Planetary Science, Solar Science and Earth Science

Planetary Science, Solar Science and Earth Science

Edited by **John Wayne**

SYRAWOOD
PUBLISHING HOUSE
New York

Published by Syrawood Publishing House,
750 Third Avenue, 9th Floor,
New York, NY 10017, USA
www.syrawoodpublishinghouse.com

Planetary Science, Solar Science and Earth Science
Edited by John Wayne

International Standard Book Number: 978-1-68286-314-5 (Hardback)

Printed in China.

Contents

Preface

Over the recent decade, advancements and applications have progressed exponentially. This has led to the increased interest in this field and projects are being conducted to enhance knowledge. The main objective of this book is to present some of the critical challenges and provide insights into possible solutions. This book will answer the varied questions that arise in the field and also provide an increased scope for furthering studies.

This book provides comprehensive insights into the field of planetary sciences, solar sciences and earth sciences through extensive discussions on various topics like scaling, criticality in turbulence, nonlinear waves, time series, pattern formations, etc. which have been thoroughly elucidated. This book is an excellent guide for those wanting to acquire advanced knowledge. It is ideal for students and academicians of planetary sciences, astronomy, physics, earth sciences and associated disciplines as it provides many interesting topics which can be taken up for research.

I hope that this book, with its visionary approach, will be a valuable addition and will promote interest among readers. Each of the authors has provided their extraordinary competence in their specific fields by providing different perspectives as they come from diverse nations and regions. I thank them for their contributions.

Editor

1

Stochastic electron motion driven by space plasma waves

G. V. Khazanov[1], **A. A. Tel'nikhin**[2], **and T. K. Kronberg**[2]

[1]NASA Goddard Space FlightCenter, Greenbelt, MD, USA
[2]Altai State University, Department of Physics and Technology, Barnaul, Russia

Correspondence to: G. V. Khazanov (george.v.khazanov@nasa.gov)

Abstract. Stochastic motion of relativistic electrons under conditions of the nonlinear resonance interaction of particles with space plasma waves is studied. Particular attention is given to the problem of the stability and variability of the Earth's radiation belts. It is found that the interaction between whistler-mode waves and radiation-belt electrons is likely to involve the same mechanism that is responsible for the dynamical balance between the accelerating process and relativistic electron precipitation events. We have also considered the efficiency of the mechanism of stochastic surfing acceleration of cosmic electrons at the supernova remnant shock front, and the accelerating process driven by a Langmuir wave packet in producing cosmic ray electrons. The dynamics of cosmic electrons is formulated in terms of a dissipative map involving the effect of synchrotron emission. We present analytical and numerical methods for studying Hamiltonian chaos and dissipative strange attractors, and for determining the heating extent and energy spectra.

1 Introduction

In this study, we treat the stochastic motion of charged particles resonantly interacting with nonlinear electromagnetic space plasma wave fields. Particular attention is given to the problem of nonlinear interaction between whistler mode waves and Earth's radiation belt electrons. Another problem of interest is the stochastic acceleration of cosmic wave electrons by space plasma waves at galactic shocks. We use a Hamiltonian formalism to treat these two problems. The Hamiltonian's equations of multiperiodical motion are assigned a measure-preserving map, the explicit form of which is defined by the closed set of nonlinear difference equations. The map is parameterized by some quantity, the control parameter, having a hard dependence on the structure of the wave packet and its power. Using topological arguments, we determine the value of the control parameter at which the system exhibits chaotic motion in the form of a strange attractor. In this case, every particle can explore the entire phase space energetically accessible to it; as a result, the upper bound of the strange attractor can be put on a one-to-one correspondence with the upper boundary of an energy spectrum whose value depends parametrically on the spectral power of the wave. The chaotic motion on the strange attractor is ergodic with mixing and as a consequence, the evolution of the distribution function and all means obeys the Fokker–Planck–Kolmogorov equation. As the wave power increases above some critical value, the phase space structure undergoes a change in topology called intermittency. The behavior is more complex, exhibiting random transitions between regular and stochastic motion. In this regime, diffusion in energy is realized through the drift of orbits in phase space. The generic Hamiltonian model is extended to include the effect of dissipation of energy associated with the synchrotron emission of relativistic electrons. It also proves possible to represent the dynamics of the system in the form of a dissipative map. For a dissipative system, the topology of attractors on which the motion appeared chaotic has the properties of fractional dimensionality. Another interesting effect in the Hamiltonian system occurs when an extrinsic noise is present. Numerical computations are presented to illustrate the methods and to give insight, and, also, to verify analytical results. We believe that acceleration mechanisms due to the nonlinear wave-particle interactions are capable of producing relativistic radiation belt electrons and galactic cosmic ray particles.

First, we will apply the model to the dynamics of Earth's radiation belts (RBs). It is suggested that wave-particle interactions, which are well correlated with the solar wind, may be responsible for the RB electrons. Energy diffusion due to resonance with chorus waves can be a viable mechanism for generating RB electrons, and the cyclotron resonances between electrons and chorus near the magnetic equator are thought to result in particles being scattered into the loss cone, forming bursts of precipitation. We show that significant heating rates and pitch angle diffusion occur for the RB electrons with energies from a few keV up to a few MeV, and the calculated timescales and the total energy input to the atmosphere from relativistic electrons are in reasonable agreement with experimental data. This is the subject of Sect. 4. We deal with the problem of acceleration of cosmic rays in Sect. 5. High-energy particles observed in cosmic rays can be regarded as tails of the particle distribution in space plasmas. The energy relations show that collective plasma processes can play an important role in the evolution of the energy spectra of cosmic rays. The observations of synchrotron emission from shell supernova remnants (SNRs) have shown that most cosmic rays are produced by SNRs. We show that electrons can be accelerated up to high energies by the Langmuir and upper-hybrid electrostatic waves at the foot of the front of galactic shocks. We follow up by applying the model to stochastic surfing electron acceleration at galactic shocks to take into account the impact of synchrotron emission on the dynamics and to interpret the synchrotron emission spectrum.

2 Dynamics of particles

We will consider the dynamics of a charged particle resonantly interacting with space plasma wave fields. Hamiltonian formalism is applied to describe the dynamics in collisionless space plasmas. As known, the Hamiltonian formalism is based on the theory of smooth manifolds and differential geometry (Schutz, 1982; Arnold, 1988, 1989). In this approach any state of a dynamic system is given by a point in phase space. Let 2n-D (dimensional) smooth manifold M on which the Hamilton function $H(q,p)$ is defined,

$$\left(q^1, q^2, \ldots, q^n, p_1, p_2, \ldots, p_n\right) = (q, p), \tag{1}$$

are canonical coordinates, be the phase space of a system. Then dynamics is determined by the Hamiltonian vector field

$$\boldsymbol{v} = \frac{\partial H}{\partial p_\mu}\frac{\partial}{\partial q^\mu} - \frac{\partial H}{\partial q^\mu}\frac{\partial}{\partial p_\mu}, \tag{2}$$

where $\partial/\partial q$ and $\partial/\partial p$ are the basis vector fields, which act as linear differential operators on smooth functions. Equations (2) are equivalent to the phase flow given by the differential equations

$$\dot{q}^\mu = \left[q^\mu, H\right] = \partial H/\partial p_\mu,$$
$$\dot{p}_\mu = \left[p_\mu, H\right] = -\partial H/\partial q^\mu, \tag{3}$$

$$H(q, p : t) = \sqrt{\left(p_\mu - A^\mu\right)\left(p^\mu - A_\mu\right) + m^2} + \varphi, \tag{4}$$

where $\boldsymbol{p}$ is the canonical momentum, $\boldsymbol{A}$ the vector potential, φ the scalar (electrostatic) potential of the wave field, H is the Hamiltonian associated with the problem, and [,] stand for the Poisson brackets. We have employed here, and throughout this paper, the system of units in which the speed of light $c = 1$ and the electron charge $|e| = 1$.

For the field variables $\boldsymbol{A}$ and φ, which are explicit functions of coordinates and time, we have to specify the coordinate system. We have chosen a Cartesian spatial coordinate system whose z axis is directed along the external magnetic field, and the plane perpendicular to the external magnetic field direction is spanned by the orthogonal coordinates x and y. Now our manifold looks locally like a linear real space $R^6, q^\mu = (x, y, z) \in R^3, p_\mu = (p_x, p_y, p_z) \in R^3$, i.e., the local topology of M is identical to that of the space. We make use of the Lorentz gauge, assuming

$$\operatorname{div} \boldsymbol{A}_{\mathrm{t}}^{\mathrm{w}} = 0, \quad \boldsymbol{A} = (-By - A_{\mathrm{t}}, 0, 0), \tag{5}$$

$$\boldsymbol{B}_{\mathrm{t}} = \operatorname{rot}\boldsymbol{A}_{\mathrm{t}}, \quad \boldsymbol{E}_{\mathrm{t}} = -\partial \boldsymbol{A}_t/\partial t, \quad \boldsymbol{E}_l = -\nabla\varphi, \tag{6}$$

where the subscripts t and l denote the transverse and longitudinal components of the electromagnetic wave field, and $-By$ is the vector potential for constant external magnetic field $\boldsymbol{B}$ (Landau and Lifshitz, 1980). Then we write the wave field in the form of a transverse longitudinal wave with slowly varying amplitude

$$\begin{pmatrix} A_t \\ \varphi \end{pmatrix} = \begin{pmatrix} A(\varepsilon t, \varepsilon z) \\ U(\varepsilon t, \varepsilon z) \end{pmatrix} \cos\left(k_z z + k_t y - \omega t\right). \tag{7}$$

The functions $A(\varepsilon t, \varepsilon z)$ and $U(\varepsilon t, \varepsilon z)$ describe the repetitive space–time structure of the envelope of the wave. Let the shape of the envelope be given by a smooth periodical function $f(\varepsilon t)$, such that

$$f(t+T, z+L) = f(t, z), \tag{8}$$

$$\varepsilon = \frac{\partial f}{\partial t}\cdot\frac{1}{\omega f} \sim (\omega T)^{-1}, \quad \frac{\partial f}{\partial z}\cdot\frac{1}{kf} \sim (kL)^{-1}, \tag{9}$$

where ε is a small parameter, the ratio of the oscillation period $2\pi/\omega$ (or $2\pi/k$) to the time (space) scale (T, L), over which the envelope varies. This ε also serves as an ordering parameter. Thus we assume the ambient magnetic field varies slowly over one wavelength $\varepsilon^2 = \frac{1}{k_z}\partial \ln B/\partial z$, and the wave field is sufficiently small, A/m, $U/m = \varepsilon$.

We now take into account the axial symmetry of the nonperturbative problem, and introduce the new variables, an action (I), and an angle (θ), by the canonical transformation,

$$y = r\sin\theta, \; p_y = (mr\omega_B)\cos\theta; \tag{10}$$

$$r = \sqrt{2m\omega_B I}/m\omega_B, \; \omega_B = B/m, \tag{11}$$

where r is the gyroradius, and ω_B is the angular gyrofrequency, respectively (Sagdeev et al., 1988). The

Hamiltonian (4) in this representation becomes

$$H(z, p_z; \theta, I; t) = H_0(p_z, I) + \left[\sqrt{2m\omega_B I}H_0^{-1}A\sum_n J_n'(k_t r) + U\sum_n J_n(k_t r)\right]\cos\psi, \quad (12)$$

$$H_0(p_z, I) = \sqrt{m^2 + p_z^2 + 2m\omega_B I}, \quad (13)$$

$$\psi = k_z z + n\theta - \omega t, \quad (14)$$

where $H_0(p_z, I)$ is the Hamiltonian of the non-perturbative system, and ψ is the phase of the particle in the wave field. In deriving (12), the representation

$$e^{ik_t r\cos\theta} = \sum_n J_n(k_t r)e^{in\theta} \quad (15)$$

has been employed, where $J_n(\cdot)$ are Bessel functions, J_n' are their derivatives with respect to the argument, and $n \in Z$, Z is the set of all integers. We consider the case of resonance between the Doppler-shifted wave frequency and the gyromotion,

$$\dot{\psi} = k_z\dot{z} + n\dot{\theta} - \omega = 0. \quad (16)$$

Resonant wave-particle interaction occurs whenever Eq. (16) holds, which is satisfied for a series of n values for particles with different momentum and energy. Sufficiently close to a resonance, ψ is slowly varying; therefore p_z and I are also slowly varying variables:

$$\varepsilon = \dot{\psi}/\omega\psi, \dot{I}/\omega I, \ \dot{p}_z/\omega p_z. \quad (17)$$

This condition means that the adiabatic approach to the problem of resonant wave-particle interaction is applicable. Choosing a particular resonance, $n = s$, we can transform the Hamiltonian (12) as follows:

$$H(z, p_z; \theta, I; t) = H_0(p_z, I) + \left[\sqrt{2m\omega_B I}H_0^{-1}\cdot A(\varepsilon t)J_s'(k_t r) + U(\varepsilon t)J_s(k_t r)\right]\cos\psi, \quad (18)$$

$$H_0(p_z, I) = \sqrt{m^2 + p_z^2 + 2m\omega_B I}, \quad (19)$$

$$\psi = k_z z + s\theta - \omega t, \quad (20)$$

where all of the perturbation terms average to zero except for $n = s$. Accordingly, we write the equations associated with the Hamiltonian (18) in the form

$$\dot{p}_z = k_z\left(\sqrt{2m\omega_B I}H_0^{-1}\cdot AJ_s'(k_t r) + UJ_s(k_t r)\right)\sin\psi, \quad (21)$$

$$\dot{I} = s\left(\sqrt{2m\omega_B I}H_0^{-1}\cdot AJ_s'(k_t r) + UJ_s(k_t r)\right)\sin\psi, \quad (22)$$

$$\dot{z} = p_z/H_0, \ \dot{\theta} = \omega_B m/H_0. \quad (23)$$

We retain in these equations only the leading terms, where the small parameter ε automatically keeps track of the ordering. Now our manifold looks locally like the real space $R^2 \times R \times S$, $(p_z, I) \in R^2$, $\dot{z} \in R$, $\theta \mod 2\pi \in S$. Because of the explicit symmetries of time translation and rotation of the phase, the phase space representation offers significant simplifications in treating the problem. Indeed, the Hamiltonian (18) is invariant under the transformations (8) and (10),

$$H = H\exp(i\psi), \ H(t+T) = H(t), \quad (24)$$

consequently, the phase flow conserves the invariant of motion,

$$sp_z - k_z I = \text{const.}, \quad (25)$$

where const. is a constant independent of time, determined by the initial condition. This leads to the restriction of dynamics onto a reduced phase space, acceptable coordinates of which appear to be a canonically conjugated pair (ψ, u), where ψ is the phase variable, and u is a new action variable. With these simplifications we can reduce this motion to quadratures only if the perturbation has a trivial form of monochromatic wave propagating along the direction of the ambient magnetic field. In the general case this non-autonomous nonlinear dynamical system is non-integrable, and the measure of its regular motions is equal to zero.

Consider the expanded phase space of the Hamiltonian system. Such a space has a natural structure of the bundle, the base of which is the discrete time $t_n = nT$, $n \in Z$, and the fiber is the $R \times S$ space (Schutz, 1982). In this case, it is sufficient to describe the motion on some time interval $(t_0, t_0 + T)$, for example, $g^1(\psi_0, u_0) = (\psi_1, u_1)$, where (ψ_0, u_0) is the initial state of the system, and g^1 is the map at one period (Arnold, 1989). In this way we have defined a map of the phase plane onto itself g^n: $R \times S \rightarrow R \times S$. The map is a diffeomorphism, which forms a one-parameter group $g^n = (g^1)^n$ of diffeomorphisms of the phase plane, and preserves the phase volume. This statement follows from Liouville's theorem. This bundle and the group g^n acting on it are shown in Fig. 1.

For the Hamiltonian (18) and its invariant of motion (25), this group can be written in the form of a set of nonlinear difference equations,

$$u_{n+1} = u_n + Q\sin\psi_n, \ \psi_{n+1} = \psi_n + F(u_{n+1}) \mod 2\pi, \quad (26)$$

where Q is the control parameter, which defines the intensity of wave-particle interaction, and the function $F(u_{n+1})$ describes the shift of phase acquired by a particle. The system (26) exhibits a rich variety of chaotic phenomena in various parameter regimes. This kind of dynamics is typically related to the appearance of a strange attractor (SA), which represents persistent chaotic motion over a global domain of phase space. In that sense, the SA is to be a paradigm for obtaining some information about the dynamics. This information is primarily geometrical and statistical, such as the fractal dimensions, spectrum of Liapunov exponents of the underlying SA and the invariant distribution on the SA. Any SA

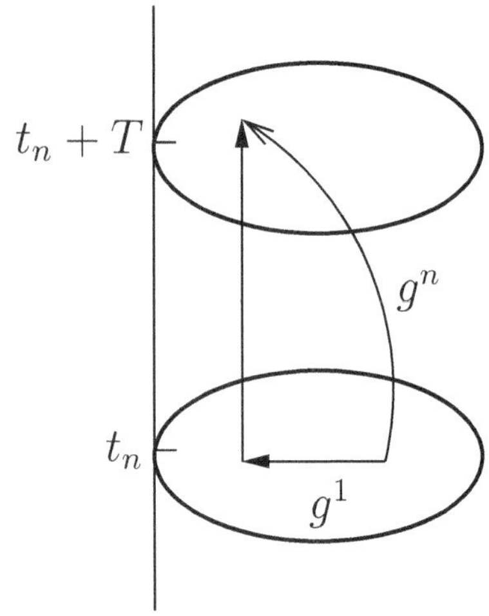

Fig. 1. Map at one period (after Arnold, 1988).

is known to be both ergodic and mixing (Arnold and Avez, 1968), and for many dynamical systems of interest, the phase variable randomizes much more rapidly than the action variable u, allowing kinetic description in terms of a standard Fokker–Planck–Kolmogorov (FPK) equation for the distribution $w(u)$ alone. For area-preserving maps, the distribution on any SA is trivially a constant. A common procedure for finding $w(u)$ consists of iterating an initial state, or of solving an FPK equation. Applying the invariant distribution, we may replace time averages by phase-space averages in calculating the steady state value of a physical observable. If $G(u)$ is an observable function in phase space, then the phase average

$$\langle G \rangle = \int du w(u) G(u) \tag{27}$$

is independent of u_0 (for almost all initial values u_0 in the basin of a given SA). We are obliged to say strange attractor; now we should wish to prove that these kinds of objects are common in the problem of interest, so we will try in the subsequent sections to develop all the proofs as explicitly as possible.

3 Dynamics of wave fields

We are going to inspect the problem of dynamics of particles in space plasma wave fields. The first question of course is how to describe these wave fields. A common technique in studying the behavior of wave fields is to Fourier analyze transform into mode amplitudes, to obtain an equation of motion for each mode. If only the N most important modes are kept, this motion is given by a set of ordinary differential equations describing the evolution in time of the mode amplitude components. This procedure of Fourier analysis followed by truncation is called the Galerkin approximation. Any stationary nonlinear wave is known to be a strongly exited state of a nonlinear medium, which emerges due to the competition between dispersion and nonlinearity. Such a wave represents in itself the bound state of a large number of harmonics. It is worthwhile emphasizing that these are not true modes of the nonlinear system, and thus interchange of energy among these quasimodes will occur even in the absence of external noise. There exists a number N that actually truncates the wave spectrum exponentially, and this N is, in fact, the constant of coupling for stationary nonlinear waves (Sagdeev et al., 1988).

Another approach to describing wave fields consists of the proper choice of basis wave models in the standard form of nonlinear wave equations. For instance, the applicability of the Korteweg–de Vriez equation for describing nonlinear space plasma phenomena is well established. In that case, the nonlinear dispersion relation, which includes the main part of information about wave dynamics, is what we chiefly need (Bernstein et al., 1957; Lutomirski and Sudan, 1966). A general approach for deriving nonlinear dispersion relations has been developed in different techniques, by Whitham (1974), Karpman (1975), and Kaup and Newell (1978). They have shown that any nonlinear dispersion relation conserves its own form even if the wave propagates in a weakly inhomogeneous and weakly nonstationary medium. This important conclusion follows from the strong stability of a nonlinear wave packet. The basic model for modulated waves in a dispersive medium appears to be the familiar nonlinear Schrödinger (NLS) equation,

$$i\left(\partial\Phi/\partial t + v_g \partial\Phi/\partial z\right) + \mu\partial^2\Phi/\partial z^2 + \left(v_g/2k_0\right)\triangle_\perp\Phi \\ -g|\Phi^2|\Phi = 0, \tag{28}$$

for the complex wave amplitude Φ (Karpman, 1975). Here the group velocity v_g of the wave packet, and the parameters of dispersion μ and nonlinearity g are determined entirely by a nonlinear dispersion relation $\omega = \omega\left(k^2, |\Phi|^2\right)$ in the usual way,

$$v_g = (\partial\omega/\partial k)_{k_0},\ \mu = \frac{1}{2}\left(\partial v_g/\partial k\right)_{k_0},\ g = \left(\partial\omega/\partial|\Phi|^2\right)_{k_0}. \tag{29}$$

The subscript denotes that we have to put these values at $k = k_0$, where k_0 is the characteristic wavenumber. The basic nonlinear mechanism in this case is the ponderomotive force that arises through the coupling of the high- and the low-frequency oscillations, and numerous possibilities of the parametric decay and the process of modulation instabilities (Sagdeev et al., 1988). Equation (28), involving the effects of self-focusing and self-compressing, describes specifically the dynamics of Langmuir, upper-hybrid mode and whistler mode waves, i.e., the very waves that usually are detected (or inferred) in space plasmas and are likely to play a major role in stochastic heating of particles. For application purposes we discuss the relevance of the representations to the whistler wave events in the Earth's magnetosphere. The spatio–temporal structure chorus will need to describe the interaction and compare obtained results with experimental data. We will rely on the data represented by the CLUSTER mission (Santolik et al., 2003), which provided temporally high-resolution measurements of the wave events. In this work, chorus emissions have been measured at a radial distance of 4.4 Earth radii within a 2000 km region located close to the Equator. The wave vector direction in this region is nearly parallel to the field lines, and the waveforms of the packet show a fine structure consisting of subpackets with amplitudes above 300 pT. In each wave packet, the frequency changes with a typical rate of a few $\mathrm{kHz\,s^{-1}}$, and the spectrum is organized into two bands between 2 and 6 kHz. The spectral power P of the wave has its peak value $P\sim 10^{-4}\,\mathrm{nT^2\,Hz^{-1}}$ at the frequency 3.8 kHz, close to the half-gyrofrequency $\omega_B/2$. The chorus emission, being a coherent whistler mode, has a slowly varying amplitude (envelope), with the time period $T=200\,\mathrm{ms}$ determining the repetitive temporal structure of the chorus.

In order to interpret these results, we will use solutions of a nonlinear Schrödinger (NLS) equation, the applicability of which to whistler mode waves is well grounded (Karpman, 1975). In the wave frame, Eq. (28), for quasi-plane waves, takes the form

$$i\Phi_t+\mu\Phi_{zz}-g|\Phi|^2\Phi=0, \tag{30}$$

with $\Phi=b\exp iS$, where $b=B^{\mathrm{w}}/B$ is the wave amplitude, normalized to the external magnetic field $\boldsymbol{B}$, and S is the eikonal of the wave. It is known (Karpman, 1975) that an NLS equation has several different classes of solutions. Of these, the most stable solutions are the so-called two-parameter envelope solitons

$$b(z-vt)=a\,\mathrm{sch}(Qz-Ft), \tag{31}$$

$$a^2=F^2/G, \tag{32}$$

$$F=Qv,\ G=\omega_0 g, \tag{33}$$

which are parameterized by the soliton velocity v and by the soliton spacescale $L=1/Q$. We will assume the value of v to be close to v_{g}. These solutions describe a nonlinear wave with a dispersion law

$$\omega(k)=\omega_0(k)+ga^2/2, \tag{34}$$

where $\omega_0(k)$ is the linear dispersion relation for whistlers

$$\omega_0(k)=\omega_B k^2c^2/\left(\omega_p^2+k^2c^2\right), \tag{35}$$

and $ga^2/2$ is the nonlinear frequency shift. The NLS equation is completely integrable and possesses an entire set of integrals of motion. One of them,

$$P=\int b^2\mathrm{d}t, \tag{36}$$

is physically significant in virtue of the relation

$$P=(1/2\pi)\int b^2(\omega)\mathrm{d}\omega, \tag{37}$$

which determines the spectral power $b^2(\omega)/(2\pi)$. Substituting Eq. (31) in the expression (36), we obtain

$$P=2a^2/F=2F/G, \tag{38}$$

and, a Fourier transform of Eq. (31) gives

$$b^2(\omega)/(2\pi)=(\pi P/4F)\,\mathrm{sch}^2(\pi\omega/2F). \tag{39}$$

The spectral representation means that the nonlinear wave can be interpreted as a nonlinear wave packet whose amplitude and width $\Delta\omega=4F/\pi$, unlike a usual wave packet, are bound with the coupling. It is known that any spatially localized perturbations to an NLS equation will convert eventually into a set of solitons. This result has been proved by the technique of inverse scattering transform (IST) (Newell, 1985). According to the IST, solitons of the NLS equations play the role of harmonic waves in linear systems, and the representation of solutions in the form of a set of solitons corresponds to a Fourier transform in linear equations. So the N-soliton solution of the NLS equation is

$$b(z,t)=\sum_{i=1}^{N_{\mathrm{s}}}a_i\,\mathrm{sch}(Q_iz-F_it), \tag{40}$$

$$a_i^2=F_i^2/G, \tag{41}$$

where N_{s} determines the total number of solitons. The IST involves quasi-classical quantization, and solutions (40) belong to the discrete spectrum of values of a and F. The solitons of the NLS equation conserve physically significant properties, so the total spectral power is the sum of spectral powers of the single solitons,

$$P=\sum_i P_i=2\Delta F/G,\ \Delta F=\sum F_i, \tag{42}$$

where ΔF is the half-width of a power spectrum. To interpret the spatio–temporal wave structure measured by Santolik et al. (2003), we apply the results of theoretical analysis to these data. We have approximated the temporal structure of chorus presented by Santolik et al. (2003) in Fig. 2 by the N-soliton solution of the type $E^{\mathrm{w}}=\sum_i E_i\mathrm{sch}(t/T_i)$ with $T_i=(14.37, 1.18, 0.89, 1.03, 0.84)\,\mathrm{ms}$, where T_i are the timescales of each subpacket. The fitted values of the waveforms were obtained from the CLUSTER data as shown in Fig. 2. Figure 2 has been added to identify the wave's fine structure. There is an apparent correspondence between the waveforms described by the envelope soliton solutions and the experimental data. This gives us confidence to assume that those signals are at least close to the waveforms having the N-soliton envelope. As a result, we will assume the representation (40) is adequate for the waveform of a single wave packet, where we have to accomplish the substitution $t\to t-nT$, to describe the quasi-periodical structure of the chorus emission. The physical reason for the chirp effect is supposed to be the nonlinear dispersion of phase velocity.

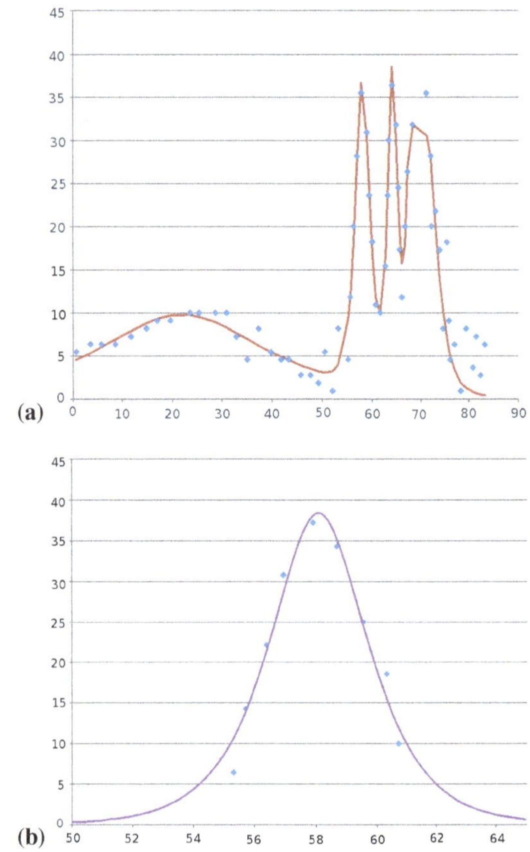

Fig. 2. (a) Single wave packet with the envelope approximated by N-soliton solution. **(b)** Smooth curve given by sch$(t/1.18\,\mathrm{ms})$ corresponds to the 2th subpacket.

Indeed, it can be shown, any gradients of physical fields will lead to the change of wave vector: $\partial k/\partial t = -\partial\omega/\partial z$. Considering this expression and the nonlinear wave dispersion (34), we can obtain the equation of transfer of wave information

$$\left(\frac{\partial}{\partial t} + v_g \frac{\partial}{\partial z}\right)\omega = 0. \tag{43}$$

As a consequence, we conclude that the nonlinear dispersion relation is always the same along the group velocity vector, $(\frac{\partial}{\partial t}, v_g \frac{\partial}{\partial z})$, even if the frequency (and wavenumber) changes slowly over the time period of the wave, $2\pi/\omega$, such that

$$\dot{\omega}2\pi/\omega \ll 1\, \dot{\omega} = \partial\omega/\partial t. \tag{44}$$

Then taking into account the nonlinear wave dispersion (34), one computes the wave drift rate, $\dot{\omega} = (g/2)\partial a^2/\partial t$. Applying to this the soliton-like solution for an envelope, the detailed evolution of the wave frequency (exponential growth of frequency on the initial stage, nonlinear dependence on the wave amplitude, etc.) can be studied. The mean and maximal rates are evaluated as $\dot{\omega} \sim ga^2/T \sim F^2/(\omega T)$, and $\dot{\omega}_{\max} \sim ga^2F/3 \sim F^3/(3\omega)$. One calculates $F = \sqrt{ga^2\omega}$, with the help of Eq. (41), then puts together the empiric data (Santolik et al., 2003); we evaluate $ga^2/2 \sim 200\,\mathrm{Hz}$, $F \sim 1,2\,\mathrm{kHz}$, and therefore, $\langle\dot{\omega}\rangle \sim 2\,\mathrm{kHz\,s^{-1}}$ and $\dot{\omega}_{\max} \sim 100\,\mathrm{kHz\,s^{-1}}$. Thus we explicitly show that the condition (44) holds. Note that longitudinal inhomogeneities of the ambient magnetic field $\boldsymbol{B}(z)$ and particle density $N_e(z)$ do not influence the effect because the fundamental mode $\omega_0 = \omega_B k^2c^2/\omega_p^2$ is proportional to B/N_e, and this ratio is in fact a constant along a force line near the equatorial plane. The rising tone of chorus is observed in the region of generation and formation of waveforms due to self-modulation and, due to the propagation of waves away from this region when the modulation instability is stabilized, this effect may vanish, as was noted by Cully et al. (2008) and Wilson et al. (2011). The spectral gap is likely due to just the same competition between nonlinearity and dispersion under conditions of the exact cyclotron and Landau resonance at $\omega = \omega_B/2$, $v_r = v_p$. Presumably the chirp-effect and spectral gap are rather the manifestation of intrinsic properties of the typical nonlinear whistler mode wave rather than a new type of emission.

The effect of diffraction on wave propagation is described by the third term of the NLS equation (28). Let AD be the angle of diffraction, $\mathrm{AD} = 2/kd$, where d is the perpendicular size of the region of wave generation, and k is the wave number. In accordance with (Santolik et al., 2003), $\mathrm{AD} \sim 0.1$ and these waves propagate from their source nearly parallel to the field line. Putting $\mathrm{AD} = 0.1$, $k = 0.25\,\mathrm{km}^{-1}$, we will have an estimate of d, $d \sim 100\,\mathrm{km}$, and the lengthscale of the wave field, $l_d = kd^2/2$, $l_d \approx 2000\,\mathrm{km}$, which is comparable with the spacescale of the envelope. At $z = 0$ the wave beam has a plane phase front and a Gaussian distribution of amplitude $b(r) = b(0)\exp\left(-r^2/d^2\right)$. The solution of the equation shows that the width of the packet grows as $d^2(z) = d^2\left(1 + (z/l_d)^2\right)$, and the radius of curvature grows as $R = z + l_d^2/z$. As $z/l_d \gg 1$; the originally plane wave becomes spherically divergent, its amplitude decreases as $kd^2b(0)/2z$, and its width grows as $d(z) \approx 2z/kd$. For example, if the wave was generated at a large L and traveled a great distance to reach Polar, this effect will lead to a lower intensity at Polar (Tsurutani et al., 2011).

As noted above, the impact of the inhomogeneity of the ambient magnetic field and density fields on the wave propagation is negligible because of a constancy of the ratio B/N_e. The solutions of an NLS equation, in particular with a parabolic profile of perturbations, are well known (Newell, 1985). According to Newell (1985), the wave power spectrum and shape remain unchanged if the envelope's size (intrinsic spacescale) is less than the spacescale of an external field. Typically, this condition is easy satisfied. Another source of perturbations can be the spacecraft itself. As long as the size of the spacecraft is much less than the wavelength, the effect of imperfection of the wave can be described as the scattering of a soliton-like packet on this defect. This effect is local, therefore it immediately appears in the wave measurements (Hastings and Garrett, 1996). It is possible that it helps explain the modification of waveforms in wave events (Santolik et al., 2003).

Another powerful tool, especially when the explicit shape of a nonlinear wave is unknown, is spectral representation in the so-called form of time- and space-like wave packets (Sagdeev et al., 1988). Let $f(t)$ be a periodic function, $f(t+T)=f(t)$, given by the convergent Fourier series

$$f(t)=\sum_{n\in Z} f_n e^{-i\omega_n t},\ \omega_n=2\pi n/T, \tag{45}$$

whose coefficients f_n satisfy the Parceval identity,

$$\sum_n |f_n|^2=\langle f^2(t)\rangle,\ \langle f^2\rangle=\frac{1}{2T}\int_{-T}^{T} f^2(t)\mathrm{d}t. \tag{46}$$

Let $f(t)$ correspond to a certain realization of a physical process. Then the measurable quantity is $\langle f^2\rangle$, and/or its spectral density $\sum_n |f_n|^2$. One assumes this function characterizes the intensity of a nonlinear wave packet. As mentioned above, each packet is a bound state of N quasimodes, and N takes part a constant of coupling. The implication of this is that as interchange of energy among these quasimodes occurs, the constant wave power, proportional to $\langle f^2\rangle$, will share evenly among the quasimodes. In virtue of Eq. (46), we obtain $\sum_n |f_n|^2 \sim N f_\mathrm{r}^2 \sim$ const., and consequently, $f_\mathrm{r}=f_0/\sqrt{N}$, where the value of f_0 would have to be extracted from a measured physical quantity, for instance, the spectral density. A more rigorous result, obtained in the theory of distributions, states the following (Richtmyer, 1978). Let $f(t)$ be a generalized function such that

$$f(\tau+1)=f(\tau),\ e^{i2\pi\tau}f(\tau)=f(\tau),\ \tau=t/T, \tag{47}$$

then, this function is found to be

$$f(\tau)=\mathrm{const.}\sum_{n\in Z}\delta(\tau-n),\ \mathrm{const.}=f_\mathrm{r}. \tag{48}$$

The wide wavepacket approach follows from these statements and the physical meaning of the theory of nonlinear waves (Sagdeev et al., 1988). Taking the structure of a wide wave packet to be of the form

$$A(\varepsilon t,\varepsilon z)=\sum_n A_n \sin(n\delta kz-n\delta\omega t), \tag{49}$$

where

$$\delta k=2\pi/L,\ \delta\omega=2\pi/T, \tag{50}$$

L, T are the length and time scales of the problem. Then we suppose that all spectral characteristics A_n are equal-amplitude, and write the envelope of wave packet in the form

$$A(\varepsilon t,\varepsilon z)=A_\mathrm{r}\sum_n \sin(n\delta kz-n\delta\omega t),$$
$$A_\mathrm{r}=A_0/\sqrt{N},\ N=\Delta\nu T, \tag{51}$$

where A_0 is the characteristic amplitude of the wave, and $\Delta\nu$ is the width of the frequency spectrum. In the limit $\delta k\to 0$, expression (51) can be transformed into

$$A(\varepsilon t,\varepsilon z)=A_\mathrm{r}\sum_{n\in Z}\delta(t/T-n). \tag{52}$$

There is a time-like representation (TLR) of a wide wave packet. In deriving it we have used the relation

$$\sum e^{\mathrm{inx}}=2\pi\sum\delta(x-2\pi n), \tag{53}$$

where $\delta(\cdot)$ is the Dirac delta function. Carrying out the same procedure with respect to Eq. (51), in another limit $\delta\omega\to 0$, the expression takes the form of a space-like (SL) wave packet

$$A(\varepsilon t,\varepsilon z)=A_\mathrm{r}\sum_{n\in Z}\delta(z/L-n). \tag{54}$$

4 Whistler-electron interaction in the Earth's radiation belts

Wave-particle interaction with chorus is believed to be the main mechanism for the acceleration of 10–100 keV radiation belt (RB) electrons to relativistic energies, a major topic in space weather physics today. In the area of investigations, progress is related to highly resolved wave measurements of intensive chorus (Santolik et al., 2003; Tsurutani et al., 2011) and oblique large amplitude whistlers (Cattell et al., 2008) as well as observations of relativistic electron microburst precipitation events (Blake et al., 1997). These results show that the usual quasilinear models of electron energizations and scattering via small amplitude incoherent waves (Summers et al., 2007) may be inadequate for understanding radiation belt dynamics. There is a need to develop a new theory for wave-particle interaction between RB electrons and coherent whistler mode waves. Some aspects of the problem have been treated by Faith et al. (1997), Roth et al. (1999), Wykes et al. (2001), and Albert (2002), who showed that the nonlinear coupling between a monochromatic whistler wave and the bouncing motion of RB electrons may lead to a random walk in energy and pitch angle. In this case, as shown by Faith et al. (1997), an electron gains energy only chaotically, when the wave field exceeds a threshold value. These results hinted that this mechanism may be promising in the many applications where the efficiency of particle heating is of prime importance. In particular, these resonant random processes may contribute to the 10- to 100-fold increase in the relativistic electron fluxes in the RBs, as well as constitute an interesting relation between wave activity and enhancements in fluxes during geomagnetic storms and other active periods (Roth et al., 1999). Finally, the concept of a strange attractor (SA) was introduced by Khazanov et al. (2007) and Khazanov et al. (2008a) in order to describe chaotic motion of RB electrons under the conditions of nonlinear resonant interaction with chorus. Strange attractors, which embody the idea of deterministic chaos, the statistical aspects of SAs have come under explicit study. The method has proved fruitful in describing the dynamics of electrons in the Earth and Jovian radiation belts.

4.1 Chaotic motion of the RB electrons driven by chorus

We are concerned with the dynamics of nonrelativistic electrons in the whistler mode chorus, which accelerates particles to relativistic energies through the resonance

$$-k_z|v_z| + \omega_B - \omega = 0. \tag{55}$$

Taking into account the relation $\omega_B I = mv_t^2/2$, we write the invariant of Hamiltonian flow (25) as

$$v_t^2 = 2(\omega_B/\omega)\, v_{\mathrm{ph}}(v_{\mathrm{r}} - |v_z|), \tag{56}$$

where $v_{\mathrm{r}} = v_{\mathrm{ph}}(\omega_B - \omega)/\omega$ is the resonance speed of an electron, and v_t is the perpendicular (to $\boldsymbol{B}$) velocity of an electron. Then the equations of motion (21)–(23) can be represented in the form

$$\dot{v}_t = \left(\omega_B^2/2\omega\right) v_{\mathrm{ph}} \left(b/\sqrt{N}\right) T \sin\psi \sum_{n\in Z} \delta(t - nT),$$
$$\dot{\psi} = \omega(v_t, \omega) = \omega_B - \omega - k_z v_{\mathrm{r}} + \left(k^2/2\omega_B\right) v_t^2. \tag{57}$$

We have also employed the invariant of motion, the value of the Bessel function $J_1'(0) = 1/2$ and the relationships

$$A/m = (\omega_B/\omega)\, v_{\mathrm{ph}} b, \;\; b = B_t/B. \tag{58}$$

After integrating these equations, the problem is formulated in terms of a G^n map, given by a closed pair of nonlinear difference equations,

$$G^n : \begin{array}{l} u_{n+1} = u_n + Q\sin\psi_n, \\ \psi_{n+1} = \psi_n + \frac{1}{4}\omega T u_{n+1}^2 \operatorname{sign} u \quad \bmod 2\pi, \end{array} \tag{59}$$

where the new variable u and the control parameter Q have been introduced by the relations

$$u = v_t/v_{\mathrm{r}}, \; Q = \left(\omega_B^2/2\omega\right)\sqrt{TP} = \omega_B T\left(\frac{\omega_B}{2\omega}\right)\frac{b}{\sqrt{N}}, \tag{60}$$

and P is the wave power normalized to B^2. The map (59) acts as the group $G^n = (G^1)^n$ of diffeomorphisms of phase space, hence the pair (M, G^n), together with the invariant of motion is equivalent to the periodic flow (21)–(23). Khazanov et al. (2008a) and Khazanov et al. (2011) have shown that the representation (52) is tantamount to that given by a solution of nonlinear wave Eq. (40). The physical explanation of this somewhat surprising result is quite clear. Any periodic, nonlinear wave can be described as wave packet consisting of some number N of quasimodes with the frequency spacing $\delta\omega \sim 1/T$, where T is the time period of the envelope. In the situation to which we refer, the half-width $\Delta\nu$ is about 1 kHz, T is no less than 100 ms, and therefore $N \sim \Delta\nu T(\sim 10^2)$. On the other hand, $\Delta\nu \sim \omega$, which indicates that $1/N \sim \varepsilon$, the small parameter given by Eq. (9). Thus we determine that the representation (52) is reasonable, so the results obtained in these approaches either coincide or differ only slightly.

To understand the physical nature of these results, we examine the dynamics of a particle resonantly interacting with some quasimode of the wave packet. According to Eq. (57), the motion is governed by the pair of closed equations

$$\dot{v}_t = \left(\omega_B^2/2\omega\right) v_{\mathrm{ph}}\left(b/\sqrt{N}\right)\sin\psi,$$
$$\dot{\psi} = \left(k^2/2\omega_B\right) v_t^2.$$

Setting $v_t = v_{\mathrm{c}} + \Delta v$, where v_{c} is the value of v_t in the resonance state, and linearizing this equation, we derive the equations

$$\ddot{\psi} + \omega_b^2 \sin\psi = 0,$$
$$\omega_b^2 = \omega\omega_B b/\sqrt{N}\left(v_{\mathrm{c}}/2v_{\mathrm{p}}\right),$$

which describes the small (bounce) oscillations of the particle with the frequency ω_{b}. Now we apply the overlapping criterion (Chirikov, 1979), $\omega_{\mathrm{b}} \geq \delta\omega\left(\sim T^{-1}\right)$, to obtain the condition of the onset of stochasticity,

$$\left(v_{\mathrm{c}}/v_{\mathrm{p}}\right) \geq (2\omega/\omega_{\mathrm{b}})\sqrt{N}/b\omega^2 T^2. \tag{61}$$

Note that the criterion can be rewritten in the form $T_{\mathrm{b}} \leq T$, where $T_{\mathrm{b}}\left(\sim \omega_{\mathrm{b}}^{-1}\right)$ is the bounce period of the particle in the wave field. In amplitude–time terms, this relation determines the condition for parametric resonance between the slow variations in the wave field and the proper oscillations of a particle in the field. Thus the basic resonances (55) and their interaction through parametric resonance play a crucial role in the appearance of stochastic motion.

Our interest now is to inspect the motion of relativistic electrons, driven by chorus. By integrating Eqs. (21)–(23) along with invariant of motion (25), we derive in the usual way the g^n map,

$$g^n : \begin{array}{l} u_{n+1} = u_n + Q\sin\psi_n, \\ \psi_{n+1} = \psi_n + (3/2)K u_{n+1}^{-2/3} \operatorname{sign} u \quad \bmod 2\pi, \end{array} \tag{62}$$

written in the notations

$$p_z/m = p, u = p^{3/2}, and K = v_{\mathrm{g}}\omega_B T. \tag{63}$$

Here the control parameter Q is given by

$$Q = 3\left(\omega_B v_{\mathrm{ph}}/2\omega\right)^{1/2} v_{\mathrm{g}}\omega_B (PT)^{1/2},$$
$$P = b^2/\Delta\nu, \; b = B_t/B. \tag{64}$$

Solutions of that map have been discussed in detail by the authors (Khazanov et al., 2008a, 2011). Let us consider a pair (M, g^n), where M is a smooth manifold, and g^n is a diffeomorphism. Denote by

$$J = \frac{\partial(u_{n+1}, \psi_{n+1})}{\partial(u_n, \psi_n)} \tag{65}$$

the Jacobi matrix of the map, the eigenvalues of which are given by the relations

$$\det J = \lambda_1 \cdot \lambda_2 = 1,$$
$$\operatorname{tr} J = \lambda_1 + \lambda_2 = 2 + KQu^{-5/3}, \tag{66}$$

where det J and tr J denote the determinant and the trace of the matrix, respectively. The Jacobian of (65) is equal to one, and therefore g^n has the structure of a differentiable area-preserving map and (ψ, u) is the canonical pair of variables. It is known (Arnold and Avez, 1968) that the the geometrical structure of M and dynamics of g^n on M are intimately closed, and that the condition

$$|\operatorname{tr} J| = 3 \tag{67}$$

corresponds to a topological modification at which the dynamics of the system becomes chaotic. In this situation, a single phase trajectory investigates all accessible phase space, which is called the strange attractor (SA). The strange attractor is the invariant set $g^n\mathrm{SA} = \mathrm{SA}, n \to \infty$ tightly embedded in phase space. The motion on any SA is random over a wide range of Q due to the global stability of the SA, on which all means (observables) are stable indifferent of any (reasonable) initial conditions. Then we apply the condition of Eq. (67) to the relation (66) to obtain the upper bound of $\{u\}$

$$u_\mathrm{b} = (KQ)^{3/5}. \tag{68}$$

Shown in Fig. 3 is the strange attractor of the pair (M, g^n). Figure 3 shows that the stochastic region extends to the values of u predicted by Eq. (68).

Finally, using the relations (63) and (68) along with invariant of motion (25), we find the Q dependence of the upper boundary of energy spectrum E_b,

$$E_\mathrm{b} = m(KQ)^{2/5}. \tag{69}$$

At typical values of parameters, $\omega_B/\omega = 2$, $\omega_B/2\pi = 8 \times 10^3\,\mathrm{Hz}$, $P = 10^{-4}\,\mathrm{nT^2\,Hz^{-1}}$, $B = 300\,\mathrm{nT}$, $T = 2 \times 10^{-1}\,\mathrm{s}$ (Santolik et al., 2003), the expression (69) yields $E_\mathrm{b} \simeq 8\,\mathrm{MeV}$. In this way as above, we can say the same about the pair (M, G^n), just changing the order of words a little. Indeed, in this case, the condition (67) proves to determine the lower bound of $\{u\}$,

$$u_\mathrm{c} = 4/\omega_B^2 T\sqrt{PT}, \tag{70}$$

which is in good agreement with the numerical solutions and results of qualitative analysis. With the values of the parameters as given above, the expression (70) yields $u_\mathrm{c} \approx 1 \times 10^{-3}$.

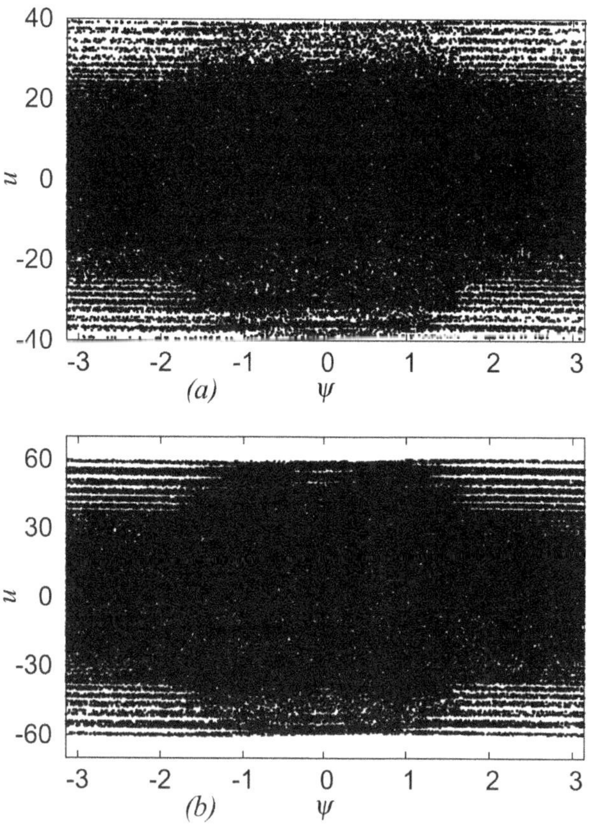

Fig. 3. Phase space for the map g^n after 2×10^6 iterations with **(a)** $Q = 0.5$ and **(b)** $Q = 0.98$.

The phase space of the map G^n is shown in Fig. 4. The overall picture of the phase space is quite different for $u < u_\mathrm{c}$ and $u > u_\mathrm{c}$. In the first case, the motion is regular. Figure 4 indicates the existence of a threshold for the initial particle velocity above which the trajectory becomes chaotic. We assume that nonlinear electron acceleration by a wave packet of whistler mode waves is always a stochastic process.

Now using the invariant of motion (56) and resonance condition (55), we have the relationship

$$e = \left(1 + u^2/4\right)^2, \tag{71}$$

where $e = E/E_\mathrm{r}$ is the particle energy normalized to that at exact resonance ($E_\mathrm{r} \simeq 25\,\mathrm{keV}$). Putting in the expression (71) $u = u_\mathrm{c}$ and $u_0 = \sqrt{2\omega_B/\omega}$, we determine the range of heating $e \in (e_\mathrm{c}, e_0)$, where

$$e_\mathrm{c} \simeq 1 + u_\mathrm{c}^2/2,\ e_0 = (1 + \omega_B/2\omega)^2, \tag{72}$$

and evaluate (72) as $E_\mathrm{c} \sim 25\,\mathrm{keV}$, $E_0 \sim 100\,\mathrm{keV}$. Lastly, the expression (72) is taken as a criterion that stochastic motion occurs, making available reliable information about the intensity of wave fields. Setting u_c be $v_\mathrm{T}/v_\mathrm{r}$, where v_T is the thermal velocity, and considering the lower bound (70) and

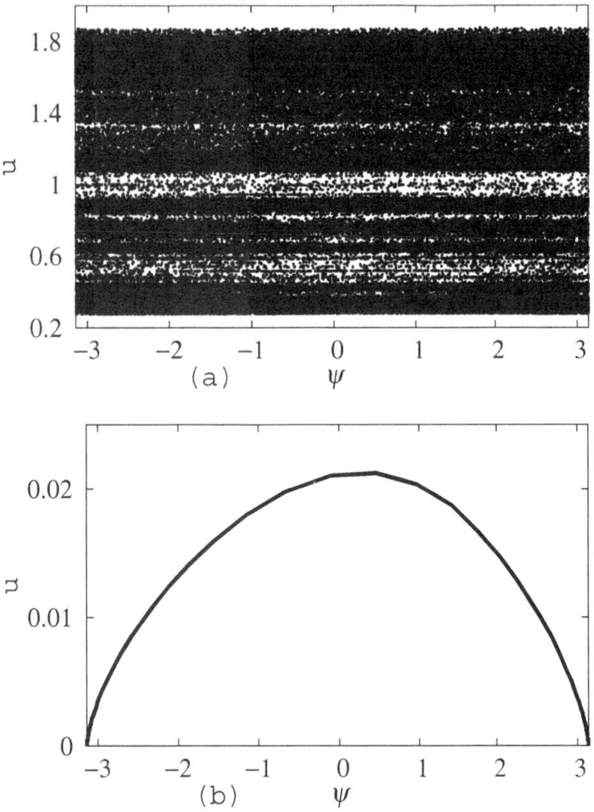

Fig. 4. Phase space for the map G^n at $Q = 2 \times 10^{-3}$. **(a)** A single trajectory of length 2×10^5. The initial point $(10^{-4}, 3 \times 10^{-1})$. **(b)** A single trajectory of length 8×10^3. The initial point $(10^{-4}, 10^{-5})$.

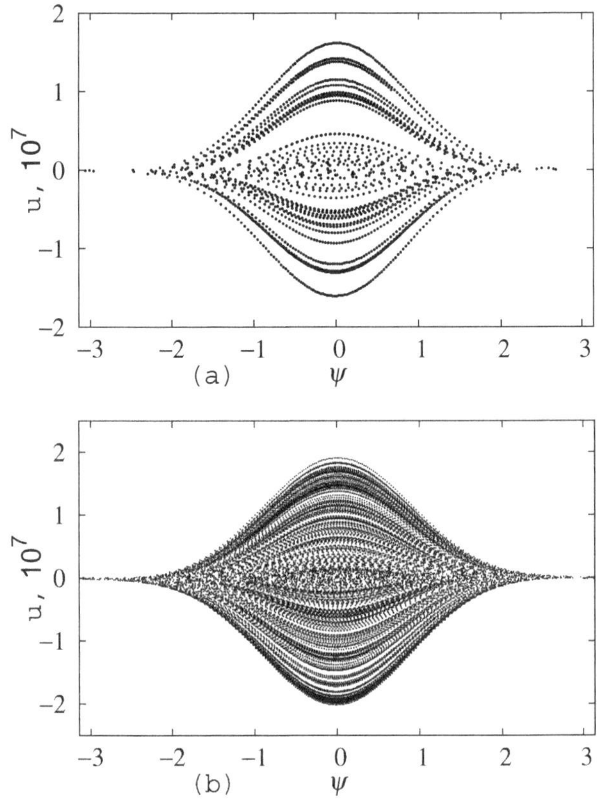

Fig. 5. Intermittent chaotic behavior of map (62) with $Q = 5 \times 10^5$. **(a)** $t = 2 \times 10^3$ and **(b)** $t = 2 \cdot 10^4$ iterations of a single initial condition.

$u_c^2 \ll 1$, we evaluate the available wave intensity,

$$P_{\text{th}} \geq 10^{-8}\,\text{nT}^2\,\text{Hz}^{-1}, \; B_{\text{th}} \geq 1\,\text{pT}. \quad (73)$$

The equations from which the SA arises are usually parameterized by some control parameter, whose variation changes the character of the dynamics. The control parameter Q, as a rule, depends directly on the intensity of the wave field. First, we consider the dynamics of g^n under conditions when the wave perturbation is larger than that observed. Figure 5 shows the evolution of the system over time at $Q = 5 \times 10^5$. The picture indicates that the system demonstrates both chaotic and regular dynamics, a so-called chaotic intermittent behavior. This behavior is more complex, exhibiting random transitions between regular and stochastic motion. In this regime, stochastic diffusion governed by the SA is realized through the drift of orbits in phase space and can be expressed in terms of a FPK equation that describes an increment in entropy rather than the diffusion in action (Khazanov et al., 2008b). The phase modification is associated with the appearance of fixed elliptic points, which occur when an initial saddle point of the attractor changes to a stable elliptic point as the control parameter Q increases above the appropriate critical value, Q_c. To determine these points, we write

$$\sin\psi_0 = 0, \left(3K/2u_0^{2/3}\right) = 2\pi, \; u_0 = u_b, \quad (74)$$

where u_b is given by expression (68). These equations are satisfied provided Q takes the value

$$Q = Q_c, \;\; Q_c = (3K/4\pi)^{5/2}\,K^{-1}. \quad (75)$$

The test does predict the phase modification rather well. We find this behavior numerically and can give an interpretation in terms of the theory developed by Khazanov et al. (2008b). Substituting Q_c in Eqs. (69) and (74), we obtain $u_0 \simeq 3\times10^3$, and therefore $E_0 \simeq 100\,\text{MeV}$. Finally, Eq. (75) yields $Q_c \simeq 10^3$, which is three orders of magnitude larger than typical values of $Q(\sim 0.5-1)$. The explicit expression for P_c, the value of the wave power at which pronounced intermittent dynamics occur, results immediately from Eq. (64):

$$P_c = (2/9)(3/4\pi)^5(\omega T)^2(B^2/\omega), \, P_c \sim 10^3\,\text{nT}^2\,\text{Hz}^{-1}. \quad (76)$$

The results show that the intermittent chaotic motion is possible only in extremely strong $(B_w/B \sim 1)$ wave fields. Under typical conditions, the long-term evolution of the system is dominated by diffusion induced by stochasticity.

4.2 Statistical properties of the dynamics

The dynamics of the system exhibits a random walk, and all points of the phase trajectory tend to a certain strange attractor (SA) at $n \to \infty$. The appearance of such an SA represents persistent chaotic motion over a global domain of phase space, as shown in Fig. 6. There the joint probability density $\rho(\psi, u, t)$ is proportional to the number of phase points in the element of phase space. We see that after a few iterations the action distribution remains localized while phase randomization occurs, and that the final distribution is random. Thus, we have good evidence for statistical properties in the region for which the system exhibits chaotic motion. Once the phase variable becomes rapidly varying, the evolution of the coarse-graining function

$$w(u,t) = \frac{1}{2\pi} \int_{-\pi}^{\pi} \rho(\psi, u, t)\, d\psi, \tag{77}$$

obeys the Fokker–Planck–Kolmogorov (FPK) equation for the u variable alone. In this case, the nature of the evolution to the steady state is so-called deterministic diffusion (Lichtenberg and Lieberman, 1992). As the canonical status of the phase variables has been demonstrated above, the distribution function (probability density) $w(u;t)$ is governed by the FPK equation in the standard form

$$\frac{\partial w(u;t)}{\partial t} = \frac{1}{2} \frac{\partial}{\partial u} D \frac{\partial w}{\partial u}. \tag{78}$$

Here D is the conventional diffusion coefficient in phase space,

$$D = < (u_{n+1} - u_n)^2 > T^{-1} = Q^2/2T, \tag{79}$$

where $(u_{n+1} - u_n)$ is substituted from either g^n or G^n, $<\cdot>$ denotes the phase average, and T is the timescale of the problem. The function $w(u,t)$ is a differentiable function supported in $\{U\}$ with the norming

$$\int_{u \in \{U\}} w(u,t)\, du = 1, \tag{80}$$

where $\{U\}$ is a range of the variable u.

First, by means of g^n (or G^n) we calculate the diffusion coefficient and evaluate the characteristic time for redistribution of u over the spectrum

$$t_d \simeq 2u_m^2/D = T(2u_m/Q)^2. \tag{81}$$

We exploit the FPK equation with $w(u)$ and its derivative $\partial w/\partial u$ vanishing at the upper and lower boundaries. We introduce the moment $< u^2 > = \int_{-\pi}^{\pi} du\, u^2 w(u)$, multiply equation (78) by u^2, and integrate the resulting equation over u to obtain

$$\mathrm{d} < u^2 > /\mathrm{d}t = D, \quad u \in \{U | u \leq u_m\}. \tag{82}$$

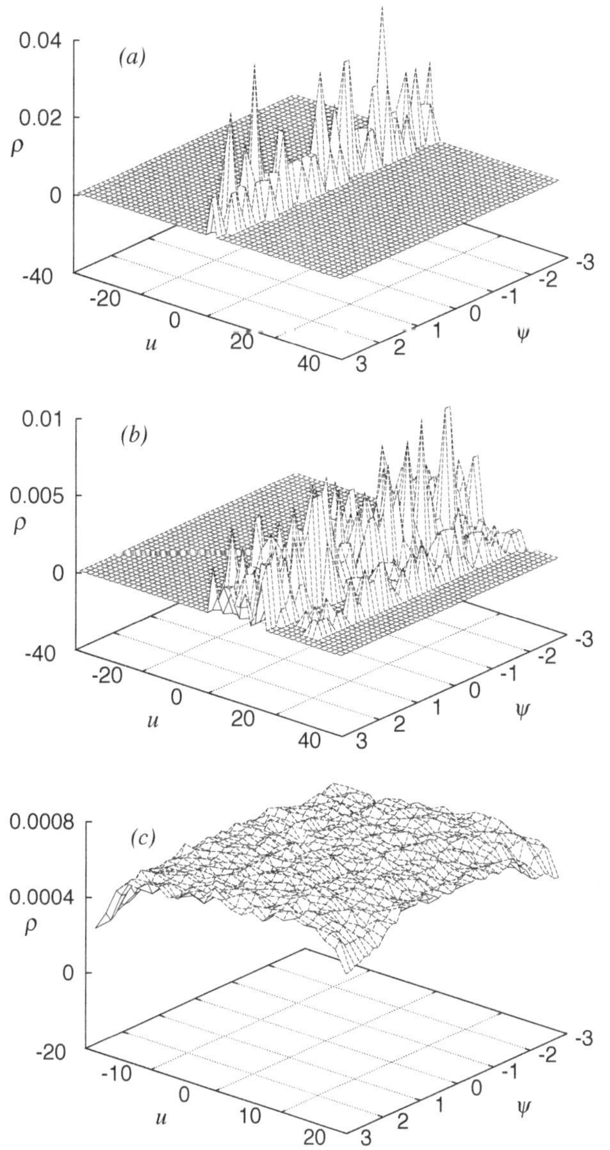

Fig. 6. Joint distribution $\rho(\psi, u)$ computed numerically via g^n at several different numbers of steps: **(a)** 10^2, **(b)** 10^3, and **(c)** 2×10^6. The parameters are the same as in Fig. 3a.

The restriction on u is needed because the spectrum is bounded from above. Then, by means of Eq. (60) along with Eq. (72), we calculate by formulas (79) and (81) the diffusion coefficient $D_u \approx 1\,\mathrm{s}^{-1}$ and the time of diffusion in u, $t_d \sim 16\,\mathrm{s}$. For the function $w(u)$ restricted to $[u_c, u_0]$, the invariant distribution can be given by

$$w(u) = (u_0 - u_c)^{-1}. \tag{83}$$

This means that the random variable u is evenly distributed on $[u_c, u_0]$. Now we evaluate the energy distribution function, $w(\varepsilon)$, in the range of non-relativistic energies. Considering $u^2 = 4(\sqrt{e} - 1)$, we note that the particle energy E is a specified function of u. Thus the measure-preserving transformation $w(e)de = w(u)du$ determines this problem completely, subject to appropriate boundary conditions. The corresponding solution for $w(e)$ is found to be

$$w(e) = w(u)/2\left[e\left(\sqrt{e}-1\right)\right]^{1/2}, \; e \in (e_c, e_0). \tag{84}$$

By means of Eq. (84) we find the mean energy $\langle e \rangle \sim 0.5e_0$. This solution relates directly to the behavior of the system near the order–chaos bifurcation transition. Equation (82) may be written as

$$\mathrm{d} < v_t{}^2 > /\mathrm{d}t = D_u v_\mathrm{r}^2. \tag{85}$$

Then considering Eq. (85) we find the rate of change of magnetic moment μ, $\dot{\mu} = v_\mathrm{r}^2 D_u/2B$, and the heating rate $\dot{E} \simeq E_0/t_d \left(\sim 6\,\mathrm{keV\,s^{-1}}\right)$. This result is nontrivial, because chaotic motion in the (u, ψ)-phase space leads to nonadiabatic magnetic moment changes and stochastic heating of plasma particles. The gyroradius is a direct measure of the perpendicular electron velocity. Thus the stochastic heating will be accompanied by a radial drift of particles in space. Indeed, considering the relation $r = v_t/\omega_B$, via (85), we get the following expression

$$\frac{\mathrm{d} < r^2 >}{\mathrm{d}t} = D_t, D_t = D_u v_t^2/\omega_B^2, \; \left(\sim 10^{10}\,\mathrm{cm^2\,s^{-1}}\right),$$

where D_t is the coefficient of collisionless diffusion across the ambient magnetic field.

Now we describe the effects associated with stochastic heating of high-energy particles. Due to one-to-one correspondence, $u = p^{3/2}$, and using $dp = 2du/3p^{1/2}$, one obtains the coefficient of diffusion in p, $D_p = 4D_u/9p$. Then the measure-preserving transformation $w_p = w_u|\partial u/\partial p|$ results in the probability density in p, $w_p = 3\sqrt{p}/2p_\mathrm{b}^{3/2}$. There is an important relationship, $w_p^2 D_p = w_u^2 D_u$. Furthermore, one can easily prove the invariant relation

$$w_v^2 D_v = w_u^2 D_u = \text{inv.}, \text{inv.} = 1/2t_d \tag{86}$$

takes place for any $v(u)$, and depends smoothly on u, the coordinate on the SA, whose distribution function $w_v = w(v)$ relates with w_u through a measure-preserving transformation, with D_v given by a quadratic form of its variation. We see that the invariant is given by the rate of diffusion on the SA. We come to know as well that the FPK equation for $w(v,t)$ can be written in divergent form

$$\frac{\partial w(v,t)}{\partial t} = \frac{\partial}{\partial v} J(v), \; J(v) = \frac{1}{2}\left(D_v \frac{\partial}{\partial v} + d_v\right) w(v,t), \tag{87}$$

where the transport coefficients are given by

$$D_v = D_u\left(\partial v/\partial u\right)^2, \; d_v = \frac{1}{2} D_u \partial/\partial v\left(\partial v/\partial u\right)^2. \tag{88}$$

The above equations have important consequences in determining the heating rate. Then, introducing the particle energy distribution $w(e,t)$, and taking into account the invariant of motion and the relation $e = E/m$ and $E = \sqrt{m^2 + p_z^2 + 2m\omega_B I}$, we find, after appropriate transformations, that the following equation for $w(e,t)$ results in

$$\frac{\partial w(e,t)}{\partial t} = \frac{\partial}{\partial e} J(e), \; J(e) = \frac{D_u}{2}\left(\frac{4}{9e}\frac{\partial}{\partial e} - \frac{2}{9e^2}\right) w(e,t). \tag{89}$$

Then it may be possible to find the heating rate

$$\frac{\mathrm{d}E}{\mathrm{d}t} \simeq \frac{8}{9}\left(\frac{e_\mathrm{b}}{e}\right)^2 \cdot \frac{E_\mathrm{b}}{t_d}, t_d = \left(\frac{4e_\mathrm{b}^3}{Q^2}\right) T, \tag{90}$$

where t_d is the relaxation time to the steady-state distribution. It is worth noting that the injection mechanism at low energies, which may be intimately related to particle heating is the important aspect of space physics. Since the rate of electron cyclotron resonance heating is proportional to E^{-2} this is, indeed the case. We put the characteristic values of these quantities at typical conditions, $\omega_B/\omega = 2$, $\omega_B/2\pi = 8\times10^3\,\mathrm{Hz}$, $P = 10^{-4}\,\mathrm{nT^2\,Hz^{-1}}$, $B = 300\,\mathrm{nT}$, $T = 2\times10^{-1}\,\mathrm{s}$, to determine, with the help of Eqs. (69) and (90), the upper boundary energy spectrum $E_\mathrm{b} \approx 8\,\mathrm{MeV}$, the time of diffusion $t_d \approx 2\times10^3\,\mathrm{s}$, and the heating rate $\dot{E} \approx 4\,\mathrm{keV\,s^{-1}}$, which are all in reasonable agreement with experimental data (Selesnick et al., 1997; Selesnick, 2006). These results lead to two important observations:

1. Nonlinear electron acceleration by a whistler wave is always a stochastic process when the wave power of the wave exceeds the threshold value given by Eq. (73).

2. In this case the related quantities have the following dependencies on the wave power,

$$E_\mathrm{b} \propto P^{1/5}, t_d \propto P^{-2/5}, \dot{E} \propto P^{3/5}, \tag{91}$$

with the power law indices determined by the type of nonlinearity in the phase advance equation.

Bursty relativistic electron precipitation (REP) events during geomagnetic storms appears to be the most striking phenomenon in the dynamics of the Earth's radiation belt. The chorus–electron coupling has been closely identified with REP events (Green et al., 2004; O'Brien et al., 2004), and Khazanov et al. (2011) have shown that the process of stochastic electron heating followed by the pitch-angle scattering is actually a viable mechanism for these REP events. The authors have investigated the problem, finding, in particular, that the energy flux in the Earth's atmosphere is directly proportional to the heating rate of a single electron. That is why the enhancement of the flux is often observed during the magnetospheric storm. The result (91) is telling in the Conclusions.

4.3 Adaptability of the Fokker–Planck–Kolmogorov equation

We have seen that diffusion on a strange attractor (SA) is described by a Fokker–Planck–Kolmogorov (FPK) equation for the u variable alone, where u is a canonical variable on the SA, proper for a certain physical situation. The FPK equation to which we refer indicates that an initial δ function distribution diffuses to a random distribution. Any measure-preserving transformation from u to some new variable $v(u)$ has been shown to preserve the invariant relation $w^2(u)D(u) = 1/2t_d$, with u replaced by v, indicating the universal nature of the diffusion process on any SA. In this case, the distribution $f(v)$ can be found by a direct transformation $f(v) = f(u)du/dv$ without solving a relative FPK equation. However, if we are, for example, concerned with the evolution of the energy spectrum, we should use the FPK in divergent form.

Quasi-linear theory (QLT) does predict the long-time evolution of RB electrons on the basis of an FPK equation. However, in the standard formulation (Lyons and Williams, 1984; Summers et al., 2007), the wave field of whistlers is treated as random, broadband and phase-incoherent. Thus, this formalism can not be directly applied to the problem of electron motion in the regular wave field of chorus. In the nonlinear approach, we apply just the same FPK equation in which phase averaging is made due to dynamics itself, and all time means of observable functions are equal to the phase space means. This approach corresponds to the standard FPK description in statistical mechanics and facilitates the applicability of kinetic models in which the RB dynamics is simulated (Khazanov, 2011). As it is known, QLTs are based on some linearized equations of motion, and the characteristic time of diffusion in energy is given by $t_d = E^2/D$ that depends on the wave power P as P^{-1} in certain energy ranges. In the nonlinear theory (NLT), the rate of diffusion hinges on the type of interaction, for instance, it obeys the scaling $t_d \sim P^{-2/5}$ for chorus-RB electron interaction. Shown in Fig. 7 is the time of diffusion as a function of wave power. Since the inequality $t_d/T \gg 1$ is a condition for applying the FPK description, these dependencies have important consequences in determining the range of values of wave power for which the FPK formalism is valid. It is just the same thing to evaluate the permissible values of chorus wave fields, P_1, P_2 and susceptible to FPK description, in QLT and NLT, respectively. Here, P_1 and P_2 are determined by Eqs. (73) and (76), respectively. They yield $P_1/P_0 = 10^{-4}$, and $P_2/P_0 = 10^7$, where $P_0 = 10^{-4}\,\mathrm{nT^2\,Hz^{-1}}$ is the typical value of wave power (Santolik et al., 2003).

Although it is normally thought that the results from simple problems give qualitative predictions of the behavior of more complicated physical systems, many questions concerning the dynamics of the system still exist. A particularly interesting aspect concerns the effects resulting from the resonance interaction with the quasi-coherent whistler waves. The effect of a magnetic mirror in modifying the motion can also play an essential role in certain physical situations.

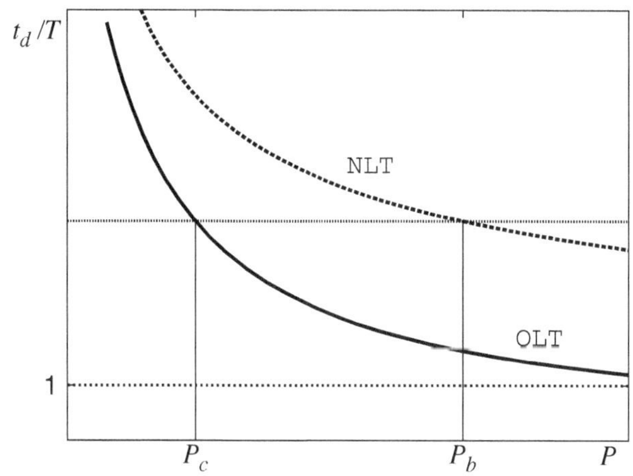

Fig. 7. Time of diffusion as a function of wave power. Here P_1 and P_2 are the critical wave powers for the applicability of the FPK equation in QLT and NLT, respectively.

Such a physical situation can emerge when RB electrons interact with the so-called quasi-coherent chorus, whose frequency spectrum contains low-intensity isotropic white noise (Tsurutani et al., 2011). The physical explanation for these wave events is quite clear. These are likely to relate to the irregular nature of the Earth's ring current injections, which generate chorus (Liemohn, 2006).

The level of background noise is relatively small; consequently, this effect can exhibit in the equations of motion only as the fluctuations of parameters ω_B and Q. Thus the control parameter is a function of the magnitude of wave field b, and therefore $Q \to Q(1+\xi), \omega_B \to \omega_B(1+\xi)$. Here, $\xi = \delta b/b$ is the random variable having a Gaussian probability density

$$p(\xi) = \frac{1}{\sqrt{2\pi}\sigma} \exp\left(-\xi^2/2\sigma^2\right),$$

the mean-square value of which is $\langle \xi_n \xi_{n'} \rangle = \sigma^2 \delta_{nn'}$, where σ^2 is the intensity of noise, and δ is the Kronecker delta.

In order to understand how the extrinsic noise modifies dynamics, we consider its influence on motion separately. Thus, we include the additional noise in G^n only in the phase advance equation, to obtain the closed pair of nonlinear stochastic equations

$$u_{n+1} = u_n + Q \sin \psi_n,$$
$$\psi_{n+1} = \psi_n + \xi_n + (1/4)\omega T u_{n+1}^2 \mod 2\pi, \tag{92}$$

where the term ξ_n plays the role of a weak stochastic force.

Numeral investigation of a related quantity, the correlation function $C(k)$,

$$C(k) = (1/N) \sum_{n \in N} \psi(n)\psi(n+k),$$

where k is the step lag, gives the results shown in Fig. 8. The motion has been shown to be δ-correlated, with $C(k) = 0$ for $k \neq 0$; this represents a very strong chaotic property, i.e., a complete decorrelation of the motion in one mapping period. Note that FPK description is valid only when the function $C(k)$ falls off rapidly with the number of mapping iterations. The effect of extrinsic noise manifests itself in the FPK equation as an additional term in the coefficient of diffusion, $D/D_u = 1\sigma^2/Q^2$. This effect does not significantly affect diffusion induced by stochasticity, due to the global stability of the SA.

The situation is quite different for ion cyclotron resonance heating (ICRH) (Lichtenberg and Lieberman, 1992). In this case the dynamics is intrinsically degenerate, and the nonlinearity arises from the finite gyroradius, which leads to a spectrum of modes in the motion, and the term proportional to Q occurs in the phase advance equation. This method has been used by the authors in conjunction with the problem of ICRH at the front of a shock (Khazanov et al., 2010).

Now we want to study how the Earth's magnetic mirror effects change the dynamics of G^n and how much impact it has on the statistical properties and long-term evolution of the system.

The drift equations for particle motion in a current-free field are given by Schmidt (1979):

$$\boldsymbol{v}_{\mathrm{d}} = v_{\parallel} \boldsymbol{B}/B + [\boldsymbol{F}, \boldsymbol{B}]/B^2, F = -\left(\mu + m v_{\parallel}^2/B\right),$$

$$\mathrm{d}v_{\parallel}/\mathrm{d}t = -(\mu/m)\,\partial B/\partial s, \tag{93}$$

where v_{d} is the drift velocity, $\boldsymbol{F}$ the free force averaged over a gyro-orbit, μ the magnetic moment, and s the coordinate along the field line.

In the drift approximation, μ is an adiabatic invariant, which is assumed to be well conserved. Since the energy of a charged particle gyrating in the Earth's magnetic mirror is strictly conserved, $v_{\parallel}$ is related to the constants of the motion μ and E through the relation $v_{\parallel} = \sqrt{2(E-\mu B)/m}$, which is easily calculated for a given function B. In the equations of motion, describing the resonant interaction with high-frequency wave fields, the forces given by Eq. (93) average to zero, as do all other off-resonant terms. All the same, residual effects are always present, which lead to the usual distortions of the phase plane near resonances of the principal frequencies, and can be physically associated with the loss of phase coherence from one passage along the field line. To show how this occurs, we first write Eq. (10) as $y = y_d + r\cos\theta$, where y_d is the guiding center position of the gyrating particle. Then, this dependence in the ψ variable gives rise to the term $k_t y_d$, and as a consequence, the term $\{k_t v_{\mathrm{d}} T\}_n - \{k_t v_{\mathrm{d}} T\}_{n+1} = \{k_t v_{\mathrm{d}} T\}$, where $\{\cdot\}$ is the fractional part, occurs in the phase advance equation. We can eliminate this term by a shift of z or t; however, small asymmetries or time variations in the magnetic field exist, which allow stochastic diffusion to occur. If these variations are random, we would write an independent equation $\{k_t v_{\mathrm{d}} T\} = \xi_n$,

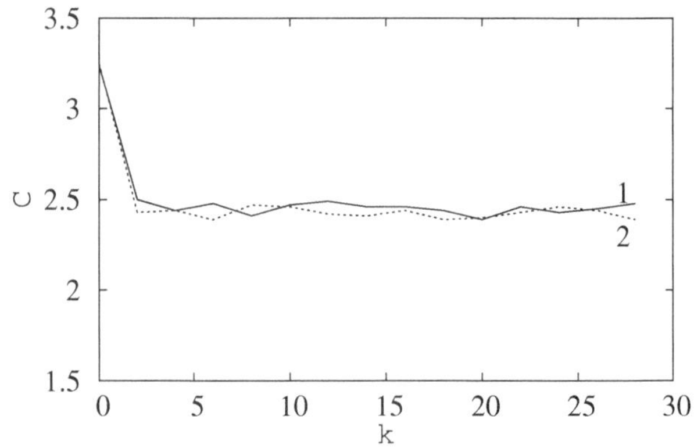

Fig. 8. Dependence of the correlation function $C(k)$ on k, corresponding (1) to the map G^n without noise and (2) with noise (92) for calculations over 2000 points.

where ξ_n is a random variable. At last, assuming that these fluctuations are white noise, we arrive at equations of motion that are similar to the map (92). Hence, the system generates diffusion analogous to that described above. Some numerical experiments seem to verify that behavior (Faith et al., 1997; Roth et al., 1999; Zheng et al., 2012, 2013). It is obvious that the effect of bouncing on the dynamics may be investigated in just the same way.

It should be noted that the term ξ_n in the system acts without significantly changing its statistical properties; moreover, numerical investigations obviously reveal that the SA remains in the sense such that we can't tell the SA of G^n from that of system (92) by whichever sign. However, if we remove the nonlinear term from the phase advance equation, then the term dominates the behavior of the system fully.

Now, we examine the problem of energization of RB electrons subject to resonant coupling with oblique large-amplitude whistler waves. The existence of such waves is confirmed by direct measurements (Cattell et al., 2008; Cully et al., 2008; Kellogg et al., 2011; Kersten et al., 2011; Wilson et al., 2011). In our calculation, we will rely on the relevant experimental data of Cattell et al. (2008) and Wilson et al. (2011). These waves have a broad range of wave normal angles ϑ with respect to the magnetic field (20–60°) and large peak-to-peak amplitudes $E_l \sim (100-200)\ \mathrm{mV\,m^{-1}}$ with $B_t \sim 1\,\mathrm{nT}$, $E_t \sim 100\,\mathrm{mV\,m^{-1}}$. The timescale of the envelope is typically $T \sim 0.2\,\mathrm{s}$, and the mean value of the frequency at peak wave power is $(0.2\text{–}0.25)\omega_B (\sim 1.5\,\mathrm{kHz})$. We have used this data, as well as the dispersion relation $\omega = \omega_B \left(k^2/\omega_{\mathrm{p}}^2\right)\cos\vartheta$ (Stix, 1992) where ω_{p} is the electron plasma frequency ($\sim 1.1 \times 10^5\,\mathrm{s^{-1}}$), to estimate the wavenumber $k \sim 2 \times 10^{-4}\,\mathrm{m^{-1}}$ and group velocity $v_{\mathrm{g}} \sim 5 \times 10^7\,\mathrm{m\,s^{-1}}$.

From Hamiltonian (18) we see that the amplitude of the interaction terms is proportional to Bessel functions, and the Bessel coefficients $J_s(k_t r)$ give the fall-off of the

Fourier amplitudes. Thus we will examine only the resonance of driving with the fundamental of gyromotion, $k_z p_z/E + \omega_B m/E - \omega = 0$. Then, because the Bessel functions tend to zero as $1/\sqrt{k_t r}$ at $k_t r \to \infty$, we have to choose $k_t r \approx 1-3$, which corresponds to the values of $J_1(\cdot)$ near a maximum in the Bessel function. This choice automatically imposes certain restrictions on the energy spectrum of particles. Indeed, taking into account the invariant of motion (25), and the relations $k_t r \sim (1-3), \omega_B/k_z \sim 1, r = \sqrt{2m\omega_B I}/m\omega_B$, we find, without much difficulty, the range of particle energies, in which the wave-particle interaction is most effective, $(k_t/k_z)\sqrt{2E/m} \sim (1-3)$, $E < 10\,\mathrm{MeV}$.

First we will study the problem provided the condition $B_t/B \gg (J_1(\cdot)U_0/J_1'm)\sqrt{E/m}$ holds. By integrating Eqs. (21)–(23) along with invariant of motion (25), we derive, in the usual way, a map that is similar to the g^n map describing the chorus–electron interaction, except that the control parameter contains the term $J_1'(k_t r)(\sim 0.1)$. This similarity means that the given dynamical system manifests itself in just the same way as the g^n map; as a consequence, the heating rate is governed by the same Eq. (90). However, the intensity of coupling, in this case, is determined by the magnitude of the wave field multiplied by the factor $J_1'(\cdot)$. It stands to reason that the expression (91) remains the same, after appropriate modifications, $\sqrt{P} \to \sqrt{P}J_1'(\cdot)$. So these waves are about four times larger in magnitude than those measured by SAMPEX (The Solar Anomalous and Magnetospheric Particle Explorer), and the heating rate of the process is comparable to that caused by chorus–electron coupling.

Let the dynamics be dominated by interaction of the particle with the electrostatic (ES) component of a wave. Taking the wave field to be in the form of a space-like (SL) wave packet (54), we convert the equations of motion into the map,

$$u_{n+1} = u_n + Q\sin\psi_n, \psi_{n+1} = \psi_n + 1/u_{n+1}, \quad \mod 2\pi, \quad (94)$$

where the variable u and the control parameter Q are defined by the relations,

$$u = (E/m\omega T), Q = J_1(\cdot)U_0/m\sqrt{N}, \quad (95)$$

and the spectral characteristic N and the ES potential are given by

$$N = \Delta\nu T, \; U_0 = E_0^l/k_z. \quad (96)$$

The map (94) belongs to the class of Fermi mappings, therefore this map possesses an SA (Lichtenberg and Lieberman, 1992). Shown in Fig. 9 is the SA of the system, whose upper bound is defined by the equations

$$u_\mathrm{b} = \sqrt{Q}, \; E_\mathrm{b} = \omega T\sqrt{Qm}, \quad (97)$$

which is borne out numerically.

The invariant distribution on the SA is a trivial constant, and the time of diffusion in energy is found to be

$$t_\mathrm{d} = (4/Q)T. \quad (98)$$

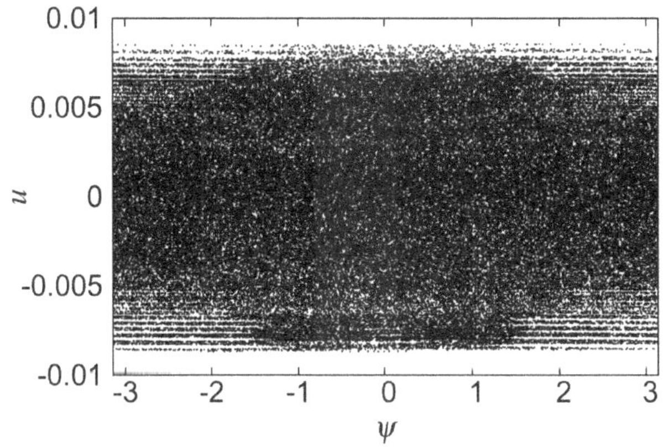

Fig. 9. Phase space $u-\psi$ for the map (94) at $Q = 0.98 \times 10^{-4}$.

We see the extent and rate of heating at this wave-particle interaction are proportional to $U_0 J_1(k_t r)$, just as it is expected. We extract from Eqs. (94) and (98) the following estimates of $E_\mathrm{b} \simeq 5\,\mathrm{MeV}$ and $t_\mathrm{d} \simeq 10^4\,\mathrm{s}$. We have seen that generic, nonlinear behavior in the interaction with oblique waves restrict the extent and rate of particle heating, which, to a certain extent, accounts for the nature of wave-particle interaction.

Electron acceleration in the radiation belts is a well-established fact. Whistler-RB electron coupling ranks among the most important accelerate mechanisms that are likely to play a major role in maintaining the entire pool of RB electrons. Acceleration of RB electrons, which are driven by whistlers up to relativistic energies, can proceed in the very short time of $\sim 1\,\mathrm{h}$ or less, and manifest itself in bursty relativistic electron precipitation events. We have used topological arguments to determine the value of the control parameter at which the dynamical system exhibits chaotic motion, in the form of a strange attractor. In this case, every particle can explore the entire phase space energetically accessible to it; as a result, the upper bound of the strange attractor can be put on a one-to-one correspondence with the upper boundary of an energy spectrum whose value depends parametrically on the spectral power of the wave. The chaotic motion on the strange attractor is ergodic with mixing, and as a consequence the evolution of the distribution function and all means obeys a FPK equation.

As it appears from our model, one thing is certain: the self-consistent nonlinear approach to the RB electron dynamics offers far more perspective than the QLT. There is every reason to believe that an FPK description has turned out to be a true one, after a fashion. The nonlinear approach suggests that it may be possible to obtain theoretically a kind of scaling law for the rate of diffusion that would be applicable to most whistler wave events. Thus, studying the model of nonlinear diffusion can lead to important results, both from the point of view of interpreting experimental data and of obtaining reliable information from the approach.

5 Stochastic acceleration of cosmic-ray electrons

It is well known that high-energy particles observed in cosmic rays can be regarded as tails of the particle distribution in space plasmas. The energy relation (the energy per unit volume of space plasma is approximately equal to the energy per unit volume of cosmic rays) shows that collective processes occurring in the plasma can play an important role in the evolution of the energy spectra of cosmic rays. In this connection, it becomes relevant to determine how the energy should be redistributed in cosmic rays so that the energy spectra could contain particles with very high energies (Kaplan and Tsytovich, 1973; Ginzburg, 1989). Supernova remnant shocks are thought to be the primary source of cosmic rays, because supernova remnants (SNRs) are able to provide the energy necessary to maintain the observed cosmic ray density. Synchrotron emission from SNRs indicates the presence of 100 TeV electrons, and inverse Compton scattering of background photons by ultrarelativistic (UR) electrons is a very plausible explanation for TeV γ-ray emission from young SNRs (Reynolds, 2001; Treumann and Terasawa, 2001; Vink, 2004; Abdo et al., 2009).

The current description of cosmic ray acceleration up to UR energies is the well-known first-order Fermi acceleration at the SNR shock front (Bell, 1978; Blandford and Ostriker, 1978). In the first-order Fermi model, wave turbulence makes the particle momenta isotropic, thus particles can cross the shock front many times. Simulations show that the efficiency of the mechanism depends on the spectra and amplitudes of MHD fluctuations, and their fluctuation amplitudes, and relativistic particle distributions are a natural consequence of the stochastic acceleration by turbulent plasma waves. As the effect of energy loss through synchrotron emission is not included in the model, the gradual high-energy cutoff has been attributed to the balance between the acceleration and escape processes, which leads to a steady-state distribution (Liu et al., 2008; Muranushi and Inutsuka, 2009). In addition to this, it should be noted, the wave-particle interactions in the model are treated in the linear approximation.

There is also another treatment of the problem relating Hamiltonian chaos to nonlinear interactions of particles with plasma waves (Sagdeev et al., 1988; Horton and Ichikawa, 1996). A dynamical system that can be represented by a measure-preserving map and that illustrates the nature of stochasticity was proposed by Fermi as a model for cosmic ray acceleration. There are, as a matter of fact, various versions of this model, all characterized by a phase advance equation that is inversely proportional to the velocity. The review of the class of Fermi maps was made by Lichtenberg and Lieberman (1992). Another important model is the relativistic standard map (RSM), which was independently proposed by Chernikov et al. (1989) and Nomura et al. (1992). The RSM is the relativistic generalization of the standard map (SM), introduced by Chirikov (1979). The standard map is well known for application to a wide class of problems, including the confinement of particles in fusion devices, radio-frequency heating, acceleration, and heating of particles by nonlinear wave packets (Lichtenberg and Lieberman, 1992). Thus, it was natural to study in terms of the RSM how relativistic effects modify the nonlinear motion described by the SM. It is worth noting that these investigations of RSM provide us with a more profound knowledge of many general properties of relativistic chaotic systems. A possible mechanism for strong heating was proposed by Papadopoulos, and the first simulation was performed by Cargill and Papadopoulos (1988). It was shown that collisionless shock waves propagating away from a supernova may be directly responsible for the 10 keV X-ray emission seen in SNRs. A sequence of plasma instabilities between the reflected ions and the background electrons at the foot of the shock front can give rise to rapid anomalous heating of electrons. A somewhat different approach to the problem was taken by Dieckmann et al. (2004). A self-consistent simulation was performed by these authors to show that ions reflected from SNR shocks can excite large-amplitude electrostatic waves through the beam-plasma instability. Another mechanism that has been investigated to explain the acceleration process includes the nonlinear coupling of the beam of cosmic rays escaping away from a supernova, with background Langmuir waves. Estimates of the rate of instability from this type of interaction were made by Ginzburg (1989), who found that the characteristic time of the instability would be about a few years.

As a consequence of this, chaotic dynamics of relativistic electrons is of particular interest. Thus, the nonlinear interaction of high-energy electrons with an extraordinary electromagnetic wave propagating across a given magnetic field was investigated by Zaslavsky et al. (1987). They estimated the rate of the diffusion and found it to be sufficient to accelerate an electron up to UR energies. Chaotic dynamics of relativistic electrons in the spectrum of Langmuir waves has been considered by Klimov and Tel'nikhin (1995) in terms of a map given by the closed pair of nonlinear difference equations. More complete calculations, including the gyroresonance effect were made by Nagornykh and Tel'nikhin (2003) who showed the role of the gyromotion in randomizing the phase of particles. Finally, Krotov and Tel'nikhin (1998) have shown that the stochastic heating of particles by the nonlinear Langmuir wave packet in space plasmas can be regarded as a possible mechanism for the formation of the energy spectrum of cosmic rays.

5.1 Stochastic surfing electron acceleration at Galactic shocks

Besides the Fermi model for cosmic ray acceleration, there is another possible mechanism for the generation of cosmic rays proposed by Teller (1971). In this approach, the shock wave emerging at a supernova explosion is capable of accelerating cosmic ions with the spectrum which quiet good corresponds to the empiric spectrum. However, this spectrum of escaping relativistic ions turns out to be unstable in regard to some plasma instabilities. The self-consistent simulation of Dieckmann et al. (2004) showed that such a beam of relativistic ions can induce a collisionless shock. This shock arises as a consequence of strong nonlinear upper-hybrid (UH) solitary waves which are driven via the beam instability by the ion beam. In this scenario, typical for collisionless shock generation (Artsimovich and Sagdeev, 1979), the shock front is formed by the dissipative process caused by particle heating, and the tail of the wave manifests itself as UH wave perturbations in the foot of the quasi-perpendicular shock. The dispersion law of the electrostatic UH wave (its wave potential is about 1 mc^2) at the foot of shock front is

$$\omega = kv_{\mathrm{f}}, v_{\mathrm{f}} \leq v_{\mathrm{b}}, \omega = \sqrt{\omega_B^2 + \omega_{\mathrm{p}}^2}, \tag{99}$$

the dispersion law of which is a typical condition for plasma modes induced by a beam (Artsimovich and Sagdeev, 1979), where ω_{p} is the electron plasma frequency, and v_{f}, v_{b} are the speed of the front and beam respectively. In this situation electrons can be accelerated up to relativistic energies through the mechanism of surfing (Sagdeev et al., 1988). The model of surfing is in general similar to the Fermi–Ulam model (Lichtenberg and Lieberman, 1992) of acceleration in which the fixed wall plays the role of Lorentz force. This force doesn't change the particle energy and results only in the drift of particles along the front (look at the picture in Fig. 10). The way the electrons gain energy is by picking up wave field energy in multiple encounters with the shock front (Sagdeev et al., 1988).

Taking the original potential of the wave to be of the form

$$\varphi(y,t) = \varphi_0 \cos(ky - \omega t), \tag{100}$$

then Eq. (22) can be written as

$$\dot{I} = \sum_n \varphi_0 n J_n(kr) \sin(n\theta - \omega t), \quad \dot{\theta} = \omega_B m/E, \tag{101}$$

$$E = \sqrt{m^2 + 2m\omega_B I}, \quad \psi = n\theta - \omega t. \tag{102}$$

In this case electron cyclotron resonance heating (ECRH) is accomplished by resonance between the gyrofrequency and the UH wave,

$$\dot{\psi} = s\omega_B m/E_{\mathrm{r}} - \omega = 0, \quad s \in Z_{\mathrm{s}} \subset Z, \tag{103}$$

which is satisfied for a series of s-values for particles of different energies.

To begin with, let us consider the dynamics in a single mode. Choosing a particle resonance $s = l$, we can write Eqs. (101)–(102) as

$$\dot{I} = \varphi_0 l J_l(kr) \sin(\psi), \quad \dot{\psi} = l\omega_B m/E - \omega. \tag{104}$$

Linearizing with Eqs. (104) in the vicinity of the exact resonance, putting of quantities at the resonance, we obtain the equation of phase oscillations

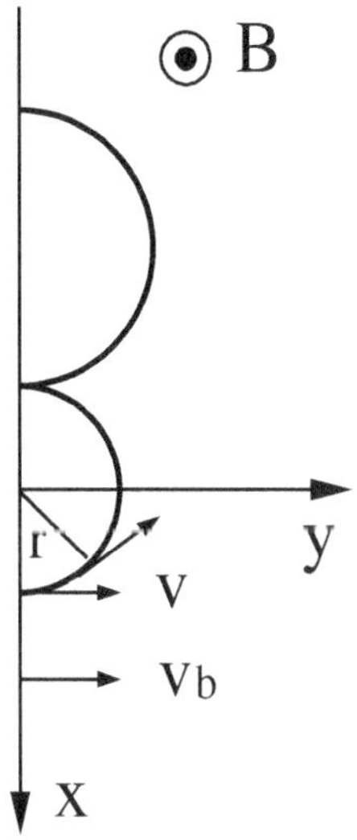

Fig. 10. Particle trajectory at the front of a shock.

$$\ddot{\psi} + \omega_{\mathrm{b}}^2 \sin\psi = 0, \quad \omega_{\mathrm{b}} = \frac{\omega_B}{e}\sqrt{\left(\frac{\omega}{\omega_B}\right)^{5/3} e^{2/3} U_0},$$
$$e = E/m, \quad U_0 = \varphi_0/m, \tag{105}$$

where ω_{b} is the bounce frequency in the wave field. In writing (105) we have taken advantage of the dispersion law (99), the relation $(kr)_{\mathrm{r}} = l$, and used the asymptotic expression of the Bessel function $J_l(l) \approx 1/l^{1/3}$, which is valid for $l \gg 1$. By requiring $\omega_{\mathrm{b}} \geq \delta\omega$, where $\delta\omega = \omega_B/e$ is the spacing between adjacent modes, we obtain the overlap condition

$$e_{\mathrm{c}} \geq (\omega_B/\omega)^{5/3} / U_0^{3/2}, \tag{106}$$

which indicates the value of wave field at which electrons with energy e can be accelerated in random manner. At last, we find from above the characteristic time of the variation of I and ψ is of order $2\pi/\omega_{\mathrm{b}} \sim (2\pi/\omega_B)e$, which essentially exceeds one period $2\pi/\omega$. The small parameter $\varepsilon = \omega_B/\omega e \approx 1/s$ serves as the condition for the applicability of the adiabatic approach, developed by the authors (Khazanov et al., 2010).

The condition of adiabaticity, $\dot{\psi} \ll \omega\psi, \dot{I} \ll \omega I$ allows us to write Eqs. (101)–(102) in the form

$$\dot{\psi} = s\omega_B e^{-1} - \omega, \quad \dot{I} = 2sJ_s(s)\varphi_0 \sin\psi T \sum \delta(t - nT),$$
$$T = 2\pi e/\omega_B. \tag{107}$$

We have carried out the following transformation here,

$$\sum \varphi_0 n J_n(\cdot) e^{in\theta - i\omega t} = \varphi_0 s J_s(s) e^{is\theta - i\omega t} \sum_{n \neq s} e^{in\omega_B t/e},$$

employed the identity $\sum e^{in\omega_B t/e} = T\sum_n \delta(t - nT)$, and taken into account that the wave-particle collisions occur twice during one gyroperiod. After integrating, Eqs. (107) reduce to a map

$$e_{n+1} = e_n + Qe_n^{2/3}\sin\psi_n,\ \ Q = 2(\omega/\omega_B)^{2/3}U_0,$$
$$\psi_{n+1} = \psi_n + 2\pi(\omega/\omega_B)e_{n+1} \mod 2\pi, \tag{108}$$

where $s = (\omega/\omega_B)e(\gg 1)$, $J_s(s) \approx 1/2\pi s^{1/3}$ have been used.

Now we have to obtain the equations of motion under conditions of strong energy losses due to synchrotron emission and inverse Compton scattering, the rate of which is given by

$$\dot{e} = -\beta_s e^2,$$

where $\beta_s \approx 0.5\times10^{-19}\,\mathrm{s}^{-1}$ at $B = 3\mu$G (Ginzburg, 1989). Calculating the loss of energy over the gyroperiod yields

$$\Delta e = -\alpha e^3,\ \alpha = \beta_s/\nu_B,\ \nu_B = \omega_B/2\pi, \tag{109}$$

which is valid if $\Delta e/e \ll 1$, $e < 10^{10}$. Then we combine Eq. (108) and Eq. (109), to obtain a dissipative map

$$e_{n+1} = e_n\left(1-\alpha e_n^2\right) + Qe_n^{2/3}\sin\psi_n,$$
$$\Delta\psi = 2\pi\frac{\omega}{\omega_B}e_{n+1} \mod 2\pi. \tag{110}$$

Then, letting $w(e,n)$ be the energy distribution, we write the FPK equation corresponding to the map (110)

$$\frac{\partial w(e,n)}{\partial n} = \frac{1}{2}\frac{\partial^2}{\partial e^2}(D(e)w) + \frac{\partial}{\partial e}\left(\alpha e^3\right)w,$$
$$D(e) = Q^2e^{4/3}/2, \tag{111}$$

where n is the iteration number. Now the evolution of observable functions can be described by the FPK equation for the moments of $w(e,n)$, $M(m) = \int e^m w(e,n)de$. In this way we have revealed the relaxation time to a stationary state with the mean value

$$\langle e\rangle = e_0 = \left(Q^2/4\alpha\right)^{3/8}, \tag{112}$$

which is equal to $t_d = 1/Q^{3/4}\alpha^{5/8}\nu_B$, and the mean rate of ECRH is determined by

$$\mathrm{d}E/\mathrm{d}t = Q^2\nu_B m/e^{2/3}. \tag{113}$$

It is worth noting that the injection mechanism at low energies, which may be intimately related to particle heating, is the important aspect of cosmic-ray physics. Since the rate of ECRH is proportional to $E^{-2/3}$, this is indeed the case. This stationary distribution, provided w, $\partial w/\partial e$ vanish at infinity, is given by

$$J(e) = \frac{Q^2}{4}\left(e^{4/3}\frac{\partial w}{\partial e} + \frac{4}{3}e^{1/3}f\right) + \alpha e^3 w = 0, \tag{114}$$

with solution

$$w(e) = \frac{\mathrm{const}}{e^{4/3}}\exp\left(-\frac{3}{8}\left(\frac{e}{e_0}\right)^{8/3}\right), \tag{115}$$

where e_0 is a characteristic energy given by Eq. (112). This equation predicts that the power law will depend on $w(e) \propto e^{-4/3}$ at low energies $e < e_0$.

Choosing the following values of quantities $B = 3\mu$G, $\nu_B = 8\,$Hz, and $\omega_p = 2\pi\cdot 8\times10^2\,\mathrm{s}^{-1}$, we evaluate $e_0 = 10^8(E_0 = 50\,\mathrm{TeV}$, $t_d \approx 5\times10^3\,$yr, $\mathrm{d}E/\mathrm{d}t \approx 100/e^{2/3}\,\mathrm{MeV\,s^{-1}})$, and the parameters $Q = 5\text{–}10$, $\alpha = 5\times10^{-21}$. Numerical investigations of Eqs. (110) at these parameters gave the results shown in Fig. 11. This figure indicates the spectrum is bounded both above and below the values of $e_b \approx 4\times10^8(E_b \approx 200\,\mathrm{TeV})$ and $E_c(\approx 14\,\mathrm{eV})$ defined by Eq. (106). In Fig. 11b, the analytical expression (115) is compared with the numerical results. The numerical points of the spectrum in the main range $E < E_0$ all fall full on the curve $w(e) = \mathrm{const}\cdot E^{-4/3}$ as predicted by (115). Then when one takes into account the geometry of problem (Fig. 10) and assume the distribution along the line of sight is isotropic, we can write the differential flux of electrons as

$$\mathrm{d}N/\mathrm{d}E\mathrm{d}\Omega = (c/2\pi)N_e(E_c)(r(E_0)/r)w(E) \tag{116}$$

with the gyroradius r proportional to E, $w(E)$ given by Eq. (115), and where $N_e(E_c)$ is the density of resonant electrons at $E = E_c$. If we substitute $w(E)$ in Eq. (116), this expression becomes

$$\mathrm{d}N/\mathrm{d}E\mathrm{d}\Omega = \mathrm{const}.E^{-7/3}\exp\left[-(3/8)(E/E_0)^{8/3}\right], \tag{117}$$

with the spectral index ~ 2.33 in the range of low energies, $E < E_0$.

Since the electrons radiating their peak spectral power at frequency ν have energies $e = \sqrt{\nu/\nu_s}$, $\nu_s = 1.26\times10^6 B$, we find that the spectral flux density of radiating electrons with the energy spectrum (117) is

$$P_\nu = \mathrm{const}.\nu^{-2/3}\exp\left[-(3/8)(\nu/\nu_0)^{4/3}\right], \tag{118}$$

where $\nu_0 = \nu_s e_0^2$ $(\approx 4\times10^{16}\,\mathrm{Hz})$. Note, the mean radio spectral index is about 0.5 (Reynolds, 2001), or 0.6 (Stevenson and Green, 2002) with a spread of order 0.2.

Inverse Compton scattering of cosmic microwave background (CMB) photons by relativistic electrons is a plausible explanation for TeV γ-ray emission from SNRs. We denote by e_p the energy of a primary CMB photon ($e_p \simeq 10^{-3}\,$eV), and E_γ as the energy of a γ-photon. It is well known (Ginzburg, 1989) that electrons with energies $e \ll m/4e_p$ can scatter only soft photons $E_\gamma = e_p e^2$ that contribute to radio and X-ray spectra, while high-energy electrons $e > m/4e_p$ will scatter photons with energies $E_\gamma \le E$, whose spectrum

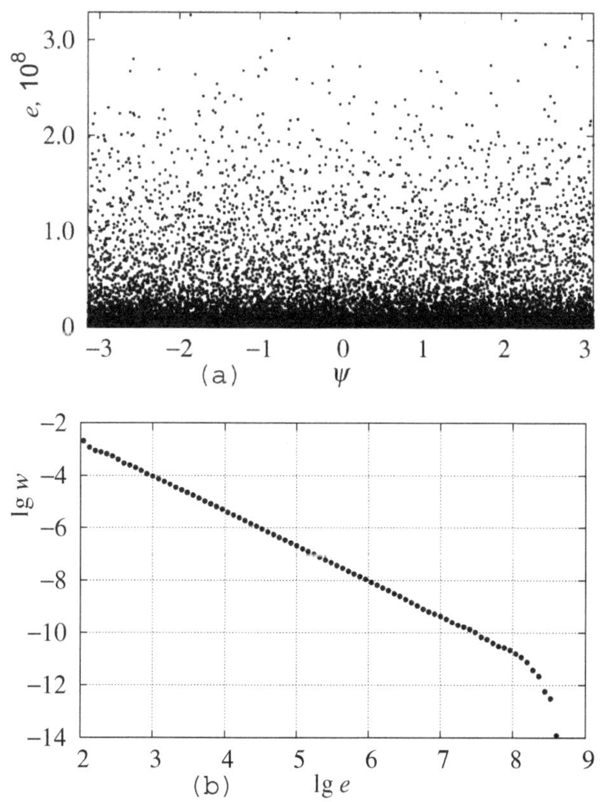

Fig. 11. **(a)** Phase space $e-\psi$ and **(b)** the energy distribution $w(e)$ for the map (110) with parameters $Q=9.2, \alpha=10^{-21}$, and $e_c=10^2$.

is similar to the energy spectrum of electrons. Note that this characteristic energy ($m/4e_{\mathrm{p}} \approx 10^8$) is comparable to the mean energy of electrons. In this case, Compton losses in fact are insensitive to the energy of electrons, and an electron loses its energy with the rate $\mathrm{d}e/\mathrm{d}t = 5\times10^{-4}\,\mathrm{s}^{-1}$ (Ginzburg, 1989). To find the spectrum of TeV γ-photons, we first modify this expression as it has been done above, then after substituting in the FPK equation, the following expression results:

$$P_\nu = \mathrm{const.}\nu^{-2/3}\exp\left[-(3/2)\left(E_\gamma/E_0\right)^{2/3}\right].$$

The tail of the equation describes TeV γ-photon spectrum. This spectrum is quite softer than that given by expression (118); the signature of the spectra observed in SNRs (Reynolds, 2001; Vink, 2004).

Our results allow the inference of an efficiency of shock acceleration of electrons via the electron cyclotron resonance heating and indicate that stochastic surfing acceleration is likely to play an important role in producing relativistic electrons at an SNR shock.

5.2 Stochastic heating of cosmic-ray electrons by a Langmuir wave packet

The interaction of relativistic particles with Langmuir waves is thought to be an important mechanism for producing cosmic rays (Kaplan and Tsytovich, 1973; Ginzburg, 1989). According to Ginzburg (1989), the rate of the Galactic beam-plasma instability, which generates Langmuir waves with the frequency $\omega_{\mathrm{p}} = 3\times10^4\,\mathrm{s}^{-1}$ and minimal wavenumber $k = 10^{-6}\,\mathrm{cm}^{-1}$ is about $10^{-8}\,\mathrm{s}^{-1}$. Thus the time–space scales of such a Langmuir wave packet are likely to be $T = 10^8\,\mathrm{s}$ and $L = 3\times10^{18}\,\mathrm{cm}$, respectively. As long as the group velocity of Langmuir waves is much more less than the speed of light, we concern ourselves with the problem of the nonlinear interaction of relativistic electrons with a spacelike (SL) wave packet provided the Cherenkov resonance condition $kp_z/E-\omega=0$ is met. A particle moving in this wave field is governed by the equations

$$\dot{p}_z = \left(k\varphi_0/\sqrt{\Delta kL}\right)\sin\psi L\sum_n \delta(z-nL),$$
$$\dot{\psi} = kp_z/E-\omega_p. \qquad (119)$$

In the usual way we reduce these equations into the map

$$E_{n+1} = E_n + \sqrt{(k/\Delta k)kL\varphi_0^2}\sin\psi_n,$$
$$\psi_{n+1} = \psi_n + kL - \omega_p L\left(E_{n+1}/\sqrt{E_{n+1}^2-m^2}\right) \mod 2\pi, \qquad (120)$$

which was introduced by Klimov and Tel'nikhin (1995) and Krotov and Tel'nikhin (1998).

In the nonrelativistic limit, these equations reduce to the well-known map

$$E_{n+1} = E_n + \sqrt{(k/\Delta k)kL\varphi_0^2}\sin\psi_n,$$
$$\psi_{n+1} = \psi_n + kL - \mathrm{const.}/\sqrt{E_{n+1}},\quad E = p^2/2m,$$

which describes the heating of particles in an SL wave packet of electrostatic waves (Sagdeev et al., 1988). In the contrary limit, $(E/m)^2 \gg 1$, map (120) in terms of the variables (ψ, u) becomes

$$g^n:\quad \begin{aligned} u_{n+1} &= u_n + Q\sin\psi_n, \\ \psi_{n+1} &= \psi_n + 1/2u_{n+1}^2\,\mathrm{sign}u \mod 2\pi, \end{aligned} \qquad (121)$$

$$u = Ep_z/|p_z|m\sqrt{kL},\quad Q=\sqrt{(k/\Delta k)U_0^2},$$
$$U_0 = \varphi_0/m. \qquad (122)$$

Numerical investigations of the map at $Q \ll 1$ indicated the appearance of the strange attractor (SA), shown in Fig. 12. The upper bound of the SA is given by

$$u_{\mathrm{b}} = Q^{1/3}, \qquad (123)$$

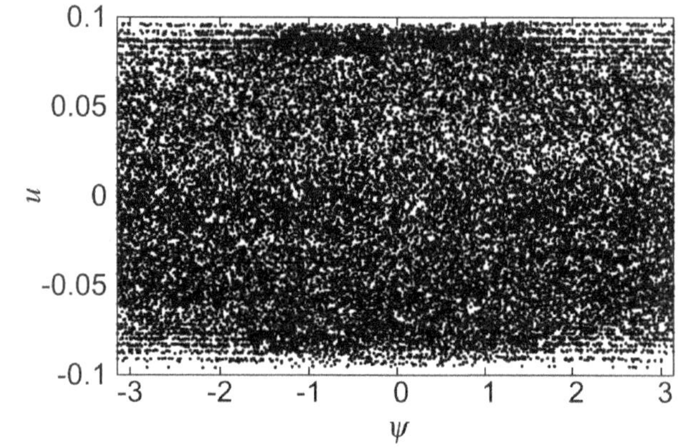

Fig. 12. Strange attractor of the map (121) at $Q = 1.1 \times 10^{-3}$.

which is borne out numerically. The energies over which heating can occur are limited to those given by $E_b = m\sqrt{kL}Q^{1/3}$ ($\sim 100\,\mathrm{GeV}$ at $Q = 10^{-3}$).

The equations from which the SA arises are usually parameterized by some control parameter whose variation changes the character of the dynamics. When the control parameter Q becomes greater than a critical value Q_c, the system exhibits intermittent chaotic behavior shown in Fig. 13. We are already familiar with this kind of dynamics; however, we have never seen such a curious behavior before. It appears that after a finite number of iterations the system demonstrates an intermittent transition to regular unbounded motion. Following the approach of Khazanov et al. (2008b), we describe the motion as the composition $G^t \circ g^n$ (Arnold, 1989), where g^n is the map (121), and G^t is the Hamiltonian flow acting on a smooth manifold M^2. The goal is to find this G^t. The action of G^t is associated with an adiabatic motion, which is realized when the following relations

$$3K/4\pi u^{2/3} \ll 1, \quad Q/u \ll 1, \quad \sin\psi = 0 \tag{124}$$

are valid. Now the evolution of the orbit's phase flow is described by the vector field

$$\boldsymbol{v} = Q\sin\psi\,\partial/\partial u + (1/2u^2)\partial/\partial\psi, \tag{125}$$

on which the family of invariant curves is given by

$$h = Q\cos\psi - 1/2u. \tag{126}$$

This h serves also as a Hamiltonian function of dynamical system (125), the proper values of which are defined by the initial values of (ψ, u) on the SA. Because the behavior of the system is completely determined by these proper values of h, we can find the density of states in the space of parameters, $\rho(h) = \tau = \mathrm{d}\Gamma/\mathrm{d}h$, where $\tau = t/T$ is the time of motion along the orbit, and $\Gamma(h) = \oint_{C_h} u(h)\mathrm{d}\psi$ is the phase area bounded by the invariant curve C_h (Arnold, 1988). Invoking

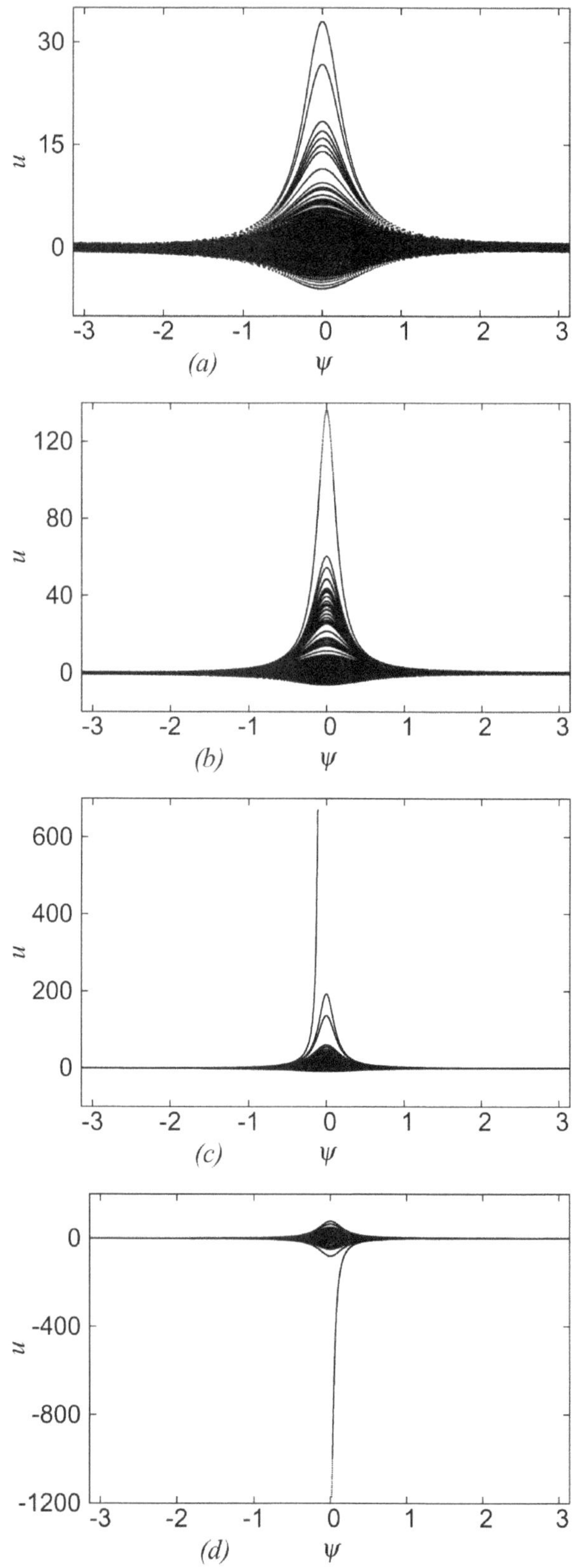

Fig. 13. Phase space u, ψ for the map (121) at $Q = 0.367$. Number of iterations increasing from **(a)–(c)** and **(d)** as $N = 4.2 \times 10^4$; 8×10^4; 2.2×10^5.

Eq. (126), we obtain

$$\rho(h) = \tau = 2\pi|h| \Big/ \left(h^2 - Q^2\right)^{3/2}, \quad |h| \geq Q. \tag{127}$$

Arising singularity in distribution (127) at $|h| = Q$, we can characterize its ε-vicinity. As the radius of the ε-vicinity tends to 0, it makes the mean density tends to infinity. As a consequence, the states $|h| = Q$ are the most likely states of the system. In finding an explicit expression for Q_c, we use Eq. (126) with $|h| = Q$ to find $|u_{\min}| = 1/4Q$. So far as all these orbits start from the SA, the $u_{\min}$ belongs to the SA. Then by equating $u_{\min}$ to the u_b given by Eq. (123), we obtain $Q_c = 1/2\sqrt{2}$, which matches quite well to the numerical values. If we substitute Q_c in (122), assuming $k/\Delta k = 100$, and $k = 10^{-6}\,\mathrm{cm}^{-1}$, these required magnitudes of wave field turn out to be $U_c \simeq 3 \times 10^{-2}$, and $(E_l)_c \sim 10^{-2}\,\mathrm{mV\,cm}^{-1}$.

Now we formulate the problem of dynamics in terms of the variables $(u, \dot{u})$. These variables allow us to obtain more fine information about the geometric structure of phase space. Carrying out the procedure, we derive the following equations:

$$W = \frac{1}{2}\dot{u}^2 + w(u), \quad w(u) = h/2u + 1/8u^2, \tag{128}$$

where W is an invariant of motion. To our surprise, we run up against the old Kepler problem here. This can be seen directly from (128) by defining the effective potential $w(u)$. It is worth noting that this problem, unlike the Kepler one, is invariant under the transformation $(-h, u) \to (h, -u)$. Until the proper values of W are less than 0, the system exhibits bounded motion; in another case, which is realized for example, at $h = -Q$ and $u_{\min} = 1/4Q$, the motion becomes unbounded. Strange as it may appear, the singularity in distribution (127) leads to a topological modification of phase space. At first, changes in topology emerge provided the conditions $Q > Q_c$ but $|h| > Q$ hold. In this case, the topological space is a direct sum of the strange attractor and torus. There is another change in topology at $|h| = Q$. As it appears from Eqs. (125)–(126), the trajectory approaches infinity at $\psi = 0$, or $\pm\pi$, and the vector field vanishes. It is known, on the other side, any vector field on a torus vanishes nowhere (Arnold, 1989); consequently this kind of dynamics is associated with a metamorphism of phase space.

Now we consider the dynamics of the system in the off-resonant case, i.e., when the phase velocities of the waves in the packet differ from the speed of light. Setting $(\omega_p T - kL)\mathrm{mod}2\pi = \Omega$, $\Omega > 0$, to be the phase shift in the phase advance equation (121), we numerically investigate the behavior of the system. Shown in Fig. 14 is the phase space evolution of the system. We cannot resist the fascination of the strange behavior. Trying to understand it, we write the phase flow associated with the Hamiltonian,

$$h = Q\cos\psi - 1/2u - \Omega, \tag{129}$$

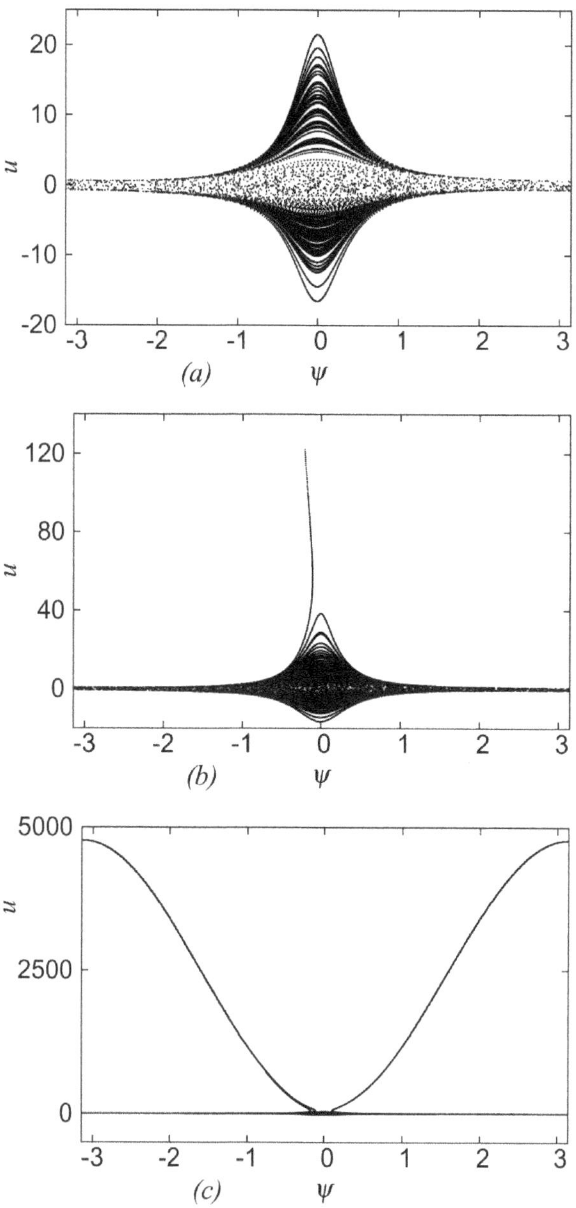

Fig. 14. Dynamics of the system (121) in the off-resonant case. Here $Q = 0.367$ and $w = \pi/2 \times 10^{-4}$. **(a)** $2 \cdot 10^4$, **(b)** 3.5×10^4, and **(c)** 8×10^4 iterations of a single initial condition.

in the form

$$\dot{u} = Q\sin\psi, \dot{\psi} = -\Omega + 1/2u^2. \tag{130}$$

The proper values of h in turn determine a family of invariant curves on the vector field

$$\boldsymbol{v} = Q\sin\psi\,\partial/\partial u + \left(1/2u^2 - \Omega\right)\partial/\partial\psi. \tag{131}$$

It is easily seen that the function h changes its sign, vector field (131) is invariant under the transformation

$$u \to -u, \ \psi' \to \psi + \pi, \tag{132}$$

and the fixed points of the vector field are given by

$$\psi_0 = 0, \ |u_0| = 1/\sqrt{2\Omega}. \tag{133}$$

One considers these dynamics in a configuration space proceeding to the variable $(u, \dot{u})$. The procedure for deriving the invariant of motion

$$W = \frac{1}{2}\dot{u}^2 + w(u), \ w(u) = h/2u + 1/8u^2 + h\Omega u + \Omega^2 u^2/2, \tag{134}$$

is the same at that used for obtaining (128). This potential is symmetric with respect to $(-h, u) \to (h, -u)$, and the turning points are situated next to $|u_{\min}| = 1/4Q$, $|u_{\max}| = 2Q/\Omega$. These $u_{\max}$ determine a monochromatic energy state $E_{\max} = 2m\sqrt{kL}Q/\Omega$ with different directions of p_z momentum. The potential $w(u)$ has a local maximum at the fixed points, and the motion near these points obeys the equation $\ddot{\psi} + (Q/u_0^3)\sin\psi = 0$, which shows just the same invariance (132), and the tangent vector at a point of fixed points changes its direction. Thus there are two allowable local coordinate systems (ψ', u') and (ψ, u), whose determinant of the transformation matrix from one to another, $\begin{vmatrix} \partial\psi'/\partial\psi & 0 \\ 0 & \partial u'/\partial u \end{vmatrix}$, is equal to -1. There is very good reason to believe that this manifold is a direct sum of the strange attractor and the Mobius strip. In the resonant case, the width of this strip approach infinity. Therefore, the off-resonant effect eliminates a possibility of unbounded motion. To describe the effect of synchrotron emission and inverse Compton scattering on the dynamics, we again apply the method developed above (Sect. 4.1, formula (109)). In this way we derive the following equations

$$\begin{aligned} u_{n+1} &= u_n(1 - \alpha|u_n|) + Q\sin\psi_n, \\ \psi_{n+1} &= 1/2u_{n+1}^2 \operatorname{sign} u \,(\operatorname{mod} 2\pi), \end{aligned} \tag{135}$$

where $\alpha = \beta_s T\sqrt{kL}\,(\sim 8\times 10^{-6})$, and all another variables and the parameter Q remain the same. When the wave perturbation is small, $Q \sim 10^{-3}$, the system exhibits chaotic behavior on a strange attractor. For the dissipative system the phase space volume contracts as the motion proceeds. In virtue of it, the topology of the SA is quite remarkable, showing the property of fractional dimensionality, $d \simeq 2 - 2\alpha Q^{1/3}/3h_k$, where h_k is the Kolmogorov entropy of the original Hamiltonian system. For the parameter range of interest, numerical investigations of the system motion are shown in Fig. 15. The figure displays the emphatic intermittent behavior. It also proves possible to represent the phase flow as

$$\begin{aligned} \dot{u} &= -\alpha u^2 + Q\sin\psi, \\ \dot{\psi} &= 1/2u^2, \end{aligned}$$

the fixed points of which are given by the equation

$$\alpha u_0^2 = Q\sin\psi_0. \tag{136}$$

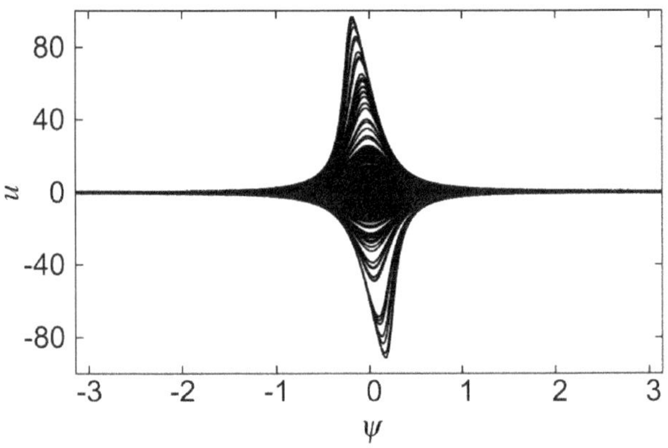

Fig. 15. Phase space for dissipative map (135) with $Q = 0.4$ and $\alpha = 8\times 10^{-6}$.

Equation (136) defines the phase shift away from the resonance, and in implicit form determines the range of accessible proper values of u, $u < \sqrt{Q/\alpha}$ (~ 200) in virtue of Eq. (122); the upper value of energy should be about 100 TeV. The estimates seem to verify this inference.

We propound the model of the stochastic dynamics of relativistic electrons driven by the Langmuir wave packet. Whether or not a model like this can be applied to the cosmic-ray electron acceleration, seemingly, will be a problem for future research, when such a kind of model will need to be compared with data at this level of precision to give confidence in the results (Reynolds, 2001).

6 Conclusions

We study the motion of relativistic electrons interacting resonantly with space-plasma waves. The nonlinear resonant interaction is shown to originate electron chaotic motion related to the profound change in topology of phase space, and the appearance of the strange attractor. In this case, every particle can explore the entire area of phase space energetically accessible to it; as a result, the upper bound of the strange attractor can be put on a one-to-one correspondence with the upper boundary of an energy spectrum whose value depends parametrically on the spectral power of the wave. The chaotic motion on the strange attractor is ergodic with mixing, and as a consequence, the evolution of the distribution function and all means obeys a FPK equation. In nonlinear theory, the rate of diffusion as a function of wave power energy hinges on the type of interaction. These dependencies have important consequences in determining the range of values of wave power for which the FPK formalism is valid. The results are applied to the problem of acceleration of cosmic-ray electrons and the Earth's radiation belt electrons.

Acknowledgements. Funding in support of this study was provided by NASA grant UPN 370-16-10, the NASA HQ POLAR project, and the NASA LWS program.

Edited by: G. Morales

References

Abdo, A. A., Ackermann, M., Ajello, M., Asano, K., Atwood, W. B., Axelsson, M., Baldini, L., Ballet, J., Barbiellini, G., Baring, M. G., Bastieri, D., Bechtol, K., Bellazzini, R., Berenji, B., Bhat, P. N., Bissaldi, E., Blandford, R. D., Bloom, E. D., Bonamente, E., Borgland, A. W., Bouvier, A., Bregeon, J., Brez, A., Briggs, M. S., Brigida, M., Bruel, P., Burgess, J. M., Burrows, D. N., Buson, S., Caliandro, G. A., Cameron, R. A., Caraveo, P. A., J., Casandjian, M., Cecchi, C., Çelik, Ö., Chekhtman, A., Cheung, C. C., Chiang, J., Ciprini, S., Claus, R., Cohen-Tanugi, J., Cominsky, L. R., Connaughton, V., Conrad, J., Cutini, S., d'Elia, V., Dermer, C. D., de Angelis, A., de Palma, F., Digel, S. W., Dingus, B. L., do Couto e Silva, E., Drell, P. S., Dubois, R., Dumora, D., Farnier, C., Favuzzi, C., Fegan, S. J., Finke, J., Fishman, G., Focke, W. B., Fortin, P., Frailis, M., Fukazawa, Y., Funk, S., Fusco, P., Gargano, F., Gehrels, N., Germani, S., Giavitto, G., Giebels, B., Giglietto, N., Giordano, F., Glanzman, T., Godfrey, G., Goldstein, A., Granot, J., Greiner, J., Grenier, I. A., Grove, J. E., Guillemot, L., Guiriec, S., Hanabata, Y., Harding, A. K., Hayashida, M., Hays, E., Horan, D., Hughes, R. E., Jackson, M. S., Jóhannesson, G., Johnson, A. S., Johnson, R. P., Johnson, W. N., Kamae, T., Katagiri, H., Kataoka, J., Kawa, N., Kerr, M., Kippen, R. M., Knödlseder, J., Kocevski, D., Komi, N., Kouveliotou, C., Kuss, M., Lande, J., Latronico, L., Lemoine-Goumard, M., Longo, F., Loparco, F., Lott, B., Lovellette, M. N., Lubrano, P., Madejski, G. M., Makeev, A., Mazziotta, M. N., McBreen, S., McEnery, J. E., McGlynn, S., Meegan, C., Mészáros, P., Meurer, C., Michelson, P. F., Mitthumsiri, W., Mizuno, T., Moiseev, A. A., Monte, C., Monzani, M. E., Moretti, E., Morselli, A., Moskalenko, I. V., Murgia, S., Nakamori, T., Nolan, P. L., Norris, J. P., Nuss, E., Ohno, M., Ohsugi, T., Omodei, N., Orlando, E., Ormes, J. F., Paciesas, W. S., Paneque, D., Panetta, J. H., Pelassa, V., Pepe, M., Pesce-Rollins, M., Petrosian, V., Piron, F., Porter, T. A., Preece, R., Rainò, S., Rando, R., Rau, A., Razzano, M., Razzaque, S., Reimer, A., Reimer, O., Reposeur, T., Ritz, S., Rochester, L. S., Rodriguez, A. Y., Roming, P. W. A., Roth, M., Ryde, F., Sadrozinski, H. F.-W., Sanchez, D., Sander, A., Saz Parkinson, P. M., Scargle, J. D., T. L., Schalk, C. Sgrò, E. J., Siskind, P. D., Smith, P., Spinelli, Stamatikos, M., Stecker, F. W., Stratta, G., Strickman, M. S., Suson, D. J., Swenson, C. A., Tajima, H., Takahashi, H., Tanaka, T., Thayer, J. B., Thayer, J. G., Thompson, D. J., Tibaldo, L., Torres, D. F., Tosti, G., Tramacere, A., Uchiyama, Y., Uehara, T., Usher, T. L., van der Horst, A. J., Vasileiou, V., Vilchez, N., Vitale, V., von Kienlin, A., Waite, A. P., Wang, P., Wilson-Hodge, C., Winer, B. L., Wood, K. S., Yamazaki, R., Ylinen, T., and Ziegler, M.: Fermi Observations of GRB 090902B: A Distinct Spectral Component in the Prompt and Delayed Emission, Astrophys. J. Lett., 706, L138 doi:10.1088/0004-637X/706/1/L138, 2009.

Albert, J. M.: Nonlinear interaction of outer zone electrons with VLF waves, Geopys. Res. Lett., 29, 116-1–116-3, doi:10.1029/2001GL013941, 2002.

Arnold, V.: Geometrical Methods in the Theory of Ordinary Differential Equations, Springer-Verlag, New York, 1988.

Arnold, V.: Mathematical Methods of Classical Mechanics, Springer-Verlag, New York, 1989.

Arnold, V. and Avez, A.: Ergodic Problems of Classical Mechanics, Benjamin, New York, 1968.

Artsimovich, L. and Sagdeev, R.: Plasma Physics for Physicists, Atom Press, Moscow, 1979 (in Russian).

Bell, A. R.: The acceleration of cosmic rays in shock fronts, Mon. Not. Roy. Astron. Soc., 182, 147–156, 1978.

Bernstein, I., Greene, J. M., and Kruskal, M. D.: Exact Nonlinear Plasma Oscillations, Phys. Rev., 108, 546–550, doi:10.1103/PhysRev.108.546, 1957.

Blake, J. B., Baker, D. N., Turner, N., Ogilvie, K. W., and Lepping, R. P.: Correlation of changes in the outer-zone relativistic-electron population with upstream solar wind and magnetic field measurements, Geophys. Res. Lett., 24, 927–929, doi:10.1029/97GL00859, 1997.

Blandford, R. and Ostriker, J.: Particle acceleration by astrophysical shocks, Astrophys. J., 221, L29–L32, 1978.

Cargill, P. J. and Papadopoulos, K.: A mechanism for strong shock electron heating in supernova remnants, Astrophys. J., 329, L29–L32, doi:10.1086/185170, 1988.

Cattell, C., Wygant, J. R., Goetz, K., Kersten, K., Kellogg, P. J., von Rosenvinge, T., Bale, S. D., Roth, I., Temerin, M., Hudson, M. K., Mewaldt, R. A., Wiedenbeck, M., Maksimovic, M., Ergun, R., Acuna, M., and Russell, C. T.: Discovery of very large amplitude whistler-mode waves in Earth's radiation belts, Geophys. Res. Lett., 35, L01105, doi:10.1029/2007GL032009, 2008.

Chernikov, A. A., Tel, T., Vattay, G., and Zaslavsky, G. M.: Chaos in the relativisti generalization of the standard map, Phys. Rev. A, 40, 4072–4076, doi:10.1103/PhysRevA.40.4072, 1989.

Chirikov, B.: A universal instability of many-dimensional oscillator systems, Phys. Rep., 52, 263–379, 1979.

Cully, C. M., Bonnell, J. W., and Ergun, R. E.: THEMIS observations of long-lived regions of large-amplitude whistler waves in the inner magnetosphere, Geophys. Res. Lett., 35, L17S16, doi:10.1029/2008GL033643, 2008.

Dieckmann, M., Eliasson, B., and Shukla, P.: Self-consistent Studies of Electron Acceleration to Ultrarelativistic Energies by Upper Hybrid Waves, Astrophys. J., 617, 1361–1370, doi:10.1086/425568, 2004.

Faith, J., Kuo, S., and Huang, J.: Electron precipitation caused by chaotic motion in the magnetosphere due to large-amplitude whistler waves, J. Geophys. Res., 102, 2233–2241, doi:10.1029/96JA01968, 1997.

Ginzburg, V.: Applications of Electrodynamics in Theoretical Physics and Astrophysics, Gordon and Breach, New York, 1989.

Green, J. C., Onsager, T. G., O'Brien, T. P., and Baker, D. N.: Testing loss mechanisms capable of rapidly depleting relativistic electron flux in the Earth's outer radiation belt, J. Geophys. Res., 109, A12211, doi:10.1029/2004JA010579, 2004.

Hastings, D. and Garrett, H.: Spacecraft-Environment Interactions, Cambridge University Press, New York, 1996.

Horton, W. and Ichikawa, Y. H.: Chaos and Structures in Nonlinear Plasmas, World Scientific, Singapore, 1996.

Kaplan, S. A. and Tsytovich, V. N.: Plasma Astrophysics, Pergamon Press, Oxford, 1973.

Karpman, V.: Nonlinear Waves in Dispersive Media, Pergamon, New York, 1975.

Kaup, D. J. and Newell, A. C.: Solitons as Particles, Oscillators, and in Slowly Changing Media: A Singular Perturbation Theory, P. R. Soc. A, 361, 413–446, doi:10.1098/rspa.1978.0110, 1978.

Kellogg, P. J., Cattell, C. A., Goetz, K., Monson, S. J., and Wilson III, L. B.: Large amplitude whistlers in the magnetosphere observed with Wind-Waves, J. Geophys. Res., 116, A09224, doi:10.1029/2010JA015919, 2011.

Kersten, K., Cattell, C. A., Breneman, A., Goetz, K., Kellogg, P. J., Wygant, J. R., Wilson III, L. B., Blake, J. B., Looper, M. D., and Roth, I.: Observation of relativistic electron microbursts in conjunction with intense radiation belt whistler-mode waves, Geophys. Res. Lett., 38, L08107, doi:10.1029/2011GL046810, 2011.

Khazanov, G., Tel'nikhin, A., and Kronberg, T.: Dynamic theory of relativistic electrons stochastic heating by whistler mode waves with application to the Earth magnetosphere, J. Geophys. Res., 113, A03207, doi:10.1029/2007JA012488, 2008a.

Khazanov, G., Tel'nikhin, A., and Kronberg, T.: Nonlinear electron motion in a coherent whistler wave packet, Phys. Plasmas, 15, 073506, doi:10.1063/1.2959121, 2008b.

Khazanov, G., Tel'nikhin, A., and Kronberg, T.: Bursty precipitation driven by chorus waves, in: Radiation Effects and Defects in Solids: Incorporating Plasma Science and Plasma Technology, Special Issue: Plasma: Fusion and Environments, 166, 795–805, doi:10.1080/10420150.2011.610319, 2011.

Khazanov, G. V.: Kinetic Theory of the Inner Magnetosheric Plasma, Astrophysics and Space Science Library 372, Springer Science, New York, 2011.

Khazanov, G. V., Tel'nikhin, A., and Kronberg, T.: Chaotic motion of relativistic electrons driven by whistler waves, Plasma Phys. Contr. F., 49, 447–466, doi:10.1088/0741-3335/49/4/008, 2007.

Khazanov, G. V., Tel'nikhin, A. A., and Krotov, A.: Stochastic ion heating by the lower-hybrid waves, in: Radiation Effects and Defects in Solids: Incorporating Plasma Science and Plasma Technology, Special Issue: Plasma Science and Plasma Technology, 165, 165–176, doi:10.1080/10420150903516684, 2010

Klimov, O. and Tel'nikhin, A.: Stochastic dynamics of a relativistic particle in the field of a wave packet, Plasma Phys. Rep., 21, 335–338, 1995.

Krotov, A. and Tel'nikhin, A.: Stochastic heating of relativistic particles in the field of a wide wave packet, Plasma Phys. Rep., 24, 767–771, 1998.

Landau, L. and Lifshitz, E.: The Classical Theory of Fields, vol. 2, Butterworth-Heinemann, Waltham, 1980.

Lichtenberg, A. and Lieberman, M.: Regular and Chaotic Dynamics, Springer, New York, 1992.

Liemohn, M. W.: Introduction to special section on "Results of the National Science Foundation Geospace Environment Modeling Inner Magnetosphere/Storms Assessment Challenge", J. Geophys. Res., 111, A11S01, doi:10.1029/2006JA011970, 2006.

Liu, S., Fan, Z., Fryer, C., Wang, J., and Li, H.: Stochastic Electron Acceleration in Shell-Type Supernova Remnants, Astrophys. J. Let.., 683, L163–L166, doi:10.1086/591793

Lutomirski, R. and Sudan, R.: Exact Nonlinear Electromagnetic Whistler Modes, Phys. Rev., 147, 156–165, doi:10.1103/PhysRev.147.156, 1966.

Lyons, L. and Williams, D.: Quantitative aspects of magnetospheric physics, Reidel, Boston, 1984.

Muranushi, T. and Inutsuka, S.: Direct Simulations of Particle Acceleration in a Fluctuating Electromagnetic Field Across a Shock, Astrophys. J., 691, L24–L26, doi:10.1088/0004-637X/691/1/L24, 2009.

Nagornykh, E. and Tel'nikhin, A.: Stochastic Motion of Relativistic Particles in the Field of a Wide Wave Packet, in: 11th International Congress on Plasma Physics: ICPP2002, 15–19 July 2002, Sydney, Australia, edited by: Falkoner, I. S., Dewar, R. L., and Khachan, J., American Institute of Physics (AIP), AIP Conference Proceedings, 669, 824–827, doi:10.1063/1.1594057, 2003.

Newell, A.: Solitons in Mathematics and Physics, SIAM, Philadelphia, 1985.

Nomura, Y., Ichikawa, Y. H., and Horton, W.: Nonlinear dynamics of the relativistic standard map, Phys. Rev. A, 45, 1103–1115, doi:10.1103/PhysRevA.45.1103, 1992.

O'Brien, T. P., Looper, M. D., and Blake, J. B.: Quantification of relativistic electron microburst losses during the GEM storms, Geophys. Res. Lett., 31, L04802, doi:10.1029/2003GL018621, 2004.

Reynolds, S.: Synchrotron Radiation from Galactic Sources: What We Can Learn About Particle Acceleration, Space Sci. Rev., 99, 177–186, doi:doi:10.1023/A:1013892931294, 2001.

Richtmyer, R.: Principles of Advanced Mathematical Physics, vol. 1, Springer-Verlag, New York, 1978.

Roth, I., Temerin, M., and Hudson, M. K.: Resonant enhancement of relativistic electron fluxes during geomagnetically active periods, Ann. Geophys., 17, 631–638, doi:10.1007/s00585-999-0631-2, 1999.

Sagdeev, R., Usikov, D., and Zaslavsky, G.: Introduction to Nonlinear Physics, Harwood, New York, 1988.

Santolik, O., Gurnett, D. A., Pickett, J. S., Parrot, M., and Cornilleau-Wehrlin, N.: Spatio-temporal structure of storm-time chorus, J. Geophys. Res., 108, 1278, doi:10.1029/2002JA009791, 2003.

Schmidt, G.: Physics of High Temperature Plasmas, Academic Press, New York, 1979.

Schutz, B.: Geometrical methods of Mathematical Physics, Cambridge University Press, Cambridge, London, 1982.

Selesnick, R. S.: Source and loss rates of radiation belt relativistic electrons during magnetic storms, J. Geophys. Res., 111, A04210, doi:10.1029/2005JA011473, 2006.

Selesnick, R. S., Blake, J. B., Kolasinski, W. A., and Fritz, T. A.: A quiescent state of 3 to 8 MeV radiation belt electrons, Geophys. Res. Lett., 24, 1343–1346, doi:10.1029/97GL51407, 1997.

Stevenson, F. and Green, D.: Historical Supernovae and Their Remnants, Clarendon Press, Oxford, 2002.

Stix, T.: Waves in Plasmas, AIP Press, New York, 1992.

Summers, D., Ni, B., and Meredith, N. P.: Timescales for radiation belt electron acceleration and loss due to resonant wave-particle interactions: 2. Evaluation for VLF chorus, ELF hiss, and electromagnetic ion cyclotron waves, J. Geophys. Res., 112, A04207, doi:10.1029/2006JA011993, 2007.

Teller, E.: Relativistic hydrodynamics in Supernovae, in: Physics of High Energy Density, edited by: Caldiroba, P. and Knoepfel, H., chap. 15, Academic Press, New York, 1971.

Treumann, R. A. and Terasawa, T.: Electron Acceleration in the Heliosphere, Space Sci. Rev., 99, 135–150, doi:10.1023/A:1013836813547, 2001.

Tsurutani, B. T., Falkowski, B. J., Verkhoglyadova, O. P., Pickett, J. S., Santolík, O., and Lakhina, G. S.: Quasi-coherent chorus properties: 1. Implications for wave-particle interactions, J. Geophys. Res., 116, A09210, doi:10.1029/2010JA016237, 2011.

Vink, J.: Shocks and particle acceleration in supernova remnants: observational features, Adv. Space Res., 33, 356–365, doi:10.1016/j.asr.2003.05.012, 2004.

Whitham, G. B.: Linear and Nonlinear Waves, Wiley, New York, 1974.

Wilson, L. B., Cattell, C. A., Kellogg, P. J., Wygant, J. R., Goetz, K., Breneman, A., and Kersten, K.: The properties of large amplitude whistler mode waves in the magnetosphere: Propagation and relationship with geomagnetic activity, Geophys. Res. Lett., 38, L17107, doi:10.1029/2011GL048671, 2011.

Wykes, W. J., Chapman, S., and Rowlands, G.: Enhanced phase space diffusion due to chaos in relativistic electron-whistler mode wave particle interactions with applications to Jupiter, Planet. Space Sci., 49, 395–404, doi:10.1016/S0032-0633(00)00160-4, 2001.

Zaslavsky, G., Natenson, M., Petrovichev, B., Sagdeev, R., and Chernikov, A.: Stochastic acceleration of relativistic particles in magnetic field, Eks. Teor. Fiz. (J. Exp. Theor. Phys.), 93, 1987.

Zheng, Q., Zheng, Y., Fok, M.-C., and Lui, A. T. Y.: Electron energy diffusion and advection due to non-linear electron-chorus wave interactions, J. Atmos. Sol.-Terr. Phy., 80, 152–160, doi:10.1016/j.jastp.2012.01.011, 2012.

Zheng, Q., Fok, M.-C., Zheng, Y., and Lui, A. T. Y.: Nonlinear whistler mode wave effects on magnetospheric energetic electrons, J. Atmos. Sol.-Terr. Phy., 102, 8–16, doi:10.1016/j.jastp.2013.03.018, 2013.

2

Brief Communication: Weibel, Firehose and Mirror mode relations

R. A. Treumann[1,2] **and W. Baumjohann**[3]

[1]Department of Geophysics and Environmental Sciences, Munich University, Munich, Germany
[2]International Space Science Institute, Bern, Switzerland
[3]Space Research Institute, Austrian Academy of Sciences, Graz, Austria

Correspondence to: R. A. Treumann (rudolf.treumann@geophysik.uni-muenchen.de)

Abstract. Excitation of Weibel magnetic fields in an initially non-magnetized though anisotropic plasma may trigger other low-frequency instabilities fed by pressure anisotropy. It is shown that under Weibel-like stable conditions the Weibel-like thermal fluctuation magnetic field allows for restricted Firehose-mode growth. In addition, low-frequency Whistlers can also propagate in the plasma under certain anisotropic conditions. When the Weibel-like mode becomes unstable, Firehose instability ceases but Mirror modes take over. This will cause bubble structures in the Weibel-like field in addition to filamentation.

1 Introduction

Historically, there are just three celebrated fundamental very low-frequency (electro-)magnetic instabilities in hot anisotropic plasmas, the well-known Firehose mode (Vedenov et al., 1961; Treumann and Baumjohann, 1997, for a plasma physics textbook), its complementary equivalent, the Mirror mode, and the Weibel instability (Weibel, 1959; Yoon and Davidson, 1987, and others). Recently, Schlickeiser et al. (2011) and Schlickeiser and Skoda (2011), applying a substantially more rigorous relativistic approach based on the theory of analytical functions, identified a much larger number of different electromagnetic low-frequency modes, weakly and strongly damped/unstable ones, which add additional dispersion channels to a magnetized plasma. Felten and Schlickeiser (2013a, b, c) and Felten et al. (2013) extended these calculations to relativistic Weibel-like (non-magnetic) conditions, again finding a large number of different dispersion channels which belong to damped or unstable modes, including very low-frequency and non-oscillating aperiodic modes.

Of the historical modes, the first is a general bulk plasma mode excited in an external magnetic field by a thermal anisotropy with larger magnetically parallel than perpendicular temperature, $T_\perp > T_\|$, resulting in Alfvén waves which radiate along the magnetic field. On the contrary, the latter Weibel-like mode acts in non-magnetized plasmas when, for some not further specified reason, the plasma exhibits a thermal (pressure) anisotropy with higher temperature (kinetic energy) in one than in the two other directions. In both the Firehose and Weibel-like cases the cause of a higher *equivalent* temperature in one direction (and thus a temperature anisotropy) can also be a fast (relative) streaming (for the physical mechanism of the Weibel-like streaming mode see Fried, 1959) of the plasma with kinetic energy exceeding the transverse thermal energy. This can be provided by beam or counter-streaming beam configurations (cf., e.g., Achterberg and Wiersma, 2007) and has been made responsible for the generation of magnetic fields under various non-dynamo conditions occurring, for instance, in shock waves (for a review see, e.g., Treumann, 2009), preferentially in relativistic shocks (for relativistic shocks see the review by Bykov and Treumann, 2011). The physical differences in the two modes are large. While Weibel-like modes provide a non-dynamo mechanism to *produce* quasi-stationary and thus non-propagating magnetic fields in an otherwise non-magnetized plasma, the Firehose mode grows on an existing field and propagates along the field at Alfvén velocity V_A thus transporting energy away from the region where it is excited, filling a large volume with magnetic fluctuations and contributing to turbulence and other effects. However, since both modes are generated by similar mechanisms though being different, having completely different properties, one expects that they will compete and possibly even act in tandem to generate magnetic fields and propagate them away from the source region.

The remaining Mirror mode, on the contrary, is an about stationary feature of the plasma with wave vector $\boldsymbol{k}$ almost perpendicular to the ambient magnetic field (cf., e.g., Baumjohann and Treumann, 2012). It generates plasma inhomogeneity at the lowest frequency of plasma turbulence. It is easily derived from anisotropic fluid theory, but its physical mechanism is attributed to trapping of particles in depleted magnetic field regions along the field. This mechanism is more complex and still not completely resolved (for a recent more complete account of the electron mirror mode which concerns us here, see, e.g., Pokhotelov et al., 2008, 2010, 2013, and references therein).

In the present note we briefly examine this situation at the example of the classical Weibel instability, showing that there indeed exists such a competition which may become important in limiting the growth of the Weibel-like modes, allowing other fluctuations to propagate on the magnetic Weibel-like background field which may contribute to distribution of magnetic fields in a larger volume. A similar analysis examining the newly identified electromagnetic modes (Schlickeiser et al., 2011; Schlickeiser and Skoda, 2011; Felten et al., 2013; Felten and Schlickeiser, 2013a, b, c), though being highly desirable, lies outside this brief communication.

2 Thermal fluctuation effects

In order to demonstrate the role of thermal magnetic fluctuations as a pathway for other modes, we restrict ourselves to the conventional Weibel-like thermal-anisotropy mode only. The following analysis could be made more complete by applying the expressions for the many dispersion channels identified in the above references to the non-magnetized plasma modes. Here we refrain from such an extension of the present communication, not least for the reason of the wealth of newly found modes but also for the complexity of the more precise expressions given there (Felten et al., 2013; Felten and Schlickeiser, 2013a, b, c), which would obscure our intentionally focussed discussion.

The thermal anisotropy-excited Weibel-like mode grows under the condition that the anisotropy is along direction z. Writing $T_z = T_\parallel$, $T_{x,y} = T_\perp$, the condition for growth is, in its simplest form, given by

$$A \equiv \frac{T_\parallel}{T_\perp} - 1 > 0. \tag{1}$$

In this case it generates a magnetic field $\boldsymbol{B}_W$ in the perpendicular direction, i.e. transverse to the direction of anisotropy. Under the opposite condition the Weibel mode is stable but still generates a zero-frequency perpendicular magnetic thermal fluctuation field (Yoon, 2007; Treumann and Baumjohann, 2012) of spectral energy density[1]

$$\frac{\langle b^2(k)\rangle}{b_0^2} = \frac{(A+1)^2 k\lambda_\mathrm{e}}{(A+2)\left[k^2\lambda_\mathrm{e}^2 - A - \mu\right]^2} \tag{2}$$

written here for the electron Weibel mode with skin depth λ_e and mass ratio $\mu = m_\mathrm{e}/m_\mathrm{i}$. Its spectral density amplitude is given by

$$\frac{b_0^2}{m_\mathrm{e}c^2} = \frac{\mu_0}{m_\mathrm{e}}\sqrt{\frac{\pi T_\perp}{m_\mathrm{e}c^2}}. \tag{3}$$

The thermal fluctuation level has a distinct dependence on wave number k. For short wavelengths $k \gg \lambda_\mathrm{e}^{-1}$ it decays like $\propto (k\lambda_\mathrm{e})^{-3}$ (confirming the result of Yoon, 2007). Due to the presence of the inert ion component it vanishes at $k \to 0$, maximizing below $k\lambda_\mathrm{e} < 1$, i.e. compared with the electron skin depth the fluctuation field is of a longer scale, providing a weakly oscillatory moderate-wavelength background magnetic field. One may note that the wavenumber is perpendicular to both, $\hat{z}$, i.e. the direction of thermal pressure anisotropy, and the Weibel-like magnetic fluctuation field.

The magnetic fluctuation field is quasi-stationary in the sense that in the final step of calculation the real part of the frequency $\omega_\mathrm{r} \approx 0$ was put to zero, which is equivalent to the – not entirely correct – assumption of purely growing/damped $\omega(\boldsymbol{k}) = i\gamma(\boldsymbol{k})$ wave modes with growth/damping rate $\gamma(\boldsymbol{k})$. In fact, $\omega_\mathrm{r} \neq 0$ for Weibel-like modes. Hence, with $k_0^2\lambda_\mathrm{e}^2 = A$ the requirement that $|\gamma(\boldsymbol{k})| < \omega_\mathrm{r}(\boldsymbol{k})$ implies a condition on the wave number

$$1 > k^2\lambda_\mathrm{e}^2 > A\left(1 - \sqrt{\frac{\pi}{4A(A+1)^3}}\right). \tag{4}$$

Under damped conditions $A = -|A|$, and from Eq. (2), one requires $1 < |A| < 2$ fixing the range of applicability of the following analysis. There is a sufficiently wide range of wavelengths longer than λ_e in the damped case. In the unstable case, shorter wavelengths around $k \sim \lambda_\mathrm{e}$ are favoured. The more precise analysis (Felten and Schlickeiser, 2013a, b) is not restricted to the condition $|\gamma| < \omega_\mathrm{r}$, but includes aperiodically growing/damped waves.

The mean magnetic fluctuation field amplitude $\bar{b} = b_0\left[\int \mathrm{d}k\langle b^2(k)\rangle\right]^{\frac{1}{2}}$ that results from the above thermal spectral density when integrating over k space, becomes

$$\frac{\bar{b}}{b_0} = \left[\frac{(A+1)^2(A+1+\mu)}{(A+2)(A+\mu)(1-A-\mu)}\right]^{\frac{1}{2}}. \tag{5}$$

[1]Clearly, under the opposite condition the direction of anisotropy switches and the Weibel instability works in the orthogonal direction. It is thus *universal, working at any thermal anisotropy in nonmagnetic plasma* and disappears only for $A \equiv 0$. Just for this reason it also makes sense to write the thermal Weibel level including a non-vanishing anisotropy, for the thermal level becomes itself anisotropic: it is thermal along the anisotropy and thermally excited transverse to it.

It serves as background magnetization on which electromagnetic waves at low frequency, much less than the electron plasma frequency $\omega \ll \omega_e$, can propagate in a non-magnetic plasma. Propagation would be inhibited otherwise, allowing only for evanescent modes in the spatial range of the electron skin depth λ_e. Actually, for the same reason the unstable Weibel-like mode can penetrate just over the electron skin depth only from its generation site into the collisionless plasma perpendicular to the direction of anisotropy. This restriction necessarily causes a pronounced magnetic filamentation of the Weibel-like unstable plasma. On the other hand, once Weibel-like thermal fluctuations provide a weak magnetic background field, spreading across the plasma becomes possible for other electromagnetic modes.

3 Secondary instability

Once the thermal fluctuation spectrum of the Weibel mode is established, the plasma behaves weakly magnetized, with the Weibel-like thermal fluctuation background field being structured at about the electron skin depth in the direction perpendicular to the anisotropy. This implies that the thermal magnetic background organizes into long magnetic filaments of a typical transverse size of a few electron skin depths, with consequences for the propagation of any secondary electromagnetic modes.

3.1 Weibel-stable case

Let us assume that the Weibel mode has been stable (see footnote), which implies that $A \lesssim 0, 1 < |A| < 2$. Under this condition we have $T_\perp > T_\parallel$, with $T_\perp$ being directed along the Weibel magnetic field, i.e. playing the role of a *magnetically parallel* anisotropy, while $T_\parallel$ is a perpendicular anisotropy. Under this condition the Firehose mode could grow only under the condition

$$\frac{\beta_\perp}{\beta_\parallel} - 1 > \frac{2}{\beta_\parallel}, \quad \beta_{\parallel,\perp} = \frac{2\mu_0 N T_{\parallel,\perp}}{\bar{b}^2}, \tag{6}$$

with $\bar{b}^2 = \int dk \langle b^2(k)\rangle$ the thermal rms magnetic field (note that indexing is with respect to the direction of the initial thermal anisotropy where the directions with respect to the magnetic field are inverted!) existing in the otherwise non-magnetic plasma. The plasma βs refer to it in the present case. The left-hand side is positive in the Weibel-stable case and, hence, the Firehose instability *can* grow under the rewritten condition

$$\beta_\perp - \beta_\parallel > 2, \quad |A| < 2, \tag{7}$$

where the second condition is imposed by the thermal fluctuation level requirement. The Firehose mode can also propagate, because its propagation direction is along the thermal Weibel-like magnetic field which is along the direction of the Weibel-like magnetic channels and therefore allowed to propagate for a parallel mode like the Firehose–Alfvén wave.

Checking for the electron Mirror mode (not including any more sophisticated effects) yields instability as long as

$$\beta_\parallel^2 - \beta_\perp > \beta_\perp\beta_\parallel \quad \text{and} \quad A = \beta_\parallel/\beta_\perp - 1 < 0. \tag{8}$$

This implies a contradiction, thus *excluding* the Mirror mode. Weibel-stable plasmas are stable against Mirror but allow for Firehose modes. These propagate on the thermal level of the magnetic field in the direction perpendicular to the Weibel-stable direction of anisotropy. This is interesting to know, as it shows that in an otherwise non-magnetic plasma where in one direction (this time the $\perp$ direction) the plasma is Weibel *unstable*, the existence of thermal magnetic fluctuations in the Weibel-stable direction causes Firehose modes to radiate away from this region if only the above Firehose condition, based on the small thermal fluctuation level of the magnetic field, is satisfied. Since the thermal level is small this will, in practice, always be the case if only $T_\perp > T_\parallel$.

In addition, if the plasma consists of a thermal background and a warm anisotropic component, the low-frequency Whistler (or Alfvén) mode can be destabilized, as it only requires that for the anisotropic component $A < 0$ and hence $\beta_\perp < \beta_\parallel$, which is given by the stability condition of the Weibel mode (note again that the indices $\parallel$, $\perp$ refer to the anisotropy frame, not to the magnetic frame!). The spectral density of the magnetic fluctuations in this range with $\beta_\parallel/\beta_\perp < 1$ in the long wavelength range $k\lambda_e < 1$ is estimated as

$$\frac{\langle b^2(k)\rangle}{b_0^2} \approx \left(\frac{T_\parallel}{T_\perp}\right)^2 k\lambda_e. \tag{9}$$

Using the Whistler dispersion relation for electrons in the low-frequency range, i.e. $k^2\lambda_e^2 \approx \omega/(\Omega_e - \omega)$, with $\Omega_e = e\bar{b}/m_e$, the electron cyclotron frequency in the quasi-stationary Weibel thermal fluctuation field yields the relation

$$\frac{\langle b^2(k)\rangle}{b_0^2} \approx \left(\frac{T_\parallel}{T_\perp}\right)^2 \sqrt{\frac{\nu}{1-\nu}}, \quad \nu = \frac{\omega}{\Omega_e} \tag{10}$$

between the Whistler frequency range $\omega(k)$ and the magnetic spectral energy density. Since we know that the magnetic spectral density vanishes at $k = 0$, and using the dispersion relation for $k\lambda_e = 1$, we find that the frequency of Whistlers is in the range

$$0 \le \nu < \frac{1}{2}, \quad \text{for} \quad 0 \le k\lambda_e < 1 \tag{11}$$

propagating along the Weibel thermal fluctuation field perpendicular to the Weibel anisotropy, i.e. in the $\perp$ direction. These are indeed very low-frequency Whistlers, since Ω_e based on the thermal fluctuation level is small.

3.2 Weibel-unstable case

The interesting domain is that of unstable Weibel modes (for instance in view of application to relativistic shocks) in which case the magnetic field grows and may become quite strong. Several mechanisms of saturation or limiting growth have been proposed (for energy arguments, quasilinear and other nonlinear mechanisms see Achterberg and Wiersma, 2007; Pokhotelov and Amariutei, 2011; Pokhotelov et al., 2010, respectively, and references therein). Here we are not interested in saturation but in the excitation of low-frequency plasma modes which may transport the field away from the region of excitation.

Since $A > 0$ for unstable Weibel-like modes growing from thermal background fluctuations, while generating a substantial magnetic field in the direction perpendicular to the direction of anisotropy, we immediately conclude that in the presence of an isotropic background plasma the Whistler mode is naturally unstable under the condition that sufficiently many resonant particles exist with large enough resonant energy. This is certainly the case when Weibel-like modes are excited by a streaming population. In this case one expects that Whistlers will be radiated along the Weibel-like field perpendicular to the exciting anisotropy direction.

What about the two remaining low-frequency electromagnetic modes, the Firehose and Mirror modes? The former is very easy to infer about. From Eq. (7) and the instability condition $A > 0$ we immediately construct the contradiction $2 < 0$, thus excluding the Firehose mode from growing in this case. The Firehose mode will not excite any Alfvén waves at the expense of the Weibel mode in this unstable case. However, the situation is different for the Mirror mode. Clearly Eq. (8) can be satisfied when the Weibel mode grows with $A > 0$ (instead of $A < 0$ as in the former Weibel-stable case). The corresponding condition can be written

$$\frac{\beta_\|}{\beta_\perp} - 1 > \frac{1}{\beta_\|} \quad \text{and} \quad A = \frac{\beta_\|}{\beta_\perp} - 1 > 0, \tag{12}$$

which yields $\beta_\| > 1/A$ for mirror instability and can clearly be satisfied. The solution is

$$\beta_\| \gtrsim \frac{1}{2}\beta_\perp \left(1 \pm \sqrt{1 + \frac{4}{\beta_\perp^2}}.\right) \tag{13}$$

Of the two solutions one is trivially satisfied. The other requires that the anisotropy $A > 1/\beta_\perp$. It is interesting to note that since the βs depend inversely on the growing magnetic field, starting from thermal level, they will decrease with Weibel-like mode growth, thereby ultimately stabilizing the second Mirror mode branch when $A \sim 1/\beta_\perp$, while the branch with the negative sign of the square root remains unaffected.

Hence, any growing Weibel-like mode will readily self-consistently evolve into a chain of Mirror structures which on their own are filled with low-frequency Whistlers (for the observational evidence and theoretical arguments see, e.g., Baumjohann et al., 1999; Treumann et al., 2000) bouncing back and force in the mirrors with some of them possibly escaping into the environment. As a consequence, the Weibel-like instability does not only lead to magnetic and current filaments of perpendicular scale of the order of the electron skin depth. In addition the magnetic Weibel-like fields evolve into a sequence of mirror structures as the lowest frequency magnetically turbulent modes which can evolve according to their own nonlinear dynamics (cf., e.g., Pokhotelov et al., 2008, 2010, 2013, and references therein).

4 Conclusions

The Weibel-like instability has frequently been made responsible for the excitation of magnetic fields in otherwise initially non-magnetized plasmas. Actually, as has been known for a long time (Landau and Lifschitz, 1960; Sitenko, 1967; Akhiezer et al., 1975, and followers), thermal fluctuations in the plasma readily lead to the spontaneous emission of magnetic fluctuations. At the lowest frequencies these form quasi-stationary magnetic fields (see, e.g., Yoon, 2007; Treumann and Baumjohann, 2012) with wavelength longer than the electron skin depth λ_e.

These fluctuations provide an initial weak magnetic background field which allows the Weibel-like magnetic field to penetrate over some distance into the otherwise magnetically nontransparent plasma, which would inhibit penetration of any magnetic field over distances longer than λ_e, effectively exponentially screening the plasma from magnetic fields outside the Weibel-like mode source region. In the presence of a pressure anisotropy, however, Firehose, Mirror and Whistler modes can grow in the plasma and compete with the Weibel-like mode. A Weibel-stable plasma allows for Firehose and Whistler modes to grow, both propagating along the magnetic thermal fluctuation field. In a Weibel-unstable plasma the Firehose instability is stable. However, Whistler and Mirror modes can grow. The latter structure the unstably generated Weibel-like magnetic field into chains of magnetic bubbles or holes and trap low-frequency Whistlers (Baumjohann et al., 1999; Treumann and Baumjohann, 2000; Treumann et al., 2000), thus contributing to magnetic structure and turbulence in addition to the known Weibel filamentation effect. This will have a profound effect on the self-consistent generation of magnetic plasma turbulence in regions of pressure anisotropy (or relative streaming) in an otherwise initially non-magnetized plasma. Not only may one expect that anisotropic or streaming plasmas will thus naturally be weakly magnetized, they will also be naturally turbulent, thus providing plenty of scattering centres for energetic particles as required in all stochastic particle acceleration scenarios (for a recent review cf., e.g., Bykov and Treumann, 2011; Schure et al., 2012, and references therein).

The present investigation has been forcefully restricted to the investigation of one quite simple low-frequency electromagnetic mode only, the conventional thermally anisotropic Weibel instability. This instability belongs to a much wider class of low-frequency electromagnetic modes which have been investigated in depth only recently (Schlickeiser et al., 2011; Schlickeiser and Skoda, 2011; Felten et al., 2013; Felten and Schlickeiser, 2013a, b). Several of these modes generate quasi-stationary magnetic fields in a similar way as the Weibel instability we were dealing with. It will be most interesting to investigate their effect on the excitation of other electromagnetic instabilities, the propagation of Alfvén and Whistler waves in the plasma that has become weakly re-magnetized by them. It should also be noted at this point that Simões Jr. et al. (2013) in a very recent paper performed particle-in-cell simulations of a thermally isotropic, quiescent and stable plasma to determine the electric and magnetic thermal fluctuation levels. As expected, both levels are different from zero. The electric fluctuations naturally map the Langmuir fluctuation branch for wavelength exceeding the Debye length, for shorter wavelengths they show a wide range of weak fluctuations extending even down below the plasma frequency. This result has been very well known for a long time (cf., e.g., Lund et al., 1996, with and without a beam) and is nicely confirmed by these simulations. The magnetic fluctuations found in this case are also expected in the whole frequency range from fluctuation theory of any thermal system (Landau and Lifschitz, 1960; Sitenko, 1967). Their absence would have been very surprising. As they should, magnetic fluctuation amplitudes maximize at lowest frequencies and longest wavelengths. This was noted by Yoon (2007) and, for vanishing anisotropy, also in Treumann and Baumjohann (2012). Both the last two papers became in this respect completely independent of the Weibel mode, indicating the general presence of magnetic fluctuations in thermally anisotropic and also thermally isotropic plasmas, thereby very softly magnetizing an initially non-magnetic plasma and permitting the propagation of low-frequency electromagnetic (mhd) modes.

Acknowledgements. This research was part of an occasional Visiting Scientist Programme in 2006/2007 at ISSI, Bern. R. A. Treumann thankfully recognizes the assistance of the ISSI librarians, Andrea Fischer and Irmela Schweizer. R. A. Treumann also thanks the two referees, P. H. Yoon and another anonymous referee, for their intriguing remarks on the paper. One of the referees has brought to our attention the series of papers on the new low-frequency relativistic electromagnetic modes, which is thankfully acknowledged.

Edited by: P.-L. Sulem

References

Achterberg, A. and Wiersma, J.: The Weibel instability in relativistic plasmas. I. Linear theory, II. Nonlinear theory and stabilization mechanism, Astron. Astrophys., 475, 1–18, doi:10.1051/0004-6361:20065365, 2007.

Akhiezer, A. I., Akhiezer, I. A., Sitenko, R. V., and Stepanov, K. N.: Nonlinear Theory and Fluctuations, in: Plasma Electrodynamics, Pregamon Press, Oxford, UK, 2, 116–142, 1975.

Baumjohann, W. and Treumann, R. A.: Basic Space Plasma Physics, Revised Edition, Imperial College Press at World Scientific, London & Singapore, 357–359, 2012.

Baumjohann, W., Treumann, R. A., Georgescu, E., Haerendel, G., Fornacon, K.-H., and Auster, U.: Waveform and packet structure of lion roars, Ann. Geophys., 17, 1528–1534, doi:10.1007/s00585-999-1528-9, 1999.

Bykov, A. M. and Treumann, R. A.: Fundamentals of collisionless shocks for astrophysical application, 2. Relativistic shocks, Astron. Astrophys. Rev., 19, 42, doi:10.1007/s00159-011-0042-8, 2011.

Felten, T. and Schlickeiser, R.: Spontaneous electromagnetic fluctuations in unmagnetized plasmas. IV. Relativistic form factors of aperiodic Lorentzian modes, Phys. Plasmas, 20, 082116, doi:10.1063/1.4817804, 2013a.

Felten, T. and Schlickeiser, R.: Spontaneous electromagnetic fluctuations in unmagnetized plasmas. V. Relativistic form factors of weakly damped/amplified thermal modes, Phys. Plasmas, 20, 082117, doi:10.1063/1.4817805, 2013b.

Felten, T. and Schlickeiser, R.: Spontaneous electromagnetic fluctuations in unmagnetized plasmas. VI. Transverse, collective mode for arbitrary distribution functions, Phys. Plasmas, 20, 104502, doi:10.1063/1.4824114, 2013c.

Felten, T., Schlickeiser, R., Yoon, P. H., and Lazar, M.: Spontaneous electromagnetic fluctuations in unmagnetized plasmas. II. Relativistic form factors of aperiodic thermal modes, Phys. Plasmas, 20, 052113, doi:10.1063/1.4804402, 2013.

Fried, B. D.: Mechanism for instability of transverse plasma waves, Phys. Fluids, 2, 337, doi:10.1063/1.1705933, 1959.

Landau, L. D. and Lifshitz, E. M.: Electrodynamics of Continuous Media, Pergamon Press, Oxford, 1960.

Lund, E. J., Treumann, R. A., and Labelle, J.: Quasi-thermal fluctuations in a beam-plasma system, Phys. Plasmas, 3, 1234–1240, doi:10.1063/1.871747, 1996.

Pokhotelov, O. A. and Amariutei, O. A.: Quasi-linear dynamics of Weibel instability, Ann. Geophys., 29, 1997–2001, doi:10.5194/angeo-29-1997-2011, 2011.

Pokhotelov, O. A., Sagdeev, R. Z., Balikhin, M. A., Onishchenko, O. G., and Fedun, V. N.: Nonlinear mirror waves in non-Maxwellian space plasma, J. Geophys. Res., 113, A04225, doi:10.1029/2007JA012642, 2008.

Pokhotelov, O. A., Sagdeev, R. Z., Balikhin, M. A., Fedun, V. N., and Dudnikova, G. I.: Nonlinear Mirror and Weibel modes: peculiarities of quasi-linear dynamics, Ann. Geophys., 28, 2161–2167, doi:10.5194/angeo-28-2161-2010, 2010.

Pokhotelov, O. A., Onishchenko, O. G., and Stenflo, L.: Physical mechanisms for electron mirror and field swelling modes, Phys. Scripta, 87, 065303, doi:10.1088/0031-8949/87/06/065303, 2013.

Schlickeiser, R. and Skoda, T.: Linear theory of weakly amplified, parallel propagating, transverse temperature-anisotropy

instabilities in magnetized thermal plasmas, Astrophys. J., 716, 1596–1606, doi:10.108870004-637X/716/2/1596, 2011.
Schlickeiser, R., Lazar, M., and Skoda, T.: Spontaneously growing, weakly propagating, transverse fluctuations in anisotropic magnetized thermal plasmas, Phys. Plasmas, 18, 012103, doi:10.1063/1.3532787, 2011.
Schure, K. M., Bell, A. R., O'C Drury, L., and Bykov, A. M.: Diffusive shock acceleration and magnetic field amplification, Space Sci. Rev., 173, 491–519, doi:10.1007/s11214-012-9871-7, 2012.
Simões Jr., F. J. R., Pavan, J., Gaelzer, R., Ziebell, L. F., and Yoon, P. H.: Particle-in-cell simulations on spontaneous thermal magnetic field fluctuations, Phys. Plasmas, 20, 100702, doi:10.1063/1.4825249, 2013.
Sitenko, A. G.: Electromagnetic Fluctuations in Plasma, Academic Press, New York, 1967.
Treumann, R. A.: Fundamentals of collisionless shocks for astrophysical application, 1. Non-relativistic shocks, Astron. Astrophys. Rev., 17, 409–535, doi:10.1007/s00159-009-0024-2, 2009.
Treumann, R. A. and Baumjohann, W.: Advanced Space Plasma Physics, Imperial College Press, London, 1997.
Treumann, R. A. and Baumjohann, W.: Collisionless mirror mode trapping, Nonlin. Processes Geophys., 7, 179–184, doi:10.5194/npg-7-179-2000, 2000.
Treumann, R. A. and Baumjohann, W.: A note on the Weibel instability and thermal fluctuations, Ann. Geophys., 30, 427–431, doi:10.5194/angeo-30-427-2012, 2012.
Treumann, R. A., Georgescu, E., and Baumjohann, W.: Lion roar trapping in mirror modes, Geophys. Res. Lett., 27, 1843–1846, doi:10.1029/2000GL003767, 2000.
Vedenov, A. A., Velikhov, E. P., and Sagdeev, R. Z.: Stability of plasma, Sov. Phys. Uspekhi., 4, 332–369, doi:10.1070/PU1961v004n02ABEH003341, 1961.
Weibel, E. S.: Spontaneously growing transverse waves in a plasma due to an anisotropic velocity distribution, Phys. Rev. Lett., 2, 83–84, doi:10.1103/PhysRevLett.2.83, 1959.
Yoon, P. H.: Spontaneous thermal magnetic field fluctuations, Phys. Plasmas, 14, 83–84, doi:10.1063/1.2741388, 2007.
Yoon, P. H. and Davidson, R. C.: Exact analytical model of the classical Weibel instability in a relativistic anisotropic plasma, Phys. Rev. A, 35, 2718–2721, doi:10.1103/PhysRevA.35.2718, 1987.

Ion acceleration by parallel propagating nonlinear Alfvén wave packets in a radially expanding plasma

Y. Nariyuki[1], T. Umeda[2], T. K. Suzuki[3], and T. Hada[4]

[1]Faculty of Human Development, University of Toyama, 3190, Toyama City, Toyama 930-8555, Japan
[2]Solar-Terrestrial Environment Laboratory, Nagoya University, Nagoya, Aichi 464-8601, Japan
[3]Department of Physics, Nagoya University, Furo-cho, Nagoya, Aichi 464-8602, Japan
[4]Department of Earth System Science and Technology, Kyushu University, 6-1, Kasuga City, Fukuoka 816-8580, Japan

Correspondence to: Y. Nariyuki (nariyuki@edu.u-toyama.ac.jp)

Abstract. The numerical simulation of the nonlinear evolution of the parallel propagating Alfvén waves in a radially expanding plasma is performed by using a kinetic-fluid model (the Vlasov–MHD model). In our study, both the nonlinear evolution of the Alfvén waves and the radial evolution of the velocity distribution function (VDF) are treated simultaneously. On the other hand, important ion kinetic effects such as ion cyclotron damping and instabilities driven by the non-equilibrium ion velocity distributions are not included in the present model. The results indicate that the steepened Alfvén wave packets outwardly accelerate ions, which can be observed as the beam components in the interplanetary space. The energy of imposed Alfvén waves is converted into the longitudinal fluctuations by the nonlinear steepening and the nonlinear Landau damping. The wave shoaling due to the inhomogeneity of the phase velocity is also observed.

1 Introduction

It is well known that nonlinear Alfvén waves are ubiquitously observed in the solar wind (Bruno and Carbone, 2013). These Alfvén waves are probably generated in the vicinity of the solar photosphere, and may play an important role in heating and acceleration of the coronal and solar wind plasmas. A lot of authors have already discussed the damping of the Alfvén waves in coronal plasmas (Ofman, 2010a; Galinsky and Shevchenko, 2013). In particular, the nonlinear damping of low-frequency, quasi-parallel propagating Alfvén waves is of great interest. This is because they can carry the momentum and energy over a long distance, since the linear collisionless damping of these waves is too small to dissipate them (Barnes, 1966).

On the other hand, it is also known that the proton beams are frequently observed in solar wind plasmas (Marsch, 2006). Recently, it was shown by Araneda et al. (2008) and Matteini et al. (2010) that the protons trapped by the envelope modulation (or corresponding longitudinal fluctuations) due to the parametric instabilities of Alfvén waves can be observed as the beam protons in the velocity space (Araneda et al., 2008). Past studies (Araneda et al., 2008; Matteini et al., 2010) assumed the uniform background plasmas and the periodic system. However, the radial inhomogeneity of the plasma density and the magnetic field may not be negligible to discuss the nonlinear evolution of low-frequency Alfvén waves in coronal plasmas, since the radial change of the nonlinearity and the non-WKB effects such as the reflection may occur (e.g., Chandran et al., 2011).

In the present paper, we firstly perform the numerical simulation of the nonlinear evolution of the parallel propagating Alfvén waves in a radially expanding plasma by using a kinetic-fluid model (the Vlasov–MHD model) (Nariyuki and Hada, 2007; Nariyuki et al., 2011, 2013). Actually, this has already been discussed by a lot of authors by using the magnetohydrodynamic (MHD) models (Suzuki and Inutsuka, 2006; Tanaka et al., 2007). In our study, the radial evolution of the Alfvén waves and the velocity distribution function (VDF) are treated simultaneously. Even if the isotropic VDFs are assumed in the Vlasov–MHD model, as far as the nonlinearity is relatively weak ($\delta b/b <\sim 0.3$) (Nariyuki and Hada, 2007), the nonlinear evolution of the Alfvén waves and the trapped ions observed in the Vlasov–MHD model agree with those in the ion hybrid simulation (Nariyuki et al., 2011). In Sect. 2, the one-dimensional (1-D) Vlasov–MHD model in the spherical coordinate and the simulation setup are introduced briefly. Section 3 contains the numerical results. We discuss and summarize the results in Sects. 4 and 5, respectively.

2 Simulation setup

We consider a radially expanding flux tube, which is measured by the heliocentric distance r. To discuss both the nonlinear evolution of the transverse fluid motion due to the Alfvén waves and the longitudinal (radial) ion kinetics, we introduce the 1-D Vlasov–MHD model in the spherical coordinate below. For simplicity, we here neglect the gravitational force term, since it is relatively small and may not play important roles in the super Alfvénic solar wind at $r > 10R_s$, where R_s is the solar radius.

The velocity distribution functions obey the Vlasov equation

$$\frac{df(\mathbf{x},\mathbf{v},t)}{dt} = \frac{\partial f}{\partial t} + \mathbf{v}\cdot\nabla_x f + \frac{d\mathbf{v}}{dt}\cdot\nabla_v f = 0, \tag{1}$$

where $f(\mathbf{x},\mathbf{v},t)$ is the single particle distribution function, and

$$\frac{d\mathbf{v}}{dt} = \mathbf{a} = \frac{q}{m}(\mathbf{e}+\mathbf{v}\times\mathbf{b}). \tag{2}$$

$\mathbf{v}=(v_x,v_y,v_z)$ is the particle velocity, $\mathbf{e}=(e_x,e_y,e_z)$ the electric field, $\mathbf{b}=(b_x,b_y,b_z)$ the magnetic field, m the rest mass, and q the charge.

In the spherical coordinate ($\mathbf{x}_s=(r,\theta,\phi)$, where $x = r\sin\theta\cos\phi$, $y = r\sin\theta\sin\phi$, $z = r\cos\theta$), $\frac{d\mathbf{v}}{dt}$ can be written as

$$\frac{d\mathbf{v}}{dt} = \mathbf{a} = \frac{dv_r}{dt}\mathbf{i}_r + \frac{dv_\theta}{dt}\mathbf{i}_\theta + \frac{dv_\phi}{dt}\mathbf{i}_\phi + \frac{d\mathbf{i}_r}{dt}v_r + \frac{d\mathbf{i}_\theta}{dt}v_\theta + \frac{d\mathbf{i}_\phi}{dt}v_\phi, \tag{3}$$

where $\mathbf{i}_j$ is the unit vector of the $j=r,\theta,\phi$ component, $\mathbf{v} = \mathbf{v}_s = (v_r,v_\theta,v_\phi) = (\frac{dr}{dt}, r\frac{d\theta}{dt}, r\sin\theta\frac{d\phi}{dt})$, and $\mathbf{a} = a_r\mathbf{i}_r + a_\theta\mathbf{i}_\theta + a_\phi\mathbf{i}_\phi$. Notice that $\mathbf{v}_s$ is the function of $\mathbf{x}$. By transforming (1) into the six-dimensional spherical coordinate ($(\mathbf{x}_s,\mathbf{v}_s)$), we find (Larson, 1969)

$$\frac{df(\mathbf{x},\mathbf{v},t)}{dt} = \frac{df}{dt} + \frac{dr}{dt}\frac{\partial f}{\partial r} + \frac{d\theta}{dt}\frac{\partial f}{\partial\theta} + \frac{d\phi}{dt}\frac{\partial f}{\partial\phi} + \frac{dv_r}{dt}\frac{\partial f}{\partial v_r} + \frac{dv_\theta}{dt}\frac{\partial f}{\partial v_\theta} + \frac{dv_\phi}{dt}\frac{\partial f}{\partial v_\phi} = 0, \tag{4}$$

where

$$\frac{dv_r}{dt} = \frac{v_\theta^2+v_\phi^2}{r} + a_r, \tag{5}$$

$$\frac{dv_\theta}{dt} = \frac{v_\phi^2\cot\theta - v_r v_\theta}{r} + a_\theta, \tag{6}$$

$$\frac{dv_\phi}{dt} = -\frac{v_\theta v_\phi\cot\theta + v_\phi v_r}{r} + a_\phi. \tag{7}$$

The relation (5)–(7) can be derived from Eq. (2).

We here define the longitudinal distribution function

$$g = \int\int f\,dv_\theta dv_\phi. \tag{8}$$

We assume the spherical symmetry ($\frac{\partial f}{\partial\theta} = \frac{\partial f}{\partial\phi} = 0$) and isotropic plasmas. By integrating Eq. (1) on v_θ and v_ϕ, we obtain

$$\frac{\partial g}{\partial t} + v_r\frac{\partial g}{\partial r} + \left(\frac{u_\perp^2+2T_i}{r} + A_r\right)\frac{\partial g}{\partial v_r} = -\frac{2gv_r}{r} + \frac{2gv_r u_\theta\cos\theta}{r}, \tag{9}$$

where u_j is the j component of the bulk velocity, and $n = \int g dv_r$, the number density. By taking the boundary condition $f\to 0$ when $v_{\theta,\phi}\to\pm\infty$,

$$\int\int a_r\frac{\partial f}{\partial v_r}dv_\theta dv_\phi = \int\int\frac{\partial a_r f}{\partial v_r}dv_\theta dv_\phi = (e_r + u_\theta b_\phi - u_\phi b_\theta)\frac{\partial g}{\partial v_r} \equiv A_r\frac{\partial g}{\partial v_r} \tag{10}$$

where

$$gu_{\theta,\phi} = \int\int v_{\theta,\phi} f\,dv_\theta dv_\phi. \tag{11}$$

Equation (11) is valid in isotropic plasmas (Nariyuki et al., 2013). We here take the ion mass $m_i = 1$, $q = 1$, and

$$gT_i \equiv \int\int (v_{\theta,\phi} - u_{\theta,\phi})^2 f\,dv_\theta dv_\phi, \tag{12}$$

where $p_i = nT_i = \int\int\int(v_r-u_r)^2 f d\mathbf{v} = \int(v_r-u_r)^2 g dv_r$.

When electrons can be treated as the massless fluid, we obtain

$$e_r + u_\theta b_\phi - u_\phi b_\theta = -\frac{1}{qn}\left(\frac{\partial p_e}{\partial r} + \frac{1}{r^2}\frac{\partial r^2 p_b}{\partial r}\right), \tag{13}$$

where $p_b = (|\mathbf{b}_\perp|^2)/2$, $\mathbf{b}_\perp = b_\theta\mathbf{i}_\theta + b_\phi\mathbf{i}_\phi$, and p_e is the electron pressure. Then, we find that

$$A_r = -\frac{1}{n}\left(\frac{\partial p_e}{\partial r} + \frac{1}{r^2}\frac{\partial r^2 p_b}{\partial r}\right). \tag{14}$$

The first term on the right-hand side of Eq. (9) comes from the sixth and seventh terms on the left-hand side Eq. (4). Notice that Eq. (9) is obtained for arbitrary f. While the second term on the right-hand side of Eq. (9) comes from the sixth term, we here neglect it by taking $\theta = \pi/2$.

$$\frac{\partial g'}{\partial t} + v_r\frac{\partial g'}{\partial r} + \left(\frac{|u'_\perp|^2+2T_i'}{r^3} + A_r\right)\frac{\partial g'}{\partial v_r} = 0, \tag{15}$$

$$\frac{\partial\rho' u'}{\partial t} = -\frac{\partial u_r\rho' u'}{\partial r} + \frac{\partial C b'}{\partial r}, \tag{16}$$

$$\frac{\partial b'}{\partial t} = -\frac{\partial}{\partial r}\left(u_r b' - B_r u'\right), \tag{17}$$

where $b = b_\theta + ib_\phi$ and $u = u_\theta + iu_\phi$ are the complex transverse magnetic field and velocity, $g = \int\int f dv_\theta v_\phi$ the longitudinal VDFs ($f(v_r, v_\theta, v_\phi)$ is the original VDF), $g' = r^2 g$, $u' = ru$, $b' = rb$, $T_i' = r^2 T_i$,

$$A_r = -\frac{1}{\rho'}\left(\frac{\partial(p_e' + p_b')}{\partial r} - \frac{2p_e'}{r}\right), \tag{18}$$

$\rho' = r^2\rho$, u_r the longitudinal velocity, p_e the electron pressure, $p_b = |b|^2/2$, and $p'_{e(b)} = r^2 p_{e(b)}$. $C = b_r r^2 = 100$, since the physical quantities are normalized to the parameters at the inner boundary as shown below. We here assume the isothermal electrons ($p_e/\rho = \text{const.}$).

By integrating Eq. (15) on v_r, we recover the continuity equations and the momentum equation of the radial component in the ideal MHD system (Suzuki and Inutsuka, 2006; Tanaka et al., 2007). The ion pressure p_i is also evaluated by the integration of g. Isotropic ions are assumed in Eqs. (15)–(17) to close the system. Note that the assumption of isotropic ions is reasonable as far as we concentrate on the low-frequency wave dynamics in low beta plasmas, although the VDFs of ions in the coronal and solar wind plasmas may be anisotropic. This is because the pressure anisotropy effects in the dispersion relation of Alfvén waves are negligibly small in low beta plasmas (see Sect. 4.2).

The simulation region of the present study is from $10R_s$ to $110R_s$. All the physical quantities are normalized to the parameters at $r = 10R_s$ as follows: the number density $n_0 = \rho_0/m_p = 2.9 \times 10^9$ [m^{-3}] (m_p is the mass of protons), the longitudinal magnetic field $b_{r0} = 1.3 \times 10^{-6}$[T], the Alfvén velocity $V_{A0} = b_{r0}/\sqrt{(\mu_0\rho_0)} = 5.0 \times 10^5$ [m s^{-1}] (μ_0 is the vacuum permeability), respectively. The space and time are normalized to the solar radius $R_s = 7 \times 10^8$ [m] and $T_0 = R_s/V_{A0} = 1.4 \times 10^3$ [s], respectively. We use the uniform mesh with 4000 grid points in the r direction and 1000 grid points in the v_r direction. The grid spacing is $\Delta r = 0.025$, $\Delta v_r = 0.01$ and the time step is $\Delta t = 2.5 \times 10^{-4}$. The transverse momentum Eq. (16) and the induction Eq. (17) are solved by the second-order rational Runge–Kutta scheme (Wambecq, 1978) for time integration and the second-order central difference method for evaluating spatial derivatives. The splitting scheme (Cheng, 1976) and the PIC (Positive Interpolation for hyperbolic Conservation laws) scheme are applied for time advancement of the Vlasov Eq. (15) (Umeda, 2009; Umeda et al., 2012). Note that no artificial dissipation term is included in the Vlasov–MHD system. The detailed numerical scheme to solve Eqs. (15)–(17) has also been described in a past study (Nariyuki et al., 2011).

Since the solar wind may become super Alfvénic flow within $r < 10$ ($r = 10$ is the inner boundary at $10R_s$), we here assume the constant radial velocity $u_{r0} = \frac{1}{\rho}\int v_r g_0 dv_r = V_{A0}$ as the initial condition, where g_0 is the initial VDF. Then, the initial density $\rho = 100/r^2$ due to the conservation of the mass flux. Thus, the critical point at which the solar wind becomes super Alfvénic flow is $r_c = 10$ in the present study. The radial magnetic field is $b_r = 100/r^2$ due to the conservation of magnetic flux. Note that $u_r > V_A = b_r/\sqrt{(\mu_0\rho)}$ at $r > 10$. Thus, the initial ion pressure p_0 is assumed to be uniform. The initial ion beta is $\beta_i = p_0/(b_r^2/2) = 0.02$ at $r = 10$. $g_0 = \frac{1}{\sqrt{\pi}v_{th}}\exp(-(v_r - u_{r0})^2/v_{th}^2)$ where $v_{th}^2 = 2T_i(t = 0) = 0.002r$.

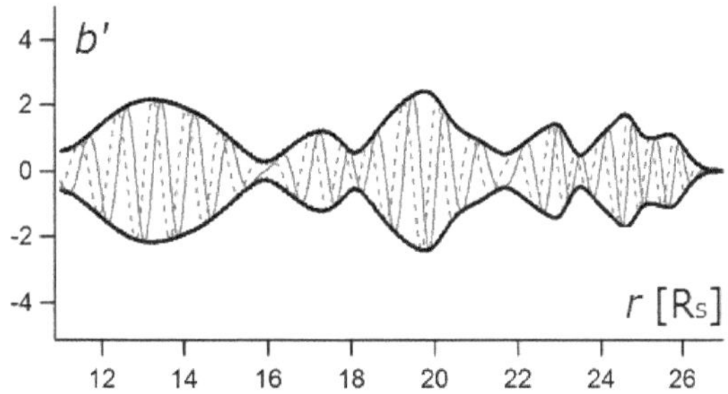

Fig. 1. The transverse magnetic field (b_θ' (gray solid lines), b_ϕ' (gray dashed lines), $\pm|b'|$ (black bold lines) at $t = 11$ (1.5×10^4 [s])).

The nonmonochromatic right-handed circularly polarized Alfveń waves

$$b(r = 10) = a\Sigma_{n=\frac{-N}{2}}^{\frac{N}{2}} b_p \exp[i(\omega_n t + \alpha_n)] \tag{19}$$

and $v(r = 10) = -b(r = 10)$ are imposed at the lower boundary, where $\omega_n = V_{A0}k_n$, $k_n = k_0 \pm n\Delta k$, $\Delta k = \pi/50$, n is an integer, and N is the number of wave modes, respectively. The phase α_n is given by a uniform random number generator with a range $[0, 2\pi]$. The inner boundary condition for g is fixed at the initial value. The open boundary condition is imposed at the upper boundary ($r = 110$). Initially, $\beta_e = 0.02$.

The imposed Alfvén waves are given as the white noise (b_p is independent of ω_n). We here set $\omega_0 = 12.566$ ($\omega_0/T_0 = 0.009$[Hz]), and $N = 60$ (0.0076[Hz] $< \omega/T_0 <$ 0.0104[Hz]) (Run A), $N = 100$ (0.0067[Hz] $< \omega/T_0 <$ 0.0113[Hz]) (Run B), $N = 140$ (0.0058[Hz] $< \omega/T_0 <$ 0.0122[Hz]) (Run C). At the early stage ($t < 10\pi/\omega_0$), the linearly increasing wave amplitude $a = t(\omega_0/10\pi)$ is set to avoid the discontinuity of the wave front. During $10\pi/\omega_0 < t < 11$, we set $a = 1$ and b_p as $< |b(r = 10)| >= 0.2$ (where $<>$ indicates time average), which is consistent with previous studies (e.g., Maneva et al., 2013). After $t > 11$, the imposed wave amplitude is diminished.

3 Results

In this section, we mainly show the results of Run B, since those of runs A and C are essentially the same as those of Run B. The difference between the runs is discussed in the last part of this section.

Figure 1 shows the transverse magnetic field (b_θ', b_ϕ', $\pm|b'|$) at $t = 11$. The imposed Alfvén wave packets have the envelope modulation corresponding to the broadband frequency spectra. Notice that the envelope modulation excites

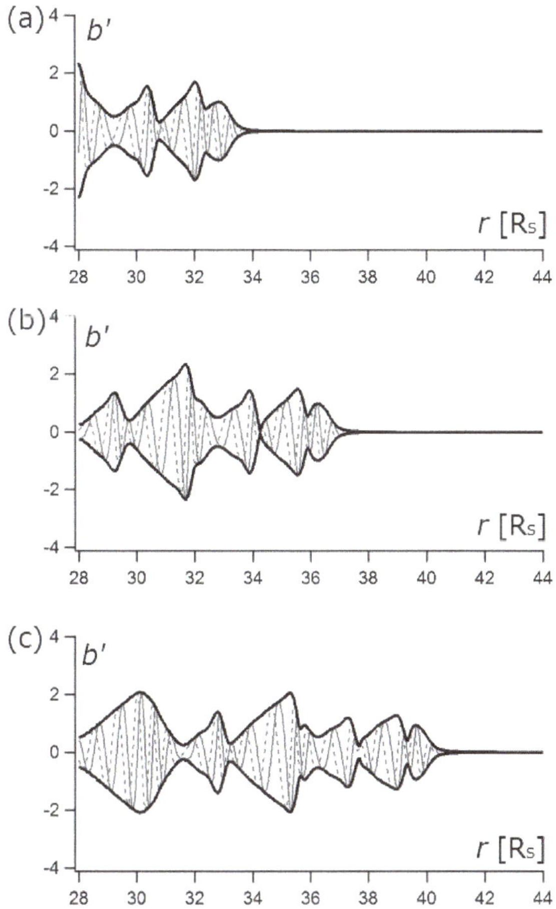

Fig. 2. The snapshot of the transverse magnetic field at **(a)** $t = 16.5$, **(b)** $t = 19.25$, and **(c)** $t = 22.0$ in Run B.

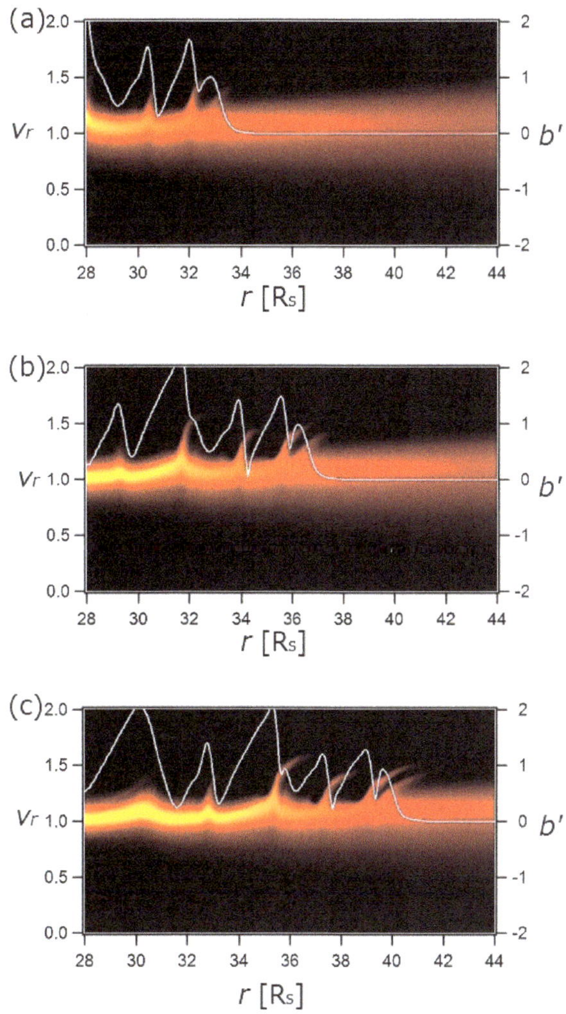

Fig. 3. The snapshot of g' at **(a)** $t = 16.5$, **(b)** $t = 19.25$, and **(c)** $t = 22.0$ in Run B. The white lines indicate $|b'|$.

the density fluctuation by the ponderomotive force (the second term of the right-hand side of Eq. 14) (Mjølhus, 1976; Mio et al., 1976). Such longitudinal fluctuations can dissipate through the nonlinear Landau damping (Nariyuki et al., 2008, 2010; Markovskii et al., 2009). Figures 2 and 3 show the snapshots of the transverse magnetic field (b') and g' at (a) $t = 16.5$, (b) $t = 19.25$, and (c) $t = 22.0$, respectively. As time elapses, nonlinear steepening of the envelope-modulated Alfvén waves (Cohen and Kulsrud, 1975) is observed (Figs. 1 and 2). In Fig. 3, the ions accelerated by the nonlinearly steepened wave front of the Alfvén waves are observed. Some ions with the higher velocity than the local phase velocity of the Alfvén waves move away from the steepened wave front and form the stretched structure in the phase space (Fig. 3b, c). It is important that the accelerated ions can be observed as the ion beams by the spacecraft observation (Fig. 4). For instance, since $V_A \simeq 10V_{A0}/r$, the relative beam speed in Fig. 4a is $v_r = 0.5 \simeq 1.8V_A(r = 36)$, which is consistent with the one observed in solar wind plasmas (Marsch, 2006).

In the one-fluid system, the bulk kinetic energy corresponding to the relative speed between core and beam protons is evaluated as the pressure and temperature parallel to the ambient magnetic field. For instance, the temperature evaluated by using the whole proton VDFs in Fig. 4a is about 1.5 times larger than the one of the core proton. On the other hand, the core proton temperature at $t = 16.5$ and $r = 36$ (Fig. 3a) is 0.0525, while the one at $t = 22.0$ and $r = 36$ is 0.0450. This suggests that the heating of ions by the nonlinear Alfvén waves in the present study corresponds to the generation of the relative bulk kinetic energy, which can be free energies to cause the beam instabilities. Actually, the present model also does not include the other important kinetic effects. The quantitative results of heating in the present study may be modified by the kinetic effects. To incorporate the instabilities into the present model through the anisotropic pressure is one of the important future issues.

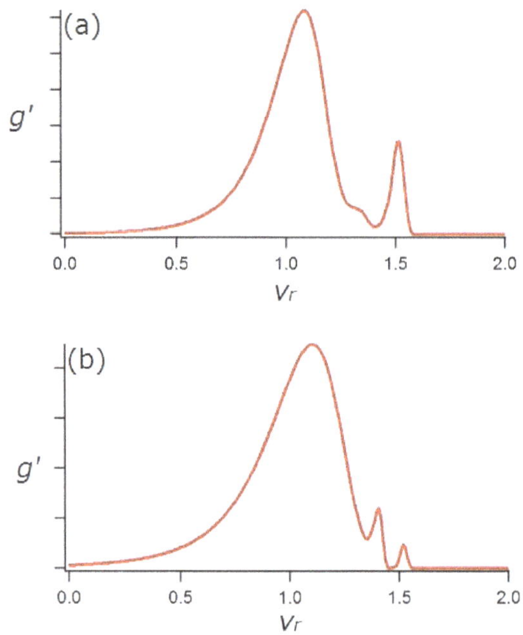

Fig. 4. The local VDFs (g' at $t = 22.0$) at **(a)** $r = 36$ and **(b)** $r = 41$ in Run B.

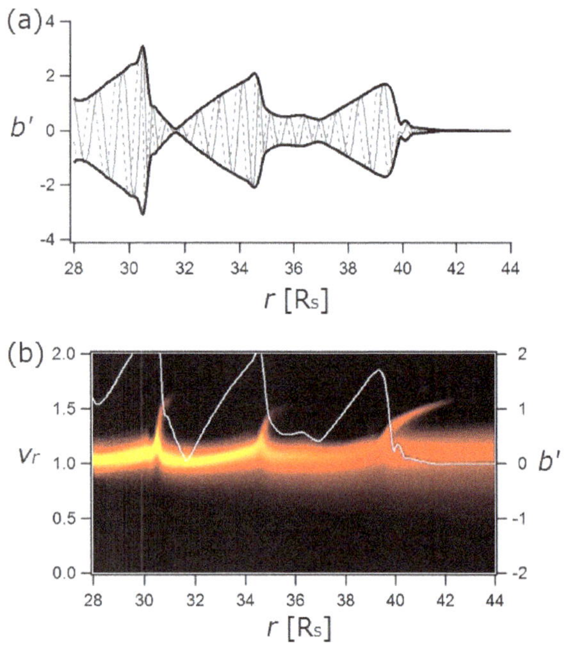

Fig. 5. The snapshot of **(a)** the transverse magnetic field and **(b)** g' at $t = 22.0$ in Run A.

Figures 5 and 6 show the snapshot of the transverse magnetic field and g' at $t = 22.0$ in runs A and C, respectively. In the resultant VDFs of all the runs (Figs. 4c, 5b, and 6b), the accelerated ions can be observed. The number of generated beams increases with increasing frequency range of the imposed Alfvén waves, since the wave length of the envelope modulation becomes smaller.

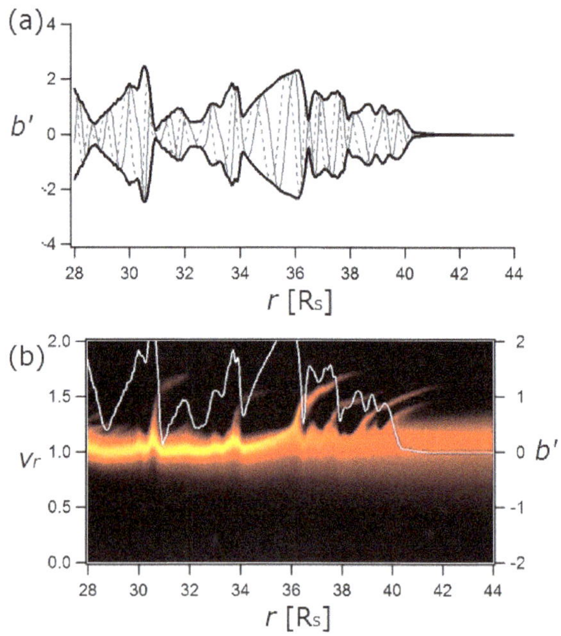

Fig. 6. The snapshot of **(a)** the transverse magnetic field and **(b)** g' at $t = 22.0$ in Run C.

Figure 7 shows the Elsasser variables $z'_p = u'_y - b'_y/\sqrt{\rho}$ and $z'_m = u'_y + b'_y/\sqrt{\rho}$ at $t = 22.0$. As shown in Fig. 7, the intensity of the sunward-propagating waves (z_m) is much less than the anti sunward-propagating waves (z_p). Namely, the non-WKB reflection by the background inhomogeneity (Suzuki and Inutsuka, 2006; Chandran et al., 2011) is weak, and the parametric decay instability of Alfvén waves (Tanaka et al., 2007) does not occur. The absence of the non-WKB reflection indicates that the WKB (Wentzel–Kramers–Brillouin) approximation is well satisfied in the present study. The absence of the parametric decay instability of Alfvén waves is discussed in the next section.

4 Discussion

4.1 Wave shoaling

In this subsection, we discuss the increase in nonlinearity in the WKB Alfvén wave packets in the solar wind. It is well known that the envelope-modulated Alfvén waves are steepened by the nonlinearity (Cohen and Kulsrud, 1975). The steepening converts the energy of transverse waves into the one of longitudinal waves (parallel ion kinetic energy), as shown in the previous section. This result suggests that the simple steepening (parallel cascading) and nonlinear ion Landau damping can play an important role in damping of Alfvén waves. On the other hand, the modulational instability does not occur due to the absence of the wave dispersion in the present system (Sakai and Sonnerup, 1983). Notice that the absence of the modulational instability does not

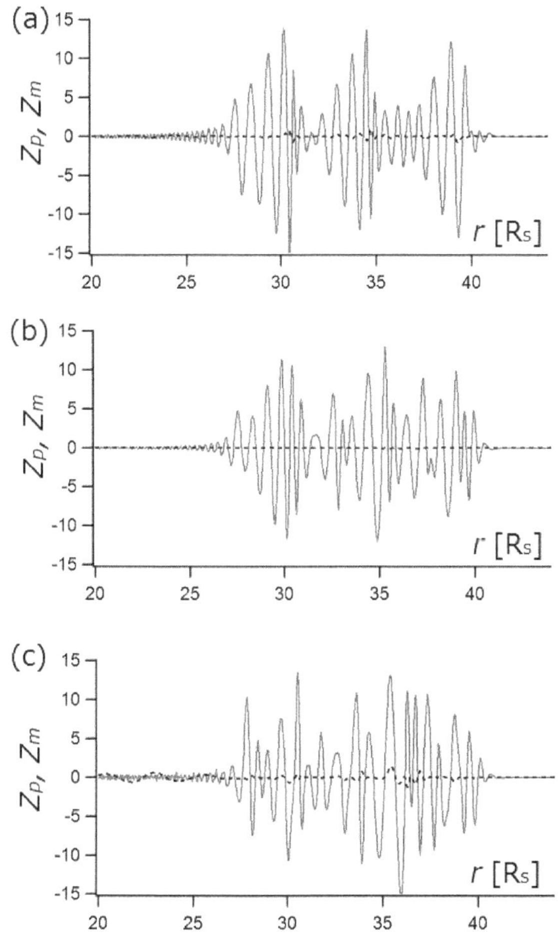

Fig. 7. The Elsasser variables $z'_p = u'_y - b'_y/\sqrt{\rho}$ (gray solid line) and $z'_m = u'_y + b'_y/\sqrt{\rho}$ (black dashed line) at $t = 22.0$ in **(a)** Run A, **(b)** Run B, and **(c)** Run C.

affect the decrease in energy of envelope-modulated Alfvén waves (Nariyuki et al., 2008). In contrast to the previous studies (Suzuki and Inutsuka, 2006; Tanaka et al., 2007) in which the MHD system was used, the decay instability is not clearly observed in the present study. This is consistent with the previous studies (Nariyuki et al., 2008; Markovskii et al., 2009) in which the ion hybrid simulations in uniform plasmas were carried out. Actually, the growth rate of the decay instability is small due to the small amplitude of each wave number mode in the present study. Moreover, the growth rate of the decay instability is weakened by the ion Landau damping of the longitudinal waves (Nariyuki and Hada, 2007).

By comparing both Figs. 1 and 2c, the contraction of the wave length ($2\pi/k$) is also observed. This is due to the conservation of the wave frequency

$$\omega = (u_r + V_A)k, \tag{20}$$

where the phase velocity $u_r + V_A$ decreases with increasing heliocentric distance. When the wave packets satisfy the WKB approximation, the nonlinearity of the waves (Barnes, 1992)

$$\left(\frac{|b|}{b_r}\right)^2 = C_0 r \left(1 + \frac{V_A}{u_r}\right)^{-2} \propto \frac{r^3}{(r + C_1)^2}, \tag{21}$$

where the constants C_0 and C_1 are determined by using the boundary condition. Equation (21) indicates that the nonlinearity of the wave packets increases with increasing heliocentric distance. The increase in the nonlinearity with the contraction of the wave length is similar to the wave shoaling of the surface gravity waves. When $u_r >> V_A$ $(r >> C_1)$, $\omega \simeq u_r k$ and $\left(\frac{|b|}{b_r}\right)^2 \propto r$. In this limit, the wave number k is the quasi-invariant and the waves can be described by using the expanding box (Grappin et al., 1993; Grappin and Velli, 1996; Liewer et al., 2001; Hellinger and Travnicek, 2005; Ofman et al., 2011). On the other hand, if $u_r << V_A$ $(r << C_1)$, $k \propto r$ and $\left(\frac{|b|}{b_r}\right)^2 \propto r^3$. Thus, the nonlinearity $\frac{|b|}{b_r}$ at a certain r becomes larger with increasing heliocentric distance from the critical point r_c. This effect of inhomogeneity promotes the nonlinear steepening and the ion acceleration by ponderomotive force.

The polarization of the imposed waves does not affect the time evolution of the system, since the wave dispersion due to the Hall effect and the finite Larmor radius effect is neglected in the present model. Notice that the nonlinear evolution of low-frequency Alfvén waves generates the higher frequency waves. The linear damping (ion cyclotron damping) and nonlinear damping (parametric instabilities) of the dispersive Alfvén waves are affected by the polarization (e.g., Liewer et al., 2001; Hellinger et al., 2005; Matteini et al., 2010). Actually, while the polarization of Alfvén waves does not affect the time evolution of the present dispersionless system, dispersive left-handed porarized waves are unstable to both decay and modulational instabilities in low beta plasmas (Longtin and Sonnerup, 1986). To incorporate the non-MHD effects (Chandran et al., 2011) such as the kinetic effects discussed in the next subsection into the present one-fluid model is an important future issue.

4.2 Ion kinetics

In the uniform and periodic system, the phase space vortex corresponding to the density fluctuations has also been observed in the nonlinear stage (Araneda et al., 2008; Matteini et al., 2010). In the present system, ions accelerated by the ponderomotive force are observed as the rolled-up VDFs in the phase space, while the vortex structures are not observed. This may be because of the asymmetrical wave form observed in Fig 2. Note that the trapped ions in the uniform system are the pseudo beam in the phase space, while the detrapped ions observed in the present study can survive as the real beam. However, the linear stability of the beams could not be discussed in the present system, since inhomogeneity of the velocity space is decoupled from the transverse wave dynamics due to the assumption of the isotropic velocity distribution function. Notice that the assumption of the isotropic VDFs in the present model is only valid for low beta plasmas such as those in the vicinity of the sun. Although the protons observed in the solar wind have the temperature anisotropy (Marsch, 2006; Bale et al., 2009; Maruca et al., 2011), the phase velocity of Alfvén waves evaluated by the Chew–Goldberger–Low (CGL) system is close to the Alfvén velocity in low beta plasmas (Nariyuki et al., 2013). The phase velocity of anti-sunward propagating Alfvén waves in CGL system is

$$v_\phi = u_r + \frac{b_r}{\sqrt{\rho\mu_0}}\left(1 - \frac{\mu_0(p_\parallel - p_\perp)}{|\mathbf{b}|^2}\right)^{\frac{1}{2}}, \tag{22}$$

where $p_\parallel$ and $p_\perp$ are the pressure parallel and perpendicular to the total magnetic field **b**. Equation (22) indicates that even if temperature (and pressure) anisotropy is large, $v_\phi \simeq u_r + V_A$ in low beta plasmas. For instance, by using an empirical law of fast solar wind plasmas in Hellinger et al. (2011), $v_\phi - u_r \simeq 1.019V_A$ $(r = 20)$, $1.020V_A$ $(r = 40)$, and $1.018V_A$ $(r = 60)$, respectively. On the other hand, as discussed below, some important kinetic effects are not included in the present model. Of course, even if the effect of temperature anisotropy on the phase velocity of Alfvén waves is negligible, the various instabilities can be caused by temperature anisotropy (Matteini et al., 2013).

In the present study, Alfvén wave packets excite the longitudinal fluctuations (deformation of the VDFs). Maneva et al. (2013) have recently confirmed the similar phase space vortex in the expanding box model. Moreover, nonresonant scattering of ions neglected in the present model can also affect the local VDFs (Araneda et al., 2008; Dong and Singh, 2013). In addition, the wave-particle resonant interactions between the Alfvén wave packets and ions become important with increasing heliocentric distance, since the ion cyclotron frequency decreases (Liewer et al., 2001). The Alfvén wave steepening observed in the present study causes the parallel cascading in the frequency space and may promote the ion cyclotron damping. It is also noteworthy that the kinetic instabilities driven by the beam components could not be described in the present system. The beam instabilities are affected by the inhomogeneity (e.g., Ofman, 2010b; Hellinger and Travnicek, 2011). To fill the gap between the past studies with the uniform system and the present study, further implementation of kinetic simulations with mesoscale models such as the expanding box models (Liewer et al., 2001; Hellinger and Travnicek, 2005; Ofman et al., 2011; Maneva et al., 2013) is required.

5 Conclusions

In the present paper, we discuss numerically the ion acceleration by nonlinear Alfvén wave packets in the radially expanding plasma. The results indicate that the steepened Alfvén wave packets outwardly accelerate ions, which can be observed as the beam components in the interplanetary space. The wave shoaling of the WKB Alfvén waves is also observed. On the other hand, as discussed in the previous section, important ion kinetic effects such as ion cyclotron damping and instabilities driven by the non-equilibrium VDFs are not included in the present model and the quantitative results of the present study may be modified by the kinetic effects. Actually, the effective heating and acceleration of ions by the Alfvén waves are very weak in the present study, while the ion beams are formed corresponding to the envelope modulation. The non-equilibrium VDFs observed in the runs still preserve the free energy to cause the instabilities. In order to address the effective ion heating of solar wind plasmas by low-frequency Alfvén waves, the kinetic effects and the other non-MHD effects such as turbulent cascading and collisions should be incorporated into the present model.

Acknowledgements. YN thanks the organizers of NWCW9 held at La Jolla, San Diego in March 2013. This work was supported by Grant-in-Aid for Young Scientists (B) no. 22740324 from MEXT of Japan. The computer simulations are performed on the DELL PowerEdge R815 supercomputer system at the Solar-Terrestrial Environment Laboratory at Nagoya University as a joint research program.

Edited by: G. S. Lakhina

References

Araneda, J. A., Marsch, E., and Vinas, A. F.: Proton Core Heating and Beam Formation via Parametrically Unstable Alfven-Cyclotron Waves, Phys. Rev. Lett., 100, 125003, doi:10.1103/PhysRevLett.100.125003, 2008.

Bale, S. D., Kasper, J. C., Howes, G. G., Quataert, E., Salem, C., and Sundkvist, D.: Magnetic fluctuation power near proton temperature anisotropy instability thresholds in the solar wind, Phys. Rev. Lett., 103, 211101, doi:10.1103/PhysRevLett.103.211101, 2009.

Barnes, A.: Collinsionless Damping of Hydrodynamic Waves, Phys. Fluids, 9, 1483, doi:10.1063/1.1761882, 1966.

Barnes, A.: Theory of magnetohydrodynamic waves: The WKB approximation revisited, J. Geophys. Res., 97, 12105, doi:10.1029/92JA00996, 1992.

Bruno, R., and Carbone, V.: The Solar Wind as a Turbulence Laboratory, Living Rev. Solar Phys., 10, 2, doi:10.12942/lrsp-2013-2, 2013.

Chandran, B. D. G., Dennis, T. J., Quataert, E., and Bale, S. D.: Incorporating kinetic physics into a two-fluid sola-wind model with temperature anisotropy and low-frequency Alfvén-wave turbulence, Astrophys. J., 743, 197, doi:10.1088/0004-637X/743/2/197, 2011.

Cheng, C. Z. and Knorr, G.: The integration of the Vlasov equation in configuration space, J. Comp. Phys., 22, 330–360, 1976.

Cohen, R. H. and Kulsrud, R. M.: Nonlinear ecolution of parallel-propagating hydro-magnetic waves, Phys. Fluids, 17, 2215, doi:10.1063/1.1694695, 1975.
Dong, C. and Singh, N.: Ion pseudoheating by low-frequency Alfvén waves revisited, Phys. Plasmas, 20, 012121, doi:10.1063/1.4789608, 2013.
Galinsky, V. L. and Shevchenko, V. I.: Induced Emission of Alfvén-Waves in Inhomogeneous Streaming Plasma: Implications for Solar Corona Heating and SolarWind Acceleration, Phys. Rev. Lett., 111, 015004, doi:10.1103/PhysRevLett.111.015004, 2013.
Grappin, R. and Velli, M.: Waves and streams in the expanding solar wind, J. Geophys. Res., 101, 425–444, doi:10.1029/95JA02147, 1996.
Grappin, R., Velli, M., and Mangeney, A.: Nonlinear wave evolution in the expanding solar wind, Phys. Rev. Lett., 70, 2190–2193, doi:10.1103/PhysRevLett.70.2190, 1993.
Hellinger, P. and Travnicek, P.: Magnetosheath compression: Role of characteristic compression time, alpha particle abundance, and alpha/proton relative velocity, J. Geophys. Res., 110, A04210, doi:10.1029/2004JA010687, 2005.
Hellinger, P. and Travnicek, P.: Proton core-beam system in the expanding solar wind: Hybrid simulations, J. Geophys. Res., 116, A11101, doi:10.1029/2011JA016940, 2011.
Hellinger, P., Velli, M., Travnicek, P., Gary, S. P., Goldstein, B. E., and Liewer, P. C.: Alfvén wave heating of heavy ions in the expanding solar wind: Hybrid simulations, J. Geophys. Res., 110, A12109, doi:10.1029/2005JA011244, 2005.
Hellinger, P., Matteini, L., Stverak, S, Travnicek, P. M., and Marsch, E.: Heating and cooling of protons in the fast solar wind between 0.3 and 1 AU: Helios revisited, J. Geophys. Res., 116, A09105, doi:10.1029/2011JA016674, 2011.
Larson, R. B.: A model for the formation of a spherical galaxy, Mon. Not. R. Astr. Soc., 145, 405–422, 1969.
Liewer, P. C., Velli, M., and Goldstein, B. E.: Alfvén wave propagation and ion cyclotron interactions in the expanding solar wind: One-dimensional hybrid simulations, J. Geophys. Res., 106, 29261, doi:10.1029/2001JA000086, 2001.
Longtin, M. and Sonnerup, B.: Modulational instability of circularly polarized Alfvén waves, J. Geophys. Res., 91, 6816–6824, 1986.
Maneva, Y. G., Vinas, A. F., and Ofman, L.: J. Geophys. Res., 118, 1–12, doi:10.1002/jgra.50363, 2013.
Markovskii, S. A., Vasquez, B. J., and Hollweg, J. V.: Proton heating by nonlinear field-aligned Alfvén waves in solar coronal holes, Astrophys. J., 695, 1413, doi:10.1088/0004-637X/695/2/1413, 2009.
Marsch, E.: Kinetic Physics of the Solar Corona and Solar Wind, Living Rev. Solar Phys., 3, 1, doi:10.12942/lrsp-2006-1, 2006.
Maruca, B. A., Kasper, J. C., and Bale, S. D.: What are the relative roles of heating and cooling in generating solar wind temperature anisotropies?, Pys. Rev. Lett., 107, 201101, doi:10.1103/PhysRevLett.107.201101, 2011.
Matteini, L., Landi, S., Velli, M., and Hellinger, P.: Kinetics of parametric instabilities of Alfvén waves: Evolution of ion distribution functions, J. Geophys. Res., 115, A09106, doi:10.1029/2009JA014987, 2010.
Matteini, L., Hellinger, P. Goldstein, B. E., Landi, S., Velli, M., and Neugebauer, M.: Signatures of kinetic instabilities in the solar wind, J. Geophys. Res., 118, 1, doi:10.1002/jgra.50320, 2013.
Mio, K., Ogino, T., Minami, K., and Takeda, S.: Modified nonlinear Schrodinger equation for Alfvén waves propagating along the magnetic field in cold plasmas, J. Phys. Soc. Japan, 41, 265–271, 1976.
Mjølhus, E.: On the modulational instability of hydromagnetic waves parallel to the magnetic field, J. Plasma. Phys., 16, 321–334, 1976.
Nariyuki, Y. and Hada, T.: Consequences of finite ion temperature effects on parametric instabilities of circularly polarized Alfvén waves, J. Geophys. Res., 112, A10107, doi:10.1029/2007JA012373, 2007.
Nariyuki, Y., Hada, T., and Tsubouchi, K.: On nonlinear evolution of Alfvénic turbulence in low beta plasmas, Phys. Plasmas, 15, 114502, doi:10.1063/1.3028316, 2008.
Nariyuki, Y., Hada, T., and Tsubouchi: Heating and acceleration of ions in nonresonant Alfvénic turbulence, Phys. Plasmas, 17, 072301, doi:10.1063/1.3449592, 2010.
Nariyuki, Y., Umeda, T., Kumashiro, T., and Hada, T.: A new numerical method for simulating the solar wind Alfvén waves: Development of the Vlasov-MHD model, Planetary and Space Science, 59, 767–771, doi:10.1016/j.pss.2011.02.017, 2011.
Nariyuki, Y., Saito, S., and Umeda, T.: A primitive kinetic-fluid model for quasi-parallel propagating magnetohydrodynamic waves, Phys. Plasmas, 20, 072115, doi:10.1063/1.4816809, 2013.
Ofman, L.: Wave modeling of the solar wind, Living Rev. Solar Phys., 7, 4, doi:10.12942/lrsp-2010-4, 2010.
Ofman, L.: Hybrid model of inhomogeneous solar wind plasma heating by Alfvén wave spectrum: Parametric studies, J. Geophys. Res., 115, A04108, doi:10.1029/2009JA015094, 2010.
Ofman, L., Viñas, A.-F., and Moya, P. S.: Hybrid models of solar wind plasma heating, Ann. Geophys., 29, 1071–1079, doi:10.5194/angeo-29-1071-2011, 2011.
Sakai, J.-I., and Sonnerup, B. U. O.: J. Geophys. Res., 88, 9069–9079, 1983.
Suzuki, T. K., and Inutsuka, S.-I.: Solar winds driven by nonlinear low-frequency Alfvén waves from the photosphere: Parametric study for fast/slow winds and disappearance of solar winds, J. Geophys. Res., 111, A06101, doi:10.1029/2005JA011502, 2006.
Tanaka, S., Ogino, T., and Umeda, T.: Parametric decay of circularly polarized Alfvén waves in the radially expanding solar wind, J. Geophys. Res., 112, A10110, doi:10.1029/2007JA012513, 2007.
Umeda, T.: A conservative and non-oscillatory scheme for Vlasov code simulations, Earth Planet. Space, 60, 773–779, 2009.
Umeda, T., Nariyuki, Y., and Kariya, D.: A non-oscillatory and conservative semi-Lagrangian scheme with fourth-degree polynomial interpolation for solving the Vlasov equation, Computer Phys. Communic., 183, 1094–1100, doi:10.1016/j.cpc.2012.01.011, 2012.
Wambecq, A.: Rational Runge-Kutta methods for solving systems of ordinary differential equations, Computing, 20, 333–342, doi:10.1007/BF02252381, 1978.

4

The comparative study of chaoticity and dynamical complexity of the low-latitude ionosphere, over Nigeria, during quiet and disturbed days

B. O. Ogunsua[1], **J. A. Laoye**[2], **I. A. Fuwape**[1], **and A. B. Rabiu**[1,3]

[1]Space Physics Laboratory, Department of Physics, Federal University of Technology, Akure, Nigeria
[2]Statistical and Nonlinear Physics Research group, Department of Physics, Olabisi Onabanjo University, Ago-Iwoye, Nigeria
[3]National Space Research and Development Agency, Abuja, Nigeria

Correspondence to: B. O. Ogunsua (iobogunsua@futa.edu.ng)

Abstract. The deterministic chaotic behavior and dynamical complexity of the space plasma dynamical system over Nigeria are analyzed in this study and characterized. The study was carried out using GPS (Global Positioning System) TEC (Total Electron Content) time series, measured in the year 2011 at three GPS receiver stations within Nigeria, which lies within the equatorial ionization anomaly region. The TEC time series for the five quietest and five most disturbed days of each month of the year were selected for the study. The nonlinear aspect of the TEC time series was obtained by detrending the data. The detrended TEC time series were subjected to various analyses for phase space reconstruction and to obtain the values of chaotic quantifiers like Lyapunov exponents, correlation dimension and also Tsallis entropy for the measurement of dynamical complexity. The observations made show positive Lyapunov exponents (LE) for both quiet and disturbed days, which indicates chaoticity, and for different days the chaoticity of the ionosphere exhibits no definite pattern for either quiet or disturbed days. However, values of LE were lower for the storm period compared with its nearest relative quiet periods for all the stations. The monthly averages of LE and entropy also show no definite pattern for the month of the year. The values of the correlation dimension computed range from 2.8 to 3.5, with the lowest values recorded at the storm period of October 2011. The surrogate data test shows a significance of difference greater than 2 for all the quantifiers. The entropy values remain relatively close, with slight changes in these values during storm periods. The values of Tsallis entropy show similar variation patterns to those of Lyapunov exponents, with a lot of agreement in their comparison, with all computed values of Lyapunov exponents correlating with values of Tsallis entropy within the range of 0.79 to 0.81. These results show that both quantifiers can be used together as indices in the study of the variation of the dynamical complexity of the ionosphere. The results also show a strong play between determinism and stochasticity. The behavior of the ionosphere during these storm and quiet periods for the seasons of the year are discussed based on the results obtained from the chaotic quantifiers.

1 Introduction

In our real world, most natural systems are nonlinear. However, a natural system can be either deterministic or stochastic. There is usually no distinction between the two in our natural system since all systems in one way or the other interact with their surroundings. Although few natural systems have been found to be low-dimensional deterministic in the sense of the theory, the concept of low-dimensional chaos has been proven to be fruitful in the understanding of many complex phenomena (Hegger et al., 1999). This is also evident in the study on magnetospheric dynamics and the ionosphere in the past decade. The study of chaos in magnetospheric index time series such as the AE index and the AL index was initially conducted by Vassiliadis et al. (1990), Shan et al. (1991), Pavlos et al. (1992) and Unnikrishnan (2008). These efforts have led to further advances in the study of chaos in the upper atmosphere. The development of this framework is based on the advances in the study of chaos in the ionosphere, revealing its complexity.

Excellent studies have been carried out relating to this framework, by some investigators like Bhattacharyya (1990), who studied chaotic behavior of ionospheric density fluctuation using amplitude and phase scintillation data, and found the existence of low-dimension chaos. Also, Wernik and Yeh (1994) further revealed the chaotic behavior of the ionospheric turbulence using scintillation data and numerical modeling of scintillation at high latitudes. They showed that the ionospheric turbulence attractor (if it exists) cannot be reconstructed from amplitude scintillation data, and their measured phase scintillation data adequately reproduce the assumed chaotic structure in the ionosphere. Also, Kumar et al. (2004) reported the evidence of chaos in the ionosphere by showing the chaotic nature of the underlying dynamics of the fluctuations of the TEC power spectrum, indicating exponential decay and the calculated positive value of the Lyapunov exponent. This is also supported by the results of the comparison of the chaotic characteristics of the time series of variations of TEC with the pseudochaotic characteristic of the colored noise time series. Also, Unnikrishnan et al. (2006a, b) analyzed the deterministic chaos at mid-latitude. Also, Unnikrishnan (2010); Unnikrishnan and Ravindran (2010) analyzed some TEC data from some Indian low-latitude stations for quiet and major storm periods, and found in their results the presence of chaos, which was indicated by a positive Lyapunov exponent and also showed storm periods with lower values compared with quiet periods.

A good number of investigators have also worked on the dynamical complexity of magnetospheric processes and the ionosphere. Balasis et al. (2008) investigated the dynamical complexity of the magnetosphere by using Tsallis entropy as a dynamical complexity measure in D_{st} time series; Balasis et al. (2009) also investigated the dynamical complexity in D_{st} further by considering different entropy measures. Coco et al. (2011), using the information theory approach, studied the dynamical changes of the polar cap potential, which is characteristic of the polar region ionosphere, by considering three cases: (i) steady IMF $B_Z < 0$, (ii) steady IMF $B_Z > 0$ and (iii) a double rotation from negative to positive and then positive to negative B_Z. They observed a neat dynamical topological transition when the IMF B_Z turns from negative to positive and vice versa, pointing to the possible occurrence of an order–disorder phase transition, which is the counterpart of the large-scale convection rearrangement and of the increase in the global coherence. Further studies of the chaotic behavior are however needed to improve our understanding of the dynamical behavior of the ionosphere of the low-latitude ionosphere, especially over Africa during quiet and storm periods for different seasons of the year, so as to be able to characterize chaoticity for different seasons.

The Nigerian subcontinent of Africa is situated in the low-latitude region of the globe within the equatorial anomaly region where the magnetic field B is almost totally parallel to the Earth's surface. In a low-latitude region such as this, off the Equator map along the F region, the eastward electric field (E) of the E region interacts with the magnetic field B during the day. This results in the electrodynamic lifting of the F region plasma over the Equator, which is known as EXB drift. The uplifted plasma over the Equator moves along the magnetic field lines in response to gravity, diffusion and pressure gradients and hence, the fountain effect. The fountain effect being controlled by the EXB drift shows the dynamics of the diurnal variation equatorial anomaly (Abdu, 1997; Unnikrishnan, 2010). There is a reduction in the F region ionization density at the magnetic Equator and much enhanced ionization density at the two anomaly crests within $\pm 15°$ of the magnetic latitude north and south of the Equator (Rama Rao et al., 2006).

The equatorial ionization anomaly and other natural processes, which include various ionization processes and recombination, the influx of solar wind, and photoionization processes among many others have a great influence on the internal dynamics of the systems of the ionosphere, and form the natural internal dynamics of the ionosphere. However, as mentioned before, there is no system in the real world without interaction with its external environment, which leads us to further study of the influence of the Sun on the ionosphere. Therefore in this study the Lyapunov exponent and Tsallis entropy from the TEC data from the ionosphere were studied for storm and quiet periods throughout the months of the year for a better understanding of the deterministic chaotic behavior and dynamical complexity of the ionosphere.

The motivation for this work is based on the fact that there is a significant level of nonlinear variability in the ionosphere that needs more investigation. Characterizing the ionosphere is of utmost importance due to numerous complexities associated with the region (Rabiu et al., 2007). The concept of chaos applied to ionospheric and magnetospheric studies of the quiet and stormy conditions is limited. The review of the works mentioned earlier shows that a good number of investigations has been carried out on the chaoticity of the upper atmosphere, especially in the magnetosphere and the ionosphere, and a number of investigations have also been carried out on the dynamical complexity of the upper atmosphere, but mainly on the magnetosphere.

To the best of our knowledge no extensive study has been carried out for all geomagnetic conditions, including quiet days, disturbed days and stormy conditions. Most investigations have only been based on quiet and storm conditions, and for all studies carried out none of the investigators has carried out a study of these concepts using the geomagnetically quiet and disturbed day classification to enable the proper characterization of the ionosphere. Secondly, this concept has not been used to study the ionosphere over the low-latitude region of Africa, and finally the study of dynamical complexity using the Tsallis entropy of the ionosphere has not been carried out using the total electron content in low-latitude Africa or any other part of the world. However, for

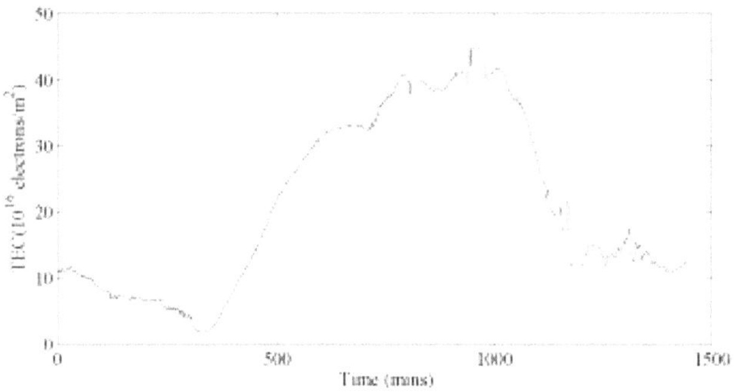

Fig. 1. A typical daily time series plot for TEC data measured from Lagos showing the typical pattern of variation.

the first time these issues mentioned are considered in this work.

2 Data and methodology

In Nigeria a number of GPS receivers are installed in different parts of the country, which falls within the equatorial ionization anomaly region. These receiver stations are called the NIGNET stations. The TEC values are gotten from these GPS receivers taking the measurement of slant TEC within a 1 m^2 columnar unit of the cross section along the ray path of the satellite and the receiver, which is given by

$$\mathrm{STEC} = \int_{\mathrm{Receiver}}^{\mathrm{Satellite}} N \mathrm{d}l. \quad (1)$$

The observation of the total number of free electrons along the ray path is derived from the frequencies L_1 (1572.42 MHz) and L_2 (1227.60 MHz) of the GPS that provide the relative ionospheric delay of electromagnetic waves travelling through the medium (Saito et al., 1998). The slant TEC is projected to vertical TEC using the thin shell model assuming a height of 350 km (Klobuchar, 1986).

$$\mathrm{VTEC} = \mathrm{STEC} \cdot \cos[\arcsin(R_\mathrm{e} \cos\Theta / R_\mathrm{e} + h_{\max})], \quad (2)$$

where $R_\mathrm{e} = 6378$ km (radius of the Earth), $h_{\max} = 350$ km (vertical height assumed in the model) and Θ = the elevation angle at the ground station.

In this study, three stations were considered: Birnin Kebbi (12°32′ N, 4°12′ E and 0.62° N geomagnetically), Enugu (6°26′ N, 7°30′ E and −3.21° N geomagnetically) and Lagos (6°27′ N, 3°23′ E and −3.07° N geomagnetically) within the low-latitude region. The TEC data obtained for January to December 2011 were considered for this study and the data are given at 1 min sampling time. The values of the five quietest days of each month were taken and the same was done for the five most disturbed days using the list of International Quiet Days (IQD) and International Disturbed Days (IDD) data provided by Geoscience Australia to compare the chaotic behavior of quiet days with that of disturbed days. The TEC data were subjected to various analyses that will be discussed in the next section.

3 Data analysis and results

3.1 Time series analysis

A given time series Sn can be defined as a sequence of scalar measurement of the same quantity taken as a series at different portions in time for a given time interval (δt). The time series describe the physical appearance of an entire system. However, it may not always describe the internal dynamics of that system. A system like the ionosphere possesses a dominant dynamics that can be seen as diurnal, so the data should be treated so as to be able to see its internal dynamics. The measured TEC time series were plotted to see the dynamics of the system. A typical plot of TEC usually has a dominant dynamics (see Fig. 1), which may be seen as the diurnal behavior. However, it can also be seen that there is a presence of fluctuations (which appear to be nonlinear) in the system as a result of the internal dynamics of the ionosphere and space plasma system due to different activities in the ionosphere. Therefore there is a need to minimize the influence of the diurnal variations, since we are more interested in the nonlinear internal dynamics of the system in this study. To do so the TEC time series was detrended by carrying out the analysis below, since it is known that there are 1440 data points in a day for daily data of 1 min sampling time, if we let a_i and $t_i, i = 1, 2, 3, \ldots, 1440$ represent the actual and observed values, respectively, of the time series of TEC. The diurnal variation reduced time is given by

$$T_i = t_i - a_j, \quad (3)$$

where $j = \mathrm{mod}\,(i, 1440)$ if $\mathrm{mod}\,(j, 1440) \neq 0$, and $j = 1440$ if $d\,(j, 1440) = 0$.

This method will give the detrended time series from the original TEC data as shown in Fig. 2. This method is similar to that used by Unnikrishnan et al. (2006a, b); Unnikrishnan (2010). Further explanations of the dynamical results can be found in Kumar et al. (2004).

3.1.1 Phase space reconstruction and nonlinear time series analysis

The chaoticity and dynamical complexity of a system are nonlinear phenomena that can describe the state of some dynamical systems. Such systems may be seen as nonlinear complex systems. The magnetosphere and the ionosphere are good examples of such systems. To be able to study such phenomena, some nonlinear time series analyses can be carried out on the time series data describing such a system. The detrended time series of TEC measurement is subjected to some

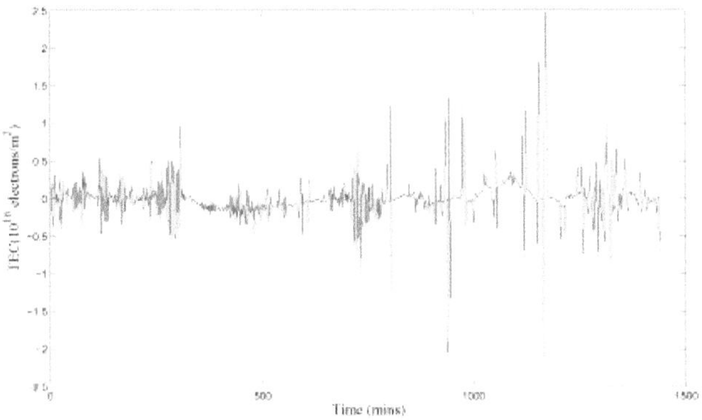

Fig. 2. A typical daily time series plot for the detrended TEC data measured from Lagos showing the typical pattern of variation.

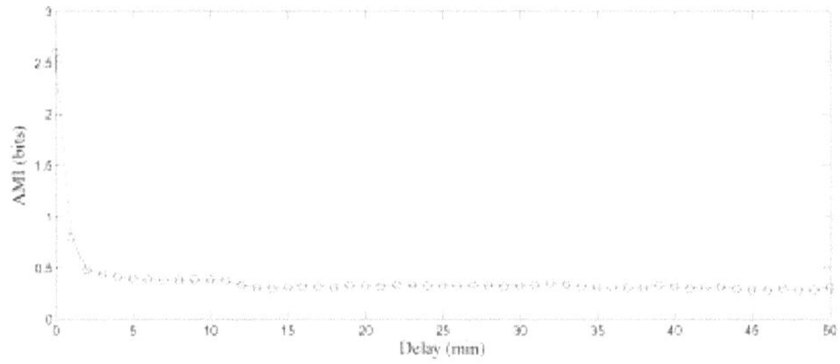

Fig. 3. Plot of mutual information against time delay obtained from the detrended TEC data measured from Lagos.

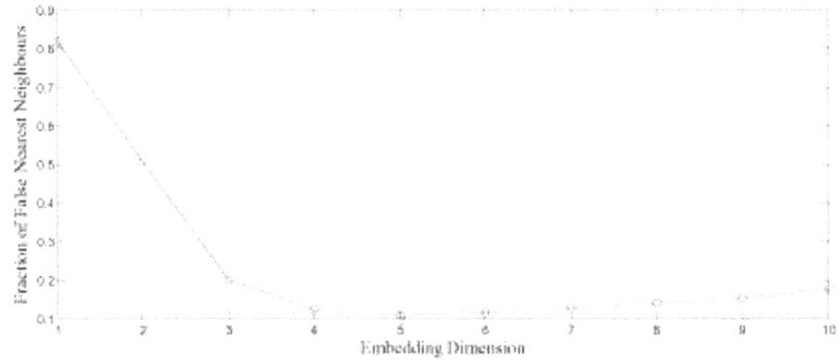

Fig. 4. Plot of fraction of false nearest neighbors against embedding dimension obtained from the detrended TEC data measured from Lagos.

nonlinear time series data analyses to obtain the mutual information and false nearest neighbors, embedding dimension and delay coordinates for the phase space reconstruction, and the evaluation of other chaotic quantifiers: Lyapunov exponents, entropy. The phase space reconstruction helps to reveal the multidirectional aspect of the system. The phase space reconstruction is based on the embedding theorem, such that the phase space is reconstructed to show the multidimensional nature as follows:

$$\boldsymbol{Y}_n = (s_{n-(m-1)\tau}, s_{n-(m-2)\tau}, \ldots, s_n), \tag{4}$$

where $\boldsymbol{Y}_{\mathrm{n}}$ are vectors in phase space. The proper choice of embedding dimension (m) and delay time (τ) is essential for phase space reconstruction (Fraser and Swinney, 1986; Kennel et al., 1992). If a plot showing the time-delayed mutual information shows a marked minimum, that value can be considered as a responsible time delay. Figure 3 shows the mutual information plotted against time delay. Likewise, the minimal embedding dimension, which corresponds to the minimum number of the nearest false neighbors as shown in Fig. 4, can be treated as the optimum value of the embedding dimension in Unnikrishnan et al. (2006a, b) and Unnikrishnan (2010). It was observed that for all the daily detrended TEC time series the choice of $\tau \geq 25$ and $m = 3$ values of delay and embedding dimension above these values are suitable for analysis of data for all stations. The choice of $\tau = 30$ and $m = 5$ was also used for further analysis for most of the stations. The reconstructed phase space trajectory is shown in Fig. 5.

3.1.2 Lyapunov exponents

The Lyapunov exponent is an important chaotic quantifier. It indicates divergence of trajectory in one dimension, or alternatively an expansion of volume, which can also be said to indicate repulsion, or attraction from a fixed point. A positive Lyapunov exponent indicates that there is evidence of chaos in a dissipative deterministic system, where the positive Lyapunov exponent indicates divergence of trajectory in one direction or expansion of values, and a negative value shows convergence of trajectory or contraction of volume along another direction. The largest Lyapunov exponent (λ_1) can be used to determine the rate of divergence as indicated by Wolf et al. (1985) where

$$\lambda_1 = \lim_{r \to \inf} \frac{1}{t} \log \frac{\Delta x(t)}{x(0)} = \lim_{r \to \inf} \frac{1}{t} \sum_{i=1}^{t} \log \frac{\Delta x(t_i)}{\Delta x(t_{i-1})}. \tag{5}$$

The Lyapunov exponent was computed for the TEC values measured from different stations. The evolution in state space was scanned with $\tau = 30, m = 5$, as shown in Fig. 6. The values of the Lyapunov exponent were computed for all stations for the five quietest and five most disturbed days of every month of the year according to the International Quiet Days (IQD) and International Disturbed Days (IDD) classification by Geoscience Australia. The difference in values was studied for these situations and compared. This was done using the implementation introduced by Rosenstein et al. (1993), and Hegger et al. (1999), both algorithms using very similar methods. The values of the Lyapunov exponent plotted for the five quietest and five most disturbed days are shown in Fig. 8a–d and show the plot values of the computed Lyapunov exponent from January to April 2011 for Birnin Kebbi station. The monthly averages of Lyapunov exponents for quiet and disturbed periods are shown in Fig. 8e.

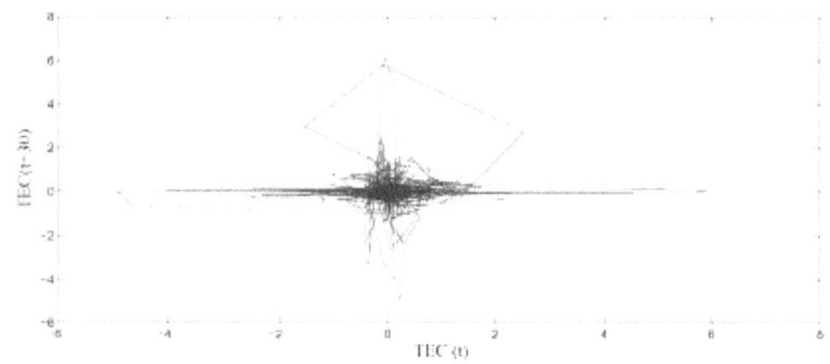

Fig. 5. Delay representation from the phase space reconstruction of the detrended time series.

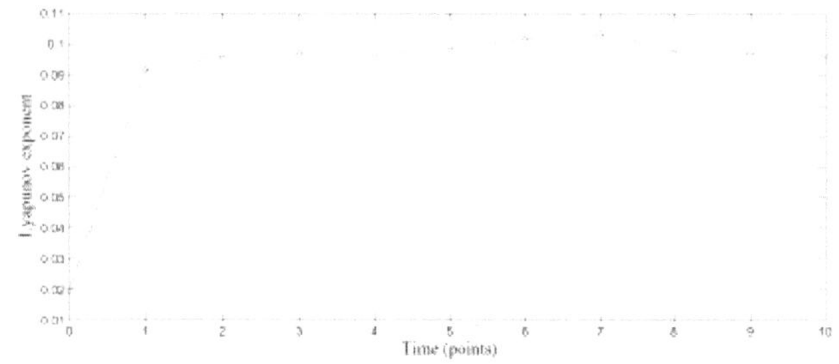

Fig. 6. Lyapunov exponent and its evolution, computed as the state space trajectory with $\tau = 30$, and $m = 5$ for detrended time series measured at Lagos with the largest Lyapunov exponent equal to 0.1035.

3.1.3 Correlation dimension

Another relevant method for studying the underlying dynamics or internal dynamics of a system is to have a good knowledge of its dimension. The correlation dimension gives a good approximation of this as suggested by Grassberger and Procaccia (1983a, b). The correlation dimension D is defined as

$$D = \lim r \to 0 \frac{\ln C(r)}{\ln r}. \tag{6}$$

The term $C(r)$ is the correlation sum for radius r where for a small radius r the correlation sum can be seen as $C(r) \sim r^d$ for $r \to 0$. The correlation sum is dependent on the embedding dimension m of the reconstructed phase space and is also dependent on the length of the time series N as follows:

$$C(r) = \frac{2}{N(N-1)} \sum_{i=1}^{N} \sum_{j=i+1}^{N} \Theta(r - ||y_i - y_j||) \tag{7}$$

where Θ is the Heaviside step function, with $\Theta(H) = 0$ if $H \leq 0$ and $\Theta(H) = 1$ if $H > 0$. The correlation dimension was computed using the Theiler algorithm approach, with Theiler window w at 180. The Theiler window was chosen to be approximately equal to the product of m and τ. A similar approach to the computation of correlation dimensions was used by Unnikrishnan and Ravindran (2010) to determine the correlation dimension of detrended TEC data for some stations in India, which lies within the equatorial region, like Nigeria. The correlation dimension for data taken for the quietest day of October 2011 and the most disturbed day of October 2011 from Lagos GPS TEC measuring station are represented by Fig. 7. The correlation dimension saturates at $m \geq 5$ for the quietest day of the month and at $m \geq 4$ for the most disturbed day. In this illustration the most disturbed day of this month falls within the storm period of October 2011. The use of quiet and disturbed day classifications in the month of October 2011 enables us to compare the quiet and storm periods together while comparing the quiet days with some relatively disturbed days.

Fig. 7. Correlation dimension for (upper panel) the quietest day of October 2011 for TEC measured at Lagos that saturates at $\tau = 30$, and $m \geq 5$ (lower panel) the most disturbed day of October 2011 for TEC measured at Lagos that saturates at $\tau = 32$, and $m \geq 4$.

3.1.4 Computation of Tsallis entropy and other entropy measures

Entropy measures are very important statistical techniques that can be used to describe the dynamical nature of a system. Tsallis entropy can be used to describe the dynamical complexity of a system and also to understand the nonlinear dynamics like chaos that may exist in a natural system. The use of entropy measure as a method to describe the state of a physical system has been employed in information theory for decades. Since entropies allow us to describe the state of disorderliness in a system, one can generalize this same concept to characterize the amount of information stored in

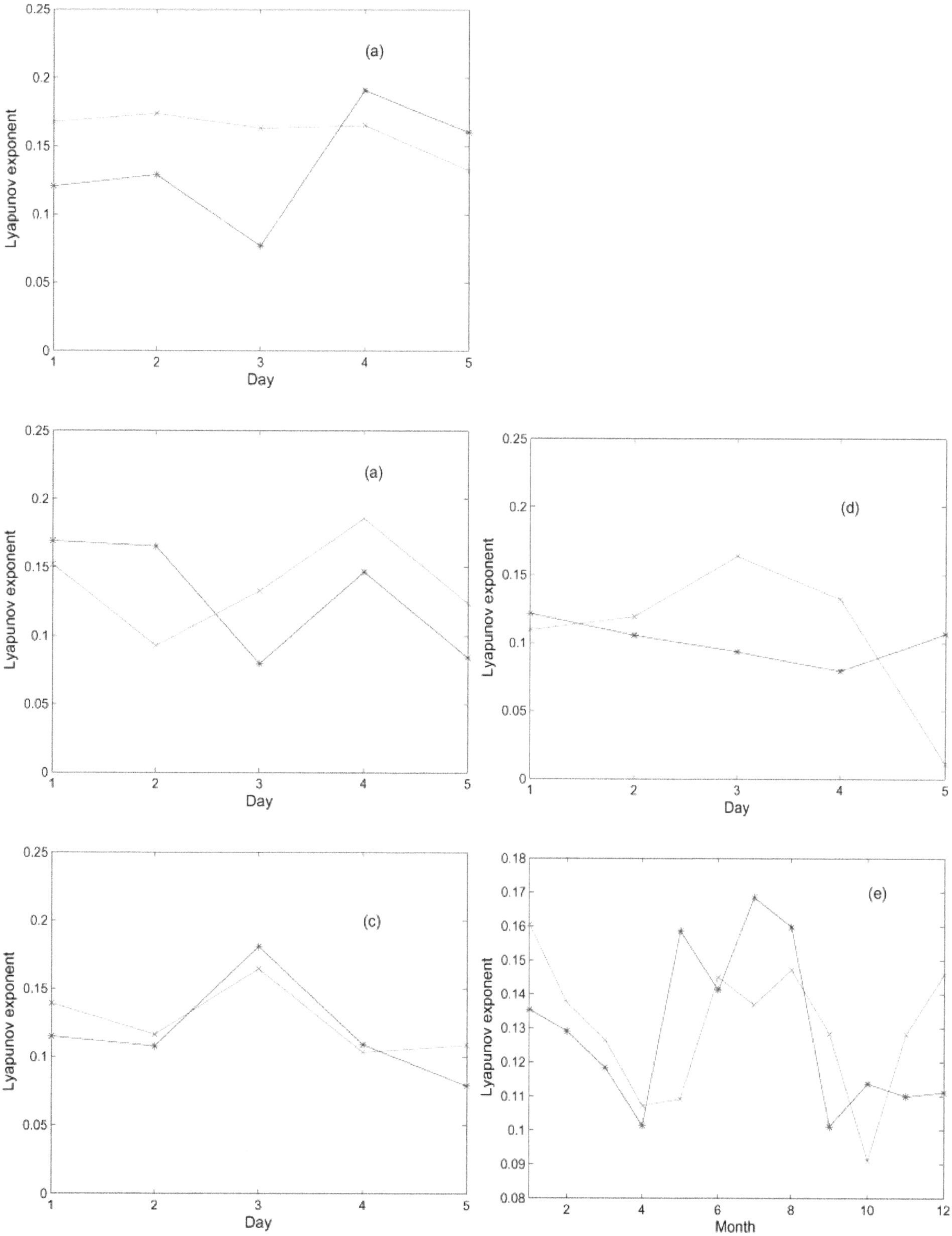

Fig. 8. Lyapunov exponents for the five quietest days (red) and the five most disturbed days (blue) for **(a)** January, **(b)** February, **(c)** March, and **(d)** April. **(e)** Monthly mean values of Lyapunov exponents for the entire year at Birnin Kebbi.

Table 1. Results of the surrogate data test for Lyapunov exponents for TEC data for the quietest days of October 2011 at Lagos station.

Original Data	Surrogate Data
0.1118	0.3397 ± 0.0325
0.0896	0.3493 ± 0.0594
0.1382	0.4091 ± 0.0684
0.1162	0.2967 ± 0.0588
0.2232	0.3914 ± 0.0207

Table 2. Results of the surrogate data test for Lyapunov exponents for TEC data for the most disturbed days of October 2011 at Lagos station.

Original Data	Surrogate Data
0.0563	0.3314 ± 0.0433
0.0370	0.1641 ± 0.0323
0.1763	0.3426 ± 0.0424
0.0872	0.2772 ± 0.0193

more general probability distributions (Kantz and Schreiber, 2003; Balasis et al., 2009). The concept of information theory is basically concerned with these principles. Information theory gives us an important approach to time series analysis, if our time series, which is a stream of numbers, is given as a source of information such that these numbers are distributed according to some probability distribution, and transitions between numbers occur with well-defined probabilities. One can deduce the same average behavior of the system at a different point and for the future.

The term entropy is used in both physics and information theory to describe the amount of uncertainty or information inherent in an object or system (Kantz and Schreiber, 2003). The state of an open system is usually associated with a degree of uncertainty that can be quantified by the Boltzmann–Gibbs entropy, a very useful uncertainty measure in statistical mechanics. However, Boltzmann–Gibbs entropy cannot describe non-equilibrium physical systems with large variability and multifractal structure such as the solar wind (Burlaga et al., 2007; Balasis et al., 2008). One of the crucial properties of the Boltzmann–Gibbs entropy in the context of classical thermodynamics is extensivity, namely proportionality with the number of elements of the system. The Boltzmann–Gibbs entropy satisfies this prescription if the subsystems are statistically (quasi-)independent, or typically if the correlations within the system are essentially local. In such cases the system is called extensive. In general, however, the situation is not of this type and correlations may be far from negligible at all scales. In such cases, the Boltzmann–Gibbs entropy is nonextensive (Balasis et al., 2008, 2009). These genralizations above were propsed by Tsallis (1988, 1998, 1999), who was inspired by the probabilistic description of multifractal geometries. He introduced an entropy measure by presenting an entropic expression characterized by an index q that leads to a nonextensive statistics,

$$S_q = k\frac{1}{q-1}(1 - \sum_{i=1}^{W} p_i^q), \tag{8}$$

where p_i are the probabilities associated with the microscopic configurations, W is their total number, q is a real number, and k is Boltzmann's constant. The value q is a measure of the nonextensitvity of the system: $q \to 1$ corresponds to the standard extensive Boltzmann–Gibbs statistics. This is the basis of the so-called nonextensive statistical mechanics, which generalizes the Boltzmann–Gibbs theory. The entropic index q characterizes the degree of nonadditivity reflected in the following pseudoadditivity rule:

$$S_q(A+B) = S_q(A) + S_q(B) + (1-q)S_q(A)S_q(B). \tag{9}$$

The cases $q > 1$ and $q < 1$ correspond to subadditivity (or subextensivity) and superadditivity (or superextensivity), respectively, and $q = 1$ represents additivity (or extensivity). For subsystems that have special theory probability correlations, extensivity is not valid for Boltzmann–Gibbs entropy, but may occur for S_q with a particular value of the index q. Such systems are sometimes referred to as nonextensive (Boon and Tsallis, 2003; Balasis et al., 2008, 2009). The parameter q itself is not a measure of the complexity of the system, but measures the degree of nonextensivity of the system. It is the time variations of the Tsallis entropy for a given $q(S_q)$ that quantify the dynamic changes of the complexity of the system. Lower S_q values characterize the portions of the signal with lower complexity. In this presentation we estimate S_q on the basis of the concept of symbolic dynamics and by using the technique of lumping (Balasis et al., 2008, 2009).

Considering the fact that Tsallis entropy has been used extensively for magnetospheric studies to obtain interesting results for the dynamical complexity by Balasis et al. (2008, 2009), we find it necessary to consider its application to the study of ionospheric dynamics. It is also necessary to compare the results obtained from the computation of Tsallis entropy with those of the Lyapunov exponent. A comparison of the relationship between the values of the Lyapunov exponent and Tsallis entropy was carried out to show their relationship as measures of complexity. This is based on the fact that Tsallis entropy has been linked to a significant degree of response to the edge of chaos and chaotic regime dynamical systems due to its non-extensive nature (Baranger et al., 2002; Anastasiadis et al., 2005), and it has been linked to weak chaos and the vanishing largest Lyapunov exponent (Kalogeropoulos, 2012, 2013). It has been established that the Lyapunov exponent varies directly as the Tsallis entropy (complexity) of a system, based on the variation of the entropic index q introduced by Tsallis (1988) and the nature of the system's dynamics. Baranger et al. (2002) were able to show that the non-extensive case of Tsallis entropy has been

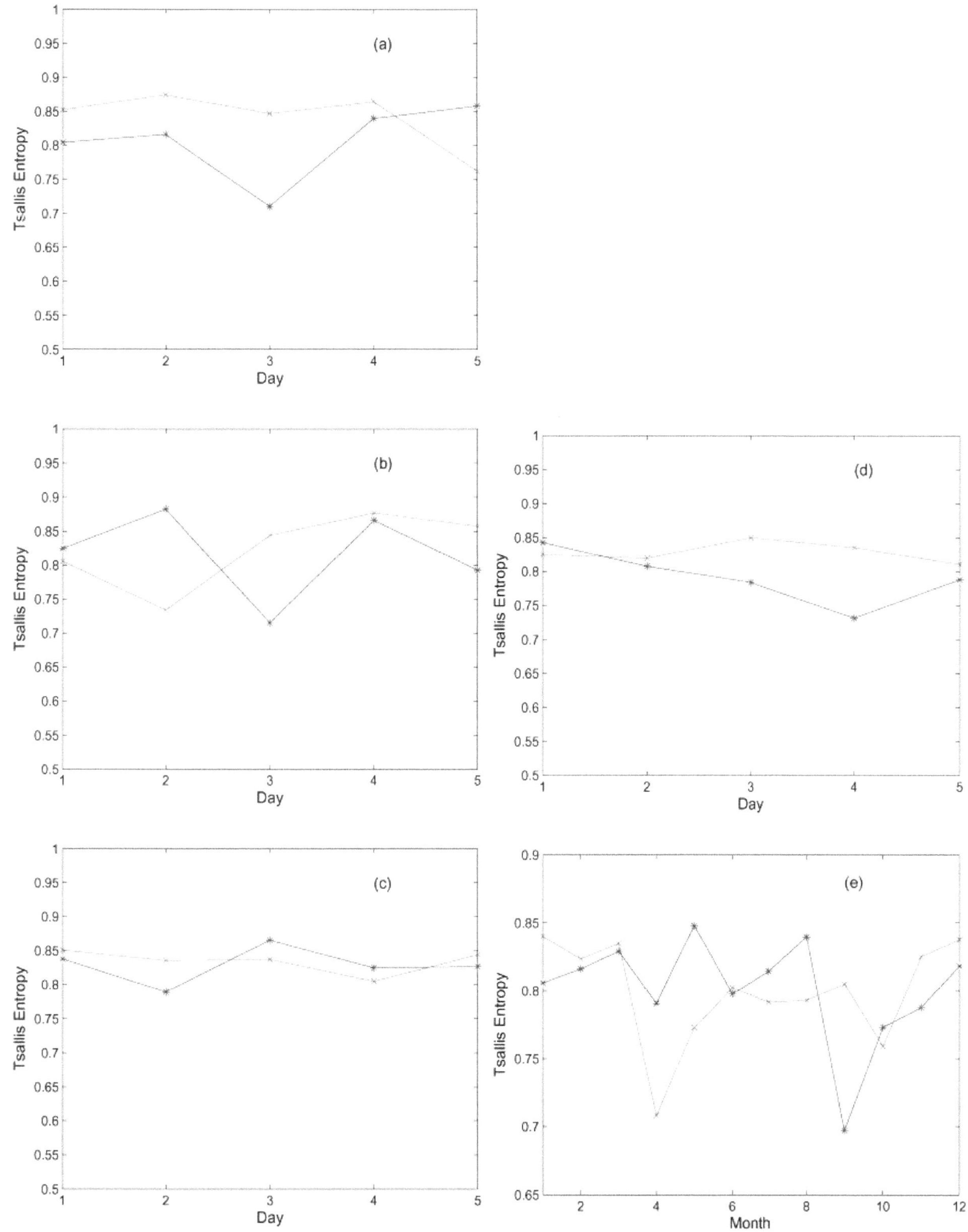

Fig. 9. Tsallis entropy for the five quietest days (red) and the five most disturbed days (blue) for **(a)** January, **(b)** February, **(c)** March, and **(d)** April. **(e)** Monthly mean values of Tsallis entropy for the entire year at Birnin Kebbi.

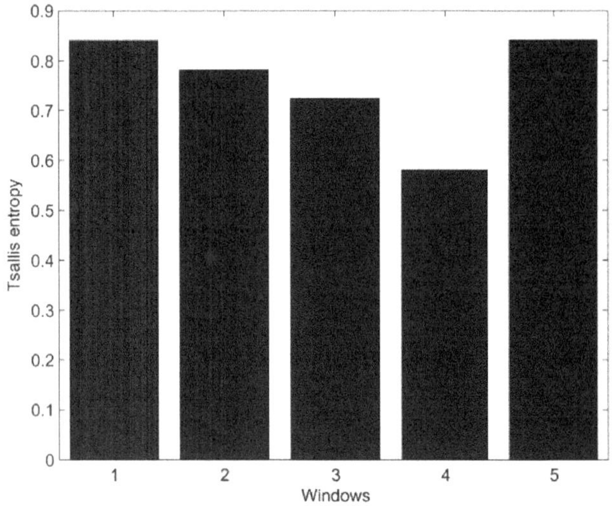

Fig. 10. Entropy values for five window intervals for the most quiet day in January 2011 from detrended TEC taken from Birnin Kebbi.

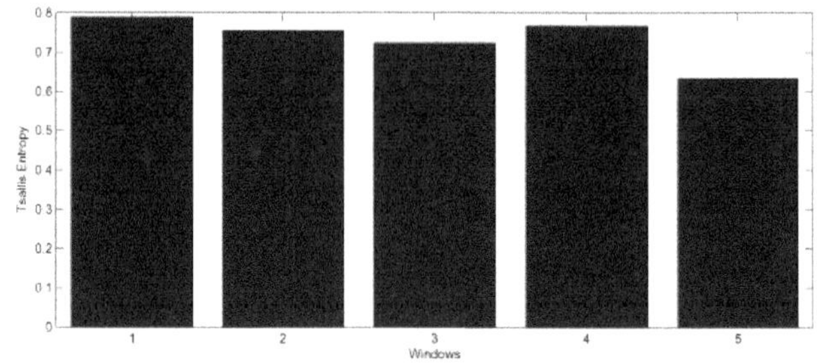

Fig. 11. Entropy values for five window intervals for the most disturbed day in January 2011 from detrended TEC taken from Birnin Kebbi.

found to vary directly as Kolmogorov–Sinai generated from Lyapunov exponents for logistic maps and dynamical systems in the threshold of chaos where $\lambda = 0$, with direct variation when $q = 1$ during a chaotic regime. They were able to show that for all cases of the positive Lyapunov exponent λ there will be an average exponential increase of any small initial distance, which can be given as

$$\xi(t) = \frac{x_t - x_t'}{x_0 - x_0'} \tag{10}$$

for $\xi(t) = \exp(\lambda t)$ where x_t and x'_t are positions of two initially closed trajectories. They were able further to relate q to the exponential increase in small distances at the edge of chaos $\lambda = 0$ as

$$\xi(t) = [1 + (1 - q)\lambda_q t]^{1/(1-q)}. \tag{11}$$

A similar Tsallis generalization was made for Lyapunov exponents in Coraddu et al. (2006), further explaining that the exponential behavior for the chaotic regime is recovered for $q \to 1$

$$\lim_{q\to 1} \exp(\lambda_q t) = \exp(\lambda t). \tag{12}$$

It was stated in their paper that a large class of generalized exponentials shows similar behavior. However, Anastasiadis et al. (2005) explored different q index values for complex networks for $\lambda < 0$ (periodic case) or $\lambda = 0$ (edge of chaos) and $\lambda > 0$ (chaotic regime), where they found $q = 2$ to be appropriate for a well-distinguished variation in Tsallis entropy between chaos and edge of chaos regimes.

The values of these entropy measures were also computed in order to study the dynamical complexity of the system under observation (the ionosphere). This will help us to obtain some more information on the state of the system. The values of Tsallis entropy were computed for the five quietest days and five most disturbed days of the month also according to the International Quiet Days (IQD) and International Disturbed Days (IDD) classification by Geoscience Australia. The computed values of Tsallis entropy representing the dynamical complexity of a system for the quiet and disturbed days are shown in Fig. 9a–d, with monthly averages for the year in Fig. 9e. The entropy measures were also computed for some daily data, which were split into five time window intervals to be able to study the changes in the pattern of the entropy values in these intervals within a day (Figs. 10 and 11).

3.1.5 Nonlinearity test using surrogate data

The test for nonlinearity using the method of surrogate data according to Kantz and Schreiber (2003) has been proven to be a good test for nonlinearity in time series describing a system. It has been accepted that the method of a surrogate data test could be a successful tool for the identification for nonlinear deterministic structure in experimental data (Pavlos et al., 1999). This method involves creating a test of significance of difference between a linearly developed surrogate and the original nonlinear time series to be tested. The test is done by carrying out the computation of the same quantity on both surrogates and the original time series and then checking for the significance of difference between the results obtained from the surrogates with the original data. Theiler et al. (1992) suggested the creation of surrogate data by using Monte Carlo techniques for accurate results. According to this method, typical characteristics of data under study are compared with those of stochastic signals (surrogates), which have the same auto-correlation function and the power spectrum of the original time series. One can safely conclude from the test of significance carried out on the surrogate and the original data that a stationary linear Gaussian stochastic model cannot describe the process under study provided that the behavior of the original data and the surrogate data are significantly different.

In this work 10 surrogate data were generated from the original data set. The geometrical and dynamical characteristics of the original data were then compared with those of the surrogates using the statistical method of significance of difference, which can be defined as

$$S = \frac{\alpha_{\text{Surr}} - \alpha_{\text{Original}}}{\sigma}. \tag{13}$$

Where α_{Surr} is the mean value of the computed quantity for the surrogate data and α_{Original} is the same quantity computed for the original TEC data, σ is the standard deviation of the same quantity computed for the surrogate data. The significance of difference considered for the null hypothesis to be rejected here is greater than 2, which enables us to reject the null hypothesis that the original TEC data describing the ionospheric system can be modeled using a Gaussian linear stochastic model with confidence greater than 95 %. The surrogate data test for all stations used in this study shows that the Lyapunov exponent of the surrogate data for the selected days in October are shown in Tables 1 and 2. The results show that the surrogate data test for the Lyapunov exponent shows a significance of difference greater than 2 for all the selected days for all the stations. Similar results were obtained for mutual information, fraction of false nearest neighbors and correlation dimension. This result gives us the confidence to reject the null hypothesis that the data used cannot be modeled using a linear Gaussian stochastic model, which shows that the system is a nonlinear system with some level of determinism. Figure 12 shows the plots comparing the mutual information plotted against time delay for the original detrended data blue with the mutual information for the surrogate data for TEC data measured at Lagos for the quietest day of March 2011, while Fig. 13 compares the fraction of false nearest neighbors for the same set of data. Table 1 shows the values of Lyapunov exponents for both original detrended and its surrogate data for TEC measured in Lagos during the quietest days of October 2011.

3.1.6 Contour/spatial plots

The contour plots for Lyapunov exponents and Tsallis entropies for quiet and disturbed days were plotted against the months of the year. This will enable us to see the behavior. The plots were done to show the variation of the values of these parameters with variations in quietness or disturbance and also to show how this varies with each month. These plots are made to show the response of the two parameters to storms and other variations describing changes in the internal dynamics of the ionosphere. These plots will also reveal the extent of the similarities in the response of these two quantifiers to ionospheric changes. The contour plots were made in this work to show the effect of the degree of quietness and monthly variation of solar activities on the variation of the two quantifiers. Figures 14 and 15 show the contour plots for Lyapunov exponents and Tsallis entropy for 2011 at Birnin Kebbi station.

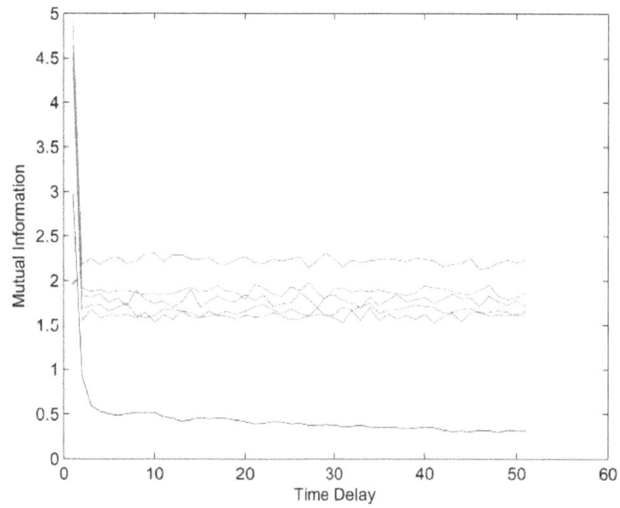

Fig. 12. The plots comparing the mutual information plotted against time delay for the original detrended data (blue curve) with the mutual information for 5 surrogate data (red curves) for TEC data measured at Lagos for the quietest day of March 2011.

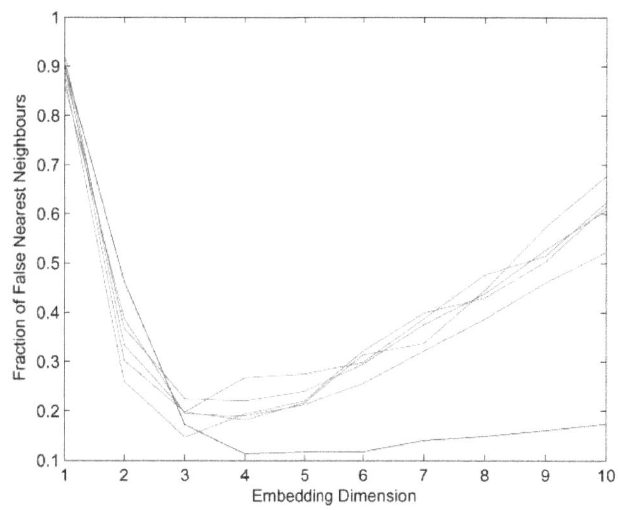

Fig. 13. The plots comparing the fraction of false nearest neighbors plotted against the embedding dimension for the original detrended data (blue curve) with the mutual information for 5 surrogate data (red curves) for TEC data measured at Lagos for the quietest day of March 2011.

4 Discussion

The results presented in this work show that the ionosphere shows a great degree of complexity for different times of the day and for different geophysical conditions. The time series plot in Fig. 1 shows the dominant characteristics of the ionosphere. The time series plot shows the rise in TEC to peak at the sunlit hours of the day; however, it can be seen that the rising to the peak of the ionosphere, which is the dominant dynamics during the day, make it impossible to see the

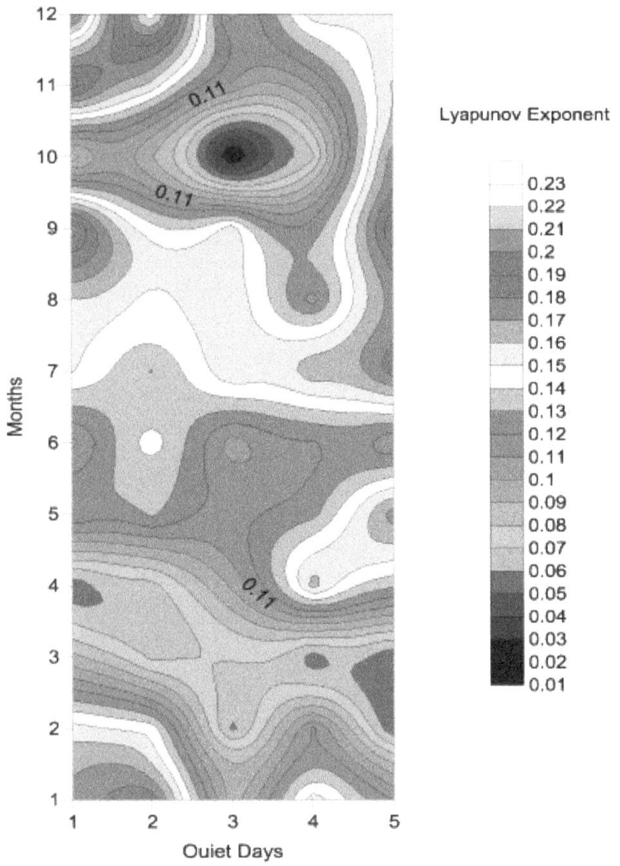

Fig. 14. Contour plot of Lyapunov exponents at Enugu station for quiet days of 2011.

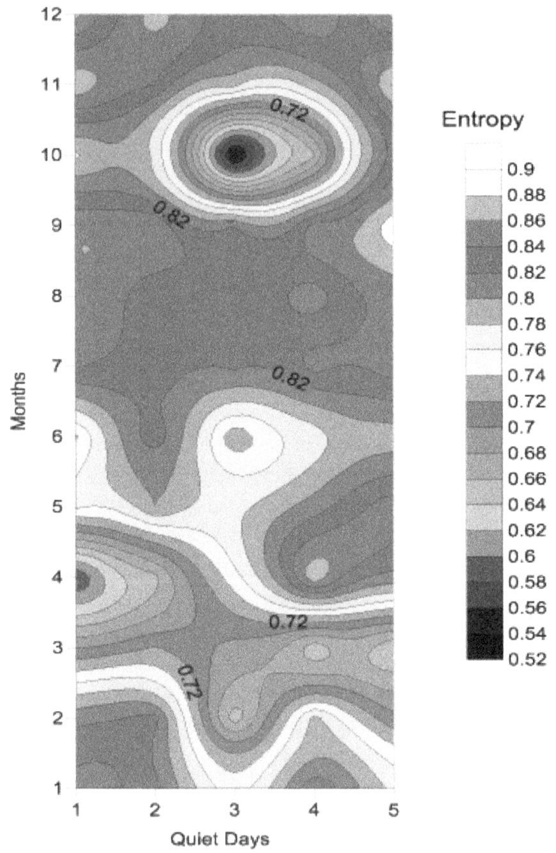

Fig. 15. Contour plot of Tsallis entropy at Enugu station for quiet days of 2011.

internal dynamics of the system from the TEC time series plot. It can be seen that the TEC time series curve is not a smooth curve with tiny variations, which probably describes a part of the internal dynamics. These visible tiny variations around the edges of the time series plot can be regarded as the rate of change of TEC, which is a phenomenon that can describe the influence of scintillations in the ionosphere. These variations are however more obvious during the night time between the 1100th and 1440th min of the day. It should be noted here that scintillations have been described as a night-time phenomenon. The detrended data shows the internal dynamics of the system more clearly, with a pattern similar to the values around the night period mentioned earlier. The values after sunset at night time in Fig. 1 show a pattern similar to the detrended TEC plot. It has been established that TEC does not decrease totally throughout the night as expected normally through the simple theory of the fact that TEC builds up during the day, but it shows some anomalous enhancements and variations and this can occur under a wide range of geophysical conditions (Balan and Rao, 1987; Balan et al., 1991; Unnikrishnan and Ravindran, 2010).

The positive values of the Lyapunov exponent indicate the presence of chaos (Wolf et al., 1985; Rosenstein et al., 1993; Hegger et al., 1999; Kantz and Schreiber, 2003). The ionosphere is a dynamic system controlled by many parameters, including acoustic motions of the atmosphere electromagnetic emission and variations in the geomagnetic field. Because of its extreme sensitivity to solar activity, the ionosphere is a very sensitive monitor of solar events. The ionosphere is that part of the upper atmosphere where free electrons occur in sufficient density to have an appreciable influence on the propagation of radio frequency electromagnetic waves. This ionization depends primarily on the Sun and its activity. Ionosphere structure and peak densities in the ionosphere vary greatly with time (sunspot cycle, seasonally and diurnally), with geographical location (polar, auroral zones, mild latitudes, and equatorial regions), and with certain solar-related ionospheric disturbances. During and following a geomagnetic storm, the ionospheric changes around the globe as observed from the ground site can appear chaotic (Fuller-Rowell et al., 1994; Cosolini and Chang, 2001; Unnikrishnan and Ravindran, 2010). The presence of chaos is indicated by the positive values of the Lyapunov exponent found in all the computations for all the TEC values obtained for the selected days for all the measuring stations used in this work. This can be expected, as it agrees with results from previous works that show that there is a reasonable presence of chaos in the ionosphere, even in the midst of the influence of stochastic drivers like solar winds (Bhattacharyya, 1990; Wernik and Yeh, 1994; Kumar et al., 2004; Unnikrishnan et al., 2006a, b; Unnikrishnan, 2010).

However, these values vary from day to day due to variations in ionospheric processes for different days on the same latitude as seen in Fig. 8a–d, with 8e showing the monthly averages for the entire year. There are also latitudinal variations due to spatial variations in the various ionospheric processes taking place simultaneously. The ionosphere is said to have a complex structure due to these varying ionospheric processes. One would expect all stations to have higher values of the Lyapunov exponent during quiet days; however, the values are sometimes higher for disturbed days, and for some stations the reverse may be the case. The values of Lyapunov exponents are however always lower for storm periods. The higher values of Lyapunov exponents during the quiet or relatively disturbed days indicate that the rate of exponential growth in infinitesimal perturbations in the ionosphere leading to chaotic dynamics might be of a higher degree during those days compared with days with lower values of Lyapunov exponents. The variation along the latitude also shows the inconsistency and complexity of the ionospheric processes. This is why for the same day of the month the values of Lyapunov exponents vary from one station to another.

The results of the correlation dimension show that the values computed are within the range of 2.8 to 3.5, with the lower values occurring mostly during the storm periods. The lower dimension during the storm periods compared with the quiet days may be due to the effect of stochastic drivers like strong solar winds and solar flares that occur during geomagnetic storms on the internal dynamics of the ionosphere; this must have been as a result of the fact that the internal dynamics must have been suppressed by the external influence. The restructuring of the internal dynamics of the ionosphere might be responsible for low-dimension chaos during storms and also the lower values of other measures like the Lyapunov exponents. The relatively disturbed day might however have a higher dimension so long as it is not a storm period, and sometimes a relatively disturbed day of the month might be a day with storms, and in this case there is usually a lower value of chaoticity and sometimes lower values of the correlation dimension as well. The surrogate data test shows a significance of difference greater than 2 for all the computed measures and we were able to reject the null hypothesis that the ionospheric system can be represented with a linear model for all the data used from the stations. However, we found that the lower significance of difference corresponds to the lower values of Lyapunov exponents during storms and extremely disturbed periods. This may be due to the rise in stochasticity during the storm period as a result of drop in values of computed quantities like Lyapunov exponents. Our ability to reject the null hypothesis for all stations shows however the presence of determinism and also that the underlying dynamics of the ionosphere is mostly nonlinear.

Tsallis entropy, which is an information theory approach derived from statistical mechanics, has been discovered to be one of the best entropy measures that can be used to describe magnetospheric dynamics, especially using D_{st} time series. The major explanation for this is the fact that Tsallis entropy is based on non-extensive statistical mechanics (Balasis et al., 2009). The ionosphere is a system that can also be described by similar dynamical processes as the magnetosphere, as both systems are also coupled (Unnikrishnan, 2010). Some of the investigations carried out on the chaoticity of both the ionosphere and the magnetosphere have yielded similar results for storm and quiet periods (Unnikrishnan, 2008, 2010). It implies that one can use similar methods to study the ionosphere and magnetosphere, and this informs the use of Tsallis entropy to study the ionosphere in this work, since it has been used to describe the magnetosphere.

Tsallis entropy is also studied for the five quietest and five most disturbed days, just as the Lyapunov exponent and the result show a similar variation to that of the Lyapunov exponent, as seen in Fig. 9a–d with the monthly averages for the entire year in Fig. 9e. The daily five window intervals for Tsallis entropy show no extensive increase in values of Tsallis entropy, but rise and fall due to changes in the internal dynamics ad dynamical complexity of the ionosphere as seen in Figs. 10 and 11. These changes can indicate some level of determinism in the ionosphere. The rise and fall in the computed values of Tsallis entropy daily five window intervals indicate the interplay between stochasticity and determinism (Unnikrishnan and Ravindran, 2010). The Tsallis entropy was able to show the deterministic behavior of the ionosphere considering its response during storm periods compared with other relatively quiet periods, as the rapid drop in values of Tsallis entropy during storms show that there is a transition from higher complexity during quiet periods to lower complexity during storms. This response in the values of Tsallis entropy is similar to the response of Lyapunov exponent values during storms. This reaction to storms of entropy computed for TEC was also described by the reaction of Tsallis entropy computed for D_{st} during storm periods (Balasis et al., 2008, 2009). This shows the deterministic nature of the ionosphere as mentioned before.

The reaction to storms may be due to the influence of stochastic drivers like strong solar winds flowing into the system as a result of solar flare or CMEs that produce the geomagnetic storms. Although there is always an influence of corpuscular radiation in the form of solar wind flowing from the Sun into the ionosphere, the influence is usually low for days without storms coming compared with days with geomagnetic storms as a result of solar flares, CMEs, etc. However, during months with high solar activities within the year, that is, during the equinoxial months, it can be seen that there is a drop in the values of Tsallis entropy that can also be clearly seen in the contour plots for quiet days. The presence of chaos even at quiet periods in the ionosphere may be due to the internal dynamics and inherent irregularities of the ionosphere, which exhibit nonlinear properties. However, this inherent dynamics may be complicated by external factors like geomagnetic storms. This may be the main reason for the drop in the values of Lyapunov exponents and Tsallis entropy during storms. According to Unnikrishnan et al. (2006a, b), geomagnetic storms are extreme forms of space weather, during which external driving forces, mainly due to solar wind, subsequent plasmasphere–ionosphere coupling, and related disturbed electric field and wind patterns, will develop. This in turn creates many active degrees of freedom with various levels of coupling among them, which alters and modifies the quiet time states of the ionosphere during a storm period. This new situation developed by a storm may modify the stability/instability conditions of the ionosphere, due to the superposition of various active degrees of freedom.

The contour plots show a lower value for March and September equinoxes in months in which storms were recorded for all stations. This can also describe the influence of solar wind (the major external factor) on the ionosphere, since a stronger influence of solar wind on the ionosphere is expected for those periods of the year. It is obvious from the results that to a great extent the influx of solar wind modifies the ionosphere or that it influences the internal dynamics of the ionosphere during geomagnetic storms (Unnikrishnan et al., 2006a, b; Unnikrishnan and Ravindran, 2010).

There are also many variations in the internal dynamics of the ionosphere that could lead to changes in chaotic behavior. The variations of Lyapunov exponents during quiet days might be a result of different variations in the intrinsic dynamics of the ionosphere. Variation patterns at different stations for the same quiet day might also be due to the same reason. It can be affirmed that the ionosphere is a complex system that varies with a short latitudinal or longitudinal interval such that even stations with one or two degrees of latitudinal differences might record different values on the same day for both quiet and disturbed periods. The same might also occur for storm periods. This is illustrated by the different pattern of variation of TEC recorded from different stations within such a close range, as used in this study.

One might experience a more sporadic rate of change in TEC as seen in the time series plots, and this is also reflected in the TEC as a result of irregularities in the internal dynamics of the ionosphere, and these might be as a result of plasma bubbles. Irregularities develop in the evening hours at F region altitudes of the magnetic Equator, in the form of depletions, frequently referred to as bubbles. The edges of these depletions are very sharp, resulting in a large time rate of TEC in the equatorial ionosphere, even during magnetically quiet conditions. The large gradient of the equatorial ionization persists in the local post-sunset hours till about 21:00 LT (DasGupta et al., 2007; Unnikrishnan and Ravindran, 2010). The TEC data for one station might experience an extremely sharp rate of change in TEC that may be due to some plasma bubbles in that region, while the TEC from the other station stays normal. These variations in the various internal dynamics like plasma bubbles leading to scintillation can cause variations in the dynamical response of the TEC. Hence, the irregular variation in the values of the Lyapunov exponent and Tsallis entropy even in quiet periods for two relatively close stations may be due to these irregularities. This might also be responsible for the quiet days in the same station having lower values of the Lyapunov exponent compared with higher values recorded for disturbed days without the external influence of storms.

As mentioned earlier, the disturbed day might be without a storm for months without a record of a geomagnetic storm, but might be the most disturbed day relative to the other days of the same month according to the IQD and IDD classification. Higher values during such a disturbed day compared with lower values of quiet days show that the variation might be due to the inherent internal dynamics causing such variations. This can also be affirmed from Unnikrishnan and Ravindran (2010), who stated that in the deterministic picture, irregularity can be generated automatically by the nonlinearity of the intrinsic dynamics, and an ionospheric scintillation is caused by irregularities of the ionospheric electron density along the signal propagation path.

The variations of these chaos and dynamical complexity parameters might also be a result of the anomalous TEC enhancements that might occur at night. This is because there can be anomalous TEC enhancements even at various geophysical conditions, as recorded by Balan and Rao (1987); Balan et al. (1991). These effects can also be seen more clearly in the Tsallis entropy values for the five window period for quiet days of January 2011, because the night-time value is higher and also shows a much higher series of fluctuations during this period compared with other periods. As mentioned in Unnikrishnan and Ravindran (2010), the irregular changes in the dynamical characteristics of TEC from the results of the Lyapunov exponent and Tsallis entropy may also be due to the collisional Raleigh–Taylor instability that may give rise to a few large irregularities in L band measurements (Rama Rao et al., 2006; Sripathi et al., 2008). All these can be seen as internal factors responsible for variations in the dynamical response of TEC, as recorded from the values of the Lyapunov exponents and Tsallis entropy completed for days without storms, which might be quiet or disturbed according to classification and which could also account for higher values of these qualifiers during disturbed days compared with quiet days. During storms however the values were much lower.

The five window interval was able to show the detailed dynamical behavior of the ionosphere within those intervals of the day. It was also able to show the response to rate of change in TEC for those intervals of the day. A drop in Tsallis entropy values shows a response to a very sharp fluctuation, showing the rate of change in TEC that can be seen for the segmented TEC plots and the detrended TEC segments. Sudden drops in Tsallis entropy values at evening and night time might also be due to night-time TEC enhancement or other interval factors like plasma bubbles as mentioned before, since this can occur for all geophysical conditions. From the results of the daily five window interval computation of the Tsallis entropy, we can infer that the enhancement and the sporadic rate of change of TEC at any time of the day (even at night), as seen in the detrended time series, can lead to sudden changes in the values of Tsallis entropy.

All the daily values of Tsallis entropy correlate positively with the values of Lyapunov exponents at values between 0.78 and 0.81. The contour plots also show similarities for both Lyapunov exponents and Tsallis entropy (Figs. 12 and 13). This indicates that both the values of Lyapunov exponents and Tsallis entropy can be used together as measures to describe the determinism and dynamical complexity of the ionosphere.

5 Conclusions

The chaotic behaviour and dynamical complexity of low-latitude ionospheric behavior over some parts of Nigeria was investigated using TEC time series measured at three different stations, namely Birnin Kebbi (12°32′ N, 4°12′ E and 0.62° N geomagnetically), Enugu (6°26′ N, 7°30′ E and −3.21° N geomagnetically) and Lagos (6°27′ N, 3°23′ E and −3.07° N geomagnetically) within the low-latitude region. The time series data obtained from the GPS data measurement were studied for chaoticity using phase space reconstruction techniques, computation of Lyapunov exponents, correlation dimension, and Tsallis entropy, and implementation of the surrogate data test was carried out. However, the Tsallis entropy was extensively used for the study of dynamical complexity of the ionospheric system described by the TEC data. The detrending analysis of the TEC data was carried out to reduce the influence of the diurnal variations so as to be able to reflect the inherent dynamics of the system for all the days considered in this work before it could be subjected to the various nonlinear time series analyses mentioned above for quantification of deterministic chaos and to show the presence of dynamical complexity. The detrended time series were subjected to further analysis for phase space reconstruction from which the choice of time delay of 30 was obtained, and an embedding dimension of 5 was considered in this study. The computed values of Lyapunov exponents show that there was the presence of chaos, since all the results are positive values. The correlation dimension computed was observed to be between the range of 2.8–3.5, where the lowest values were recorded for the storm periods, showing a transition to a lower dimension during the storm period of October 2011.

The results of Tsallis entropy show that the changes in the ionospheric response to disturbances may be due to geomagnetic storms and other phenomena like changes in the internal irregularities (like plasma bubbles or scintillations) of the ionosphere. The response of the Tsallis entropy to various changes in the ionosphere also shows the deterministic nature of the system. The results of the Tsallis entropy show a lot of similarities to those of the ionosphere. Both results show that equinoxial months and other months with geomagnetic storms indicate low values of the Lyapunov exponent, and the same was observed for Tsallis entropy. The results of the values of the Lyapunov exponent were expected to be lower for the days of the months in which a storm was recorded, and it was found to be the same, which agrees with previous works by other investigators. A similar pattern of results was obtained for the computed values of Tsallis entropy. However, the values of the two quantifiers for most disturbed days of some months from the IDD classification were higher than those of the quietest from the IQD classification. This shows that a more relatively disturbed day of the month without storms might be more chaotic.

The random variations in the values of chaoticity in the detrended TEC describing the internal dynamics of the ionosphere as seen in the result obtained from both the Lyapunov exponent and Tsallis entropy show that the ionosphere is not totally deterministic but also has some elements of stochasticity. The variation in the dynamical complexity can be seen clearly from the rise and fall in the values of Tsallis entropy, which demonstrates an inconsistent pattern. It can also be seen that the results of Tsallis entropy following the same pattern with the Lyapunov exponent show that both quantifiers can be used simultaneously and comparatively as measures of chaos and dynamical complexity, as the correlation of all the values obtained for both quantities gives values between 0.78 and 0.81, and observation of their similarities in response to changes in the ionosphere can also be deduced from the contour plots.

Considering the response of the chaotic quantifiers and entropy to the changes in the geophysical conditions, namely quiet, disturbed and storm periods, it can be seen that the ionospheric internal dynamic is chaotic, with some measure of complexity. However, change in the ionospheric response due to external influence mainly occurs due to stochastic drivers like solar wind. The intermittent variations between stochasticity and determinism from the sudden rise and fall in the value of the chaotic quantifiers and entropy measures show the dynamical complexity of the ionosphere, and these changes may occur not only due to changes in the three geophysical conditions, namely the quiet, disturbed and storm periods, but also due to changes as a result of the irregularities in the internal system of the ionosphere.

The knowledge of being able to characterize the ionospheric behavior using Lyapunov exponents and Tsallis entropy comparatively in work and the similarities in their response to the dynamical changes in the ionosphere show their ability to measure levels of determinism when used to complement each other. The ability of these two quantifiers to describe the ionospheric response to changes in geophysical conditions and also their ability to describe the changes in the internal dynamics of the ionosphere indicate that these quantifiers might be useful as indices for the description of the state of the ionospheric conditions in the near future. The relationship between these two quantifiers as established from this work calls for more research in using the two qualifiers to enable proper description and characterization of the state of the ionosphere. Therefore there is a need for further work on the use of these two parameters as indices that can describe the state of the ionosphere from time to time.

Acknowledgements. The authors would like to thank Professor A. C. L. Chian, the anonymous referees and the editorial team for contributions which lead to the final form of this manuscript. The NIGNET GPS data was made available by the office of the Surveyor General of Nigeria.

Edited by: A. C. L. Chian

References

Abdu, M. A.: Major Phenomena of the equatorial ionosphere thermosphere system under disturbed conditions, J. Atmos. Solar-Terr. Phys., 59, 1505–1519, 1997.

Anastasiadis, A., Costa, L., Gonzales, C., Honey, C., Szeliga, M., and Terhesiu, D.: Measures of Structural Complexity in Networks, in: Complex Systems Summer School 2005, Santa Fe, 2005.

Balan, N. and Rao, P. B.: Latitudinal variations of nighttime enhancements in total electron content, J. Geophys. Res., 92, 3436–3440, 1987.

Balan, N., Bailey, G. J., and Balachandian, N. . R.: Solar and Magnetic effects on the latitudinal variations of nighttime TEC enhancement, Ann. Geophys., 9, 60–69, 1991, http://www.ann-geophys.net/9/60/1991/.

Balasis, G., Daglis, I. A., Papadimitrou, C., Kalimeri, M., Anastasiadis, A., and Eftaxias, K.: Dynamical complexity in D_{st} time series using non-extensive Tsallis entropy, Geophys. Res. Lett., 35, L14102, doi:10.1029/2008GL034743, 2008.

Balasis, G., Daglis, I. A., Papadimitrou, C., Kalimeri, M., Anastasiadis, A., and Eftaxias, K.: Investigating Dynamical complexity in the magnetosphere using various entropy measures, J. Geophys. Res., 114, A00D06, doi:10.1029/2008JA014035, 2009.

Baranger, M., Latora, V., and Rapisarda, A.: Time evolution of thermodynamic entropy for conservative and dissipative chaotic maps, Chaos Soliton Fract, 12, 471–478, 2002.

Bhattacharyya, A.: Chaotic behavior of ionosphere turbulence from scintillation measurements, J. Geophys. Res., 17, 733–738, 1990.

Boon, J. and Tsallis, C. (Eds.): Nonexistensive statistical mechanics: New trends, new perspectives, Cambridge University Press, 2003.

Burlaga, L. F., Vinas, A. F., and Wang, C.: Tsallis distribution of magnetic field strength variations in the heliosphere: 5 to 90 AU, J. Geophys. Res., 112, A07206, doi:10.1029/2006JA012213, 2007.

Coco, I., Consolini, G., Amata, E., Marcucci, M. F., and Ambrosino, D.: Dynamical changes of the polar cap potential structure: an information theory approach, Nonlin. Processes Geophys., 18, 697–707, doi:10.5194/npg-18-697-2011, 2011.

Coraddu, M., Lissia, M., and Tonelli, R.: Statistical descriptions of nonlinear systems at the onset of chaos arXiv:cond-mat/0511736v1, Physica A, 365, 252–257, 2006.

Cosolini, G. and Chang, T.: Magnetic field topology and criticality in geotail dynamics relevance to substorm phenomena, Space Sci. Rev., 95, 309–321, 2001.

DasGupta, A., Paul, A., and Das, A.: Ionospheric total electron content (TEC) studies with Gps in the equatorial region, Indian J. Radio Space, 36, 278–292, 2007.

Fraser, A. M. and Swinney, H. L.: Independent coordinates for strange attractors from mutual information, Phys. Rev. A, 33, 1134–1141, 1986.

Fuller-Rowell, T. J., Codrescu, M. V., Moffett, R. J., and Quegan, S.: Response of the magnetosphere and ionosphere to geomagnetic storms, J. Geophys. Res., 99, 3893–3914, 1994.

Grassberger, P. and Procaccia, I.: Characterization of strange attractors, Phys. Rev. Lett., 50, 346–349, 1983a.

Grassberger, P. and Procaccia, I.: Measuring the strangeness of strange attractors, Physica D, 9, 189–208, 1983b.

Hegger, R., Kantz, H., and Shrieber, T.: Practical implementation of nonlinear time series method: The Tisean package, Chaos, 9, 413–430, 1999.

Kalogeropoulos, N.: Weak chaos from Tsallis entropy, Qscience Connect, 12, doi:10.5339/connect.2012.12, 2012.

Kalogeropoulos, N.: Vanishing largest Lyapunov exponent and Tsallis entropy, Qscience Connect, 26, doi:10.5339/connect.2013.26, 2013.

Kantz, H. and Schreiber, T.: Nonlinear Time Series Analysis, Cambridge University Press, 2003.

Kennel, M. B., Brown, R., and Abarbanel, H. D. I.: Determining minimum embedding dimension using a geometrical construction, Phys. Rev. A, 45, 3403–3411, 1992.

Kumar, K. S., Kumar, C. V. A., Renuka, B. G. G., and Venugopal, C.: Analysis of the fluctuations of the total electron content, measured at Goose Bay using tools of nonlinear methods, J. Geophys. Res., 10, A02308, doi:10.1029/2002/A009768, 2004.

Pavlos, G. P., Kyriakov, G. A., Rigas, A. G., Liatsis, P. I., Trochoulos, P. C., and Tsonis, A. A.: Low dimensionless chaos in magnetospheric activity from AE time series, Ann. Geophys., 10, 309–315, 1992, http://www.ann-geophys.net/10/309/1992/.

Pavlos, G. P., Athanasiu, M. A., Kugiumtzis, D., Hatzigeorgiu, N., Rigas, A. G., and Sarris, E. T.: Nonlinear analysis of magnetospheric data Part I. Geometric characteristics of the AE index time series and comparison with nonlinear surrogate data, Nonlin. Processes Geophys., 6, 51–65, doi:10.5194/npg-6-51-1999, 1999.

Rabiu, A. B., Mamukuyomi, A. I., and Joshua, E. O.: Variability of equatorial ionosphere inferred from geomagnetic field measurements, Bull. Astr. Soc. India, 35, 607–615, 2007.

Rama Rao, P. V. S., Gopi Krishna, S., Niranjan, K., and Prasad, D. S. V. V. D.: Temporal and spatial variations in TEC using simultaneous measurements from the Indian GPS network of receivers during the low solar activity period of 2004–2005, Ann. Geophys., 24, 3279–3292, doi:10.5194/angeo-24-3279-2006, 2006.

Rosenstein, M. T., Collins, J. J., and DeLuca, C. J. A.: A practical method for calculating largest Lyapunov exponents from small data sets, Physica D, 65, 117–134, 1993.

Saito, A., Fukao, S., and Mayazaki, S.: High resolution mapping of TEC perturbations with the GSI GPS network over Japan, Geophys. Res. Lett., 25, 3079–3082, 1998.

Shan, H., Hansen, P., Goertz, C. K., and Smith, K. A.: Low dimensionless chaos in magnetospheric activity from AE time series, J. Geophys. Res., 18, 147–150, 1991.

Sripathi, S., Bose, S., Patra, A. K., Pant, T. K., Kakad, B., and Bhattacharyya, A.: Simultaneous observations of ESF irregularities over Indian region using radar and GPS, Ann. Geophys., 26, 3197–3213, doi:10.5194/angeo-26-3197-2008, 2008.

Theiler, J., Eubank, S., Longtin, A., Galdrikian, B., and Farmerm, J. D.: Testing for nonlinearity in time series: The method of surrogate data, Physica D, 58, 77–81, 1992.

Tsallis, C.: Possible generalization of Boltzmann-Gibbs statistics, J. Stat. Phys., 52, 487–497, 1988.

Tsallis, C.: Generalised entropy-based criterion for consistent testing, Phys. Rev. E, 58, 1442–1445, 1998.

Tsallis, C.: Nonextensive statistics: theoretical, experimental and computational evidences and connections, Braz. J. Phys., 29, 1–35, 1999.

Unnikrishnan, K.: Comparison of chaotic aspects of magnetosphere under various physical conditions using AE index time series, Ann. Geophys., 26, 941–953, doi:10.5194/angeo-26-941-2008, 2008.

Unnikrishnan, K.: A comparative study on chaoticity of equatorial/low latitude ionosphere over Indian subcontinent during geomagnetically quiet and disturbed periods, Nonlin. Processes Geophys., 17, 765–776, doi:10.5194/npg-17-765-2010, 2010.

Unnikrishnan, K. and Ravindran, S.: A study on chaotic behavior of equatorial/ low latitude ionosphere over Indian subcontinent, using GPS TEC time series, J. Atmos. Solar-Terr. Phys., 72, 1080–1089, 2010.

Unnikrishnan, K., Saito, A., and Fukao, S.: Differences in magnetic storm and quiet ionospheric deterministic chaotic behavior. GPS TEC Analyses, J. Geophys. Res., 111, A06304, doi:10.1029/2005JA011311, 2006a.

Unnikrishnan, K., Saito, A., and Fukao, S.: Differences in day and night time ionosphere deterministic chaotic behavior : GPS TEC Analyses, J. Geophys. Res., 111, A07310, doi:10.1029/2005JA011313, 2006b.

Vassiliadis, D. V., Sharma, A. S., Eastman, T. E., and Papadopoulos, K.: Low dimensionless chaos in magnetospheric activity from AE time series, Geophys. Res. Lett., 17, 1841–1844, 1990.

Wernik, A. W. and Yeh, K. C.: Chaotic behavior of ionosphere turbulence from scintillation measurements, Radio Sci., 29, 135–139, 1994.

Wolf, A., Swift, J. B., Swinney, H. L., and Vastano, J. A.: Determining Lyapunov exponents from a time series, Physica D., 16, 285–317, 1985.

5

Turbulence in the interstellar medium

D. Falceta-Gonçalves[1,2], G. Kowal[2], E. Falgarone[3], and A. C.-L. Chian[4,5,6]

[1]SUPA, School of Physics & Astronomy, University of St Andrews, North Haugh, St Andrews, Fife, KY16 9SS, UK
[2]Escola de Artes, Ciencias e Humanidades, Universidade de Sao Paulo, Rua Arlindo Bettio 1000, CEP 03828-000 Sao Paulo, Brazil
[3]LERMA/LRA, CNRS, Ecole Normale Supérieure and Observatoire de Paris, 24 rue Lhomond, 75231 Paris Cedex, France
[4]Observatoire de Paris, LESIA, CNRS, 92190 Meudon, France
[5]National Institute for Space Research (INPE) and World Institute for Space Environment Research (WISER), P.O. Box 515, 12227-010 São José dos Campos-SP, Brazil
[6]School of Mathematical Sciences, University of Adelaide, Adelaide, SA 5005, Australia

Correspondence to: D. Falceta-Gonçalves (dfalceta@usp.br)

Abstract. Turbulence is ubiquitous in the insterstellar medium and plays a major role in several processes such as the formation of dense structures and stars, the stability of molecular clouds, the amplification of magnetic fields, and the re-acceleration and diffusion of cosmic rays. Despite its importance, interstellar turbulence, like turbulence in general, is far from being fully understood. In this review we present the basics of turbulence physics, focusing on the statistics of its structure and energy cascade. We explore the physics of compressible and incompressible turbulent flows, as well as magnetised cases. The most relevant observational techniques that provide quantitative insights into interstellar turbulence are also presented. We also discuss the main difficulties in developing a three-dimensional view of interstellar turbulence from these observations. Finally, we briefly present what the main sources of turbulence in the interstellar medium could be.

1 Introduction: the basics of turbulence

Turbulence is characterised by chaotic motions in a fluid (Rempel et al., 2004; He and Chian, 2005; Chian et al., 2006, 2007, 2010) that lead to diffusion of matter and dissipation of kinetic energy. It is to be stressed that not all chaotic motions in a fluid may be called "turbulent". Because of its chaotic nature turbulence can only be studied and modelled in terms of statistical quantities. Long-term deterministic local properties of a turbulent fluid are unpredictable.

For nearly incompressible and unmagnetised fluids, the temporal evolution of the fluid velocity field is given by the Navier–Stokes equation

$$\frac{\partial \boldsymbol{u}(\boldsymbol{x},t)}{\partial t} + \boldsymbol{u}(\boldsymbol{x},t)\cdot\nabla\boldsymbol{u}(\boldsymbol{x},t) = -\frac{\nabla p(\boldsymbol{x},t)}{\rho(\boldsymbol{x},t)} + \nu\nabla^2\boldsymbol{u}(\boldsymbol{x},t) + \boldsymbol{F}(\boldsymbol{x},t), \quad (1)$$

where $\boldsymbol{u}(\boldsymbol{x},t)$ represents the velocity field, p the pressure, ν the kinematic viscosity, and $\boldsymbol{F}$ an external force normalised by the local density. ρ is the gas mass density and is set constant in the incompressible case (with $\nabla\cdot\boldsymbol{u}=0$). Even in this simplified mathematical description the fluid dynamics is not a trivial solution. Equation (1) is non-linear, as seen from the advective term on the left-hand side, and non-local – in the sense that the local properties of the fluid are related to all the other regions – through the pressure term. The incompressibility condition results in an infinite sound speed, and in an instantaneous propagation of any perturbation in the fluid.

Burgers (1939) modelled the time evolution of the simplified version of the Navier–Stokes equation by considering $\nabla p = 0$. This equation has exact solutions, which may sound interesting, but it results in non-universal "turbulence", even though Burgers turbulence models have gained increasing interest due to their ability to describe the statistics of shock-induced structures and many other applications (see review by Bec and Khanin, 2007).

In the full Navier–Stokes equation, perturbations in $\boldsymbol{u}(\boldsymbol{x},t)$ are expected to have their distribution changed due to non-linear terms. These instabilities may drive local vorticity and result in the fragmentation of large-amplitude eddies into smaller ones, creating a turbulent pattern. As imagined by Richardson (1922), *big whirls have little whirls that feed on their velocity, and little whirls have lesser whirls, and so on to viscosity*. This statement represents one of the first conceptual descriptions of the energy cascade in turbulent flows. The shear drives unstable motions at large scales, which are broken and fragmented into smaller vortices, down to the smallest scales where they are damped, e.g. due to viscosity. In an incompressible viscous fluid this damping scale is that at which the timescale for viscous damping is of the order of the turnover dynamical time. At that scale, the eddy kinetic energy is transferred to internal energy due to viscosity. Turbulence is naturally developed over larger range of scales if viscosity is small, i.e. with a large Reynolds number ($Re = UL/\nu \gg 1$), the characteristic velocity U being injected at a length scale L.

Kolmogorov (1941, hereafter K41) realised that it would be possible to solve the Navier–Stokes equation for a turbulent flow if $\boldsymbol{u}(\boldsymbol{x},t)$ is considered a stochastic distribution. One of the key assumptions in the K41 theory is that the energy transfer rate ϵ should be constant at all scales. It is defined as $\epsilon \simeq \delta u_l^2/\tau_l$, where δu_l is the velocity fluctuation amplitude at length scale l, and $\tau_l = \tau_{\mathrm{eddy}} = l/\delta u_l$ its dynamical timescale[1]. Therefore, one obtains

$$\delta u_l \simeq (\epsilon l)^{\frac{1}{3}}. \tag{2}$$

Equation (2) means that turbulence can be modelled by scaling laws. This would be true within the so-called *inertial range of scales*, i.e. the scales where the energy transfer rate is constant, generally between the energy injection and the viscous damping scales. The velocity power spectrum $P_u(k)$ is defined[2] here by $\int_{k=1/l}^{\infty} P_u(k')\,dk' = \delta u_l^2$, from which we obtain the standard Kolmogorov power spectrum for the velocity field:

$$P_u(k) \propto \epsilon^{2/3} k^{-5/3}. \tag{3}$$

In other words, it is possible to reinterpret Kolmogorov's idea in Fourier space in terms of non-linear interaction between similar wavenumbers. This theory is a result of the so-called *locality*, i.e. similar wavenumbers, $k = 2\pi/\lambda$, of the non-linear wave–wave interaction that result in the energy cascade through smaller scales (Kraichnan, 1965a). From the spectral form of the Navier–Stokes equation, the three-wave interactions follow the selection rule $k_3 = k_1 + k_2$. The extrema are found at $k_3 \to 0$ and $k_1 = k_2$, which is the locality assumed in Kolmogorov's theory, resulting in $k_3 = 2k_1$.

The theory also predicts the scaling laws for the moments of velocity spatial increments, known as *velocity structure functions*, defined as

$$S_{\mathrm{p}}(l) = \left\langle \{[\boldsymbol{u}(\boldsymbol{r}+\boldsymbol{l}) - \boldsymbol{u}(\boldsymbol{r})]\cdot \boldsymbol{l}/l\}^p \right\rangle = C(p)\epsilon^{p/3} l^{p/3}, \tag{4}$$

where p is a positive integer representing the moment order and $\boldsymbol{l}$ is the spatial increment vector. In incompressible fluids, if the turbulence is considered *homogeneous*, *isotropic* and *self-similar*, i.e. scale invariant, then

$$S_{\mathrm{p}}(l) = C(p)\epsilon^{p/3} l^{p/3}, \tag{5}$$

where $C(p)$ was initially assumed by Kolmogorov to be constant with p.

One of the main successes of the Kolmogorov–Obukhov turbulence theory is the explanation of the empirical determination of the diffusion coefficient by Richardson (1926), done more than a decade before K41. The diffusion coefficient is related to the time evolution of the separation between Lagrangian points (e.g. particles dragged by the flow) in a turbulent medium. The probability distribution function Φ of pairs of points separated by a distance $\boldsymbol{r}$ may be described as

$$\frac{\partial \Phi(\boldsymbol{r},t)}{\partial t} = \frac{1}{r^2}\frac{\partial}{\partial r} r^2 K(r) \frac{\partial \Phi(\boldsymbol{r},t)}{\partial r}, \tag{6}$$

where $K(r)$ represents the diffusion coefficient. It is easy to determine, from dimensional analysis, that if $\dot{r} = u(r) \propto r^{1/3}$ as in the Kolmogorov scaling, the diffusion coefficient for the inertial range will be $K(r) = k_0 \epsilon^{1/3} r^{4/3}$, the scaling proposed by Richardson (1926). This diffusion coefficient for the inertial range substituted in Eq. (6) then results in

$$\Phi(\boldsymbol{r},t) = \frac{A}{(k_0 t)^3 \epsilon} \exp\left(-\frac{9r^{2/3}}{4k_0 \epsilon^{1/3} t}\right), \tag{7}$$

where A is a normalisation coefficient. The Richardson distribution is therefore non-Gaussian. Several experiments and numerical models have shown the validity of the turbulent diffusion scaling (Elliott Jr. and Majda, 1996; Fung and Vassilicos, 1998; Zouari and Babiano, 1994; Boffetta and Sokolov, 2002), as has also been recently used in the predictions of stochastic magnetic reconnection[3] (Lazarian et al., 2012).

This theory of turbulence has been quite successful in reproducing most of the experimental data, and there is

[1]Note that we distinguish τ_l and τ_{eddy} here, since τ_l represents the timescale for energy transfer at scale l, while τ_{eddy} is the eddy turnover timescale. In the K41 theory both timescales are the same, but this is not true for other cases, e.g. as in some magnetised cases.

[2]The power spectrum is defined as the one-dimensional spectrum in Fourier space, while the energy spectrum, generally defined as $E_u(k) = k^2 P_u(k)$, is the three-dimensional spectrum. For the sake of simplicity, we use the term *power spectrum* to represent the latter.

[3]This term accounts for the magnetic reconnection that is induced by turbulent motions near the current sheet – the separation layer between fields with components of opposed directions – which would then result in reconnection rates as a function of the stochastic motions of the fluid.

a flourishing literature with hundreds of works available, e.g. Armstrong et al., 1995; Leamon et al., 1998; Bale et al., 2005; Koga et al., 2007; Bourras et al., 2009; Chian and Miranda, 2009; Chepurnov and Lazarian, 2010; Sahraoui et al., 2010; Chian and Muñoz, 2011; Chang et al., 2012; Hurricane et al., 2012; Miranda et al., 2013, just to mention a few. Naturally, many authors criticised the fact that $C(p)$ is a constant in Kolmogorov's initial theory, given the breakdown of self-similarity at small scales and the possible non-universality of turbulence (given its "memory" related to the energy injection). These criticisms have been addressed later in the Kolmogorov–Obukhov turbulence theory (Kolmogorov, 1962; Obukhov, 1962), including the effects of *intermittency*. Intermittency results from rare and large local fluctuations in the velocity field which break the similarity condition (Frisch, 1995). One of the effects of intermittency is observed in the probability distribution function (PDF) of velocity longitudinal increments $\delta u_l = [\boldsymbol{u}(\boldsymbol{r}+\boldsymbol{l}) - \boldsymbol{u}(\boldsymbol{r})]\cdot\hat{l}$, which shows large deviations from the Gaussian distribution at small scales, with large-amplitude tails and peaked distributions at $\delta u_l \sim 0$ (see Fig. 1). Kraichnan (1991) pointed that sharp shocks could, for instance, result in more regions with smooth fluid flows and also more regions with sharp transitions in velocities, compared to the standard picture of the self-similar K41 turbulence. We would then expect non-Gaussian PDFs at both small and large scales.

Many authors have attempted to determine theoretically the scalings of turbulence with intermittency. One of the most successful approaches is the multifractal description of the energy dissipation field proposed by She and Levêque (1994). This theory results in $S_p(l) \propto l^{\zeta(p)}$, with

$$\zeta(p)=\frac{p}{3}\left(1+\frac{2}{3}\right)+(3-D')\left[1-\left(1-\frac{2}{3(3-D')}\right)^{p/3}\right], \quad (8)$$

where D' represents the dimensionality of the dissipation structures. In Kolmogorov–Obukhov theory, structures of highest dissipation are filamentary, better described then by $D' \sim 1$, while recent numerical simulations reveal a dominance of two-dimensional intermittent structures at small scales (e.g. Moisy and Jimenez, 2004; Kowal et al., 2007; Kowal and Lazarian, 2007; Boldyrev and Perez, 2012), which is also supported by experimental data (e.g. Fredriksen et al., 2003; Thess and Zikanov, 2007). Multifractal analysis of Voyager 1 and 2 in situ data has also shown intermittent features in the magnetic turbulence at the solar wind and the termination shock (Macek and Szczepaniak, 2008; Macek et al., 2011, 2012). On the theoretical side, Birnir (2013) derived a statistical solution of the stochastic Navier–Stokes equation from the linear Kolmogorov–Hopf differential equation, accounting for the She–Levêque intermittency corrections. His results satisfactorily reproduce the PDFs built on observations and numerical simulations of turbulent flows. Compressibility and coupling between magnetic fields and the plasma flow – both present in the dynamics of the

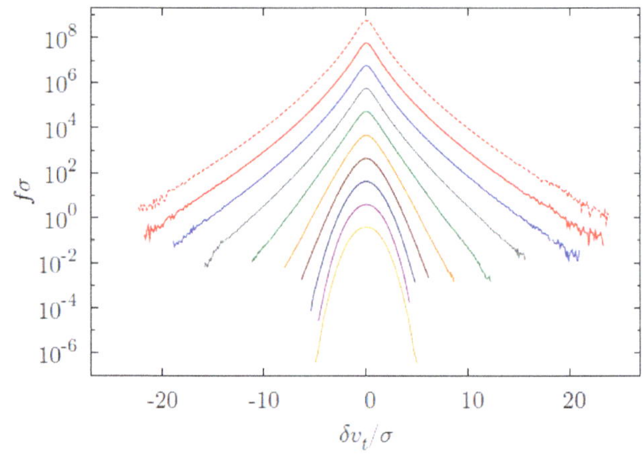

Fig. 1. PDF of velocity increments as a function of the lag length $|l|$, from small (top) to large scales (bottom) (extracted from Wilczek, 2010). The non-Gaussianity is clear for velocity increments at small scales.

interstellar medium (ISM) – make the description of the interstellar turbulence even more complex.

1.1 Supersonic turbulence

Compressible plasmas are of great interest in astrophysics, and particularly in the case of interstellar turbulence. Compressibility in turbulent flows results in the formation of a hierarchy of density structures, viewed as dense cores nested in less dense regions, which are in turn embedded in low-density regions and so on. Such a hierarchical structure was described by von Weizsäcker (1951) as

$$\frac{\rho_\nu}{\rho_{\nu-1}}=\left(\frac{l_\nu}{l_{\nu-1}}\right)^{-3\alpha}, \quad (9)$$

where ρ_ν represents the average density of a structure at hierarchical level ν, at a length scale l, and α the compressibility degree, assumed to be the same at each level. The dimensionality of the system is obtained by $D' = 3 - 3\alpha$. Therefore, the average mass within each substructure must follow the relation $M_l \propto l^{3-3\alpha}$.

The density hierarchy as described above must then be coupled to the local turbulent motions. The energy density transfer rate must now be rewritten as $\epsilon_l = \rho_l\, \delta u_l^3/l$ to account for the density changes at different scales (Lighthill, 1955). If, once again, one assumes the constancy of the energy transfer rate across scales within the inertial range (Fleck, 1996), one obtains the scaling of the amplitude of the velocity fluctuations

$$\delta u_l \propto l^{\frac{1}{3}+\alpha}, \quad (10)$$

and the velocity power spectrum is then given by

$$P_u(k) \propto k^{-5/3-2\alpha}. \quad (11)$$

Note that for stationary energy distribution solutions in compressible turbulence, $\alpha > 0$, which results in steeper velocity power spectra compared to the standard K41 scaling. The density power spectrum, on the other hand, instead of following the velocity field as a passive scalar would do, presents a distinct power spectrum given by

$$P_\rho(k) \propto k^{6\alpha-1}, \tag{12}$$

i.e. for $\alpha \sim 1/6$, the power spectrum of the density field becomes flat in the inertial range. One of the most striking results of the hierarchical model for the density field in compressible turbulence is its ability to recover the standard Kolmogorov scalings for the density-weighted velocity field $\boldsymbol{v} \equiv \rho^{1/3}\boldsymbol{u}$ (Fleck, 1996).

Numerical simulations of compressible turbulence have confirmed the scalings described above for $\alpha \simeq 0.15$ (Kritsuk et al., 2007; Kowal and Lazarian, 2007), close to $\alpha = 1/6$, for which the density power spectrum becomes flat. The velocity power spectrum on the other hand becomes $P_u(k) \propto k^{-2}$. Remarkably, this is the exact slope obtained for Burger's turbulence, despite the different framework of that theory.

1.2 Magnetized turbulence

Magnetic fields introduce further complexity in the plasma dynamics that can be described by the magnetohydrodynamic (MHD) equations in the fluid approximation and assuming perfect coupling between the field and the plasma:

$$\frac{\partial \boldsymbol{u}}{\partial t} + \boldsymbol{u}\cdot\nabla\boldsymbol{u} = -\frac{\nabla p}{\rho} + \nu\nabla^2\boldsymbol{u} + \frac{(\nabla\times\boldsymbol{B})\times\boldsymbol{B}}{4\pi\rho} + \boldsymbol{F}, \tag{13}$$

$$\frac{\partial \boldsymbol{B}}{\partial t} = \nabla\times(\boldsymbol{u}\times\boldsymbol{B}) + \eta\nabla^2\boldsymbol{B}, \tag{14}$$

where $\boldsymbol{B}$ is the magnetic field and η the plasma resistivity ($\eta = 0$ for ideal plasmas).

Let us first consider an external uniform magnetic field B_0. Any perturbation in the fluid velocity field will be coupled to the magnetic field. The magnetic tension/pressure results in a decrease in the non-linear growth of perturbations, but only in those perpendicular to the magnetic field lines. This complex coupling between the flow and the magnetic field makes the modelling of turbulence in magnetised plasmas an interesting task[4].

1.2.1 The Iroshnikov–Kraichnan model

A useful simplification of the equations above is made by considering $\boldsymbol{B} = \boldsymbol{B}_0 + \delta\boldsymbol{B}$ and using the Elsässer variable $z^{\pm} = \boldsymbol{u} \pm \delta\breve{\boldsymbol{B}}$, where $\breve{B} = B/(4\pi\rho)^{1/2}$. This was derived independently by Iroshnikov (1964) and Kraichnan (1965a, b) (IK hereafter). From this change in variables, Eqs. (13) and (14) result in (see Schekochihin and Cowley, 2007)

[4]More details on MHD turbulence may be found in Biskamp (2003).

$$\frac{\partial z^{\pm}}{\partial t} \mp v_A \nabla_{||} z^{\pm} + z^{\mp}\cdot\nabla z^{\pm} = -\nabla p + \frac{\nu+\eta}{2}\delta z^{\pm} + \frac{\nu-\eta}{2}\delta z^{\mp} + \boldsymbol{F}, \tag{15}$$

where $v_A = B_0/\sqrt{4\pi\rho}$ is the Alfvén velocity and $\nabla_{||}$ is the spatial derivative parallel to the direction of the mean magnetic field.

In their model, Iroshnikov and Kraichnan proposed that incompressible magnetised turbulence results from the non-linear interactions of counter-propagating wave packets. The timescale for the two wave packets to cross each other is of the order of the Alfvén time $\tau_A \sim l_{||}/v_A$, where $l_{||}$ is the length scale of the wave packet parallel to the mean magnetic field. In their phenomenological description of the MHD turbulence, the interactions between the wave packets are *weak*, i.e. $|z^{\pm}| \ll \breve{B}_0$, or the field perturbations are much smaller than B_0. Notice that, superimposed on the magnetic fluctuations, the fluid is also perturbed and the dynamical timescale of a fluid "eddy" is $\tau_{\rm eddy} \equiv l/\delta u_l$. The different wave modes (mechanical and magnetic perturbations) thus interact with each other. For the interaction between modes to be weak the Alfvén time must be much smaller than the dynamical timescale, i.e. $\tau_A \ll \tau_{\rm eddy}$. The non-linear decay of the wave packets in such weak interactions, and subsequently the turbulent cascade, can only occur after several interactions. Since interactions are random, the wave packet amplitude changes in a random walk fashion, i.e. $N = (\tau_l/\tau_A)^{1/2}$ interactions are needed for the wave packet to change significantly. At the same time, N is also defined by the number of crossings in a decay timescale $N = \tau_l/\tau_{\rm eddy}$, which results in

$$\tau_l \sim \frac{\tau_{\rm eddy}^2}{\tau_A} \sim \frac{l^2 v_A}{l_{||}\delta u_l^2}. \tag{16}$$

Therefore, the turnover time at scale l is longer by a large factor and, as expected, the non-linear cascade proceeds much more slowly.

The second major assumption in the IK theory of weak turbulence is its isotropy, i.e. $l_{||} \sim l$. Substituting this scaling into the relation $\epsilon = \delta u_l^2/\tau_l$, one obtains

$$\delta u_l \sim (\epsilon v_A)^{1/4} l^{1/4} \text{ and } P_u(k) \sim (\epsilon v_A)^{1/2} k^{-3/2}. \tag{17}$$

There is evidence of an IK cascade in the solar wind and interplanetary medium (e.g. Bamert et al., 2008; Ng et al., 2010). However, many observations of the solar wind turbulence also suggest a more Kolmogorov-like turbulence, i.e. $\propto k^{-5/3}$ (e.g. the early studies of Coleman, 1968, 1968; Matthaeus and Goldstein, 1982; and the more recent papers by Alexandrova et al., 2008; Chian and Miranda, 2009; Sahraoui et al., 2010; Li et al., 2011; Chian and Muñoz, 2011; Kozak et al., 2012; Hellinger et al., 2013). It is possible though that a mix of both cascades may occur, as pointed out by e.g. Salem et al. (2009) and Alexakis (2013), which

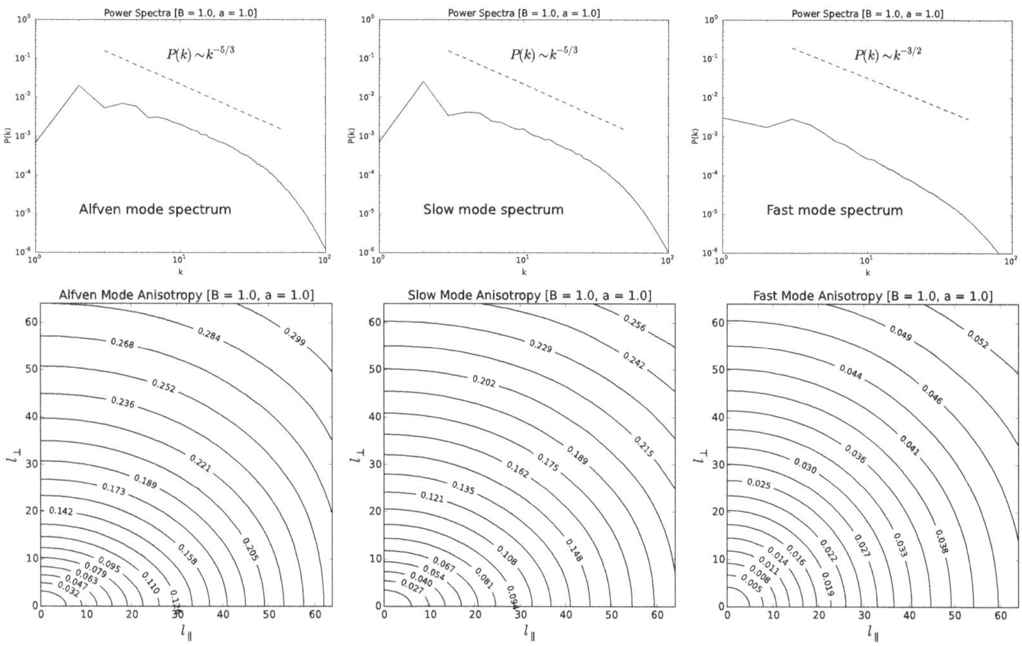

Fig. 2. Spectra and second-order structure function anisotropy of dispersion (δv) of the different wave modes in MHD turbulence. The Alfvén and slow modes present the K41 power cascade and strong anisotropy of dispersion of velocity at small scales, while the fast waves present the IK cascade and are basically isotropic at all scales. Data from a 1024^3 isothermal, sub-Alfvénic and subsonic turbulence model.

showed a mix of K41 and IK cascades for the magnetic and velocity field fluctuations, respectively. Moreover, most of these data also reveal the solar wind turbulence to be highly anisotropic (i.e. $\delta u_l^{||} \neq \delta u_l^{\perp}$) with respect to the local magnetic field (Horbury et al., 2008, 2012).

As pointed out by Goldstein (2001), one of the main issues raised by the solar wind is *why is the power spectrum of this anisotropic, compressible magnetofluid often Kolmogorov-like?*

1.2.2 The Goldreich–Sridhar model

Marsch (1990) remarked that if, instead of an Alfvén time, the timescale for the waves to interact non-linearly with each other was the regular eddy turnover time, i.e. $\tau_l \simeq \tau_{\rm eddy} \sim l_{||}/\delta u_l$, one would get a K41 cascade for the magnetised turbulence. This would be true also for the case of strong turbulence, $|z^{\pm}| > \breve{B}_0$. The isotropy condition was retained, which raised a problem, most of the observational data mentioned above revealing strongly anisotropic turbulence.

Hereafter Goldreich and Sridhar (1995) proposed a turbulent model based on anisotropic fluctuations, with strong coupling between the wave modes. Strictly speaking the GS95 model assumes a critical balance between mechanical and Alfvénic modes in such a way that $l_{\perp}/\delta u_l \simeq l_{||}/v_A$. Therefore

$$l_{||} \sim v_A \epsilon^{-1/3} l_{\perp}^{2/3} \text{ and } P_u(k) \propto k^{-5/3}. \tag{18}$$

From Eq. (18), not only is the magnetized turbulence anisotropic, but it is *local* in the sense that the anisotropy is measured in the reference frame of the local magnetic field. Such an anisotropy is expected to occur in both the dispersion of velocity (δv) and wave vectors $\boldsymbol{k}$, though it is easier to observe velocity dispersion anisotropies from the interstellar medium, as discussed below. Therefore, statistically, a large number of eddies with local fields randomly distributed in space result in an average zero anisotropy (even at small scales). In the strong magnetised cases though, the anisotropy would be more clearly detected in experiments and observations.

Several direct numerical simulations of magnetised turbulence in a quasi-incompressible regime have been performed in the past decade. Many numerical experiments reveal that MHD turbulence indeed has a large part of its energy cascade close to a K41 distribution. However, as shown by Cho and Lazarian (2002, 2003), Cho et al. (2002) and Kowal and Lazarian (2010), the decomposition of the different modes in MHD turbulence actually reveals that, although Alfvén and slow modes behave as a K41 type of turbulence and are anisotropic, the fast modes are isotropic and follow IK statistics (see Fig. 2).

The effects of imbalanced (or cross-helicity) turbulence in the cascade and statistics of the local fields have also been addressed in the past few years (Lithwick et al., 2007; Beresnyak and Lazarian, 2008, 2010; Wicks et al., 2011; Markovskii and Vasquez, 2013, and references therein). Imbalanced turbulence occurs when waves travelling in opposite directions along the mean magnetic field are of unequal amplitudes, i.e. carry different energy fluxes to small length scales, so that $z_l^+/z_l^- \neq 1$ and $\epsilon_l^+/\epsilon_l^- \neq 1$. The imbalance may arise in MHD turbulence since the interaction timescales between waves z_l^+ and z_l^- are different, and the cascade generally occurs faster for z_l^-. This is understood as the number of interactions (N) being much larger for counter-propagating wave packets, resulting in $\epsilon_l^+/\epsilon_l^- > 1$. In such a scenario, numerical simulations show that the anisotropy is not equal for the different wave modes.

Locality of scales for wave–wave interactions has also been the subject of recent studies in turbulence (Carati et al., 2006; Alexakis, 2007; Mininni et al., 2008; Aluie and Eyink, 2009; Beresnyak and Lazarian, 2010). Magnetic fields are responsible for long-range interactions, from the Lorentz force acting over the whole fluid frozen to it. Therefore, different wavelengths may interact with each other non-linearly. Bispectra of fluctuations of density are discussed in Burkhart et al. (2009), and the non-local interactions appear to be important in MHD and supersonic turbulence models. A similar approach is used for studying the non-local interactions of Elsässer modes (Cho, 2010), resulting in a substantial fraction of non-local interactions in MHD turbulence. The role of the non-local interactions in the turbulent cascade is still not clear though. Turbulence in magnetised collisionless plasmas has been also studied in the past few years (e.g. Hellinger et al., 2006; Schekochihin et al., 2008; Bale et al., 2009, and others) in order to determine the role of collisionless plasma instabilities in the dynamics of plasma turbulence. Simulations by Kowal et al. (2011) and Santos-Lima et al. (2013) reveal that the statistics are still dominantly Kolmogorov-like, though strong asymmetries that may also arise due to instabilities (firehose, mirror and cyclotron instabilities) are small scale.

2 Signatures of a turbulent ISM

In the previous section some theoretical aspects of turbulence have been presented. Its direct comparison to the dynamics of the interstellar gas is not trivial, as we discuss in the following. However, we will present here some observational evidence of a turbulent ISM, and discuss the possible turbulent regimes that may be inferred from these.

The recognition of a turbulent interstellar medium dates back to the 1950s with the work of von Weizsäcker (1951) on the spatial distribution of dense structures in the plane of the sky. He recognised the hierarchy of structures and suggested its turbulent origin. The identification of turbulent motions was provided shortly after it was measured from velocity dispersions (von Hoerner, 1951). Later on, the observational and theoretical supports for a turbulence-dominated ISM have grown considerably (see reviews by Elmegreen and Scalo, 2004; Mac Low and Klessen, 2004; Hennebelle and Falgarone, 2012, and references therein), causing a major shift in the understanding of the ISM nature, from a thermal pressure-dominated system, as thought before, to a very dynamic multi-phase system.

2.1 Density distributions

As mentioned above, one of the main signatures of the turbulent character of the ISM is related to the density distribution of its contents. Up to now, tracers of the gas density distributions of the ISM at large scales have been predominantly indirect[5]. They rely on spectral lines and continuum emissions from the different phases of the ISM: the hot and fully ionised (HIM), the warm and fully/partially ionised medium (WIM/WNM), and the cold weakly ionised (CNM). These emissions being integrated along lines of sight and projected in the plane of the sky, sophisticated inversion methods have to be implemented. It is the statistical approach of the temporal and spatial variability of these emission fluxes that are the readily accessible observational techniques for studying interstellar turbulence.

With hydrogen being the most abundant element in the universe, the λ 21 cm line of neutral hydrogen is a key diagnostic. Its line-integrated emission is proportional to the bulk of the hydrogen column density, since its opacity remains low over most of the ISM. Statistics of the HI intensity spatial distributions have therefore been used to probe interstellar turbulence, but the results are far from homogeneous. Green (1993) studied the power law of the spatial power spectrum of the HI emission from different fields in our galaxy. He obtained power spectra with slopes between -2.2 and -2.8 at a scale range between 35 and 200 pc. From the HI 21 cm absorption towards Cas A. Roy et al. (2009) derived a power law with index -2.7, consistent with Kolmogorov turbulence in the diffuse interstellar medium. However, Miville-Deschênes et al. (2003) find an impressive power law in the nearby ISM at high galactic latitude with the same slope of -3.6 over two orders of magnitude in scales (between 0.1 and 25 pc). Similar studies have been performed since then, including other density tracers such as the CO and ^{13}CO line emission of molecular clouds, and power laws have also been inferred (e.g. Bensch et al., 2001; Hill et al., 2008). A review of the scatter of the power-law slopes measured is given in Hennebelle and Falgarone (2012). The scatter of the slope values is certainly affected by projection effects: one would expect a 2-D power spectrum $k^{-8/3}$ for an intrinsic Kolmogorov scaling. However, the integration along lines of sight crossing often large amounts of turbulent ISM with different properties tends to blur such a simple law. Moreover, the different tracers originate in truly different phases of the ISM, with varying amounts of small-scale structure that may affect the power spectrum of the density distributions (i.e. in many cases, like supersonic turbulence, density fluctuations

[5] In situ data have been obtained at the nearby interstellar plasma by Voyager 1 (Gurnett et al., 2013), though no direct study of the local turbulence has been discussed yet.

are not simply advected by turbulence as passive scalars, see e.g. Audit and Hennebelle, 2005). Indeed, as seen in Fig. 10 of Hennebelle and Falgarone (2012), many studies (including the power spectrum of the dust thermal emission) give power-law indices close to -2.7. It is not possible though to presume that a Kolmogorov-like cascade operates in the ISM, with scalings given by Eqs. (2) and (3). Even though compressibility seems to have little effect on the statistics of the ISM, except for small scales ($\sim$ pc scales) and cold and dense regions, magnetisation effects may be important, as we discuss further below.

Armstrong et al. (1995) used another tracer of density fluctuations, the scintillation of the background radiation (i.e. changes in the refraction index due to the turbulent motions in the ionised components of the ISM) in order to obtain the density spectrum along the line of sight. As a complementary method, fluctuations of the Faraday rotation measurements (RM) in the plane of sky are also used to estimate density fluctuations (once the magnetic field is known) in the line of sight (e.g. Minter and Spangler, 1996). The combined data provide the density fluctuations along the line of sight, but for different length scales, as seen in Fig. 3. The turbulence probed by both methods (scintillations and RM) presents a most impressive spectrum, with a unique Kolmogorov-like slope across more than ten orders of magnitude in wavenumber.

Similar works have been done for external galaxies. Turbulence has been characterised based on similar techniques for the Small Magellanic Cloud (see Stanimirovic et al., 1999; Stanimirovic and Lazarian, 2001; Chepurnov et al., 2008; Burkhart et al., 2010) and has revealed spatial variations in HI morphology. Dutta et al. (2013) calculated an HI intensity fluctuation power spectrum for a sample of 18 spiral galaxies and found slopes in the range of -1.9 to -1.5. Shallower spectra, compared to K41, could be evidence of two-dimensional eddy-dominated turbulence at scales larger than the disk thicknesses.

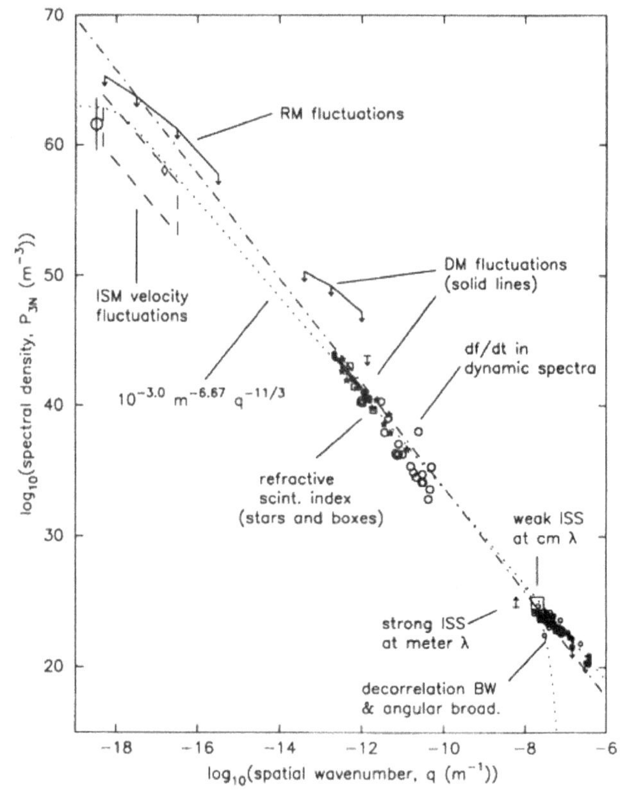

Fig. 3. Power spectrum of density along the line of sight from different data sets, and the dashed line as a reference for a Kolmogorov-like spectrum for one dimension ($k^{-11/3}$) (extracted from Armstrong et al., 1995).

2.2 Velocity fields

2.2.1 Direct statistical analysis

Spectral lines of several species observed with high spectral resolution may be used to infer the turbulence velocity distributions in the different phases of the ISM, such as hydrogen lines (mostly) and some ions for the diffuse ISM (e.g. Bowen et al., 2008), and molecular spectral lines (^{12}CO and ^{13}CO in most surveys) for the molecular clouds. The early surveys of Larson (1981) and Solomon et al. (1987) revealed the universal line-width and mass-distribution scalings among molecular clouds. Notably, both works pointed to a velocity dispersion relation $\sigma_v \propto l_v^\alpha$, with $\alpha \sim 0.5$ (see Fig. 4, left panel). Many similar studies were carried out to study the velocity distribution in molecular clouds, such as the work by Goldsmith et al. (2008), Yoshida et al. (2010), Qian et al. (2012) and Heyer and Brunt (2012) in the Taurus Molecular Cloud; Gustafsson et al. (2006) and Liu et al. (2012) for the Orion complex, and many others.

More recent studies confirmed the same scaling relation although with slopes varying significantly (Heyer and Brunt, 2004; Qian et al., 2012). Qian et al. (2012) for instance used the variance of the velocity difference of cores in molecular clouds, instead of the line width, and obtained $\alpha_v \sim 0.7$. On the other hand, massive cores are known to exhibit shallower slopes compared to what is frequently assumed (i.e. $\alpha_v < 0.5$).

Recently, Ballesteros-Paredes et al. (2011) compiled different observational surveys and concluded that, while in general terms the typical CO clouds observed by Heyer et al. (2009) lie close to Larson's relation, this is clearly not the case for the dense and massive cores, which exhibit large velocity dispersions for their relatively small sizes (Fig. 4). Those authors propose that the large dispersions observed at small scales are related to increased velocities as the clouds become gravitationally bound. However, the increased dispersion at small scales has already been reported in Falceta-Gonçalves et al. (2010a), based on numerical simulations without self-gravitating objects. For these authors the large

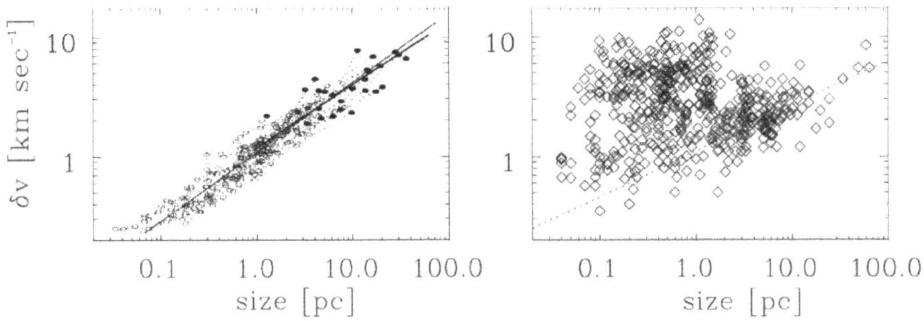

Fig. 4. Velocity dispersion relations from different surveys by Heyer and Brunt (2004) (left panel) and compilation from several surveys done by Ballesteros-Paredes et al. (2011) (right panel). As the later authors point out, while large CO clouds from the survey by Heyer et al. (2009) exhibit the typical Larson relationship, denser structures show a larger dispersion of velocity. This fact has been interpreted by those authors as being due to gravity in collapsing cores, while Falceta-Gonçalves et al. (2010b) argued for projection effects and compressibility.

dispersion observed at small scales is an intrinsic feature of the turbulent gas. The broad dispersion of the scaling relation indicates a turbulent regime dominated by compressible motion at small scales, as discussed in Sect. 1.1, though regular incompressible turbulence dominates at larger scales. Compressibility, as described in Fleck's model (see Eq. 10), naturally gives larger slopes for the dispersion relation, with a value of $\alpha \sim 0.16$ favoured by observations. It is not clear though what the actual role of gravity is in the statistics of the molecular cloud emissions.

At the large-scale end of the cascade, the apparent uniqueness of the scaling of the velocity dispersion with size scale suggests a universal source (or mixture of sources) of energy for the molecular gas turbulence in our galaxy. Chepurnov and Lazarian (2010) presented a statistical analysis of high-latitude HI turbulence in the Milky Way based on the velocity coordinate spectrum (VCS) technique. They found a velocity power spectrum $P_u(k) \propto k^{-3.8}$ and an injection scale of $\sim 140 \pm 80$ pc. The slightly steeper slope, compared to K41, can be the result of shock-dominated (compressible) turbulence, with averaged sonic Mach numbers $\sim$7–8 (see Sect. 1.1 above).

Two-point statistics are also used, but, since in situ measurements are not yet available, one easily accessible observable turns out to be the variations in the plane of the sky of the line-of-sight centroid velocity of spectral lines. Lis et al. (1996) showed that they trace the plane-of-the-sky projection of the vorticity. Using a sample of about one million independent CO spectra in a diffuse field, Hily-Blant and Falgarone (2009) identified, on statistical grounds, the ensemble of positions at which vorticity departs from a Gaussian distribution. These form coherent elongated structures at the parsec scale that are found to harbour sub-structures of most intense velocity shears down to the milliparsec scale (Falgarone et al., 2009). These coherent structures are proposed to be the manifestations of the intermittency of turbulent dissipation in diffuse molecular clouds (see the review of Hennebelle and Falgarone, 2012), which may be compared to Eq. (8) above.

Li and Houde (2008) studied the scaling relations of the velocity dispersions from different neutral and ionised molecular species, namely HCN and HCO^+, in the region of M17. As occurs in many other star-forming regions, the ionised molecules systematically present a smaller dispersion of velocity compared to the neutral molecules. Such a difference arises as turbulent energies dissipate differently for the species due to ambipolar diffusion. Falceta-Gonçalves et al. (2010b) showed that the dispersion for ions is typically smaller than that for the neutral species basically due to the damping of the ion turbulence at the ambipolar diffusion scales ($\simeq 0.01$ pc).

The direct comparison between statistics of observational data and the theory must be done with caution. Column density projections, or in another sense emission maps, are influenced by projection effects. Different structures projected onto the same line of sight, but decorrelated at a given length scale, may be observed as a single structure on the projected emission map. Some deconvolution is possible though once the velocity profile is known, with a high spectral resolution.

2.2.2 Indirect access to the velocity field via maser emission

The low surface brightness of the above tracers and projection effects make the direct analysis of turbulent flows in the ISM difficult. Maser spots that are bright point sources and are transported by turbulence as passive scalars (because they are tiny and low-mass structures) turn out to be powerful tracers of the turbulent velocity field. Maser radiation in molecular lines appear in dense regions where population level inversion can be generated by radiative pumping, for instance in the dense molecular gas of star-forming regions (SFRs) associated with ultra-compact HII regions, embedded IR sources, hot molecular cores, Herbig–Haro objects, and outflows (Litvak, 1974; Reid and Moran, 1981; Elitzur, 1992; Lo, 2005). Maser emissions are often characterised by high brightness temperatures and high degrees of polarisation. Intense maser emission is detected in the molecular lines of hydroxyl (OH), water (H_2O), silicon monoxide (SiO), ammonia (NH_3), methanol (CH_3OH), among others.

Walker (1984) used the *very long baseline interferometry* (VLBI) maps of the H_2O maser source in W49N to demonstrate that both two-point velocity increments and two-point spatial correlation functions exhibit power-law dependencies on the maser spot separation, which is indicative of a turbulent flow. Gwinn (1994) performed statistical analysis of VLBI data for W49N to confirm the power-law dependence of the velocity dispersion and spatial density of masing spots on a spatial scale, and interpreted this observation as evidence of turbulence. Imai et al. (2002) reported sub-milliarcsec structures of H_2O masers in W3 IRS 5. A cluster of maser spots (emission spots in individual velocity channels) displays velocity gradients and a complicated spatial structure. Two-point spatial correlation functions for the spots can be fitted by the same power laws in two very different spatial ranges. The Doppler-velocity difference of the spots as a function of spot separation increases as expected in Kolmogorov-type turbulence. Strelnitski et al. (2002) used VLBI data to investigate the geometry and statistical properties of the velocity field traced by H_2O masers in five star-forming regions. In all sources the angular distribution of the H_2O maser spots shows approximate self-similarity over almost four orders of magnitude in scale. The lower order structure functions for the line-of-sight component of the velocity field can be fitted by power laws, with the exponents close to the Kolmogorov value. Similar results were also obtained for other regions (e.g. Richards et al., 2005; Strelnitski, 2007; Uscanga et al., 2010).

2.3 Turbulent magnetic fields

The magnetic fields in our galaxy are modelled as a superposition of different components: (i) a large-scale field, following a spiral structure in the plane of the galactic disk, and extending high above the plane into the galactic halo, and (ii) a complex component of locally disturbed magnetic fields, which are related to molecular clouds and star formation regions. The spiral pattern in the disk aligns with the spiral arms (e.g. Han, 2006). This is expected, since the shear of gaseous motion around the centre of the galaxy stretches the field lines in this direction (see the review of mean field dynamo by Beck et al., 1996).

There are four main methods of studying the fluctuations in the ISM magnetic field, namely the polarisation of dust thermal emission (both in emission in the far-infrared (FIR) and absorption in the visible and near-IR), the Zeeman effect of spectral lines, Faraday rotation and polarisation of the synchrotron emission. Polarised synchrotron emission can also be mapped in order to provide the geometry of the field lines in the plane of the sky. Faraday rotation and synchrotron polarisation measurements excel in probing the magnetic field of the diffuse ionised medium of the ISM, i.e. they are excellent tools for studying the large-scale fields of galaxies in general. More extensive reviews both of magnetic fields in star-formation regions and galactic-scale magnetic fields are given in Crutcher (2012) and Han (2006), respectively.

As mentioned earlier, synchrotron emission polarisation can be used for mapping the large-scale structure of the magnetic fields in galaxies (see review by Beck, 2009). The fields traced by the polarised synchrotron emission present intensities of the order of ~ 10–$15\,\mu$G. However, the synchrotron emission probes the ionised medium only, which is less useful in determining the turbulence properties of the star formation regions, dominated by the dense and neutral components of the ISM. Therefore, a magnetic field with intensity ~ 10–$15\,\mu$G is supposed to thread most of galactic disk, except the dense regions of the arms where the local properties of the plasma and stellar feedback may dramatically change the field properties.

Oppermann et al. (2012) compiled an extensive catalogue of Faraday rotation measure (RM) data of compact extragalactic polarised radio sources in order to study the angular distribution of the all-sky RMs. The authors found an angular power spectrum $P(k) \propto k^{-2.17}$ for the Faraday depth, which is given by the product of the line-of-sight magnetic field component $B_{\rm LOS}$ and the electron number density n_e. The combination of the RM and polarisation vectors of the synchrotron emission allows one to reconstruct the three-dimensional structure of galactic magnetic fields. Such angular fluctuations of the Faraday depth are thought to be related to the turbulent ISM. However, the relationship between the fluctuations of the RM and the local fluctuations of electron density and magnetic fields is not clear yet. This, for instance, is an interesting subject for further comparisons with simulations (as in Gaensler et al., 2011).

Possibly the most direct method for estimating the magnetic field intensity in the dense and cold ISM relies on the detection of the Zeeman effect (see Robishaw et al., 2008, for details). For instance, Sarma et al. (2002) detected and studied the Zeeman effect in H_2O masers in several SFRs and determined line-of-sight magnetic field strengths ranging from 13 to 49 mG. They found a close equilibrium between the magnetic field energy and turbulent kinetic energy in masing regions.

Alves et al. (2012) showed that shock-induced H_2O masers are important magnetic-field tracers of very high-density gas in low-mass protostellar core IRAS 16293-2422. They investigated whether the collapsing regime of this source is controlled by magnetic fields or other factors such as turbulence, and concluded that the magnetic field pressure derived from data is comparable to the ram pressure of the outflow dynamics. This indicates that the magnetic field is energetically important for the dynamical evolution of the protostellar core.

Due to its brightness, maser emission is better for probing magnetic fields, but they are rare and limited in extent. The Zeeman effect in non-masing regions has been detected for HI, OH, and CN lines for which the turbulent broadening is typically larger than the Zeeman splitting in frequency. The compilation by Crutcher (1999) and recent CN Zeeman observations in SFRs by Falgarone et al. (2008) show that the turbulent motions within the SFRs and molecular clouds are supersonic but sub-Alfvénic. The *upper limit* magnetic field intensity scales with density, estimated from a Bayesian analysis, as $B \propto n^{\kappa}$, with $\kappa \sim 0.47$ (Crutcher et al., 2010). Collapsed structures along the mean field would produce $\kappa \rightarrow 0$, while shock compressions perpendicular to the field lines result in $\kappa \rightarrow 1$. The observed relation with $\kappa \sim 0.47$ is expected, for instance, in Alfvénic perturbations and is in agreement with MHD simulations (e.g. Burkhart et al., 2009). It was also claimed in that work that, despite its relative importance in the overall dynamics of clouds, the uniform magnetic fields in these clouds are in general not strong enough to prevent gravitational collapse based on the mass-to-flux (M/Φ) ratios observed. Other major compilations of Zeeman measurements in molecular clouds are given, for example, in Bourke et al. (2001) and Troland and Crutcher (2008) with similar results.

Polarisation maps of molecular clouds

Radiation may be polarised due to a preferred direction for emission/absorption from aspherical dust grains, as well as by some molecules and atoms. The ISM in known to be populated by a complex distribution of grain sizes and shapes. Depending on its composition an aspherical rotating dust particle may align with the magnetic field line. The orientation of the polarisation of radiation is then linked to the orientation of the magnetic field itself (see review by Lazarian, 2007). Many observational data have been made available in the past decade both on absorption and emission dust polarisation (e.g. Heiles, 2000; Chapman et al., 2011).

The strengths of magnetic fields can be estimated from polarisation maps by the Chandrasekhar and Fermi (1953) (CF) technique. The CF method is based on the assumption that the magnetic and turbulent kinetic pressures are the dominant ones within the cloud, and that the fluid motions are coupled to the magnetic field lines. In this sense, any perturbation from the fluid turbulence will result in a change in the orientation of the field lines. Major improvements in the CF technique are given e.g. by Falceta-Gonçalves et al. (2008), Hildebrand et al. (2009) and Houde et al. (2009). If the velocity dispersion $\delta v_{\rm los}$ is known, e.g. from spectral lines, the mean magnetic field in the plane of sky can be estimated as (Falceta-Gonçalves et al., 2008)

$$B_{\rm sky}^{\rm uniform} \simeq \delta v_{\rm los} (\pi < \rho >)^{1/2} \left[1 + \tan^{-1}(\delta\phi)\right], \qquad (19)$$

where $\delta\phi$ represents the dispersion in the polarisation angle.

From the equation above, the ratio $\delta B/B_{\rm sky}$ – assumed to be $\sim \tan^{-1}\delta\phi$ – is directly related to the Alfvénic Mach number of the turbulence. Notice that the dependence of the projected $\delta B/B_{\rm sky}$ on the actual 3-D MHD turbulence may be removed from higher order statistical analysis, as proposed in Falceta-Gonçalves et al. (2008).

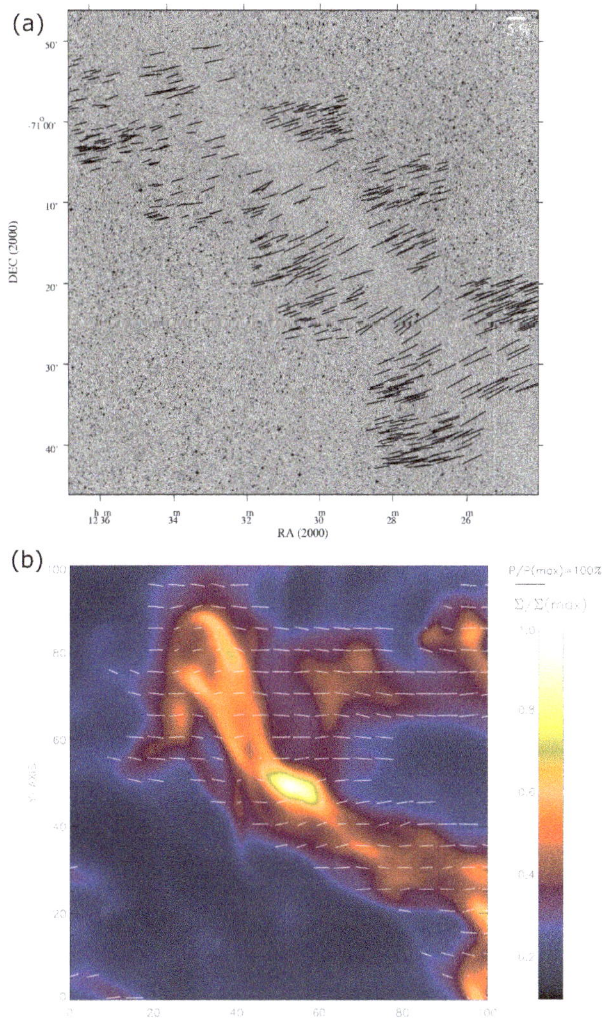

Fig. 5. Left panel: optical polarisation map of the Muska dark cloud (extracted from Pereyra and Magalhães, 2004). Right panel: simulated polarisation map from a three-dimensional simulation of MHD sub-Alfvénic turbulence (extracted from Falceta-Gonçalves et al., 2008).

The left image of Fig. 5 presents the polarisation map of the Muska dark cloud (Pereyra and Magalhães, 2004) in the optical wavelengths as a result of dust absorption. Vectors represent the magnetic field orientation. The filamentary morphology of the dark cloud is perpendicular to the external field, which is very uniform, indicating a sub-Alfvénic turbulent regime. On the right-hand side of Fig. 5, the polarisation map is overplotted on the column density projection of a 3-D MHD numerical simulation of sub-Alfvénic turbulence (Falceta-Gonçalves et al., 2008). Such comparisons between MHD numerical simulations and measurements of magnetic fields in the ISM are important in unveiling the physics of

MHD turbulence, and its role in other phenomena such as star formation.

Spatial dispersion of magnetic fields in molecular clouds from polarisation maps may be used to characterise the power spectrum of magnetised turbulence in the inertial and dissipation ranges. Houde et al. (2011) found a power law inertial range for the magnetic field spatial distribution that is $\propto k^{-2.9\pm0.9}$, and a cutoff at scales ~ 0.009 pc, which is claimed by the authors to be related to the ambipolar diffusion scales.

Again, as another issue in a proper modelling of the statistics of velocity, gravity is claimed to interfere with the statistics of the observed polarisation maps (Koch et al., 2012a, b). Gradients in emission towards the cores of molecular clouds have been shown to be associated with gradients in the polarisation angles. A transition from a magnetically subcritical to a supercritical state[6] could then explain the trend, and this technique could provide an independent way of estimating the local magnetic force compared to gravity.

Heyer and Brunt (2012) showed that the turbulence in the densest regions of the Taurus molecular cloud is super-Alfvénic, while the reverse is true in the surrounding lower density medium, threaded by a strong magnetic field. This observational result is in agreement with the transition expected between scales as dense structures are formed, e.g. by shocks, in a supersonic but sub-Alfvénic large-scale turbulence (see the discussion in Falceta-Gonçalves et al., 2008; Heyer et al., 2008; Burkhart et al., 2009).

Similar to the synchrotron radiation case, by combining dust polarisation maps with Zeeman measurements in molecular clouds one can determine the three-dimensional structure of the magnetic field. Poidevin et al. (2013) recently succeeded in testing this approach for a number of objects of the SCUBA Polarimeter Legacy (SCUPOL) data catalogue. The authors were able to determine the orientation of the mean field with respect to the line of sight, as well as to estimate the turbulence regime within several molecular clouds. The authors also claimed that all observed clouds seem to present a universal large-scale turbulence that is supersonic ($M_s \sim 6$–8) and sub-Alfvénic ($M_a \sim 0.5$–0.9), at scales as large as 50pc.

In terms of comparing these data with basic theories of magnetised turbulence, most observations point towards a magnetically dominated turbulence at scales larger than a few tenths of parsecs. Heyer et al. (2008) also showed some of the first evidence of anisotropic turbulence in molecular clouds, with respect to the large-scale magnetic field orientation. The observations of the Taurus Molecular Cloud revealed a significant anisotropy in the dispersion of velocity (δv), being larger for lags perpendicular to the mean large-scale field lines. Even though a Goldreich–Sridhar similarity relation is not obtained, the anisotropy observed is a strong indication of strong coupling between MHD wave modes in the insterstellar turbulence, as predicted by the GS95 model. We could extrapolate a bit and say that a GS95 model combined with fractal density distributions, as given in Fleck (1996), is favoured for the ISM turbulence based on current observations.

3 Origins of interstellar turbulence

Surveys of different atomic and molecular line emissions have shown us that the diffuse ISM is turbulent at scales > 150 pc, with $\delta v \geq 50\,\mathrm{km\,s^{-1}}$. This results in a specific energy transfer rate[7] of $\epsilon \simeq m_H n_H \delta v_L^3 / L \sim 10^{-25} - 10^{-24}\,\mathrm{erg\,cm^{-3}\,s^{-1}}$. Brunt et al. (2009) estimated the driving scales of turbulence for molecular clouds by comparing observed and synthetic CO velocity dispersions from numerical simulations. They found that only models with large-scale sources of turbulence, such as supernovae-driven outflows (SNe) and galactic dynamics, fit well to the observed data.

Supernovae have been claimed as main turbulence drivers by many authors (e.g. Norman and Ferrara, 1996; Mac Low and Klessen, 2004; Avillez and Breitschwerdt, 2004, 2005; Joung and Mac Low, 2006; Hill et al., 2012). They certainly correspond to an important driving mechanism for turbulence in starburst regions and small galaxies (e.g. Falceta-Gonçalves et al., 2010a; Ruiz et al., 2013). However, their impact on galactic turbulence, in a more generalized sense, is still a matter of debate.

One issue is that numerical simulations of SNe-driven turbulence create superbubbles that are far too hot and diffuse (see Avillez and Breitschwerdt, 2005; Joung and Mac Low, 2006; Hill et al., 2012). Other critical arguments disfavour SNe as a main driver mechanism as well, at least for our galaxy. Zhang et al. (2001) analysed the CO emission lines from the Carina complex and obtained a turbulent energy flux per mass density unit cascading over scales $\sim 10^{-7}\,(\mathrm{km\,s^{-1}})^2\,\mathrm{yr^{-1}}$ that could not be explained from stellar feedback, but which is in rough agreement with the injection rate of energy from the gravitational interaction of the ISM gas and the galactic spiral arms. Sánchez-Salcedo et al. (2007) also showed that HI mapping of our galaxy is consistent with a turbulence injection rate that is not directly related to the star formation rate, but is about constant with respect to the galactocentric radius. Also, the correlation lengths related to SNe turbulence are strongly dependent on local properties (such as local density and temperature) (see Leão et al., 2009). Such local dependence also occurs with respect to the height related to the galactic plane, since SNe energy is easily released outwards (e.g. Melioli et al., 2009; Marasco et al., 2013). The universality of the observed properties of

[6] i.e. the system becomes supercritical once gravity overcomes the magnetic pressure.

[7] This estimate is at least one order of magnitude larger than that of Mac Low and Klessen (2004), since these authors considered a lower injection velocity at the largest scales ($\delta v_L = 10\,\mathrm{km\,s^{-1}}$).

turbulence in our galaxy, together with the extremely large injection scales (> 100 pc), suggest a galactic-scale driving source, which is later amplified, as second-order effects, by local stellar feedback. Qian et al. (2012), for instance, obtained similar core and ambient turbulent statistics, which suggested that molecular cores condense from more diffuse gas, and that there is little (if not no) additional energy from star formation into the more diffused gas.

Turbulence driven by galactic dynamics models, such as driven by velocity shears in the galactic disk, were posed long ago (e.g. Fleck, 1981). Instabilities such as the magneto-rotational instability have also been proposed (e.g. Sellwood and Balbus, 1999; Kim and Ostriker, 2002). Interactions between the arms of the galaxy and the disk gas also generate perturbations, as large as $20\,\mathrm{km\,s^{-1}}$ (Gómez and Cox, 2002), that could explain most of the injection of energy into turbulent motions. It is not clear yet which of these mechanisms (SNe or galactic dynamics) is more important for the observed turbulence in the ISM. Certainly, it is a promising subject for studies in the upcoming years, both from theoretical and observational sides.

4 Conclusions

In this work, we briefly reviewed part of the current understanding of incompressible, compressible and/or magnetised turbulence, which can be applied to characterise the interstellar medium. There is a vast literature available for each of these and a complete review of turbulence is out of the scope of this work. We discussed the recent theoretical improvements made in the modelling and characterisation of the different turbulent regimes. Multifractal description, statistics of probability functions, and spectral analysis are just a few that have been currently employed to characterise spatial and temporal variations of plasma properties associated with turbulent motions.

Phenomenological descriptions of turbulence in Fourier space, such as that of Kolmogorov–Obukhov, are particularly simple and still very useful in the diagnostics of interstellar turbulence. Since scaling relations for compressible, incompressible and magnetised turbulence of these theories may differ among each other, observations can be used to determine the turbulent regime of the ISM.

Spectroscopy has been long used to probe the velocity distributions along the line of sight. The observed amplitudes of the turbulent motions indicate that the ISM transits from a supersonic turbulent regime at scales of tens to hundreds of parsecs, at which the turbulence is driven, to subsonic at smaller scales. The scales where turbulence is subsonic depend on the "phase" of the ISM plasma. Dense molecular clouds present lower temperatures, which result in subsonic turbulence only at very small scales ($\ll 1$ pc). The warmer and more diffuse media, such as the warm neutral medium and the warm diffuse medium, present subsonic flows at scales of a few parsecs due to the larger local sound speeds. It is interesting to mention that this transition is deeply related to the origin of the dense molecular clouds. These objects originate either due to the large-scale compressible motions of the gas (e.g. Williams et al., 2000), or at small scales due to other mechanisms, such as thermal instabilities. Current observations favour the first, given their length scales and internal dynamics (see Poidevin et al., 2013). Spatial gas distributions over the plane of the sky are also provided observationally. The filamentary structure observed reveals a compressible dominated turbulent regime, at least at most of the observed scales. Observations also reveal a magnetised ISM. All these ingredients combined result in compressible and magnetised ISM turbulence, challenging theorists to provide a phenomenological description of the combined effects of supersonic flows and strong magnetic fields. Despite the good agreement between observations and the Goldreich–Sridhar model for magnetised turbulence, and Fleck's model for compressible ones, such as spectral slopes, scaling relations and multifractal analysis applied to emission maps, a complete unified theory has yet to be developed.

One of the major problems in comparing statistics of observed quantities with theories of turbulence relies on projection effects. Observations are spatially limited in the sense that all statistics are done either along the line of sight (e.g. scintillation, velocity dispersion from spectral lines, Faraday rotation) or in the plane of the sky. In addition, even the plane-of-the-sky maps are related to integrated quantities (e.g. emission lines, column density, Stokes parameters for the polarisation maps). One must therefore be careful when comparing these with theories of three-dimensional turbulence.

Other effects may also make a direct comparison between theory and observations challenging. Self-gravity of dense gas and stellar feedback, for instance, have been neglected in this paper. These processes are responsible for extra sources of energy and momentum, but are not easily linked to the turbulent cascade. Despite their obvious importance for the process of e.g. star formation, their role in the statistics of the turbulence is not completely clear. Naturally, fragmentation and clumping would be enhanced if self-gravity is considered (Vázquez-Semadeni et al., 1996; Ballesteros-Paredes et al., 2011; Cho and Kim, 2011); however, its role in the cascade itself and in intermittency is unknown.

Future studies from the theoretical side are possibly to be focused on the understanding of combined processes, such as magnetic fields, gravity, compressibility and radiation, and on the energy transfer among scales. The formation of coherent structures and how their statistics relate to the bulk of the fluid are vital for theories of star formation. New data are also expected for the upcoming years. Although the Herschel mission ended in early 2013 its data have not yet been fully explored. Other major observational facilities, such as the Planck[8] satellite and the Atacama Large Millimeter Array (ALMA), will provide complementary data at radio and microwave frequencies with very large sensitivity, thereby going "deeper" than the sensitivity reached by other instruments. Also, Gurnett et al. (2013) recently presented the first in situ measurement of the interstellar plasma as Voyager 1 crossed the heliopause and started to probe the nearby interstellar plasma. This opens new possibilities in studying interstellar turbulence locally. It is clear that the future of interstellar turbulence science is going to be very exciting.

Acknowledgements. D. Falceta-Gonçalves thanks the European Research Council (ADG-2011 ECOGAL) and Brazilian agencies CNPq (no. 300382/2008-1), CAPES (3400-13-1) and FAPESP (no. 2011/12909-8) for financial support. G. Kowal acknowledges support from FAPESP grant no. 2013/04073-2. A. C.-L. Chian thanks the European Commission for the award of a Marie Curie International Incoming Fellowship, CNPq for support, and the Paris Observatory for the kind hospitality.

Edited by: G. S. Lakhin

References

Alexakis, A.: Non-local Phenomenology for Anisotropic Magnetohydrodynamic Turbulence, Astrophys. J., 667, L93–L96, 2007.

Alexakis, A.: Large-Scale Magnetic Fields in Magnetohydrodynamic Turbulence, Phys. Rev. Lett., 110, 084502, 2013.

Alexandrova, O., Lacombe, C., and Mangeney, A.: Spectra and anisotropy of magnetic fluctuations in the Earth's magnetosheath: Cluster observations, Ann. Geophys., 26, 3585–3596, doi:10.5194/angeo-26-3585-2008, 2008.

Aluie, H. and Eyink, G. L.: Localness of energy cascade in hydrodynamic turbulence, II. Sharp spectral filter, Phys. Fluids, 21, 115108–115108-16, 2009.

Alves, F. O., Vlemmings, W. H. T., Girart, J. M., and Torrelles, J. M.: The magnetic field of IRAS 16293-2422 as traced by shock-induced H_2O masers, Astron. Astrophys., 542, A14, doi:10.1051/0004-6361/201118710, 2012.

Armstrong, J. W., Rickett, B. J., and Spangler, S. R.: Electron density power spectrum in the local interstellar medium, Astrophys. J., 443, 209–221, 1995.

Audit, E. and Hennebelle, P.: Thermal condensation in a turbulent atomic hydrogen flow, Astron. Astrophys., 433, 1–13, 2005.

Avillez, M. A. and Breitschwerdt, D.: Volume filling factors of the ISM phases in star forming galaxies, I. The role of the disk-halo interaction, Astron. Astrophys., 425, 899–911, 2004.

Avillez, M. A. and Breitschwerdt, D.: Global dynamical evolution of the ISM in star forming galaxies, I. High resolution 3D simulations: Effect of the magnetic field, Astron. Astrophys., 436, 585–600, 2005.

[8]The Planck mission's main goal is to observe the comic microwave background emission for cosmological purposes; however, the foreground ambient is the ISM, and proper modelling of the ISM structure and magnetic fields will be mandatory.

Bale, S. D., Kellogg, P. J., Mozer, F. S., Horbury, T. S., and Reme, H.: Measurement of the Electric Fluctuation Spectrum of Magnetohydrodynamic Turbulence, Phys. Rev. Lett., 94, 215002, doi:10.1103/PhysRevLett.94.215002, 2005.

Bale, S. D., Kasper, J. C., Howes, G. G., Quataert, E., Salem, C., and Sundkvist, D.: Magnetic Fluctuation Power Near Proton Temperature Anisotropy Instability Thresholds in the Solar Wind, Phys. Rev. Lett., 103, 211101, doi:10.1103/PhysRevLett.103.211101, 2009.

Ballesteros-Paredes, J., Hartmann, L. W., Vázquez-Semadeni, E., Heitsch, F., and Zamora-Avilés, M. A.: Gravity or turbulence? Velocity dispersion-size relation, Mon. Notice. Roy. Astron. Soc., 411, 65–70, 2011.

Bamert, K., Kallenbach, R., le Roux, J. A., Hilchenbach, M., Smith, C. W., and Wurz, P.: Evidence for Iroshnikov-Kraichnan-Type Turbulence in the Solar Wind Upstream of Interplanetary Traveling Shocks, Astrophys. J., 675, L45–L48, 2008.

Bec, J. and Khanin, K.: Burgers turbulence, Phys. Rep., 447, 1–66, 2007.

Beck, R.: Galactic and extragalactic magnetic fields – a concise review, Astrophys, Space Sci. Trans., 5, 43–47, 2009.

Beck, R., Brandenburg, A., Moss, D., Shukurov, A., and Sokoloff, D.: Galactic Magnetism: Recent Developments and Perspectives, Ann. Rev. Astron. Astrophys., 34, 155–206, 1996.

Bensch, F., Stutzki, J., and Ossenkopf, V.: Quantification of molecular cloud structure using the Delta-variance, Astron. Astrophys., 366, 636–650, 2001.

Beresnyak, A. and Lazarian, A.: Strong Imbalanced Turbulence, Ap. J. Astrophys. J., 682, 1070–1075, 2008.

Beresnyak, A. and Lazarian, A.: Scaling Laws and Diffuse Locality of Balanced and Imbalanced Magnetohydrodynamic Turbulence, Astrophys. J., 722, L110–L113, 2010.

Birnir, B.: The Kolmogorov-Obukhov Statistical Theory of Turbulence, J. Nonlin. Sci., 23, 657–688, 2013.

Biskamp, D.: Magnetohydrodynamical Turbulence, Cambridge University Press, Cambridge, 2003.

Boffetta, G. and Sokolov, I. M.: Relative Dispersion in Fully Developed Turbulence: The Richardson's Law and Intermittency Corrections, Phys. Rev. Lett., 88, 094501, doi:10.1103/PhysRevLett.88.094501, 2002.

Boldyrev, S. and Perez, J. C.: Spectrum of Kinetic-Alfvén Turbulence, Astrophys. J., 758, L44–L49, 2012.

Bourke, T. L., Myers, P. C., Robinson, G., and Hyland, A. R.: New OH Zeeman Measurements of Magnetic Field Strengths in Molecular Clouds, Astrophys. J., 554, 916–932, 2001.

Bourras, D., Weill, A., Caniaux, G., Eymard, L., Bourlès, B., Letourneur, S., Legain, D., Key, E., Baudin, F., Piguet, B., Traullé, O., Bouhours, G., Sinardet, B., Barrié, J., Vinson, J. P., Boutet, F., Berthod, C., and Clémençon, A.: Turbulent air-sea fluxes in the Gulf of Guinea during the AMMA experiment, JGRC J. Geophys. Res.-Oceans, 114, C04014, doi:10.1029/2008JC004951, 2009.

Bowen, D. V., Jenkins, E. B., Tripp, T. M., Sembach, K. R., Savage, B. D., Moos, H. W., Oegerle, W. R., Friedman, S. D., Gry, C., Kruk, J. W., Murphy, E., Sankrit, R., Shull, J. M., Sonneborn, G., and York, D. G.: The Far Ultraviolet Spectroscopic Explorer Survey of OVI Absorption in the Disk of the Milky Way, Astrophys. J. Suppl. Ser., 176, 59–163, 2008.

Brunt, C. M., Heyer, M. H. and Mac Low, M. M.: Turbulent driving scales in molecular clouds, Astron. Astrophys., 504, 883–890, 2009.

Burgers, J. M.: Mathematical examples illustrating relations occurring in the theory of turbulent fluid motion, Proc. Acad. Sci. Amst., 17, 1–53, 1939.

Burkhart, B., Falceta-Gonçalves, D., Kowal, G., and Lazarian, A.: Density Studies of MHD Interstellar Turbulence: Statistical Moments, Correlations and Bispectrum, Astrophys. J., 693, 250–266, 2009.

Burkhart, B., Stanimirovic, S., Lazarian, A., and Kowal, G.: Characterizing Magnetohydrodynamic Turbulence in the Small Magellanic Cloud, Ap. J. Astrophys. J., 708, 1204–1220, 2010.

Carati, D., Debliquy, O., Knaepen, B., Teaca, B., and Verma, M.: Energy transfers in forced MHD turbulence, J. Turbulence, 7, N51, doi:10.1080/14685240600774017, 2006.

Chandrasekhar, S. and Fermi, E.: Magnetic Fields in Spiral Arms, Ap. J. Astrophys. J., 118, 113–115, 1953.

Chang, K., Bewley, G. P., and Bodenschatz, E.: Experimental study of the influence of anisotropy on the inertial scales of turbulence, J. Fluid Mech., 692, 464–481, 2012.

Chapman, N. L., Goldsmith, P. F., Pineda, J. L., Clemens, D. P., Li, D., and Krco, M.: The Magnetic Field in Taurus Probed by Infrared Polarization, Astrophys. J., 741, 21, doi:10.1088/0004-637X/741/1/21, 2011.

Chepurnov, A. and Lazarian, A.: Extending the Big Power Law in the Sky with Turbulence Spectra from Wisconsin H Mapper Data, Astrophys. J., 710, 853–858, 2010.

Chepurnov, A., Lazarian, A., Gordon, J., and Stanimirovic, S.: Topology of Neutral Hydrogen within the Small Magellanic Cloud, Ap. J. Astrophys. J., 688, 1021–1028, 2008.

Chian, A. C.-L. and Miranda, R. A.: Cluster and ACE observations of phase synchronization in intermittent magnetic field turbulence: a comparative study of shocked and unshocked solar wind, Ann. Geophys., 27, 1789–1801, doi:10.5194/angeo-27-1789-2009, 2009.

Chian, A. C.-L. and Muñoz, P. R.: Detection of current sheets and magnetic reconnections at the turbulent leading edge of interplanetary coronal mass ejection, Astrophys. J. Lett., 733, L34, doi:10.1088/2041-8205/733/2/L34, 2011.

Chian, A. C.-L., Kamide, Y., Rempel, E. L., and Santana, W. M.: On the chaotic nature of solar-terrestrial environment: interplanetary Alfvén intermittency, J. Geophys. Res.-Space., 111, A07S03, doi:10.1029/2005JA011396, 2006.

Chian, A. C.-L., Santana, W. M., Rempel, E. L., Borotto, F. A., Hada, T., and Kamide, Y.: Chaos in driven Alfvén systems: unstable periodic orbits and chaotic saddles, Nonlin. Processes Geophys., 14, 17–29, doi:10.5194/npg-14-17-2007, 2007.

Chian, A. C.-L., Miranda, R. A., Rempel, E. L., Saiki, Y., and Yamada, M.: Amplitude-phase synchronization at the onset of permanent spatiotemporal chaos, Phys. Rev. Lett., 104, 254102, doi:10.1103/PhysRevLett.104.254102, 2010.

Cho, J.: Non-locality of Hydrodynamic and Magnetohydrodynamic Turbulence, Astrophys. J., 725, 1786—1791, 2010.

Cho, J. and Lazarian, A.: Compressible Sub-Alfvńic MHD Turbulence in Low-Plasmas, Phys. Rev. Lett., 88, 245001, doi:10.1103/PhysRevLett.88.245001, 2002.

Cho, J. and Lazarian, A.: Compressible magnetohydrodynamic turbulence: mode coupling, scaling relations, anisotropy, viscosity-damped regime and astrophysical implications, Mon. Notice. Roy. Astron. Soc., 345, 325–339, 2003.

Cho, J., Lazarian, A., and Vishniac, E.: Simulations of Magnetohydrodynamic Turbulence in a Strongly Magnetized Medium, Astrophys. J., 564, 291–301, 2002.

Cho, W. and Kim, J., Enhanced core formation rate in a turbulent cloud by self-gravity, Mon. Notice. Roy. Astrono. Soc., 410, L8–L12, 2011.

Coleman, P. J.: Turbulence, Viscosity, and Dissipation in the Solar-Wind Plasma, Astrophys. J., 153, 371–388, doi:10.1086/149674, 1968.

Crutcher, R. M.: Magnetic Fields in Molecular Clouds: Observations Confront Theory, Astrophys. J., 520, 706–713, 1999.

Crutcher, R. M.: Magnetic Fields in Molecular Clouds, Ann. Rev. Astron. Astrophys., 50, 29–63, 2012.

Crutcher, R. M., Wandelt, B., Heiles, C., Falgarone, E., and Troland, T. H.: Magnetic Fields in Interstellar Clouds from Zeeman Observations: Inference of Total Field Strengths by Bayesian Analysis, Astrophys. J., 725, 466–479, 2010.

Dutta, P., Begum, A., Bharadwaj, S., and Chengalur, J. N.: Probing interstellar turbulence in spiral galaxies using H I power spectrum analysis, New Astronomy, 19, 89–98, 2013.

Elitzur, M.: Astronomical masers, Ann. Rev. Astron. Astrophys., 30, 75–112, 1992.

Elliott Jr., F. W. and Majda, A. J.: Pair dispersion over an inertial range spanning many decades, Phys. Fluids, 8, 1052–1060, 1996.

Elmegreen, B. and Scalo, J.: Interstellar Turbulence I: Observations and Processes, Ann. Rev. Astron. Astrophys., 42, 211–273, 2004.

Falceta-Gonçalves, D., Lazarian, A., and Kowal, G.: Studies of Regular and Random Magnetic Fields in the ISM: Statistics of Polarization Vectors and the Chandrasekhar-Fermi Technique, Astrophys. J., 679, 537–551, 2008.

Falceta-Gonçalves, D., de Gouveia Dal Pino, E. M., Gallagher, J. S., and Lazarian, A.: Turbulence and the Formation of Filaments, Loops, and Shock Fronts in NGC 1275, Astrophys. J., 708, L57–L60, 2010.

Falceta-Gonçalves, D., Lazarian, A., and Houde, M.: Damping of Magnetohydrodynamic Turbulence in Partially Ionized Gas and the Observed Difference of Velocities of Neutrals and Ions, Astrophys. J., 713, 1376–1385, 2010.

Falgarone, E., Troland, T. H., Crutcher, R. M., and Paubert, G.: CN Zeeman measurements in star formation regions, Astron. Astrophys., 487, 247–252, 2008.

Falgarone, E., Pety, J., and Hily-Blant, P.: Intermittency of interstellar turbulence: extreme velocity-shears and CO emissi Astron. Astrophys., 507, 355–368, 2009.

Fleck, R. C.: On the generation and maintenance of turbulence in the interstellar medium, Astrophys. J., 246, L151–L154, 1981.

Fleck, R. C.: Scaling Relations for the Turbulent, Non-Self-gravitating, Neutral Component of the Interstellar Medium, Astrophys. J., 458, 739–741, 1996.

Fredriksen, Å, Riccardi, C., Cartegni, L., and Pecseli, H.: Coherent structures, transport and intermittency in a magnetized plasma, Plasma Phys. Control. Fusion, 45, 721–733, 2003.

Frisch, U.: Turbulence. The legacy of A. N. Kolmogorov, Cambridge University Press, Cambridge, UK, 1995.

Fung, J. C. H. and Vassilicos, J. C.: Two-particle dispersion in turbulentlike flows, Phys. Rev. E, 57, 1677–1690, 1998.
Gaensler, B. M., Haverkorn, M., Burkhart, B., Newton-McGee, K. J., Ekers, R. D., Lazarian, A., McClure-Griffiths, N. M., Robishaw, T., Dickey, J. M., and Green, A. J.: Low-Mach-number turbulence in interstellar gas revealed by radio polarization gradients, Nature, 478, 214–217, 2011.
Goldreich, P. and Sridhar, S.: Toward a theory of interstellar turbulence, 2: Strong alfvenic turbulence, Astrophys. J., 438, 763–775, 1995.
Goldsmith, P. F., Heyer, M., Narayanan, G., Snell, R., Li, D., and Brunt, C.: Large-Scale Structure of the Molecular Gas in Taurus Revealed by High Linear Dynamic Range Spectral Line Mapping, Astrophys. J., 680, 428–445, 2008.
Goldstein, M.: Major Unsolved Problems in Space Plasma Physics, Astrophys. Space Sci., 277, 349–369, 2001.
Gómez, G. C. and Cox, D. P.: Three-dimensional Magnetohydrodynamic Modeling of the Gaseous Structure of the Galaxy: Setup and Initial Results, Astrophys. J., 580, 235–252, 2002.
Green, D. A.: A power spectrum analysis of the angular scale of Galactic neutral hydrogen emission towards $L = 140$ deg, $B = 0$ deg, Mon. Notice. Roy. Astron. Soc., 262, 327–342, 1993.
Gurnett, D. A., Kurth, W. S, Burlaga, L. F., and Ness, N. F.: In Situ Observations of Interstellar Plasma with Voyager 1, Science, 341, 1489–1492, 2013.
Gustafsson, M., Field, D., Lemaire, J. L., and Pijpers, F. P.: Probing turbulence with infrared observations in OMC1, Astron. Astrophys., 445, 601–616, 2006.
Gwinn, C. R.: Hypersonic acceleration and turbulence of H_2O masters in W49N, Astrophys. J., 429, 241–252, 1994.
Han, J. L.: Magnetic Fields in Our Galaxy: How much do we know? III. Progress in the Last Decade, Chin. J. Astron. Astrophys., 6, 211–217, 2006.
He, K. and Chian, A. C.-L.: Nonlinear dynamics of turbulent waves in fluids and plasmas, Nonlin. Processes Geophys., 12, 13–24, doi:10.5194/npg-12-13-2005, 2005.
Heiles, C.: 9286 Stars: An Agglomeration of Stellar Polarization Catalogs, Astronom. J., 119, 923–927, 2000.
Hellinger, P., Trávnícek, P., Kasper, J. C., and Lazarus, A. J.: Solar wind proton temperature anisotropy: Linear theory and WIND/SWE observations, Geophys. Res. Lett., 33, L09101, doi:10.1029/2006GL025925, 2006.
Hellinger, P., Trávnícek, P. M., Stverák, S., Matteini, L., and Velli, M.: Proton thermal energetics in the solar wind: Helios reloaded, J. Geophys. Res.-Atmos., 118, 1351–1365, 2013.
Hennebelle, P. and Falgarone, E.: Turbulent molecular clouds, Astron. Astrophys. Rev., 20, 55, doi:10.1007/s00159-012-0055-y, 2012.
Heyer, M. H. and Brunt, C. M.: The Universality of Turbulence in Galactic Molecular Clouds, Astrophys. J., 615, L45–L48, 2004.
Heyer, M. H. and Brunt, C. M.: Trans-Alfvénic motions in the Taurus molecular cloud, Mon. Notice. Roy. Astron. Soc., 420, 1562–1569, 2012.
Heyer, M. H., Gong, H., Ostriker, E., and Brunt, C.: Magnetically Aligned Velocity Anisotropy in the Taurus Molecular Cloud, Astrophys. J., 680, 420–427, 2008.
Heyer, M. H., Krawczyk, C., Duval, J., and Jackson, J. M.: Re-Examining Larson's Scaling Relationships in Galactic Molecular Clouds, Ap. J., 699, 1092–1103, 2009.
Hildebrand, R. H., Kirby, L., Dotson, J. L., Houde, M., and Vaillancourt, J.: Dispersion of Magnetic Fields in Molecular Clouds. I, Ap. J., 696, 567–573, 2009.
Hill, A. S., Benjamin, R. A., Kowal, G., Reynolds, R. J., Haffner, M., and Lazarian, A.: The Turbulent Warm Ionized Medium: Emission Measure Distribution and MHD Simulations, Astrophys. J., 686, 363–378, 2008.
Hill, A. S., Joung, M. R., Mac Low, M. M., Benjamin, R. A., Haffner, L. M., Klingenberg, C., and Waagan, K.: Vertical Structure of a Supernova-driven Turbulent, Magnetized Interstellar Medium, Astrophys. J., 750, 104, 2012.
Hily-Blant, P. and Falgarone, E.: Intermittency of interstellar turbulence: parsec-scale coherent structure of intense, velocity shear, Astron. Astrophys., 500, 29–32, 2009.
Horbury, T. S., Forman, M., and Oughton, S.: Anisotropic Scaling of Magnetohydrodynamic Turbulence, Phys. Res. Lett., 101, 175005, doi:10.1103/PhysRevLett.101.175005, 2008.
Horbury, T. S., Wicks, R. T., and Chen, C. H. K: Anisotropy in Space Plasma Turbulence: Solar Wind Observations, Space Sci. Rev., 172, 325–342, 2012.
Houde, M., Vaillancourt, J. E., Hildebrand, R. H., Chitsazzadeh, S., and Kirby, L.: Dispersion of Magnetic Fields in Molecular Clouds. II., Astrophys. J., 706, 1504–1516, 2009.
Houde, M., Rao, R., Vaillancourt, J. E., and Hildebrand, R. H.: Dispersion of Magnetic Fields in Molecular Clouds. III., Astrophys. J., 733, 109, doi:10.1088/0004-637X/733/2/109, 2011.
Hurricane, O. A., Smalyuk, V. A., Raman, K., Schilling, O., Hansen, J. F., Langstaff, G., Martinez, D., Park, H.-S., Remington, B. A., Robey, H. F., Greenough, J. A., Wallace, R., Di Stefano, C. A., Drake, R. P., Marion, D., Krauland, C. M., and Kuranz, C. C.: Validation of a Turbulent Kelvin-Helmholtz Shear Layer Model Using a High-Energy-Density OMEGA Laser Experiment, Phys. Rev. Lett., 109, 155004, doi:10.1103/PhysRevLett.109.155004, 2012.
Imai, H., Deguchi, S., and Sasao, T.: Microstructure of Water Masers in W3 IRS 5, Astrophys. J., 567, 971–979, 2002.
Iroshnikov, P. S.: Turbulence of a Conducting Fluid in a Strong Magnetic Field, Soviet Astron., 7, 566–572, 1964.
Joung, M. K. R. and Mac Low, M.-M.: Turbulent Structure of a Stratified Supernova-driven Interstellar Medium, Astrophys. J., 653, 1266–1279, 2006.
Kim, W.-T. and Ostriker, E. C.: Formation and Fragmentation of Gaseous Spurs in Spiral Galaxies, Astrophys. J., 570, 132–151, 2002.
Koch, P., Tang, Y.-W., and Ho, P. T. P.: Magnetic Field Strength Maps for Molecular Clouds: A New Method Based on a Polarization-Intensity Gradient Relation, Astrophys. J., 747, 79, doi:10.1088/0004-637X/747/1/79, 2012a.
Koch, P., Tang, Y.-W., and Ho, P. T. P.: Quantifying the Significance of the Magnetic Field from Large-scale Cloud to Collapsing Core: Self-similarity, Mass-to-flux Ratio, and Star Formation Efficiency, Astrophys. J., 747, 80, doi:10.1088/0004-637X/747/1/80, 2012b.
Koga, D., Chian, A. C.-L., Miranda, R. M., and Rempel, E. L.: Intermittent nature of solar wind turbulence near the Earth's bow shock: Phase coherence and non-Gaussianity, Phys. Rev. E, 75, 046401, doi:10.1103/PhysRevE.75.046401, 2007.

Kolmogorov, A. N.: The Local Structure of Turbulence in Incompressible Viscous Fluid for Very Large Reynolds' Numbers, Dokl. Akad. Nauk SSSR, 30, 301–305, 1941.

Kolmogorov, A. N.: A refinement of previous hypotheses concerning the local structure of turbulence in a viscous incompressible fluid at high Reynolds number, J. Fluid Mech., 13, 82–85, 1962.

Kowal, G. and Lazarian, A.: Scaling Relations of Compressible MHD Turbulence, Astrophys. J. Lett., 666, L69–L72, 2007.

Kowal, G. and Lazarian, A.: Velocity Field of Compressible Magnetohydrodynamic Turbulence: Wavelet Decomposition and Mode Scalings, Astrophys. J., 720, 742–756, 2010.

Kowal, G., Lazarian, A., and Beresnyak, A.: Density Fluctuations in MHD Turbulence: Spectra, Intermittency, and Topology, Astrophys. J., 658, 423–445, 2007.

Kowal, G., Falceta-Gonçalves, D. A., and Lazarian, A.: Turbulence in collisionless plasmas: statistical analysis from numerical simulations with pressure anisotropy, New J. Phys., 13, 053001, doi:10.1088/1367-2630/13/5/053001, 2011.

Kozak, L. V., Savin, S. P., Budaev, V. P., Pilipenko, V. A., and Lezhen, L. A.: Character of turbulence in the boundary regions of the Earth's magnetosphere, Geomagnet. Aeron., 52, 445–455, 2012.

Kraichnan, R. H.: Inertial-Range Spectrum of Hydromagnetic Turbulence, Phys. Fluids, 8, 1385–1387, 1965a.

Kraichnan, R. H.: Lagrangian-History Closure Approximation for Turbulence, Phys. Fluids, 8, 575–598, 1965b.

Kraichnan, R. H.: Turbulence and Stochastic Processes, edited by: Hunt, J. C. R., Phillips, O. M., and Williams, D., Philos. T. Roy. Soc. A, 434, 65–78, doi:10.1098/rspa.1991.0080, 1991.

Kritsuk, A. G., Norman, M. L., Padoan, P., and Wagner, R.: The Statistics of Supersonic Isothermal Turbulence, Astrophys. J., 665, 416–431, 2007.

Larson, R. B.: Turbulence and star formation in molecular clouds, Mon. Notice. Roy. Astron. Soc., 194, 809–826, 1981.

Lazarian, A.: Tracing magnetic fields with aligned grains, J. Quant. Spectrosc. Ra., 106, 225–256, 2007

Lazarian, A., Eyink, G. L., and Vishniac, E. T.: Relation of astrophysical turbulence and magnetic reconnection, Phys. Plasmas, 19, 012105–012105-8, doi:10.1063/1.3672516, 2012.

Leamon, R. J., Smith, C. W., Ness, N. F., Matthaeus, W. H., and Wong, H. K.: Observational constraints on the dynamics of the interplanetary magnetic field dissipation range, J. Geophys. Res., 103, 4775–4787, doi:10.1029/97JA03394, 1998.

Leão, M. R. M., de Gouveia Dal Pino, E. M., Falceta-Gonçalves, D., Melioli, C., abd Geraissate, F. G.: Local star formation triggered by supernova shocks in magnetized diffuse neutral clouds, Mon. Notice. Roy. Astron. Soc., 394, 157–173, 2009.

Li, H.-B. and Houde, M.: Probing the Turbulence Dissipation Range and Magnetic Field Strengths in Molecular Clouds, Astrophys. J., 677, 1151–1156, 2008.

Li, G., Miao, B., Hu, Q., and Qin, G.: Effect of Current Sheets on the Solar Wind Magnetic Field Power Spectrum from the Ulysses Observation: From Kraichnan to Kolmogorov Scaling, Phys. Rev. Lett., 106, 125001, doi:10.1103/PhysRevLett.106.125001, 2011.

Lighthill, M. J.: The Effect of Compressibility on Turbulence, in: IAU Symp. 2, Gas Dynamics of Cosmic Clouds, Amsterdam, 121–130, 1955.

Lis, D. C., Pety, J., Phillips, T. G., and Falgarone, E.: Statistical Properties of Line Centroid Velocities and Centroid Velocity Increments in Compressible Turbulence, Astrophys. J., 463, 623–629, 1996.

Lithwick, Y., Goldreich, P., and Sridhar S.: Imbalanced Strong MHD Turbulence, Astrophys. J., 655, 269–274, 2007.

Litvak, M. M.: Coherent molecular radiation, Annu. Rev. Astron. Astrophys., 12, 97–112, 1974.

Liu, T., Wu, Y., and Zhang, H.: Molecular Environments of 51 Planck Cold Clumps in the Orion Complex, Astrophys. J. Suppl., 202, 4, doi:10.1088/0067-0049/202/1/4, 2012.

Lo, K. Y.: Mega-Masers and Galaxies, Annu. Rev. Astron. Astrophys., 43, 625–676, 2005.

Mac Low, M. M. and Klessen, R. S.: Control of star formation by supersonic turbulence, Rev. Modern Phys., 76, 125–194, 2004.

Macek, W. M. and Szczepaniak, A.: Generalized two-scale weighted Cantor set model for solar wind turbulence, Geophys. Res. Lett., 35, L02108, doi:10.1029/2007GL032263, 2008.

Macek, W. M., Wawrzaszek, A., and Carbone, V.: Observation of the multifractal spectrum at the termination shock by Voyager 1, Geophys. Res. Lett., 38, L19103, doi:10.1029/2011GL049261, 2011.

Macek, W. M., Wawrzaszek, A., and Carbone, V.: Observation of the multifractal spectrum in the heliosphere and the heliosheath by Voyager 1 and 2, J. Geophys. Res.-Space, 117, A12101, doi:10.1029/2012JA018129, 2012.

Marasco, A., Marinacci, F., and Fraternali, F.: On the origin of the warm-hot absorbers in the Milky Way's halo, Mon. Notice. Roy. Astron. Soc., 433, 1634–1647, 2013.

Markovskii, S. A. and Vasquez, B. J.: Magnetic Helicity in the Dissipation Range of Strong Imbalanced Turbulence, Astrophys. J., 768, 62, doi:10.1088/0004-637X/768/1/62, 2013.

Marsch, E.: Turbulence in the solar wind, in: Reviews in Modern Astronomy, edited by: Klare, G., Springer, Berlin, p. 43, 1990.

Matthaeus, W. and Goldstein, M.: Measurement of the rugged invariants of magnetohydrodynamic turbulence in the solar wind, J. Geophys. Res., 87, 6011–6028, 1982.

Melioli, C., Brighenti, F., a DÉrcole, A., de Gouveia Dal Pino, E. M.: Hydrodynamical simulations of Galactic fountains – II. Evolution of multiple fountains, Mon. Notice. Roy. Astron. Soc., 399, 1089–1105, 2009.

Mininni, P. D., Alexakis, A., and Pouquet, A.: Nonlocal interactions in hydrodynamic turbulence at high Reynolds numbers: The slow emergence of scaling laws, Phys. Rev. E, 77, 036306, doi:10.1103/PhysRevE.77.036306, 2008.

Minter, A. H. and Spangler, S. R.: Observation of Turbulent Fluctuations in the Interstellar Plasma Density and Magnetic Field on Spatial Scales of 0.01 to 100 Parsecs, Astrophys. J., 458, 194–214, doi:10.1086/176803, 1996.

Miranda, R. A., Chian, A. C.-L., and Rempel, E. L.: Universal scaling laws for fully-developed magnetic field turbulence near and far upstream of the Earth's bow shock, Adv. Space Res., 51, 1893–1901, 2013.

Miville-Deschênes, M.-A., Joncas, G., Falgarone, E., and Boulanger, F.: High resolution 21 cm mapping of the Ursa Major Galactic cirrus: Power spectra of the high-latitude H I gas, Astron. Astrophys., 411, 109–121, 2003.

Moisy, F. and Jimenez, J.: Geometry and clustering of intense structures in isotropic turbulence, J. Fluid Mech., 513, 111–133, 2004.

Ng, C. S., Bhattacharjee, A., Munsi, D., Isenberg, P. A., and Smith, C. W.: Kolmogorov versus Iroshnikov–Kraichnan spectra: Consequences for ion heating in the solar wind, J. Geophys. Res., 115, A02101, doi:10.1029/2009JA014377, 2010.

Norman, C. A. and Ferrara, A.: The Turbulent Interstellar Medium: Generalizing to a Scale-dependent Phase Continuum, Astrophys. J., 467, 280–291, doi:10.1086/177603, 1996.

Obukhov, A. M.: Some Specific Features of Atmospheric Turbulence, J. Fluid Mech., 13, 77–81, 1962.

Oppermann, N., Junklewitz, H., Robbers, G., Bell, M. R., Enßlin, T. A., Bonafede, A., Braun, R., Brown, J. C., Clarke, T. E., Feain, I. J., Gaensler, B. M., Hammond, A., Harvey-Smith, L., Heald, G., Johnston-Hollitt, M., Klein, U., Kronberg, P. P., Mao, S. A., McClure-Griffiths, N. M., O'Sullivan, S. P., Pratley, L., Robishaw, T., Roy, S., Schnitzeler, D. H. F. M., Sotomayor-Beltran, C., Stevens, J., Stil, J. M., Sunstrum, C., Tanna, A., Taylor, A. R., and Van Eck, C. L.: An improved map of the Galactic Faraday sky, Astron. Astrophys., 542, A93, doi:10.1051/0004-6361/201118526, 2012.

Pereyra, A. and Magalhães, A. M.: Polarimetry toward the Musca Dark Cloud, I. The Catalog, Astrophys. J., 603, 584–594, 2004.

Poidevin, F., Falceta-Gonçalves, D., Kowal, G., de Gouveia Dal Pino, E. M., and Magalhães, A. M.: Magnetic Field Components Analysis of the SCUPOL 850 m Polarization Data Catalog, Astrophys. J., 777, 112, doi:10.1088/0004-637X/777/2/112, 2013.

Qian, L., Li, D., and Goldsmith, P. F.: ^{13}CO Cores in the Taurus Molecular Cloud, Astrophys. J., 760, 147, doi:10.1088/0004-637X/760/2/147, 2012.

Reid, M. J. and Moran, J. M.: Masers, Annu. Rev. Astron. Astrophys., 19, 231–276, 1981.

Rempel, E. L., Chian, A. C.-L., Preto, A. J., and Stephany, S.: Intermittent chaos driven by nonlinear Alfvén waves, Nonlin. Processes Geophys., 11, 691–700, doi:10.5194/npg-11-691-2004, 2004.

Richards, A. M. S., Cohen, R. J., Crocker, M., and Lekht, E. E.: Merlin and Puschino Observations of H_2O Masers in Outer Galactic SFR S128N, Astrophys. Space Sci., 295, 19–25, 2005.

Richardson, L. F.: Weather prediction by numerical process, Cambridge University Press, 1922.

Richardson, L. F.: Atmospheric Diffusion Shown on a Distance-Neighbour Graph, P. Roy. Soc. Lond. A, 110, 709–737, 1926.

Robishaw, T., Quataert, E., and Heiles, C.: Extragalactic Zeeman Detections in OH Megamasers, Astrophys. J., 680, 981–998, 2008.

Roy, N., Chengalur, J. N., Bharadwaj, S., and Dutta, P.: ASP Conference Series, Vol. 407, proceedings of the conference held 8–12 December 2008, at National Centre for Radio Astrophysics (NCRA), TIFR, Pune, India, edited by: Saikia, D. J., Green, D. A., Gupta, Y., and Venturi, T., Astronomical Society of the Pacific, San Francisco, p. 272, 2009.

Ruiz, L. O., Falceta-Gonçalves, D., Lanfranchi, G. A., and Caproni, A.: The mass-loss process in dwarf galaxies from 3D hydrodynamical simulations: the role of dark matter and starbursts, Mo. Notice. Roy. Astron. Soc., 429, 1437–1449, 2013.

Sahraoui, F., Goldstein, M. L., Belmont, G., Canu, P., and Rezeau, L.: Three Dimensional Anisotropic k Spectra of Turbulence at Subproton Scales in the Solar Wind, Phys. Rev. Lett., 105, 131101, doi:10.1103/PhysRevLett.105.131101, 2010.

Salem, C., Mangeney, A., Bale, S. D., and Veltri, P.: Solar Wind Magnetohydrodynamics Turbulence: Anomalous Scaling and Role of Intermittency, Astrophys. J., 702, 537–553, 2009.

Sánchez-Salcedo, F. J., Santillán, A., and Franco, J.: Turbulence in the Outskirts of the Milky Way, J. Korean Astro. Soc., 40, 171–177, 2007.

Santos-Lima, R., de Gouveia Dal Pino, E. M., Kowal, G., Falceta-Gonçalves D., Lazarian, A., and Nakwacki, M. S.: Magnetic Field Amplification and Evolution in Turbulent Collisionless Magnetohydrodynamics: An Application to the Intracluster Medium, Astrophys. J., 781, 84, doi:10.1088/0004-637X/781/2/84, 2013.

Sarma, A. P., Troland, T. H., Crutcher, R. M., and Roberts, D. A.: Magnetic Fields in Shocked Regions: Very Large Array Observations of H_2O Masers, Astrophys. J., 580, 928–937, 2002.

Schekochihin, A. A and Cowley, S. C.: Magnetohydrodynamics – Historical Evolution and Trends, edited by: Molokov, S., Moreau, R., and Moffatt, H. K., Springer, Berlin, p. 85, 2007.

Schekochihin, A. A., Cowley, S. C., Kulsrud, R. M., Rosin, M. S., and Heinemann, T.: Nonlinear Growth of Firehose and Mirror Fluctuations in Astrophysical Plasmas, Phys. Rev. Lett., 100, 081301, doi:10.1103/PhysRevLett.100.081301, 2008.

Sellwood, J. A. and Balbus, S. A.: Differential Rotation and Turbulence in Extended H I Disks, Astrophys. J., 511, 660–665, 1999.

She, Z. and Levêque, E.: Universal scaling laws in fully developed turbulence, Phys. Rev. Lett., 72, 336–339, 1994.

Solomon, P. M., Rivolo, A. R., Barrett, J., and Yahil, A.: Mass, luminosity, and line width relations of Galactic molecular clouds, Astrophys. J., 319, 730–741, 1987.

Stanimirovic, S. and Lazarian, A.: Velocity and Density Spectra of the Small Magellanic Cloud, Astrophys. J. Lett., 551, L53–L56, 2001.

Stanimirovic, S., Staveley-Smith, L., Dickey, J. M., Sault, R. J., and Snowden, S. L.: The large-scale HI structure of the Small Magellanic Cloud, Mon. Notice. Roy. Astron. Soc., 302, 417–436, 1999.

Strelnitski, V.: Masers as probes of supersonic turbulence, edited by: Chapman, J. M. and Baan, W. A., Proceedings IAU Symposium No. 242, International Astronomical Union, Cambridge Journals Online, 25–31, doi:10.1017/S1743921307012513, 2007.

Strelnitski, V., Alexander, J., Gezari, S., Holder, B. P., Moran, J. M., and Reid, M. J.: H_2O Masers and Supersonic Turbulence, Astrophys. J., 581, 1180–1193, 2002.

Thess, A. and Zikanov, O.: Transition from two-dimensional to three-dimensional magnetohydrodynamic turbulence, J. Fluid Mech., 579, 383–412, doi:10.1017/S0022112007005277, 2007.

Troland, T. H. and Crutcher, R. M.: Magnetic Fields in Dark Cloud Cores: Arecibo OH Zeeman Observations, Astrophys. J., 680, 457–465, 2008.

Uscanga, L., Cantó, J., Gómez, J. F., Anglada, G., Torrelles, J. M., Patel, N. A., Raga, A. C., and Curiel, S.: Statistical Analysis of Water Masers in Star-forming Regions: Cepheus A and W75 N, Astrophys. J., 715, 132–142, 2010.

Vázquez-Semadeni, E., Passot, T., and Pouquet, A.: Influence of Cooling-induced Compressibility on the Structure of Turbulent Flows and Gravitational Collapse, Astrophys. J., 473, 881–893, 1996.

von Hoerner, S.: A method for studying the turbulence of the interstellar matter, Z. Astrophysik, 30, 17–64, 1951.

von Weizsäcker, C. F.: The Evolution of Galaxies and Stars, Astrophys. J., 114, 165–186, doi:10.1086/145462, 1951.

Walker, R. C.: H_2O in W49N, II – Statistical studies of hyperfine structure, clustering, and velocity distributions, Astrophys. J., 280, 618–628, 1984.

Wicks, R. T., Horbury, T. S., Chen, C. H. K., and Schekochihin, A. A.: Anisotropy of Imbalanced Alfvénic Turbulence in Fast Solar Wind, Phys. Res. Lett., 106, 045001, doi:10.1103/PhysRevLett.106.045001, 2011.

Wilczek M.: Statistical and Numerical Investigations of Fluid Turbulence, Ph.D. Thesis, Westfalische Wilhelms Universitat, Munster, Germany, 2010.

Williams, J. P., Blitz, L., and McKee, C. F.: The Structure and Evolution of Molecular Clouds: from Clumps to Cores to the IMF, Protostars and Planets IV, University of Arizona Press, Tucson, p. 97, 2000.

Yoshida, A., Kitamura, Y., Shimajiri, Y., and Kawabe, R.: Dependence of the Turbulent Velocity Field on Gas Density in L1551, Astrophys. J., 718, 1019–1035, 2010.

Zhang, X., Youngung, L., Bolatto, A., and Stark, A. A.: CO(J = 4–3) and [C I] Observations of the Carina Molecular Cloud Complex, Astrophys. J., 553, 274–287, 2001.

Zouari, N. and Babiano, A.: Derivation of the relative dispersion law in the inverse energy cascade of two-dimensional turbulence, Physica D, 76, 318–328, 1994.

6

Alfvén waves in space and astrophysical dusty plasmas

V. Jatenco-Pereira[1], **A. C.-L. Chian**[2,3,4], **and N. Rubab**[5]

[1]Instituto de Astronomia, Geofísica e Ciências Atmosféricas (IAG/USP), Rua do Matão, 1226, 05508-090 São Paulo, SP, Brazil
[2]Paris Observatory, LESIA, CNRS, 92190 Meudon, France
[3]National Institute for Space Research (INPE) and World Institute for Space Environment Research (WISER), P.O. Box 515, 12227-010 São José dos Campos, SP, Brazil
[4]School of Mathematical Sciences, University of Adelaide, Adelaide, SA 5005, Australia
[5]Department of Space Science, Institute of Space Technology (IST), Islamabad, Pakistan

Correspondence to: V. Jatenco-Pereira (vera.jatenco@iag.usp.br)

Abstract. In this paper, we present some results of previous works on Alfvén waves in a dusty plasma in different astrophysical and space regions by taking into account the effect of superthermal particles on the dispersive characteristics. We show that the presence of dust and superthermal particles sensibly modify the dispersion of Alfvén waves. The competition between different damping processes of kinetic Alfvén waves and Alfvén cyclotron waves is analyzed. The nonlinear evolution of Alfvén waves to chaos is reviewed. Finally, we discuss some applications of Alfvén waves in the auroral region of space plasmas, as well as stellar winds and star-forming regions of astrophysical plasmas.

1 Introduction

The dusty plasma has been actively investigated in the past few decades and a number of its applications have been found in laboratory systems (e.g., cold atmospheric plasma for cancer therapy, Keidar et al., 2013; Köritzer et al., 2013; Walk et al., 2013) as well as in space and astrophysical systems (e.g., auroral regions, stellar winds and star-forming regions). The aim of this paper is to review some recent advances in the study of parallel propagation of Alfvén waves and kinetic Alfvén waves in space and astrophysical dusty plasmas as well as nonlinear evolution of Alfvén waves to chaos. The layout of the manuscript is organized as follows: an overview of the theory of Alfvén waves and dispersion characteristics in dusty plasmas is presented in Sect. 2.1. In Sect. 2.2, we discussed the damping of kinetic Alfvén waves. Section 2.3 is devoted to a short overview of the Lorentzian particle velocity distribution function and its effects on Alfvén waves. The nonlinear evolution of large-amplitude Alfvén waves to chaos is analyzed in Sect. 2.4. Various space and astrophysical applications of the interactions of Alfvén waves with dust plasma particles are discussed in Sects. 3 and 4, respectively. Finally, a brief summary is given in Sect. 5.

2 Alfvén waves in dusty plasmas

Dusty plasmas are ionized gases in the presence of microparticles and often are termed as "complex plasmas" or "colloidal plasmas". These dust particles, or dust grains, are highly massive and usually negatively charged due to the higher mobility and the large flux of incoming electrons, and therefore an isolated dust grain in a plasma acts like a floating probe. Orbital motion limited (OML) is one of the ways to determine the charging cross-section on a grain surface pertaining to the laws of conservation of energy and momentum. For an isolated dust grain it can be formulated as $a \ll \lambda_D \ll d$ where a is the dust grain size, d is the the grain average separation ($d = (3/4\pi n_{d0})^{1/3}$, where n_{d0} is the grain number density) and the dusty plasma Debye length of β species particles ($\lambda_{D\beta 0} = (k_B T_\beta / 4\pi n_{\beta 0} q_\beta^2)^{1/2}$, where k_B is the Boltzmann constant). In the case under consideration, we always have $a \ll \lambda_{D\beta 0}$ and for dusty plasma $d \leq \lambda_{D\beta 0}$ and the dust particles can be treated as massive point particles where the effect of neighboring particles is expressive and we can expect collective behavior (de Angelis, 1992). Dusty plasmas are present in many astrophysical environments, as for example the rings of Saturn, cometary tails or interstellar clouds (Goertz, 1989; Verheest, 2000). The grain size is not uniform in many space and astrophysical dusty plasmas and can vary from sub-micron to centimeter scale; therefore, one can consider different dust size distributions. It is well known that the presence of dust particles significantly modifies the electrostatic/electromagnetic modes due to quasi-neutrality and is responsible for some new modes that are associated with dusty dynamics, and the coupling of these tiny dust grains to the magnetic field is very important to determine the wave propagation in the planetary and interstellar media, i.e., ions and dust micro-particulates provide an effective coupling between Alfvén wave and neutral gas in the molecular clouds (Elmegreen and Fiebig, 1993).

2.1 Dispersion relation of Alfvén waves in a dusty plasma

It is well known that the properties of Alfvén and magnetoacoustic waves are radically modified in a dusty medium, especially when the charge on the dust grains fluctuates; when the charged grains acquire a circular motion around the field lines that is defined by the dust cyclotron frequency; and when the waves and the grains experience the same frequency, a resonance occurs that leads to the damping of the waves.

Various modes of propagation in a dusty plasma arise from the fact that frequencies associated with dust particles are smaller than those for electrons and ions. These waves are ultra-low frequency dust modes (Rao, 1996) and are associated with the dust particle inertia. Here, charged dust grains have a collective behavior and take part in the wave dynamics. As examples of these oscillation modes, we could mention dust acoustic waves (DAW) (Rao et al., 1990; D'Angelo, 1995) and electrostatic dust-cyclotron waves (EDCW) (Sorasio and Rosenberg, 2001). Magnetized dusty plasmas support additional electrostatic low-frequency waves involving the dynamics of magnetized/unmagnetized dust grains and magnetized electrons/ions, i.e., dust lower hybrid (DLH) waves with $\omega \ll \omega_{ci}$, which propagate nearly perpendicular to the external magnetic field.

The dispersion relation of Alfvén waves, with frequencies smaller than the ion cyclotron frequency, assuming constant charged particles in a neutral and cold dusty plasma, is given by (Cramer et al., 2002)

$$k_z^2 = u_1 \pm u_2, \tag{1}$$

where

$$u_1 = \frac{\omega^2 \Omega_i^2}{v_{Ai}^2 \left(\Omega_i^2 - \omega^2\right)} + \frac{\omega^2 \omega_{max}^2}{s v_{Ad}^2} \int_1^{a_m} \frac{f(\mathcal{R})\, d\mathcal{R}}{\mathcal{R}\left(\omega_{max}^2/\mathcal{R}^4 - \omega^2\right)}, \tag{2}$$

and

$$u_2 = \frac{\omega^3 \Omega_i}{v_{Ai}^2 \left(\Omega_i^2 - \omega^2\right)} + \frac{\omega^3 \omega_{max}}{s v_{Ad}^2} \int_1^{a_m} \frac{\mathcal{R} f(\mathcal{R})\, d\mathcal{R}}{\omega_{max}^2/\mathcal{R}^4 - \omega^2}. \tag{3}$$

In the equations ω is the angular wave frequency, Ω_i is the ion cyclotron frequency, v_{Ai} and v_{Ad} are the Alfvén speeds of the ions and dust, respectively, $s = c_4 \ln(a_m)$ where $a_m = a_{max}/a_{min}$ (a_{min} and a_{max} are the minimum and the maximum dust radii, respectively), $c_4 = 3/\left(1 - a_m^{-3}\right)$ and c is the speed of light. Also, $\omega_{max} = q_{min}B/(m_{min}c)$ is the maximum dust cyclotron frequency (i.e., for the smallest dust grain with mass m_{min} and charge q_{min}). The need for m_{min} and q_{min} comes from the fact that considering a Maxwellian velocity distribution for electrons and ions, in the equilibrium situation where the dust charge does not vary with time, the electric currents on the dust surface must be identical (Vladimirov, 1997). As a result, for grain radius $a > 10^{-7}$ cm, the dust charge increases linearly with the grain size.

The right-hand polarized wave (positive sign in Eq. 1) is the mode damped by the dust resonance. In the case of $\omega_{min} < \omega < \omega_{max}$, the integral in the particle radii has singularities, whose residues give the complex part of the wave number ($k_z \equiv k_{\mathcal{R}} + i k_{\mathcal{C}}$) and lead to the dust-cyclotron damping of the waves, with damping length $L \equiv L(\omega) = 2\pi/k_{\mathcal{C}}(\omega)$. If we consider a distribution of grain sizes, the waves are damped in a broad band of resonance (Vidotto and Jatenco-Pereira, 2006).

2.2 Damping of kinetic Alfvén waves

Alfvén waves are a fundamental physical phenomenon in magnetized plasma that contributes to a wide variety of physical processes in space plasmas, for example, turbulence, plasma heating, acceleration along the field lines, wave particle interactions and generation of geomagnetic perturbations. The interaction of the wave with thermal ions is one of the important features in many astrophysical regions, such as solar corona/solar wind, for transporting thermal energy (Yukhimuk et al., 2000). Kinetic Alfvén waves (KAWs) are small-scale dispersive Alfvén waves, differ drastically from MHD Alfvén waves and appear when the cross field scales are comparable to ion gyroradius.

A kinetic Alfvén wave is an important electromagnetic wave that transports electromagnetic energy in many space and astrophysical regions (Nakano et al., 2002; Voitenko et al., 2003). The KAW can transfer the wave energy to electrons via Landau damping, which is the collisionless damping of low-frequency waves that occurs when thermal electrons travel along the magnetic field lines. In this process the particles gain kinetic energy at the expense of the wave if the distribution function has a negative slope in the range $v \sim \omega/k$ and it results in the heating of plasmas or acceleration of electrons along the magnetic field direction (Cramer, 2001).

In many space and astrophysical regions, the wave properties such as the wave dispersion and the onset of Landau/cyclotron damping are dependent on the plasma beta (Gary and Borowsky, 2004).

The wave-particle interaction in KAWs depends sensitively on the plasma β parameter (Wu et al., 1996). Hasegawa and Cheng (1976) showed that in a low beta plasma $1 > \beta > m_e/m_i$, the wave is weakly damped due to ion Landau damping leading to ion heating (Hasegawa and Cheng, 1976), whereas if electron temperature is comparable to ion temperature, nonlinear Landau damping occurs. For some parameter regimes where $\beta \sim m_e/m_i$, the strong electron Landau damping will lead to electron heating (Wu and Fang, 1999).

The dissipation of KAWs by electron Landau damping has been suggested to play a significant role in the damping of interplanetary magnetic field fluctuations and to explain the slope of the magnetic turbulence spectrum where the spectral break is believed to cause the collisionless damping at short scales, which converts the turbulent energy into thermal energy.

Zhao et al. (2011) discussed the damping mechanism of KAWs in the dense plume structure of coronal holes and showed that KAWs are strongly damped in this region due to Landau damping when $\lambda_e k_\perp < 0.3$.

Wu and Fang (1999, 2003, 2007) and Wu et al. (1996) showed that the parallel electric field of the KAW can lead to effective electron heating due to kinetic damping or Ohmic dissipation, which can further dissipate the KAW energy to sustain the temperature of the coronal loop plasmas. This dissipation occurs due to the parallel electric field of the KAW, which helps the local electron heating and acceleration mechanism to balance the additional radiative loss of the bright plume.

The presence of dust grains in many space/astrophysical regions may also play an important role in the dynamics of Alfvén waves, i.e., breaking down of frozen approximation in field conditions (Kamaya and Nishi, 2000). The coupling of charged dust grains to plasma through electric or magnetic fields can affect the spectrum of low-frequency waves through dust inertia and cyclotron resonance. The damping of KAWs in dusty plasma has been studied rigorously by many authors. Das et al. (1996) studied the KAW by retaining the finite Larmor radius of dust grains and showed that unlike MHD Alfvén waves, the KAW propagate across the magnetic field and because of its coupling to electrostatic modes, i.e., dust acoustic mode, and it undergoes Landau damping for each plasma species (Das et al., 1996)

$$\omega^2 = k_\parallel^2 V_{\mathrm{DA}}^2 \left[1 + k_\perp^2 \rho_{\mathrm{d}}^2 \left(\frac{3}{4} + \frac{T_{\mathrm{e}}}{T_{\mathrm{d}}} \right) \pm \frac{i\pi n_{\mathrm{d}0} \omega_{\mathrm{cd}}^2 \left(\frac{r_{\mathrm{d}}^2}{\lambda_{\mathrm{Dio}}^2} + \frac{r_{\mathrm{d}}^2}{\lambda_{\mathrm{Deo}}^2} \right) \left(1 + \frac{3}{4} k_\perp^2 \rho_{\mathrm{d}}^2 \right)}{\omega_{\mathrm{pd}}^2 k_\parallel \left(\omega - k_\parallel v_\parallel \right)} \right], \quad (4)$$

where $V_{\mathrm{DA}} = c\omega_{\mathrm{cd}}/\omega_{\mathrm{pd}}$ is the dust Alfvén speed, r_{d} is the radius of dust grains, $\rho_{\mathrm{d}} = C_{\mathrm{D}}/\omega_{\mathrm{cd}}$, $\left(C_D = \sqrt{T_{\mathrm{e}}/m_{\mathrm{d}}}\right)$ is the dust Larmor radius and $\omega_{\mathrm{cd}} = Q_{\mathrm{d}0} B_0 / m_{\mathrm{d}}$ is the dust cyclotron frequency with $Q_{\mathrm{d}0}$ being the mean charge of the dust particles.

Rubab et al. (2009) showed that in a dusty Lorentzian plasma the Alfvén wave is coupled to the dust acoustic wave and the damping due to dust charge fluctuation is insensitive to the form of distribution function in the ultra low-frequency regime

$$\omega^2 = k_\parallel^2 V_{\mathrm{DA}}^2 \left[1 + \frac{2\kappa - 3}{2\kappa - 1} \left(1 + \frac{\omega_{\mathrm{pd}}^2}{c^2 k_\perp^2} \right) k_\perp^2 \rho_{\mathrm{d}}^2 \pm i\pi n_{\mathrm{d}0} \frac{r_{\mathrm{d}}^2}{k_\parallel} \right], \quad (5)$$

where $\rho_{\mathrm{d}} = C_D/\omega_{\mathrm{cd}}$, $\left(C_D = \sqrt{T_{\mathrm{eff}}/m_{\mathrm{d}}}\right)$, and $T_{\mathrm{eff}} = n_{\mathrm{d}0} Z_{\mathrm{d}}^2 \left(T_{\mathrm{e}} n_{\mathrm{i}0} + T_{\mathrm{i}} n_{\mathrm{e}0} \right) / n_{\mathrm{i}0} n_{\mathrm{e}0}$.

However, Gaelzer et al. (2010) focused on high-frequency regimes with superthermal character of electrons and reported a significant modification of the dispersive and absorption characteristics. The absorption coefficient increases due to kappa-distributed electrons, which in turn have a large effect on the Whistler mode.

Gaelzer et al. (2009) studied extensively the properties of oblique Alfvén waves in a Maxwellian plasma with dust charge fluctuation effects using kinetic theory. They focused on the competition between various damping mechanisms, like Landau damping and damping due to charge variations where inelastic collisional frequency between the dust and plasma species plays a vital role in the wave absorption and hence the damping rates.

Dust grains play a crucial role in most space/astrophysical regions and are found to affect the growth/damping of many electromagnetic waves drastically. In a collisional dusty plasma the damping rate is found to be proportional to the collision frequency between dust and ions (Yang et al., 2012). It has been shown by many authors (Khan et al., 2013; Zubia et al., 2007) that the collisionless Landau damping is negligibly small for nearly perpendicular propagation (small $k_\parallel$), while dust charge fluctuation damping is the main mechanism for the damping of kinetic Alfvén waves, which could be a reason for plasma heating. de Juli et al. (2005) also explained that the Landau damping is negligible for the transverse ordinary Alfvén waves propagating parallel to the magnetic field.

2.3 Alfvén waves in Lorentzian plasmas

Velocity distribution functions with suprathermal tails in space and astrophysical plasmas are ubiquitous and persistent features and have been confirmed by many spacecraft missions. In kinetic and thermodynamical equilibrium the Maxwell–Boltzmann distributions come out as a solution of the Boltzmann collision term. Nevertheless, in a collisionless system exhibiting strong correlation and interactions on large scales, the Fokker–Planck approach happens to be invalidated. Treumann (1999), in persuit of the same goal, established the kinetic foundation of Lorentzian statistical mechanics to show that power-law velocity space distributions are a particular thermodynamic equilibrium state. According to the Lorentzian statistical mechanics, the power or the spectral index contains the information about the tangible turbulent system whose entropy is believed to increase slowly due to internal dissipation processes to maintain a stationary state. When the correlations in the system are absent, the nonthermal distribution approaches the classical Boltzmann distribution.

In most of the theoretical investigations reported so far, the particle distribution has been assumed to be Maxwellian. Spacecraft measurements of electron energy spectra and in situ observations in the near-Earth space plasma clearly indicate the presence of suprathermal particle populations. These particles may arise due to the effect of external forces acting on the natural space environments or due to wave-particle interaction. The measurements of velocity space and energy distribution functions in most of the realistic situations differ significantly from Fokker–Planck solutions. These observed distributions contain a plentiful supply of suprathermal particles, i.e., particles that move faster than the thermal speed and are represented by the family of kappa or Lorentzian distribution, which obeys a power law in particle speed with high-energy tails. Such a distribution gives a best fit to the observed Maxwellian-like distributions. The presence of suprathermal particles has prompted a number of investigations (Treumann, 1999; Collier, 1993; Vasyliunas, 1968). The mathematical form of isotropic kappa distribution in the literature is

$$f^{\kappa}(r,v) = \frac{n}{2\pi\left(\kappa v_{\text{th}}^2\right)}\frac{\Gamma(\kappa+1)}{\Gamma\left(\kappa-\frac{1}{2}\right)\Gamma\left(\frac{3}{2}\right)}\left(1+\frac{v^2}{\kappa v_{\text{th}}^2}\right)^{-\kappa-1}, \quad (6)$$

where $v_{\text{th}}^2 = \left(\frac{2\kappa-3}{\kappa}\right)\frac{kT}{m}$ is the thermal velocity of the particles, m is the mass of the particles of species, n their number density, T their equivalent temperature, and Γ is the Gamma function. The free parameter κ ($\kappa > 3/2$) describes the deviation of the turbulent nonthermal distribution function from the Boltzmann distribution and is responsible for the shape of the distribution function and hence the slope of the energy spectrum of the suprathermal particles forming the tail of the velocity distribution function. A low value of κ describes a power-law distribution with a strongly accelerated and hard spectrum collisionless plasma. For the large values κ, the function degenerates into a Maxwellian. Such deviations from Maxwellian distributions are expected to exist in low-density plasma where binary collisions are sufficiently rare.

In natural (space and astrophysical) or fusion plasmas, non-Maxwellian distributions originate from wave–plasma interactions, runaway electrons, turbulence, and the fusion of Alpha particles. Suprathermal radiation field and highly energetic beam sources are used for heating. Super-diffusion processes have been proposed to explain the suprathermal tails observed in many space plasmas (Treumann, 1997).

The presence of a high-energy tail component in a kappa distribution effectively changes the rate of resonant energy transfer between particles and plasma waves or by the action of plasma wave fluctuations in the process of transferring free energy to the wave and hence generating instabilities (Lazar et al., 2011). Wave absorption can increase or decrease with respect to the equivalent Maxwellian distribution function. In addition, plasma stability can either be enhanced or degraded.

The dispersion properties of low-frequency waves in a magnetized plasma by using kappa distribution has been derived by a number of authors. For instance, Rubab et al. (2009, 2010, 2011) employed kappa distribution to study kinetic Alfvén waves by showing that the cross field instability is found to be insensitive to the form of distribution function, while for parallel propagation ($V_0 \| B_0$) has a strong influence on the growth rates, i.e., for low values of kappa, the instability reduces, while by increasing the spectral index ($\kappa > 5$), the amplitude of the unstable region increases. They also showed that the parallel phase velocity of the dispersion relation remains unchanged, while nonthermality is found to be more effective in the perpendicular direction, which can be shown from the following relation:

$$\omega^2 = k_{\|}^2 V_A^2\left[1+\frac{2\kappa-3}{2\kappa-1}k_{\perp}^2\rho_{\text{d}}^2 \pm i\pi n_{\text{d}0}\frac{r_{\text{d}}^2}{k_{\|}}\right]. \quad (7)$$

The same has been investigated by Watt and Rankin (2007) by showing that the parallel electron acceleration due to inertial Alfvén wave remains insensitive to parallel phase velocity by using kappa distribution and enumerated that wave particle interactions are enhanced, indicating that Landau damping increases for suprathermal particles. Furthermore, electron energy flux is also found to be dependent on kappa (Pierrard et al., 2001). Kinetic Alfvén solitary waves in a double layer structure have been studied by Gogoi and Khan (2010) using the kappa distribution function and it is found that the spectral index plays a significant role in determining the shape and size of solitons in double layers.

Observational data from the FREJA wave experiment showed that the Earth auroral plasma energization observed nonlinear soliton-type structures, which can be interpreted as solitary kinetic Alfvén waves (Wahlund et al., 1994; Louarn et al., 1994). Further, the particle distribution function in the auroral source regions with high-energy tails $\kappa < 7$ has been observed (Christon et al., 1988; Kletzing et al., 2003). On the basis of theory and observations the data of solar winds suggest that such distributions give best fit curves. The existence of suprathermal distributions of charged particles at different heliospheric distances in the solar wind and in other space and planetary environments has been regularly confirmed by Interball/Aurora probes (Ermakova and Antonova,

2006), Helios, (Prested et al., 2008) Ulysses (Zouganelis, 2008), Cassini (Schippers et al., 2008), and Wind observations (Salem et al., 2006). The presence of suprathermal charged particles has been vindicated in the magnetosphere of other planets such as Jupiter's megnetosheath (Krimigis et al., 1981; Mauk et al., 2004), Mercury (Christon, 1987), Neptune (Krimigis et al., 1989), and Saturn (Sittler Jr. et al., 1983). Rudakov et al. (2011) showed that the non-Maxwellian distribution function has a significant effect on the dynamics of solar wind plasmas. In particular, they showed that linear Landau damping leads to the formation of a plateau in the parallel electron distribution function, which diminishes the Landau damping rate significantly, and generalized nonlinear scattering of waves by plasma particles to short wavelengths. The incorporation of these effects leads to the steepening of the wave spectrum between the inertial and the dissipation ranges with a spectral index between 2 and 3, as observed in the solar wind plasmas.

2.4 Alfvén chaos

Nonlinear dust Alfvén waves were investigated by Verheest (1994), where the dust dynamics is modeled by a number of cold, highly negatively charged and very massive fluids, in addition to electrons and protons; he showed that several low-frequency modes occur that are typical for the dust components, some of them described by the derivative nonlinear Schrödinger equation (DNLS) for electromagnetic waves. Verheest and Meuris (1996) showed that fluctuations in the charges of dust grains, due to capture or liberation of electrons and protons, induce density and momentum losses or gains in dusty plasmas, giving rise to wave damping or growth; in an intermediate regime the dust charging influences the nonlinear development of dust Alfvén waves, leading to a DNLS with a source term. Shukla and Verheest (2003) showed that nonlinear dynamics of modulated dust Alfvén wave packets is governed by a DNLS that admits localized envelope solitons; for dusty plasma parameters relevant to interstellar molecular clouds, the soliton width is on the order of a fraction of a parsec. Verheest (2004) applied the reductive perturbation theory to derive a DNLS to describe the nonlinear evolution of parallel propagating electromagnetic waves in multispecies plasmas; he showed that the DNLS cannot have stationary solitary waves, only envelope solitons. Verheest and Cattaert (2004) derived a DNLS for parallel electromagnetic modes in plasmas containing polydisperse charged dust; they showed that when modeling the charged dust by a power-law distribution in planetary rings and other astrophysical systems, it depends very much on the power-law index whether the smaller or the larger grains are more important. Pandey et al. (2008) studied the nonlinear wave propagation in collisional dusty plasma when electrons and ions are highly magnetized so that the Lorentz force acting on the plasma particles dominates its collision with the dust and charged dust remains weakly magnetized; they showed that large-amplitude waves can be easily excited in such a collisional dusty medium and can be described by the DNLS, whose soliton solutions explain the parsec scale structures in astrophysical plasmas.

The nonlinear evolution of a large-amplitude Alfvén wave propagating along an ambient magnetic field in the x direction is described by the following derivative nonlinear Schrödinger equation (Hada et al., 1990; Chian et al., 1998, 2006)

$$\partial_t b + \alpha \partial_x (|b|^2 b) + i\,(\mu + i\eta)\,\partial_x^2 b = S(x,t)\,, \tag{8}$$

where $b = b_y + i b_z$ is the complex transverse wave magnetic field normalized to the constant ambient magnetic field B_0, η is a dissipative scale length, time t is normalized to the inverse of the ion cyclotron frequency $\omega_{\rm ci} = eB_0/m_{\rm i}$, the space coordinate x is normalized to $c_A/\omega_{\rm ci}$, $c_A = B_0/(\mu_0\rho_0)^{1/2}$ is the Alfvén velocity, $\alpha = 1/[4(1-\beta)]$, $\beta = c_S^2/c_A^2$, $c_S = (P_0/\gamma\rho_0)^{1/2}$ is the acoustic velocity and μ is the dispersive parameter. The external driving force $S(x,t) = A\exp(ik\phi)$ is a monochromatic left-hand circularly polarized wave with a wave phase $\phi = x - Vt$, where V is a constant wave velocity, and A and k are arbitrary constants.

Low-dimensional chaos of DNLS has been studied by Hada et al. (1990), Buti (1997), Chian et al. (1998, 2006, 2007) and Rempel et al. (2007) by carrying out a dynamical analysis of the stationary solutions of Eq. (8) in the driver frame. This system exhibits a wealth of information on the nonlinear dynamics of Alfvén waves such as the routes from order to chaos via local (period-doubling and saddle-node) bifurcations and global bifurcations (interior crisis and boundary crisis). Although Eq. (8) was derived for a two-fluid plasma, its mathematical structure is very similar to the DNLS in dusty plasmas. Hence, the results of chaos in DNLS in two-fluid plasmas are relevant for DNLS in dusty plasmas. Buti (1997) demonstrated that even a small fraction of charged dust grains can suppress chaos in Alfvén systems that are chaotic in the absence of dust particles. Note, however, that this analysis was performed in the absence of dissipation and noise. Since a dynamical system is sensitive to the variation of the physical parameters, the presence of dissipation and noise may alter the result of Buti (1997) significantly.

The effects of noise on Alfvén chaos were investigated by Rempel et al. (2006, 2008). By adding a Gaussian noise to the stationary solutions of the DNLS, Rempel et al. (2006) studied the occurrence of extrinsic transients and attractor hopping in a multistable regime of Alfvén waves. The complex dynamics of Alfvén waves described by DNLS was studied by Rempel et al. (2008) when a nonlinear Alfvén wave is driven towards an intermittent regime by the addition of noise. The effects of Gaussian and non-Gaussian noise were compared. In the intermittent regime, a nonlinear Alfvén wave exhibits random qualitative changes in its dynamics as the result of competition between three co-existing attractors and a chaotic saddle embedded in the fractal basin boundary.

Low-dimensional chaos of DNLS can also be studied by considering Alfvén modulational wave coupling processes (Sánchez-Arriaga et al., 2009; Miranda et al., 2012). Sánchez-Arriaga et al. (2009) studied a truncation model of DNLS with three resonant traveling waves. They showed that the modulational instability criterion is in agreement with the existence of two hyperbolic fixed points joined by a heteroclinic orbit that forces the exchange of energy among the three waves. In the presence of dissipative and growth terms, the system exhibits very complex dynamics including multi-attractors and period-doubling bifurcation to chaos. Miranda et al. (2012) studied the truncated three-wave model of DNLS involving a linearly unstable Alfvén wave and two linearly damped Alfvén waves with different damping rates. The maximum Lyapunov exponent was computed as a function of the damping rates in a two-parameter phase space, and shrimp-shaped self-similar structures in the parameter phase space were identified. By varying the damping rate of the low-frequency wave, a bifurcation diagram was constructed to show saddle-node bifurcation and interior crisis associated with a periodic window. Chaotic saddles and their stable and unstable manifolds were detected, and it was shown that the interaction between two chaotic saddles via the coupling of unstable periodic orbits can lead to crisis-induced intermittency.

The aforementioned studies on Alfvén chaos elucidate how Alfvén intermittency appears in space and astrophysical plasmas, such as the magnetic intermittent turbulence observed in the solar wind (Koga et al., 2007; Sahraoui, 2008; Chian and Miranda, 2009; Chian and Muñoz, 2011), due to nonlinear dynamical phenomena such as the type-I intermittency induced by a saddle-node bifurcation, the crisis-induced intermittency induced by an interior crisis, and the extrinsic intermittency induced by a noise (Hada et al., 1990; Chian et al., 1998, 2006, 2007; Rempel et al., 2006, 2008; Sánchez-Arriaga et al., 2009; Miranda et al., 2012).

3 Alfvén waves: space plasma applications

Alfvén waves are responsible for a variety of physical processes in space plasmas as described in Sect. 2. Hence space plasmas provide an ideal laboratory to study the properties of this mode. As an illustration, we review briefly some observations of Alfvén waves in the Earth's auroral region.

Kinetic Alfvén waves play an important role in auroral particle acceleration (Chaston et al., 2000, 2006). Field aligned currents (FACs) are the major source of ionosphere–magnetosphere coupling and the formation of auroras, as the change in the auroras is associated with the change in the auroral currents. The auroral acceleration not only occurs due to the closure of FACs (Paschmann et al., 2003), but also in the case of empowering Alfvénic auroras through the action of free energy on a small scale (Chaston et al., 2008).

Chaston and Seki (2010) simulated a 3-D model to structure small-scale auroras by considering two different scenarios, where at the electron inertial scale slippage between plasma and geomagnetic field lines becomes significant, leading the thin auroral sheet towards Kelvin–Helmholtz instability and hence the generation of vortices, while the other setup involves a resistive layer between ionosphere and magnetosphere.

Measurements from the FAST satellite (Chaston et al., 2000, 2006) and the POLAR spacecraft (Wygant et al., 2000) have speculated the existence of small-scale Alfvén waves that carry a large net Poynting flux along magnetic field lines towards the Earth. These are the intense and short-lived bursts of precipitating suprathermal electron flux ($E <$ 1 keV). Chaston et al. (2003) developed numerical simulations that confirmed that the downward energy flux of electrons accelerated by inertial Alfvén waves is responsible for exciting visible auroras.

The phenomenon associated with the Alfvén waves in auroral region can also be examined by in situ measurements where the fluctuating electric and magnetic fields and their correlation for the propagation of Alfvén waves can be determined. The parallel current that closes in the ionosphere is expected to be due to these fields in quasi-static structures. The electrons accelerated through such structures, where the field line potential is small as compared to the variation in potential, are shifted in energy, which show first a maximum in energy spectrum and then a decrease, following an inverted-V feature (Frank and Ackerson, 1972). It has been shown by Stasiewicz et al. (2000) that for quasi-static auroral arc-like structures, the magnetic field has a direct relation to field aligned currents, given by

$$\frac{1}{\mu_0}\frac{\partial b_\perp}{\partial s} = J_\perp = \sigma_{\mathrm{p}} E_\perp, \tag{9}$$

where σ_{p} is the Pedersen conductivity in the ionosphere.

Reimei spacecraft observations by Asamura et al. (2009) revealed spatial–temporal evolution of small-scale Alfvén wave generation in auroral acceleration regions. They observed the dispersive electron burst, auroral structures driven by inverted-V electrons and the distribution function responsible for driving them. They concluded that inverted-V electron population is responsible for auroral luminosity, while low-energy electrons assemble information from the regions where inverted-V events take place.

Observational data from the FREJA wave experiment showed a close connection between Alfvén and ion acoustic wave activity within auroral energization regions. Wahlund et al. (1994) found Alfvénic signatures that appear more electrostatic-like. The waveforms contain features with scales on the order of the ion gyroradius where the kinetic (thermal) effects are expected to play a role in high-frequency electrostatic turbulence up to 15 kHz. They also observed that the transverse scale of the spikes is on the order of the ion inertial scale that is comparable to a discrete auroral arc and showed that the Earth auroral plasma energization is associated with nonlinear soliton-type structures that can be interpreted as solitary kinetic Alfvén waves. Based on this data a number of authors identified electromagnetic waveforms having characteristics of Alfvén waves and suprathermal electron bursts (Wahlund et al., 1998; Gary et al., 1998).

The particle distribution function in the auroral source regions with high-energy tails ($\kappa < 7$) is also observed (Christon et al., 1988; Kletzing et al., 2003); therefore, it may be useful to study the Lorentzian inertial/kinetic Alfvén waves in this region.

4 Alfvén waves: astrophysical plasma applications

In this section we present models/results related to Alfvén wave dynamics in stellar winds and star formation where dust grains are embedded.

4.1 Stellar winds

Alfvén waves might play an important role in driving stellar winds in many regions of the HR diagram. For cool giant and supergiant stars, the most promising mechanism is an outward directed flux of Alfvén waves. This mechanism can explain both the high mass-loss rate of $\sim 10^{-8} - 10^{-6}\,\mathrm{M_\odot\,yr^{-1}}$ and the low terminal velocities of $\sim 20 - 80\,\mathrm{km\,s^{-1}}$ of these winds. It is a characteristic of cool winds to have terminal velocities u_∞ less than the escape velocity at the stellar surface, v_{e0}, and the observations show that $u_\infty \sim v_{e0}/2$. The waves can produce significant mass loss even in stars that do not have hot coronas or strong radiative fluxes. This treatment was previously developed by Hollweg (1973); Hartmann and MacGregor (1980); Hartmann and Avrett (1984); Jatenco-Pereira and Opher (1989), for example.

In the solar corona the presence of Alfvén waves was confirmed observationally (Tomczyk et al., 2007). There is direct evidence for Alfvén waves in the solar wind: waves are observed as large perturbations in the magnetic field and as negligible perturbations in density (Chian et al., 1998, 2006), and if there are oscillations in the magnetic field at the base of the wind, Alfvén waves will be generated. These waves propagate outward, dissipate their energy and transfer the momentum to accelerate the wind. It is known that the coronal holes are the source of the high-speed solar wind streams at the Earth's orbit. The measurement of particle fluxes mapped back to coronal holes implies mass fluxes at the base of a coronal hole that are 5–10 times greater than the mass flux obtained by assuming a spherically symmetric solar wind (Hundhausen, 1977). For coronal holes in the Sun, the solid angle for a radial geometry, Ω_0, is observed to suffer an expansion of from Ω_0 at the surface (i.e., at $r = R_\odot$) to $\sim 7\,\Omega_0$ at $r = 3R_\odot$ (Munro and Jackson, 1977). Beyond $3R_\odot$, the edges of the coronal hole are radial and a net non-radial expansion factor $F = \Omega/\Omega_0 = 7.26$ is found. For cool giant and supergiant stars it is assumed that $F = 10$ (Jatenco-Pereira and Opher, 1989).

It was shown by Cramer et al. (2002) that the dispersion relation of Alfvén waves is modified in a dusty plasma, because charged grains in a magnetized plasma are highly coupled to the waves due to cyclotron resonances. Since it is observed that the winds of late-type supergiant stars contain significant amounts of dust, the Alfvén waves can be dissipated in these winds. In order to evaluate how this damping mechanism can accelerate the wind, Vidotto and Jatenco-Pereira (2006) proposed a model that considers an outward flux of Alfvén waves generated at the stellar surface that is damped only due to the interaction with dust grains. They studied a K5 supergiant star with the following magnetic field configuration: from the surface, the magnetic field configuration is divergent and after a transition radius r_t, it becomes radial. If $A(r)$ is the cross section of the geometry at a radial distance r, then

$$A(r) = \begin{cases} A(r_0)(r/r_0)^S & \text{if } r \le r_t \\ A(r_0)(r_t/r_0)^S\,(r/r_t)^2 & \text{if } r > r_t, \end{cases} \tag{10}$$

where S is a parameter that determines the divergence of the geometry. The mass, energy, and momentum equations, which were solved simultaneously by taking into account both the radiative losses and the wave heating in the energy equation, determine the temperature.

Vidotto and Jatenco-Pereira (2006) assumed that the wave energy density is dissipated as follows:

$$\epsilon(\omega) = \epsilon_0(\omega)\frac{M_0}{M}\left(\frac{1+M_0}{1+M}\right)^2 \exp\left\{-\int_{r_0}^{r}\frac{1}{L}\,\mathrm{d}r'\right\}, \tag{11}$$

where $M_0 = u_0/v_{A0}$ is the initial Alfvén–Mach number, $M = u/v_A$ is the Alfvén–Mach number, the Alfvén speed is given by $v_A = (B/\sqrt{4\pi\rho})$, and L is the damping length. They also assumed that $\epsilon_0(\omega)$, the initial wave energy density, obeys a power law as in the Sun (Tu et al., 1989)

$$\epsilon_0(\omega) = \epsilon_0(\omega_0)\left(\frac{\omega}{\omega_0}\right)^{-0.6}, \tag{12}$$

and the following initial parameters for a K5 supergiant star: $r_0 = 400\,\mathrm{R_\odot}$, $M_\star = 16\,\mathrm{M_\odot}$, initial Alfvén wave flux $\phi_{A_0} = 5.5\times10^6\,\mathrm{erg\,cm^{-2}\,s^{-1}}$, $\rho_0 = 1.07\times10^{-13}\,\mathrm{g\,cm^{-3}}$, $B_0 = 10\,\mathrm{G}$, $T_0 = 3500\,\mathrm{K}$, and $S = 5.0$. They also considered a distribution function of dust grain sizes, as first proposed by Mathis et al. (1977) for grains in the interstellar medium

$$f(a) = \frac{p-1}{1-a_m^{1-p}}\,a^{-p}, \tag{13}$$

where $a_m = a_2/a_1$ with a_1 being the minimum dust grain radius, and a_2 the maximum dust grain radius. $a_1 = 5.0\times10^{-7}$

and $a_2 = 2.5 \times 10^{-5}$ cm (Rodgers and Glassgold, 1991) and $p = 4$ were adopted.

Vidotto and Jatenco-Pereira (2006) obtained the wind velocity and temperature profiles of a K5 supergiant star compatible with the observations (e.g., the terminal velocity lower than the surface escape velocity and a rise in the wind temperature as the distance increases from the stellar surface similar to the solar wind), which shows that the dust-cyclotron damping mechanism can be very promising in order to accelerate the wind.

4.2 Star formation regions

The literature contains numerous studies comprising theoretical and numerical simulations and concerning discs around the young stellar objects. Usually the central object provides all of the gravitational force on the disc and the orbit of the material in the disc is close to circular (Keplerian) and generally characterized as differential. The ambient of the cloud that gives rise to the process of star formation contains turbulence, magnetic field and dust that must be present in the later stages of pre-main-sequence stellar evolution (e.g., Schulz, 2005).

The magneto-rotational instability (MRI, Balbus and Hawley, 1991) is probably the mechanism responsible for magneto-hydrodynamic (MHD) turbulence that leads to disc accretion: it provides a necessary torque for the disc's particles to lose angular momentum and to move towards the central object. The concept of a quiescent layer incapable of allowing the operation of the MRI in a protostar disc was proposed by Gammie (1996). In this *dead zone*, the disc is not sufficiently ionized, making it difficult for the gas to be magnetically coupled to the field lines. However, if the ionization degree is sufficiently high, the MRI can operate and the disc becomes turbulent (Adams, 2010).

In a series of papers, Turner et al. (2007) and Turner and Sano (2008) used the minimum-mass solar nebula as the disc model to study under which conditions the dead zone can be eradicated, leaving the entire disc active. They showed that magnetic stresses can occur in the dead zone and that the interior of the disc retains a weak coupling to the magnetic field. Jatenco-Pereira (2013) investigated the damping of Alfvén waves as an extra non-thermal source of heating that could increase the midplane disc temperature. Although this non-thermal heating mechanism had already been suggested by Vasconcelos et al. (2000), it was not considered in the context when the dust particles are taken into account in the plasma. The waves are damped while propagating through the disc along the magnetic field lines that are taken to be perpendicular to the disc midplane. While they propagate, the waves are damped and transfer energy to the disc (Jatenco-Pereira, 2013).

Jatenco-Pereira (2013) assumed that the disc is geometrically thin, optically thick, and is in the Keplerian rotation while adopting the Shakura and Sunyaev (1973) model with some changes proper for a cold ambient: the radiation pressure term in the equation of state was neglected and the opacity was given by Bell and Lin (1994). The α prescription (Shakura and Sunyaev, 1973) suggests that the viscosity is set by the interaction of turbulent eddies: the maximum size of a turbulent cell cannot exceed the disc scale height (H) and its velocity is less than the sound speed (c_s). Hence the maximum kinematic viscosity is then $\nu_{max} = Hc_s$. It is considered that when a disc is irradiating like a black body, its effective temperature is obtained by setting the equilibrium between the energy flux liberated by the viscous dissipation ($\mathcal{F}_v$) and the black-body energy flux (σT_{eff}^4). When an additional source of energy for the discs is included, the equilibrium between the energy gained and the energy dissipated (black body) is given by Vasconcelos et al. (2000)

$$\mathcal{F}_v(H) + \mathcal{F}_A(H) = \sigma T_{eff}^4, \tag{14}$$

where $\mathcal{F}_A(H)$ is the energy flux dissipated by the Alfvén waves. It was assumed that the wave energy flux per frequency range can be described by a power law as in Sect. 4.1.

The results for a typical protostar are characterized by $M_\star = 0.7 M_\odot$ and $R_\star = 2.5 R_\odot$ and the accretion disc with a mean molecular weight $\mu = 2.33$, internal radius $R_i = 2R_\star$, accretion rate $\dot{M} = 10^{-7}\, M_\odot\, yr^{-1}$, $\alpha = 10^{-2}$, the minimum and maximum grain sizes are $a_{min} = 0.001\,\mu m$ and $a_{max} = 0.25\,\mu m$, respectively, of the effective temperature T_{eff} for two different locations of the disk as a function of the initial flux of Alfvén waves, parameterized by f ($\Phi_A = f^2 v_A^3 \rho$) compared to a purely α model are shown in Table 1.

Table 1. Effective temperature T_{eff} of the disk as a function of the parameterized initial flux of Alfvén waves ($\Phi_A = f^2 v_A^3 \rho$).

Model	f	T_{eff} ($r = 0.1$ AU) (K)	T_{eff} ($r = 1$ AU) (K)
M-1	–	660	133
M-2	0.20	917	213

We show that the damping of Alfvén waves by the dust cyclotron mechanism is more efficient than the "anomalous" viscosity mechanism and can increase the effective temperature of the disc, allowing the MRI to take place in a large part of the disc (Jatenco-Pereira, 2013).

5 Summary

We reviewed Alfvén waves in space and astrophysical dusty plasmas by introducing some basic physical concepts of the dusty environments. It was shown that the dispersion relation of Alfvén waves is modified significantly in a dusty plasma since the charged grains in a magnetized plasma are highly coupled to the waves due to cyclotron resonances. The charged dust particles introduce a cutoff in the Alfvén wave at the dust cyclotron frequency. If a distribution of grain sizes is considered, a large band of resonance frequencies appear instead of a single frequency. We discussed the damping of kinetic Alfvén waves, the Lorentzian particle velocity distribution functions, and the nonlinear evolution of large-amplitude Alfvén waves to chaos. We also reviewed some applications in space and astrophysical plasmas, showing the importance of Alfvén waves in these studies. Finally we would like to add some future issues that we intend to develop: (i) the role of damping of KAWs in coronal holes and the consequences for solar wind acceleration; (ii) the study of Lorentzian inertial/kinetic Alfvén waves in aurora regions to elucidate the role of particle distribution function; (iii) the contribution of damping of KAWs for acceleration of stellar winds, among other studies.

Acknowledgements. VJP would like to thank the Brazilian agency CNPq (under grant 304184/2010-1) for partial support. ACLC thanks CNPq for support, the European Commission for the award of a Marie Curie International Incoming Fellowship, and the Paris Observatory for hospitality.

Edited by: T. Hada

References

Adams, F. C.: The birth environment of the solar system, Annu. Rev. Astron. Astr., 48, 47–85, doi:10.1146/annurev-astro-081309-130830, 2010.

Asamura, K., Chaston, C. C., Itoh, Y., Fujimoto, M., Sakanoi, T., Ebihara, Y., Yamazaki, A., Hirahara, M., Seki, K., Kasaba, Y., and Okada, M.: Sheared flows and samll-scale Alfvén wave generation in the auroral acceleration region, Geophys. Res. Lett., 36, L05105, doi:10.1029/2008GL036803, 2009.

Balbus, S. A. and Hawley, J. F.: A powerful local shear instability in weakly magnetized disks. I. Linear analysis, Astrophys. J., 376, 214–222, 1991.

Bell, K. R. and Lin, D. N. C.: Using FU Orionis outbursts to constrain self-regulated protostellar disk models, Astrophys. J., 427, 987–1004, 1994.

Buti, B.: Control of chaos in dusty plasmas, Phys. Lett. A, 235, 241–247, 1997.

Chaston, C. C. and Seki, K.: Small-scale auroral current sheet structuring, J. Geophys. Res., 115, A11221, doi:10.1029/2010JA015536, 2010.

Chaston, C. C., Carlson, C. W., Ergun, R. E., and McFadden, J. P.: Alfvén Waves, Density Cavities and Electron Acceleration Observed from the FAST Spacecraft, Phys. Scripta, T84, 64–68, 2000.

Chaston, C. C., Peticolas, L. M., Bonnell, J. W., Carlson, C. W., Ergun, R. E., McFadden, J. P., and Strangeway, R. J.: Width and brightness of auroral arcs driven by inertial Alfvén waves, J. Geophys. Res., 108, 1091, doi:10.1029/2001JA007537, 2003.

Chaston, C. C., Genot, V., Bonnell, J. W., Carlson, C. W., McFadden, J. P., Ergun, R. E., Strangeway, R. J., Lund, E. J., and Hwang, K. J.: Ionospheric erosion by Alfvén waves, J. Geophys. Res., 111, A03206, doi:10.1029/2005JA011367, 2006.

Chaston, C. C., Salem, C., Bonnell, J. W., Carlson, C. W., Ergun, R. E., Strangeway, R. J., and McFadden, J. P.: The turbulent Alfvénic aurora, Phys. Rev. Lett., 100, 175003, doi:10.1103/PhysRevLett.100.175003, 2008.

Chian, A. C.-L. and Miranda, R. A.: Cluster and ACE observations of phase synchronization in intermittent magnetic field turbulence: a comparative study of shocked and unshocked solar wind, Ann. Geophys., 27, 1789–1801, doi:10.5194/angeo-27-1789-2009, 2009.

Chian, A. C.-L. and Muñoz, P. R.: Detection of current sheets and magnetic reconnections at the turbulent leading edge of interplanetary coronal mass ejection, Astrophys. J. Lett., 733, L34, doi:10.1088/2041-8205/733/2/L34, 2011.

Chian, A. C.-L., Borotto, F. A., and Gonzalez, W. D.: Alfvén intermittent turbulence driven by temporal chaos, Astrophys. J., 505, 993–998, 1998.

Chian, A. C.-L., Kamide, Y., Rempel, E. L., and Santana, W. M.: On the chaotic nature of solar-terrestrial environment: interplanetary Alfvén intermittency, J. Geophys. Res., 111, A07S03, doi:10.1029/2005JA011396, 2006.

Chian, A. C.-L., Santana, W. M., Rempel, E. L., Borotto, F. A., Hada, T., and Kamide, Y.: Chaos in driven Alfvén systems: unstable periodic orbits and chaotic saddles, Nonlin. Processes Geophys., 14, 17–29, doi:10.5194/npg-14-17-2007, 2007.

Christon, S. P.: A comparison of the Mercury and Earth magnetosphere measurements and substorm timescales, Icarus, 71, 448–471, doi:10.1016/0019-1035(87)90040-6, 1987.

Christon, S. P., Mitchell, D., Williams, D., Frank, L., Huang, C., and Eastman, T.: Energy spectra of plasma sheet ions and electrons from about 50 eV/e to about 1 MeV during plasma temperature transitions, J. Geophys. Res., 93, 2562–2572, 1988.

Collier, M. R.: On generating kappa-like distribution functions using velocity space Levy flights, Geophys. Res. Lett., 20, 1531–1534, 1993.

Cramer, N.: The Physics of Alfvén Waves, Wiley, Berlin, 2001.

Cramer, N., Verheest, F., and Vladimirov, S.: The Alfvén resonance in a dusty plasma with a distribution of grain sizes, Phys. Plasmas, 9, 4845–4850, 2002.

D'Angelo, N.: Coulomb solids and low-frequency fluctuations in RF dusty plasmas, J. Phys. D Appl. Phys., 28, 1009–1010, 1995.

Das, A. C., Misra, A. K., and Goswami, K. S.: Kinetic Alfvén wave in three-component dusty plasmas, Phys. Rev. E, 53, 4051–4055, 1996.

de Angelis, U.: The physics of dusty plasmas, Phys. Scripta, 45, 465–474, 1992.

de Juli, M. C., Schneider, R. S., Ziebell, L. F., and Jatenco-Pereira, V.: Effects of dust-charge fluctuation on the damping of Alfvén waves in dusty plasmas, Phys. Plasmas, 12, 052109, doi:10.1063/1.1899647, 2005.

Elmegreen, B. G. and Fiebig, D.: On the minimum length for magnetic waves in molecular clouds, Astron. Astrophys. 270, 397–400, 1993.

Ermakova, N. O. and Antonova, E. E.: On the role of non-Maxwellian forms of distribution functions in the process of acceleration of auroral particles, Int. Conf. Substorms, 8, 61–64, 2006.

Frank, L. A. and Ackerson, K. L.: Local-time survey of plasma at low altitudes over the auroral zones, J. Geophys. Res., 77, 4116–4127, doi:10.1029/JA077i022p04116, 1972.
Gaelzer, R., de Juli, M. C., Schneider, R. S., and Ziebell, L. F.: Obliquely propagating Alfvén waves in a Maxwellian dusty plasma, Plasma Phys. Contr. F., 51, 015011, doi:10.1088/0741-3335/51/1/015011, 2009.
Gaelzer, R., de Juli, M. C., and Ziebell, L. F.: Effect of superthermal electrons on Alfvén wave propagation in the dusty plasmas of solar and stellar winds, J. Geophys. Res., 115, A09109, doi:10.1029/2009JA015217, 2010.
Gammie, C. F.: Layered accretion in T Tauri disks, Astrophys. J., 457, 355–362, 1996.
Gary, J. B., Zanetti, L. J., Anderson, B. J., Potemra, T. A., Clemmons, J. H., Winningham, J. D., and Sharber, J. R.: Identification of auroral oval boundaries from in situ magnetic field measurements, J. Geophys. Res., 103, 4187–4197, doi:10.1029/97JA02395, 1998.
Gary, S. P. and Borovsky, J. E.: Alfvén-cyclotron fluctuations: Linear Vlasov theory, J. Geophys. Res., 109, A06105, doi:10.1029/2004JA010399, 2004.
Goertz, C.: Dusty plasmas in the solar system, Rev. Geophys., 27, 271–292, 1989.
Gogoi, R. and Khan, M.: Arbitrary amplitude kinetic Alfvén solitary waves and double layers in a non Maxwellian plasma, Phys. Plasmas, 17, 112311, doi:10.1063/1.3506820, 2010.
Hada, T., Kennel, C. F., Buti, B., and Mjølhus, E.: Chaos in driven Alfvén systems, Phys. Fluids B, 2, 2581–2590, 1990.
Hartmann, L. and Avrett, E. H.: On the extended chromosphere of Alpha Orionis, Astrophys. J., 284, 238–249, 1984.
Hartmann, L. and MacGregor, K. B.: Momentum and energy deposition in late-type stellar atmospheres and winds, Astrophys. J., 242, 260–282, 1980.
Hasegawa, A. and Cheng, L.: Kinetic processes in plasma heating by resonant mode conversion of Alfvén wave, Phys. Fluids, 19, 1924–1932, 1976.
Hollweg, J. V.: Alfvén waves in a two-fluid model of the solar wind, Astrophys. J., 181, 547–566, 1973.
Hundhausen, A.: Plasma Flow from the Sun, in: The Solar Output and its Variation, Proceedings, edited by: White, O. R., Colorado Associated University Press, Boulder, USA, 36 pp., 1977.
Jatenco-Pereira, V.: Damping of Alfvén waves as a heating source in protostellar accretion discs, Mon. Not. R. Astron. Soc., 431, 3150–3158, 2013.
Jatenco-Pereira, V. and Opher, R.: Effect of diverging magnetic fields on mass loss in late-type giant stars, Astron. Astrophys., 209, 327–336, 1989.
Kamaya, H. and Nishi, R.: Frozen condition for the charged particles in molecular clouds, Astrophys. J., 543, 257–270, 2000.
Keidar, M., Shashurin, A., Volotskova, O., Stepp, M. A., Srinivasan, P., Sandler, A., and Trink, B.: Cold atmospheric plasma in cancer therapy, Phys. Plasmas, 20, 057101, doi:10.1063/1.4801516, 2013.
Khan, M. S. A., Sultana, S., and Samilullah, M.: Nature of electromagnetic wave in uniform dusty plasma, International Journal of Science and Research (IJSR), 2, 132–136, 2013.
Kletzing, C. A., Scudder, J. D., Dors, E. E., and Curto, C.: Auroral source region: Plasma properties of the high-latitude plasma sheet, J. Geophys. Res., 108, 1360, doi:10.1029/2002JA009678, 2003.
Koga, D., Chian, A. C.-L., Miranda, R. M., and Rempel, E. L.: Intermittent nature of solar wind turbulence near the Earth's bow shock: Phase coherence and non-Gaussianity, Phys. Rev. E, 75, 046401, doi:10.1103/PhysRevE.75.046401, 2007.
Köritzer, J., Boxhammer, V., Schafer, A., Shimizu, T., Klämptl, T. G., Li, Y.-F., Welz, C., Schwenk-Zieger, S., Morfill, G. E., Zimmermann, J. L., and Schlegl, J.: Restoration of sensitivity in chemo – resistant glioma cells by cold atmospheric plasma, PLoS ONE, 8, e64498, doi:10.1371/journal.pone.0064498, 2013.
Krimigis, S. M., Carbary, J. F., Keath, E. P., Bostrom, C. O., Axford, W. I., Gloeckler, G., Lanzerotti L. J., and Armstrong, T. P.: Characteristics of hot plasma in the Jovian magnetosphere: Results from Voyager spacecraft, J. Geophys. Res., 86, 8227–8257, doi:10.1029/JA086iA10p08227, 1981.
Krimigis, S. M., Armstrong, T. P., Axford, W. I., Bostrom, C. O., Cheng, A. F., Gloeckler, G., Hamilton, D. C., Keath, E. P., Lanzerotti, L. J., Mauk, B. H., and Van Allen, J. A.: Hot plasma and energetic Particles in Neptune's Magnetosphere, Science, 246, 4936, doi:10.1126/science.246.4936.1483, 1989.
Lazar, M., Poedts, S., and Schlickeiser, R.: Instability of the parallel electromagnetic modes in Kappa distributed plasmas – I. Electron whistler–cyclotron modes, Mon. Not. R. Astron. Soc., 410, 663–670, doi:10.1111/j.1365-2966.2010.17472.x, 2011.
Louarn, J. E., Wahlund, T., Chust, H., de Feraudy, A., Roux, B., Holback, P. O., Dovnor, A., Ericksson, I., and Holmgren, G.: Observation of kinetic Alfvén waves by the FREJA spacecraft, Geophys. Res. Lett., 21, 1847–1850, 1994.
Mathis, J., Rumple, W., and Nordsiek, K.: The size distribution of interstellar grains, Astrophys. J., 217, 425–433, 1977.
Mauk, B. H., Mitchell, D. G., McEntire, R. W., Paranicas, C. P., Roelof, E. C., Williams, D. J., Krimigis, S. M., and Lagg, A.: Energetic ion characteristics and neutral gas interactions in Jupiter's magnetosphere, J. Geophys. Res., 109, A09S12, doi:10.1029/2003JA010270, 2004.
Miranda, R. A., Rempel, E. L., and Chian, A. C.-L.: Chaotic saddles in nonlinear modulational interactions in a plasma, Phys. Plasmas, 19, 112303, doi:10.1063/1.4766472, 2012.
Munro, R. H. and Jackson, B. V.: Physical properties of a polar coronal hole from 2 to 5 solar radii, Astrophys. J., 213, 874–886, 1977.
Nakano, T., Nishi, R., and Umebayashi, T.: Mechanism of magnetic flux loss in molecular clouds, Astrophys. J., 573, 199–214, 2002.
Pandey, B. P., Vladimirov, S. V., and Samarian, A.: Nonlinear waves in collisional dusty plasma, Phys. Plasmas, 15, 053705, doi:10.1063/1.2918341, 2008.
Paschmann, G., Haaland, S., and Treumann, R.: Auroral Plasma Physics, Kluwer Academic, Dordrecht, p. 94, 2003.
Pierrard, V., Maksimovic, M., and Lemaire, J.: Core, Halo and Strahl Electrons in the Solar Wind, Astrophys. Space Sci., 277, 195–200, doi:10.1023/A:1012218600882, 2001.
Prested, C., Schwadron, N., Passuite, J., Randol, B., Stuart, B., Crew, G., Heerikhuisen, J., Pogorelov, N., Zank, G., Opher, M., Allegrini, F., cComas, M. Reno, D. J., Roelof, E., Fuselier, S., Funsten, H., Moebius, E., and Saul, L.: Implications of solar wind suprathermal tails for IBEX ENA images of the heliosheath, J. Geophys. Res., 113, A06102, doi:10.1029/2007JA012758, 2008.

Rao, N. N.: Advances in dusty plasmas, edited by: Shukla, P. K., Mendis, D. A., Desai, T., World Scientific, Singapore, 1996.

Rao, N. N., Shukla, P. K., and Yu, M. Y.: Dust-acoustic waves in dusty plasmas, Planet. Space Sci., 38, 543–546, 1990.

Rempel, E. L., Wanderson, S. M., and Chian, A. C.-L.: Alfvén multistability: transient and intermittent dynamics induced by noise, Phys. Plasmas, 13, 032308, doi:10.1063/1.2186527, 2006.

Rempel, E. L., Chian, A. C.-L., and Miranda, R. A.: Chaotic saddles at the onset of intermittent spatiotemporal chaos, Phys. Rev. E, 76, 056217, doi:10.1103/PhysRevE.76.056217, 2007.

Rempel, E. L., Chian, A. C.-L., Koga, D., Miranda, R. A., and Santana, W. M.: Alfvén complexity, Int. J. Bifurcation Chaos, 18, 1697–1703, 2008.

Rodgers, B. and Glassgold, A. E.: The temperature of the circumstellar envelope of Alpha Orionis, Astrophys. J., 382, 606–616, 1991.

Rubab, N., Erkaev, N. V., and Biernat, H. K.: Dust kinetic Alfvén wave and acoustic waves in a Lorentzian plasma, Phys. Plasmas, 16, 103704, doi:10.1063/1.3244625, 2009.

Rubab, N., Erkaev, N. V., Langmayr, D., and Biernat, H. K.: Kinetic Alfvén wave instability in a Lorentzian dusty magnetosplasma, Phys. Plasmas, 17, 103704, doi:10.1063/1.3491336, 2010.

Rubab, N., Erkaev, N. V., Biernat, H., K., and Langmayr, D.: Kinetic Alfvén wave instability in a Lorentzian dusty plasma: Non–resonant particle approach, Phys. Plasmas, 18, 073701, doi:10.1063/1.3599600, 2011.

Rudakov, L., Mithaiwala, M., Ganguli G., Crabtree, C.: Linear and nonlinear Landau resonance of kinetic Alfvén waves: Consequences for electron distribution and wave spectrum in the solar wind, Phys. Plasmas, 18, 012307, doi:10.1063/1.3532819, 2011.

Sahraoui, F.: Diagnosis of magnetic structures and intermittency in space-plasma turbulence using the technique of surrogate data, Phys. Rev. E, 78, 026402, doi:10.1103/PhysRevE.78.026402, 2008.

Salem, C., Hubert, D., Bale, S. D., and Lin, R. P.: Nature and origin of the electron distribution functions in the slow and fast solar wind at 1 AU: Wind observations, 36th COSPAR Scientific Assembly, Beijing, China, 16–23 July 2006.

Sánchez-Arriaga, G., Hada, T., and Nariyuki, Y.:, The truncation model of the derivative nonlinear Schrodinger equation, Phys. Plasmas, 16, 042302, doi:10.1063/1.3093383, 2009.

Schippers, P., Blanc, M., André, N., Dandouras, I., Lewis, G. R., Gilbert, L. K., Persoon, A. M., Krupp, N., Gurnett, D. A., Coates, A. J., Krimigis, S. M., Young, D. T., and Dougherty, M. K.: Multi-instrument analysis of electron populations in Saturn's magnetosphere, J. Geophys. Res., 113, A07208, doi:10.1029/2008JA013098, 2008.

Schulz, N. S.: From Dust To Stars, in: Studies of the Formation and Early Evolution of Stars, Springer Praxis Books in Astrophysics and Astronomy, Praxis Publishing Ltd, ISBN 3-540-23711-9, 147 pp., 2005.

Sittler Jr., E. C., Ogilvie, K. W., and Scudder, J. D.: Survey of low-energy plasma electrons in Saturn's magnetosphere: Voyagers 1 and 2, J. Geophys. Res., 88, 8847–8870, 1983.

Shakura, N. I. and Sunyaev, R. A.: Black holes in binary systems. Observational appearance, Astron. Astrophys., 24, 337–355, 1973.

Shukla, P. K. and Verheest, F.: Dust Alfvén envelope solitons in astrophysical dusty plasmas, Astron. Astrophys., 401, 849–850, 2003.

Sorasio, G. and Rosenberg, M.: Instability of higher-harmonic electrostatic dust cyclotron waves, J. Plasma Phys., 65, 319–329, 2001.

Stasiewicz, K., Bellan, P., Chaston, C., Kletzing, C., Lysak, R., Maggs, J., Pokhotelov, O., Seyler, C., Shukla, P., Stenflo, L., Streltsov, A., and Wahlund, J.-E.: Small scale Alfvénic structure in the aurora, Space Sc. Rev., 92, 423–533, 2000.

Tomczyk, S., McIntosh, S. W., Keil, S. L., Judge, P. G., Schad, T., Seeley, D. H., and Edmondson, J.: Alfvén Waves in the Solar Corona, Science, 317, 1192–1196, 2007.

Treumann, R. A.: Theory of super-diffusion for the magnetopause, Geophys. Res. Lett., 24, 1727–1730, 1997.

Treumann, R. A.: Kinetic theoretical foundation of Lorentzian statistical mechanics, Phys. Scripta, 59, 19, doi:10.1238/Physica.Regular.059a00019, 1999.

Tu, C.-Y., Marsch, E., and Thieme, K. M.: Basic properties of solar wind MHD turbulence near 0.3 AU analyzed by means of Elsaesser variables, J. Geophys. Res., 94, 11739–11759, 1989.

Turner, N. J. and Sano, T.: Dead zone accretion flows in protostellar disks, Astrophys. J., 679, L131–L134, 2008.

Turner, N. J., Sano, T., and Dziourkevitch, N.: Turbulent mixing and the dead zone in protostellar disks, Astrophys. J., 659, 729–737, 2007.

Vasconcelos, M. J., Jatenco-Pereira, V., and Opher, R.: Alfvénic heating of protostellar accretion disks, Astrophys. J., 534, 967–975, 2000.

Vasyliunas, V. M.: Low-energy electrons on the day side of the magnetosphere, J. Geophys. Res., 73, 7519–7523, 1968.

Verheest, F.: Nonlinear dust modes, Space Sci. Rev., 68, 109–114, 1994.

Verheest, F.: Waves in dusty space plasmas, Astrophys. Space Sc. Library, Kluwer Academic Publishers, Dordrecht, Vol. 245, 2000.

Verheest, F.: Nonlinear evolution equations for parallel electromagnetic multispecies plasma modes: Reductive perturbation theory in the linear wave frame and electric field generation, Phys. Scripta, T113, 92–96, 2004.

Verheest, F. and Cattaert, T.: Nonlinear electromagnetic modes in astrophysical plasmas with dust distributions, Astron. Astrophys., 421, 17–21, 2004.

Verheest, F. and Meuris, P.: Nonlinear electromagnetic modes in plasmas with variable dust charges, Phys. Lett. A, 210, 198–201, 1996.

Vidotto, A. A. and Jatenco-Pereira, V.: The effects of Alfvén waves and radiation pressure in dusty winds of Late-Type stars. II. Dust-cyclotron damping, Astrophys. J., 639, 416–422, 2006.

Vladimirov, S. V.: Plasma Collective Effects in the Presence of Dust, Astrophys. Space Sc., 256, 85–105, 1997.

Voitenko, Y., Goossens, M., Sirenko, O., and Chian, A. C.-L.: Nonlinear excitation of kinetic Alfvén waves and whistler waves by electron beam-driven Langmuir waves in the solar corona, Astron. Astrophys., 409, 331–345, 2003.

Wahlund, J. E., Louarn, P., Chust, T., de Feraudy, H., Roux, A., Holback, B., Dovner, P. O., and Holmgren, G.: On ion acoustic turbulence and the nonlinear evolution of kinetic

Alfvén waves in aurora, Geophys. Res. Lett., 21, 1831–1834, doi:10.1029/94GL01289, 1994.

Wahlund, J. E., Eriksson, A. I., Holback, B., Boehm, M. H., Bonnell, J., Kintner, P. M., Seyler, C. E., Clemmons, J. H., Eliasson, L., Knudsen, D. J., Norqvist, P., and Zanetti, L. J.: Broadband ELF plasma emission during auroral energization, 1, Slow ion acoustic waves, J. Geophys. Res., 103, 4343–4376, doi:10.1029/97JA02008, 1998.

Walk, R. M., Snyder, J. A., Scrivasan, P., Kirch, J., Diaz, S. O., Blanco, F. C., Shashurin, A., Keidar, M., and Sandler, A. D.: Cold atmospheric plasma for the ablative treatment of neuroblastoma, J. Pediatr. Surg., 48, 63–73, 2013.

Watt, C. E. J. and Rankin, R.: Electron acceleration due to inertial Alfvén waves in a non-Maxwellian plasma, J. Geophys. Res., 112, A04214, doi:10.1029/2006JA011907, 2007.

Wu, D. J. and Fang, C.: Tw–fluid motion of plasma in Alfvén waves and the heating of solar coronal loops, Astrophys. J., 511, 958–964, 1999.

Wu, D. J. and Fang, C.: Coronal plume heating and kinetic dissipation of kinetic Alfvén aves, Astrophys. J., 596, 656–652, 2003.

Wu, D. J. and Fang, C.: Sunspot chromospheric heating by kinetic Alfvén waves, Astrophys. J., 659, L181–L184, 2007.

Wu, D. J., Huang, G. L., Wang, D. Y., and Fälthammar, C. G.: Solitary kinetic Alfvén waves in the two-fluid model, Phys. Plasmas, 3, 2879–2884, 1996.

Wygant, J. R., Keiling, A., Cattell, C. A., Johnson, M., Lysak, R., Temerin, M., Mozer, F. S., Kletzing, C. A., Scudder, J. D., Peterson, W., Russell, C. T., Parks, G., Brittnacher, M., Germany, G., and Spann, J.: Polar spacecraft based comparisons of intense electric fields and Poynting flux near and within the plasma sheet–tail lobe boundary to UVI images: An energy source for the aurora, J. Geophys. Res., 105, 18675–18692, 2000.

Yang, X., Wang, C.-L., Liu, C.-B., Zhang, J.-R., Shi, Y.-R., Duan, W.-S., and Yang, L.: The collision effect between dust grains and ions to the dust ion acoustic waves in a dusty plasma, Phys. Plasmas, 19, 103705, doi:10.1063/1.4757637, 2012.

Yukhimuk, A., Fedun, V., Sirenko, O., and Voitenko, Yu.: Excitation of fast and slow magnetosonic waves by kinetic Alfvén waves, AIP Conf. Proc., 537, 311, doi:10.1063/1.1324955, 2000.

Zhao, J. S., Wu, D. J., and Lu, J. Y.: Kinetic Alfvén waves excited by oblique magnetohydrodynamic Alfvén waves in coronal holes, Astrophys. J., 735, 114, doi:10.1088/0004-637X/735/2/114, 2011.

Zouganelis, I.: Measuring suprathermal electron parameters in space plasmas: Implementation of the quasi-thermal noise spectroscopy with kappa distributions using in situ Ulysses/URAP radio measurements in the solar wind, J. Geophys. Res., 113, A08111, doi:10.1029/2007JA012979, 2008.

Zubia, K., Rubab, N., Shah, H. A., Salimullah, M., and Murtaza, G.: Kinetic Alfvén waves in a homogeneous dusty magnetoplasma with dust charge fluctuation effects, Phys. Plasmas, 14, 032105, doi:10.1063/1.2710457, 2007.

7

Provision of boundary conditions for a convection-permitting ensemble: comparison of two different approaches

C. Marsigli, A. Montani, and T. Paccagnella

ARPA-SIMC, HydroMeteoClimate Service of ARPA Emilia-Romagna, Bologna, Italy

Correspondence to: C. Marsigli (cmarsigli@arpa.emr.it)

Abstract. The current resolution of the operational global models favours the possibility of driving convection-permitting limited-area model (LAM) simulations directly, sparing the necessity for an intermediate step with a coarser-resolution LAM. Though the resolution of global ensemble systems is generally lower than that of deterministic ones, it is also possible to consider this opportunity in the field of ensemble forecasting. The aim of this paper is to investigate the effect of this choice for driving a convection-permitting ensemble based on the COSMO model, for a specific application, namely the forecast of intense autumn precipitation events over Italy. The impact of the direct nesting in the ECMWF global ensemble is compared to a two-step nesting, which makes use of a LAM ensemble system with parametrised convection. Results show that the variability introduced in the geopotential field by the direct nesting is usually contained within the uncertainty described by the standard ensemble, and differences between pairs of members following different nesting approaches are generally smaller than the ensemble error, computed with respect to analysis. The relation between spread and error is even improved by the direct nesting approach. In terms of precipitation, it is found that the forecasts issued by members with different nesting approaches generally have differences at spatial scales between 16 and 180 km, depending on the case, hence not negligible. Nevertheless, the skill of the LAM ensemble precipitation forecasts, evaluated by means of an objective verification, is comparable. Therefore, the overall quality of the 2.8 km ensemble for the specific application is not deteriorated by the provision of lower resolution lateral boundary conditions directly from the global ensemble.

1 Introduction

The current resolution of operational global circulation models (GCM) paves the way towards driving convection-permitting limited-area model (LAM) integrations directly, with lateral boundary conditions (LBCs) provided by a global model. Presently, high-resolution LAM runs are generally nested in coarser-resolution integrations of the same model. The intermediate step with a coarser-resolution run is usually performed to ensure a ratio of spatial resolutions between nested models in the range of 2 : 1–5 : 1. The aim is to reduce the loss of information due to nesting which would affect the nested model in the case of a larger resolution gap (see for example Warner et al., 1997 and Denis et al., 2001).

With the resolution increase of the GCMs, the intermediate step with a coarser-resolution LAM could be avoided, with some potential benefits. First of all, removing one step in the nesting procedure would decrease the time needed to perform the whole cascade up to the convection-permitting integration, with a consequent reduction of the computational costs. This would be most beneficial in an operational environment, requiring timely delivery of products. Secondly, one shortcoming of the intermediate LAM run is that nowadays this usually runs in the so-called "grey zone", where convection is partly resolved and partly sub-grid. This is recognised to be a delicate issue for many models (see e.g. Gerard et al., 2009). On the other hand, the coarser spatial and temporal resolution of the LBCs provided by a global model could introduce errors in the forecast, especially if the LAM domain is small, as is often the case for convection-permitting models. Some studies have addressed this issue.

As for the spatial resolution of LBCs, de Elia et al. (2002) showed that the quality of the LAM simulation is rather insensitive to the resolution of the driving data. However, this is true if boundaries are set far from the region of interest, which is usually not true for convection-permitting simulations. Temporal resolution is also an issue, since LBCs are often produced at much lower frequency than the time step used in the LAM. If information changes rapidly at the boundary, then the LBCs might not reflect the actual changes in the state of the atmosphere (Termonia et al., 2009). Termonia (2003) found that the use of a coupling update interval of 3 h has a detrimental effect on the forecast quality of a very intense storm, affecting significantly the depth of the low-pressure system. Davies (2014) showed that, with an 11 km resolution model, 3-hourly LBCs already lead to a loss of information with respect to hourly LBCs at a 12 h forecast range. Amengual et al. (2007) found that for regional climate model (RCM) runs the update frequency of LBCs has a larger impact on the downscaling results than their spatial resolution. Nevertheless, the ratio between the two factors is dependent on the actual resolutions involved and on the forecast range. As for the relation between the two effects and their impact on LAM ensembles, Nutter et al. (2004) showed that the impact of coarsely resolved LBCs or temporal interpolation of LBCs on error growth is quite similar. Both effects act to remove small-scale features from the external fields passing through the lateral boundary, thereby constraining ensemble dispersion. The impact of coarsely resolved LBCs has been identified as stronger than the one of temporal interpolation when the LBC update frequency is reasonably high.

In this work, the issue of LBC resolution is considered in the framework of ensemble forecasting. The aim of this paper is to investigate the effect of the provision of LBCs to a LAM ensemble run at a convection-permitting resolution by a global ensemble, compared with providing LBCs from an intermediate LAM ensemble run at coarser resolution. The analysis is confined to a specific meteorological situation, namely autumn precipitation cases over the Mediterranean.

Aiming at the development of an ensemble system at the convection-permitting scale over Italy, ARPA-SIMC has implemented an experimental ensemble based on the COSMO model (Steppeler et al., 2003) during SOP (Special Observation Period) 1.1 of the Hymex Project (HYMEX, 2010–2020). The ensemble, named COSMO-H2-EPS (COSMO Hymex 2.8 km Ensemble Prediction System, Marsigli et al., 2013), consists of 10 runs of the COSMO model at 2.8 km horizontal resolution, with 50 levels in the vertical, and receives perturbed ICs and LBCs from the first 10 members of COSMO-LEPS (COSMO Limited-area Ensemble Prediction System, Montani et al., 2011). COSMO-LEPS, running at 7 km horizontal resolution with parametrised convection, in turn receives initial and boundary conditions from the global ensemble of ECMWF, referred to as ENS. ENS is currently running with an approximate horizontal resolution of 32 km, which is planned to increase to about 20 km in 2015. This may enable direct use of LBCs from ENS members to drive ensemble systems for the convection-permitting scale (2–3 km).

In this work, it is studied how the performance of the COSMO-H2-EPS ensemble varies if the 2.8 km runs receive ICs and BCs from the ENS members directly, skipping the intermediate step with COSMO-LEPS at 7 km. The aim is to evaluate the effect of removing this step, heading for the future operational set-up of the ensemble, which will benefit from the resolution increase of ENS, up to 20 km. Though in the present configuration the resolution gap between driving model (32 km, retrieved on a grid of 0.25° mesh size) and driven model (2.8 km) is quite high, it is believed that an assessment of the difference between the two nesting procedures on the specific ensemble application can provide scientific guidance useful for defining the operational set-up. It is also worth mentioning that, in its operational configuration, the 2.8 km ensemble will not use as ICs the downscaled ENS analyses, since high-resolution perturbed initial conditions will be provided by a LETKF scheme developed in the COSMO consortium (KENDA system, Reich et al., 2011), which is currently under testing.

The paper is organised as follows: in the next section, the configuration of the ensembles is described. Then, in Sect. 3 results are commented on, divided into an analysis of fields of geopotential at 500 hPa and precipitation. Finally, conclusions are drawn in Sect. 4.

2 Configuration of the ensemble systems

COSMO-H2-EPS is an ensemble system specifically designed for the Hymex Project, which was run regularly during SOP 1.1, taking place from 5 September to 6 November 2012. The ensemble consists of 10 runs of the COSMO model at 2.8 km horizontal resolution, with 50 levels in the vertical, over a domain including most of the target areas of the project. In particular, it covers a large part of Italy, Switzerland and the French Mediterranean coast (Fig. 1). The model domain contains 399×412 grid points. The ensemble was run once per day, starting at 12:00 UTC, for a forecast range of 36 h.

Initial and boundary conditions are provided by the first 10 members of COSMO-LEPS. COSMO-LEPS receives, in turn, ICs and BCs from 16 members of the ECMWF ENS, selected as the most representative members of the full ensemble through an ensemble reduction algorithm (Molteni et al., 2001; Marsigli et al., 2001). Therefore, perturbed ICs of COSMO-H2-EPS are also derived from ECMWF ENS ones by means of a two-step interpolation: first from the ENS archiving grid to the COSMO-LEPS grid at 7 km and then from it to the COSMO-H2-EPS grid at 2.8 km. This implies that no data assimilation is performed for the two ensembles; initial conditions are simply downscaled from global analyses. Model perturbations are also applied to the 2.8 km runs, since few parameters of the COSMO physics schemes (turbulence, microphysics, soil scheme) are set to values different from the default ones in the 10 COSMO runs. It should be highlighted that this same model perturbation technique is adopted in COSMO-LEPS, with some differences in the parameters which are perturbed. These pertubations affect the model run starting from the first time step, but do not affect the analysis, therefore ICs are not changed due to model perturbation. For a comprehensive description of the two systems the reader is referred to Marsigli et al. (2013) for COSMO-H2-EPS and to Montani et al. (2011) for COSMO-LEPS.

A second experimental ensemble, named EXP-H2-EPS, has been implemented for the present study. It has been run for 21 cases of the SOP 1.1 period, selected as events of moderate to intense (observed) precipitation. A description of the meteorological situation during the SOP and of a few severe events can be found in Ducrocq et al. (2013) and Ferretti et al. (2013) (the latter focussing on Italy).

The ensemble configuration of EXP-H2-EPS is identical to that of COSMO-H2-EPS in terms of resolution, domain, model set-up, and parameter perturbation. The only difference is in the way ICs and BCs are provided: for EXP-H2-EPS they are derived directly from 10 members of ENS, the same members that have been selected to drive the first 10 COSMO-LEPS operational runs. Therefore, pairs of members of EXP-H2-EPS and COSMO-H2-EPS can be compared directly: the ith member of the former ensemble is driven by an ENS member, while the ith member of the latter ensemble is driven by the ith COSMO-LEPS member, which was in turn driven by this same ENS member. It should be underlined that the temporal resolution of the LBCs is also different: hourly BCs are provided to COSMO-H2-EPS by COSMO-LEPS, while 3-hourly BCs are provided to EXP-H2-EPS by ENS.

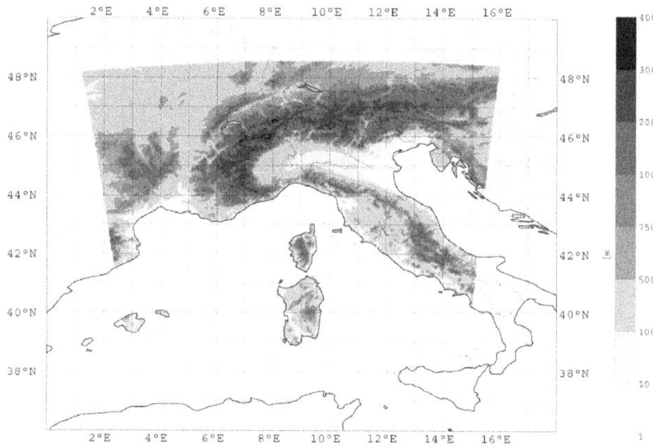

Fig. 1. Orography of the COSMO model at 2.8 km horizontal resolution, also showing the extension of the integration domain of COSMO-H2-EPS.

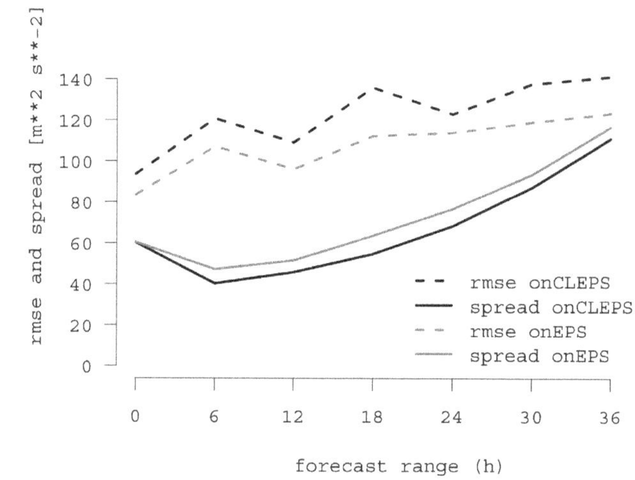

Fig. 2. RMSE of the ensemble mean (dashed lines) and ensemble spread (solid lines) for COSMO-H2-EPS (black, labelled "onCLEPS") and EXP-H2-EPS (grey, labelled "onEPS"), in terms of geopotential at 500 hPa.

3 Comparison of the two ensembles

3.1 Analysis of 500 hPa geopotential

A quantitative assessment of the differences between the geopotential fields at 500 hPa predicted by the 2.8 km ensemble runs, when driven by COSMO-LEPS members and when driven by ENS members, is provided here.

First, the spread and the root-mean-square error (RMSE) of the ensemble mean in terms of geopotential at 500 hPa are computed for the 21 cases and over the whole domain. Ensemble spread has been computed as the root-mean-square (RMS) distance of all the ensemble members from the ensemble mean. RMSE of the ensemble mean has been computed against the operational ECMWF analyses interpolated on the 2.8 km grid. Since analyses are available only every 6 h, RMSE has also been computed every 6 h, while spread has been computed hourly. The relation between spread and error is usually regarded as an indicator of the capability of the ensemble to represent the forecast error through the dispersion of the members. The spread matching the error guarantees that the ensemble variability allows one to find, on average, the true atmospheric state among the states predicted by the ensemble (e.g. Buizza et al., 2005).

The spread and error of the two systems are compared in Fig. 2. EXP-H2-EPS has slightly more spread than COSMO-H2-EPS, with a similar increase with the forecast range. The coincidence of the two spread values at initial time is due to the fact that the initial conditions of the two ensembles are the same ECMWF ENS perturbed analyses, downscaled at 2.8 km with an interpolation performed either in one step (for EXP-H2-EPS) or in two steps (for COSMO-H2-EPS). The relatively large value may be ascribed to some noise due to the interpolation itself (visible in the z500 maps, not shown).

The RMSEs of the ensemble mean of the two systems increase with lead time, with lower error values for EXP-H2-EPS with respect to COSMO-H2-EPS. Therefore, the direct nesting allows one to improve the spread–error relation in terms of geopotential at 500 hPa, evaluated over the full period and over the whole domain.

In order to check whether the direct nesting causes problems in particular meteorological situations, error and spread have also been computed case by case. First, the RMS difference between the geopotential field forecasted by each member of the COSMO-H2-EPS ensemble and by the correspondent member of the EXP-H2-EPS ensemble is computed. This is performed for each case separately, for each forecast hour (from 0 to 36 h) and over the whole integration domain. RMS differences are shown in Fig. 3 as thin solid lines, while the thick dotted line is the average RMS difference, computed between all pairs of members. Results are shown for 4 cases only, selected as representative of the whole sample.

In order to provide a reference against which to evaluate the magnitude of these differences, the spread of the two ensembles has also been computed for each case. The spread is regarded as a measure of the uncertainty expressed by each ensemble. Therefore a difference between pairs of members smaller than the spread would not be detected in the standard ensemble configuration. The spread is plotted as thick solid lines, black for COSMO-H2-EPS and grey for EXP-H2-EPS. Finally, the RMSE of the ensemble mean is also computed for each case, and it is shown in Fig. 3 as a dashed line, black for COSMO-H2-EPS and grey for EXP-H2-EPS. RMSE provides the ultimate limit for the RMS differences between pairs of members, since it quantifies the error of the ensemble: until the difference induced by the different nesting approach is smaller than the error of the ensemble, it is not actually possible to establish if this difference is detrimental to the forecast.

In all cases, the spread of each ensemble generally lies among the larger RMS differences between pairs of members and is larger than the average RMS difference. This indicates that the variability introduced by nesting the 2.8 km model on ENS directly is usually included within the uncertainty described by the reference ensemble (COSMO-H2-EPS). Furthermore, the RMS differences between pairs of members are generally smaller than the RMS error of the ensemble mean, with the few exceptions discussed below. This indicates that in most cases the impact of the different nesting is lower than the forecast error of the reference system.

Nevertheless, some differences can also be identified. The behaviour of the major part of cases (15 cases) is close to the one shown in Fig. 3a, relative to the 2012092512 case. The spread is smaller than the error of the ensemble mean for both ensembles, the two spread curves remaining quite close together for the entire forecast range. The differences between pairs of members are generally small, and do not give rise to additional uncertainty with respect to that represented by the COSMO-H2-EPS ensemble. In a few cases among these 15, the RMS differences can increase up to the level of the RMSE for a few hours during the forecast range.

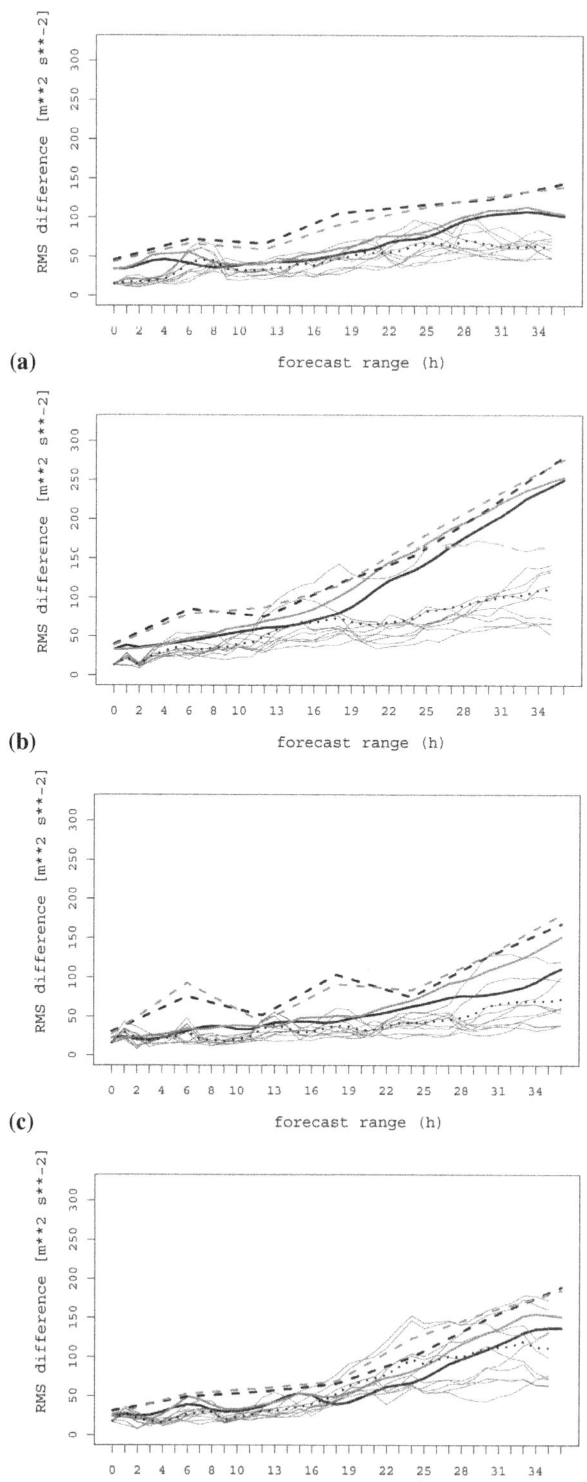

Fig. 3. RMS differences between pairs of correspondent members (thin solid), average RMS difference between all pairs (thick dotted), RMS spread (thick solid) and RMSE of the ensemble mean (thick dashed), as a function of the forecast range. RMS spread and error lines are black for COSMO-H2-EPS and grey for EXP-H2-EPS. Plots are relative to four cases only: 2012092512 **(a)**, 2012101412 **(b)**, 2012102512 **(c)**, and 2012103012 **(d)**.

In the 2012101412 case (Fig. 3b), the spread increases quite sharply after 18–21 h of simulation, gaining distance from the RMS differences, which become negligible, with the exception of one pair of members. The spread curves roughly match the error curves. In this case, a trough enters the domain from the northwest corner after about 18 h of integration. The z500 field forecasted by member 1 of EXP-H2-EPS (not shown) reproduces the observed trough and its evolution quite well, at difference with the corresponding member of COSMO-H2-EPS. The other pairs of members do not differ that much, but provide less skillful predictions. In the 2012102512 case (Fig. 3c), the RMS differences between members are generally smaller than the spread and quite below the RMSE values. Also in this case, a marked spread increase is observed for EXP-H2-EPS after 20 h, leading to a better match to the RMSE. The meteorological situation is similar to the previous one, with one member of EXP-H2-EPS performing better than the others and better than the corresponding COSMO-H2-EPS member. In both the 2012101412 and 2012102512 cases, additional spread is gained with the direct nesting approach, with almost no increase in the RMSE, mainly due to the better performance of one EXP-H2-EPS member. A similar feature has been detected in two other cases (not shown).

The 2012103012 case (Fig. 3d) offers the most notable exception to the behaviour identified so far. The spread increases with the forecast range, especially after 18 h, in agreement with the increase in the forecast error. Nevertheless, some members have differences larger than the ensemble spread, so that the variability introduced by the direct nesting can be detected as an "outlier" with respect to the variability described by the COSMO-H2-EPS ensemble. Furthermore, two pairs of members have differences larger than the RMSE. Since the RMSE of EXP-H2-EPS is higher that the RMSE of COSMO-H2-EPS, it is concluded that for this case additional error is introduced by the direct nesting. In this case, a deep low-pressure system over the western Mediterranean develops and enters the domain approximately at the 18 h forecast range. Three members of EXP-H2-EPS provide a poor description of the position and intensity of the low, unlike the correspondent COSMO-H2-EPS members. For the other ensemble members, differences are smaller. Similar behaviour is found on the following day, for the 2012103112 case (not shown).

The same analysis has also been performed in terms of mean sea level pressure (Fig. 4), in order to assess the robustness of the previous analysis.

Results are generally in good agreement with those obtained in terms of geopotential at 500 hPa. In the 2012092512 case (Fig. 4a), few RMS differences reach values close to or even above the RMSE. For most of the cases similar to this, though, RMS differences are well below these values. In the 2012103012 case (Fig. 4d), RMS differences of a few pairs of members are also quite high in terms of MSLP, confirming

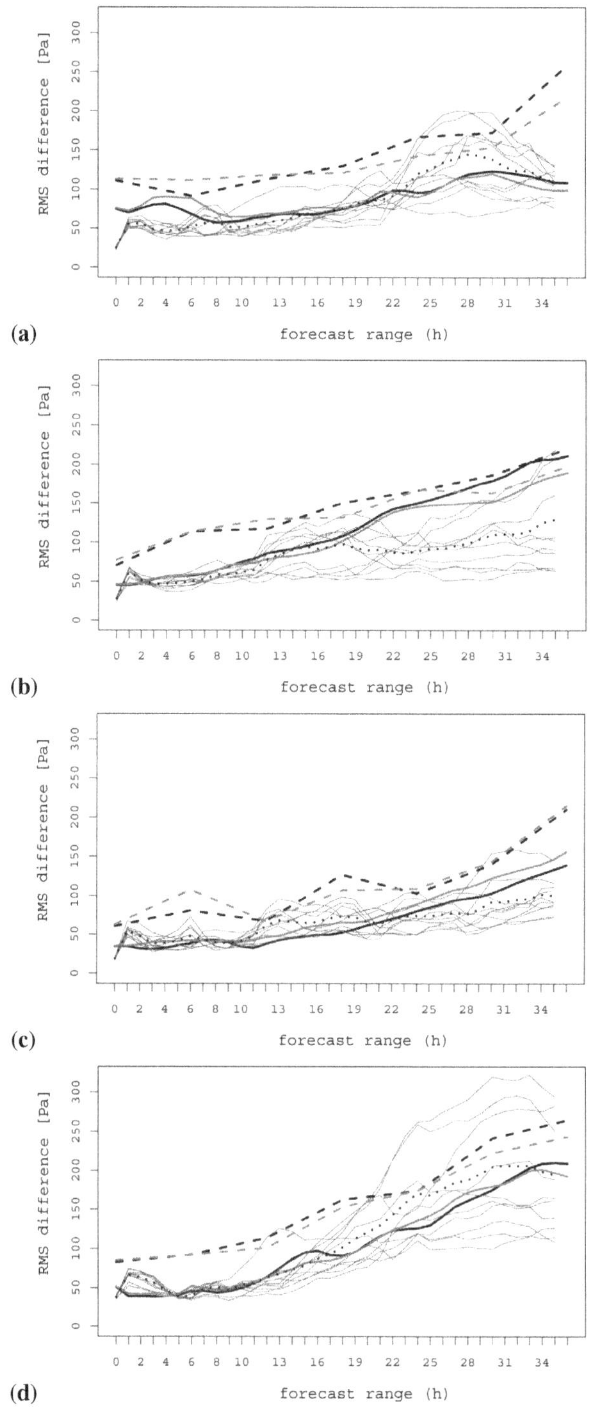

Fig. 4. The same as in Fig. 3 but for mean sea level pressure.

the problem already detected in terms of z500 and ascribed to the evolution of the deep low-pressure system.

Focussing on the 2012102512 case only, the behaviour of the two ensembles is also discussed by analysing the spectra of the perturbations (Fig. 5). Perturbations are computed as the difference between each ensemble member and the ensemble mean in terms of geopotential field at 500 hPa.

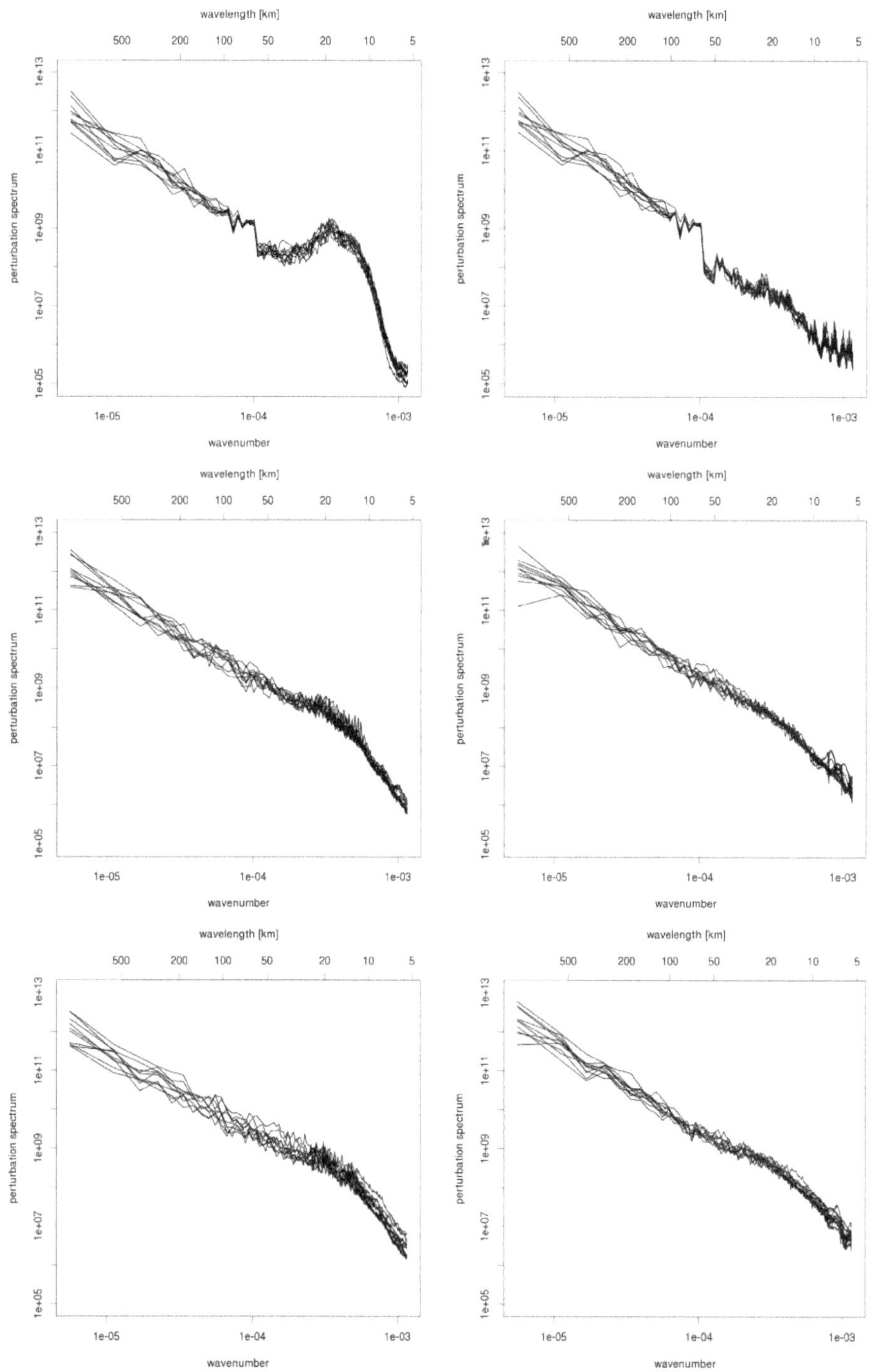

Fig. 5. Perturbation spectra of the field of geopotential at 500 hPa for each member of COSMO-H2-EPS (left panels) and EXP-H2-EPS (right panels) at the initial time (top), and at +6 h (middle) and +12 h (bottom) forecast range.

Initial state perturbations come, for both ensembles, from the ENS perturbed analyses, to which only the effect of the interpolation is added (Fig. 5, top panels). It is recalled here that the initial conditions of EXP-H2-EPS are directly interpolated from the ECMWF ENS ones, from a 0.25 to 0.025 degree grid, while the initial conditions of COSMO-H2-EPS are interpolated from the COSMO-LEPS initial conditions, which are in turn interpolated from the ECMWF ENS ones, with a two-step interpolation: the former from a 0.25 to 0.0625 degree grid, the latter from the 0.0625 grid to the 0.025 deg grid. The spectra are identical up to a wavenumber around 1e-04, equivalent to a wavelength of about 60 km, approximately equal to the scale resolved by the global ensemble (estimated as double the grid spacing), which provides all the larger-scale signals. For smaller scales, the spectra behave quite differently. With the direct interpolation (top right), perturbation amplitude is low on scales smaller than 60 km. The magnitude of the perturbations decreases quite steadily, with some noise especially at the wavelengths between 10 and 5 km, close to the resolution of the 2.8 km grid. However, with the intermediate step (top left) there is a peak of magnitude in the perturbations at wavelengths between 20 and 10 km, approximately at the scale resolved by the 7 km model, followed by a sharp decrease and with little signal on the smaller scales. The peak at the 20–10 km wavelength is the part of the perturbations introduced by the interpolation on the 7 km grid. After 6 h of forecast (middle panels), the difference between the two spectra are significantly reduced, both at the 20–10 and 10–5 km scale. The two spectra become quite similar after 12 h (bottom panels), both being characterised by a decrease in the perturbation magnitude for wavelengths below 10 km. Therefore the geopotential perturbations behave similarly from the point of view of how the signal propagates at the different scales. Similar behaviour has been detected for the other cases (not shown).

3.2 Analysis of the precipitation fields

The impact of the different nesting approaches on the precipitation forecast is first evaluated by analysing the precipitation patterns, then by performing an objective verification of the ensemble performances against observations.

The difference between 6-hourly precipitation fields forecasted by the members of COSMO-H2-EPS and of EXP-H2-EPS has been computed, summing up the RMS differences between pairs of members. This computation has been performed at first on the original grid, then it has been repeated after an upscaling of the fields over boxes which contain an increasing number of grid points: $2 \times 2, 4 \times 4, \ldots$, up to 256×256 points. Finally, the RMS difference has been divided by the spread of the COSMO-H2-EPS ensemble, to provide an indication of its significance with respect to the variability of the reference ensemble. Computation has been performed for each case and for each forecast range separately. In Fig. 6 the ratio between RMS difference and spread is shown for six cases, as a function of the horizontal scale of upscaling, for forecast ranges from 0–6 up to 30–36 h. It should be noted that the vertical scale is not the same in all the plots. This choice has been made to improve readability and should not affect the analysis of the results, since, as described, RMS difference is normalised by the correspondent spread. In order to identify the scale characterising the difference in the precipitation patterns, the curve decrease is analysed: a marked decrease in the curve (referred to as a "knee" in the curve) indicates which level of upscaling is needed to make the precipitation patterns significantly more similar to each other, compared with the spread of the ensemble. This level of upscaling, expressed in km of the upscaling box, detects the minimum spatial scale of the differences. When it is not possible to find a "knee", the scale at which the ratio between RMS difference and spread halves is adopted as a quantitative indicator of the scale of the differences. Here only 6 cases are shown, selected in order to represent the full spectrum of different behaviours of the 21 cases.

For the 2012092512 case (Fig. 6a), the differences are quite constant up to a horizontal scale of 8×8 grid points, then they start to decrease. A more marked decrease takes place for most forecast ranges at scales between 16 and 32 grid points. Therefore, the precipitation patterns forecasted by the two systems have a minimum scale of differences of the order of 24 km, but differences remain relatively high up to 90 km.

For the 2012093012 case (Fig. 6b), different behaviour is detected. The differences decrease quite steadily from the beginning, with a more marked decrease starting from the 4x4 grid point scale. Therefore, it is difficult to detect a minimum scale for the differences. The scale at which the ratio halves with respect to its original value is around 32×32 grid points, therefore about 90 km.

For the 201210612 case (Fig. 6c), the curves behave similarly to those of the first case. For all the forecast ranges a sharp decrease is observed, either at 16×16 or 32×32 grid point scale. Therefore, the minimum scale of the differences is in the order of 45 or 90 km, depending on the range considered. The 6 h forecast range should not be considered, since only very light and localised precipitation was forecast and observed.

For the 2012101712 case (Fig. 6d), the behaviour is intermediate: the curves decrease already from the beginning, but the decrease is faster than for the 2012093012 case. The scale at which the ratio halves is between 16 and 32 grid points (45 and 90 km) for most of the forecast ranges, reducing to about 24 km only for the 12 h forecast range.

For the 2012102212 case (Fig. 6e), the curve decrease is initially very slow, then more pronounced. The scale at which the ratio halves is quite large, about 64 points (180 km) for the forecast ranges 18, 24, and 30 h, while it is smaller for the 12 (about 90 km) and 6 (about 45 km) h forecast range.

Finally, the 2012103012 case (Fig. 6f) has behaviour similar to that of the 2012101712 case for the first 4 forecast ranges: the curves decrease quite rapidly, especially from the 8×8 grid point scale, reaching values of half the initial ratio at scales between 45 and 90 km.

In general, two main behaviours are found: for some cases and some forecast ranges it is possible to detect a "knee" in the curve of normalised RMS difference, indicating that a minimum scale exists at which the difference between the precipitation patterns takes place. This scale is of variable size, from about 16 (in one case, not shown) up to about 90 km. For other cases and other forecast ranges, the curve already decreases from the beginning, either quite slowly or more rapidly. In these cases, the scale at which the ratio halves has been considered: values are between 45 and 180 km. Overall, it is found that the difference in the

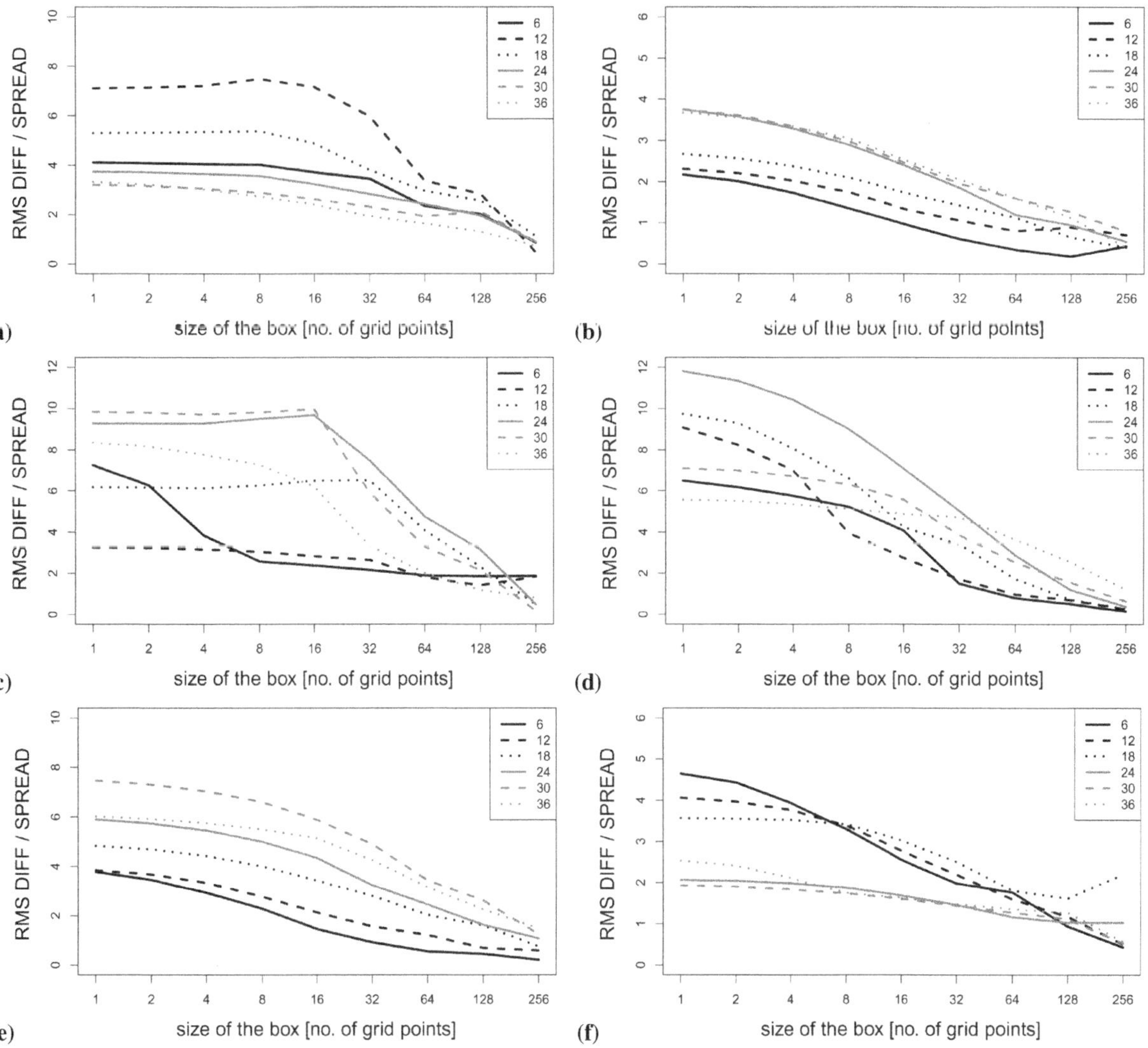

Fig. 6. Ratio between the RMS difference between all the members of the two ensembles and the spread of COSMO-H2-EPS as a function of the horizontal scale of upscaling (indicated as the size, in grid points, of each upscaling box) in terms of 6 h accumulated precipitation for the different forecast ranges (from 0–6 to 30–36 h). Plots are for the cases (from top left to bottom right): 2012092512, 2012093012, 2012101612, 2012101712, 2012102212, 2012103012.

precipitation patterns forecasted by pairs of members is not negligible, even if in some cases it might be difficult to detect it properly with a synop observation network. It is not possible to draw conclusions about the influence of the forecast range, since the cases have been examined separately, and precipitation forecast in the different ranges is actually mainly dependent on the evolution of the phenomenon rather than on the forecast range itself.

Since differences are not negligible, an objective verification has also been performed, in order to assess the quality of the precipitation forecasts issued by the two ensembles. The verification period extends from 25 September to 6 November 2012. As already mentioned, the experimental ensemble has only been run for 21 cases in this period, with a 36 h forecast range. Therefore, the verification sample is slightly smaller than 1 month.

Verification was first performed over the entire domain, in terms of precipitation accumulated over 12 h compared against synop observations (about 400 stations in the domain), considering the forecast value at the nearest grid point for each observation location.

In Table 1 the values of the ranked probability score (RPS) and skill score (RPSS) are shown for each system and for each forecast range (6–18 and 18–30 h). The skill of the two ensembles is comparable, with a little outperformance of EXP-H2-EPS over COSMO-H2-EPS.

In Fig. 7 the Brier skill score (BSS) as a function of the precipitation threshold is shown. The scores are quite close to each other for all thresholds and for both forecast ranges,

Table 1. Ranked probability score and skill score of the two ensembles computed in terms of 12 h accumulated precipitation against SYNOP data.

Ensemble system	6–18 h		18–30 h	
	RPS	RPSS	RPS	RPSS
COSMO-H2-EPS	0.21	0.49	0.23	0.40
EXP-H2-EPS	0.20	0.51	0.23	0.41

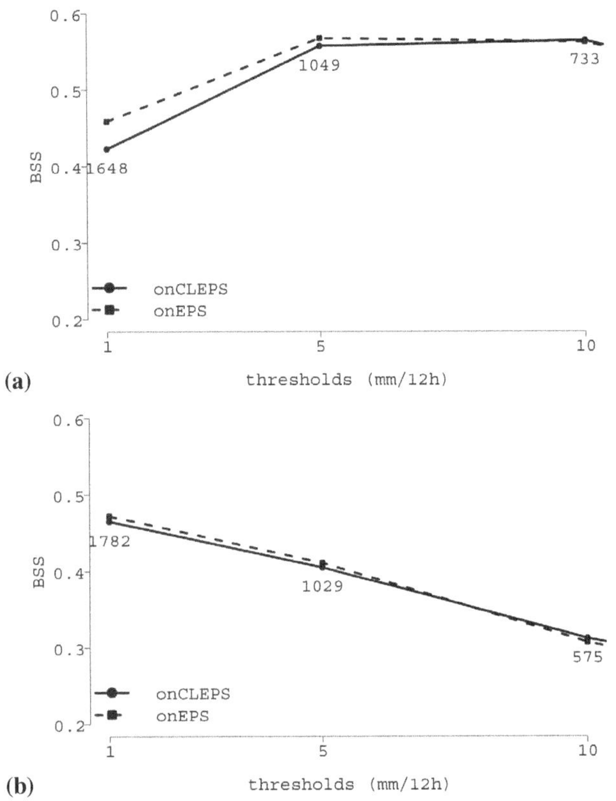

Fig. 7. Brier skill score of COSMO-H2-EPS (solid line, labelled "onCLEPS") and EXP-H2-EPS (dashed line, labelled "onEPS") as a function of the threshold for the 6–18 h **(a)** and 18–30 h **(b)** forecast range. Verification is performed for 12-hourly precipitation. Numbers indicate the observed occurrences of each event.

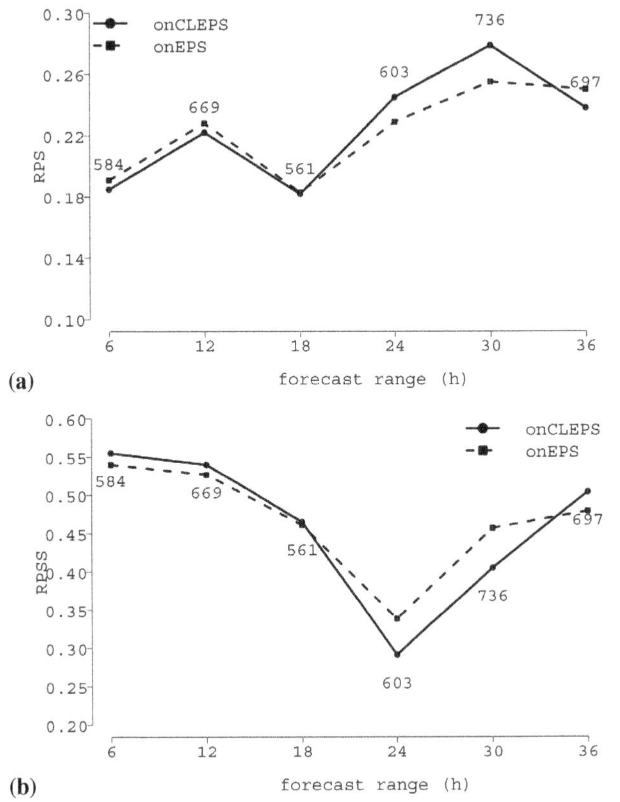

Fig. 8. Ranked probability score **(a)** and skill score **(b)** of COSMO-H2-EPS (solid line, labelled "onCLEPS") and EXP-H2-EPS (dashed line, labelled "onEPS") as a function of the forecast range. Verification is performed for 6-hourly precipitation over northern and central Italy.

EXP-H2-EPS outperforming COSMO-H2-EPS for light precipitation at the first forecast range (6–18 h). The number of events for each threshold is also reported in the plots.

Since some heavy precipitation events took place over northern and central Italy during the analysed period, a different verification has also been performed, aimed at evaluating the impact of the LBC diversity focussing on intense precipitation cases. Precipitation accumulated over 6 h has been considered and scores have been computed against observations collected by a dense network covering northern and central Italy (about 850 stations). In order to perform a verification more suitable for high-resolution forecast of intense events, a different methodology has been adopted, consisting in an upscaling of both forecast and observed values (Marsigli et al., 2008). The verification area has been covered with boxes of 0.2 degrees × 0.2 degrees, then the averages (one for each member) of the values forecasted on the grid points falling in each box are compared with the average of the values observed on the stations falling in the same box, provided that a box contains at least 3 stations.

Both RPS and RPSS indicate that the skill of the two ensembles is comparable (Fig. 8), EXP-H2-EPS slightly outperforming COSMO-H2-EPS at the 24 and 30 h forecast ranges. This dependence on the forecast range is ascribed to the performance of the two systems over a few intense events, which took place at these ranges, rather than on the forecast range itself. In order to check the dependency on the precipitation intensity, the Brier skill score is also shown, for the 1, 5, and 10 mm thresholds (Fig. 9).

While for the 1 mm threshold the scores are quite comparable (Fig. 9a), for the higher thresholds EXP-H2-EPS has higher BSS than COSMO-H2-EPS at the 24 and 30 h forecast ranges, confirming that the outperformance is mainly due to a better forecast of high precipitation.

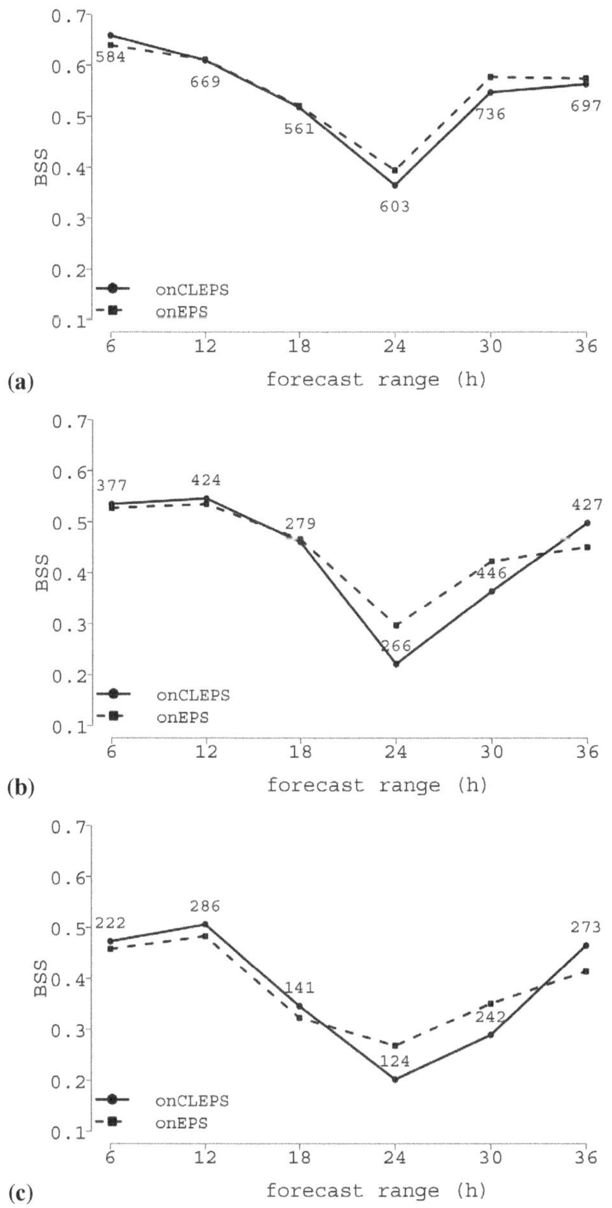

Fig. 9. The same as in Fig. 8 but for the Brier skill score. Thresholds are: 1 mm **(a)**, 5 mm **(b)**, and 10 mm **(c)**.

4 Summary and conclusions

An analysis has been undertaken of the performance of two different nesting approaches for convection-permitting ensemble forecasting. The reference approach, adopted in the COSMO-H2-EPS ensemble, consists of providing ICs and LBCs to the 2.8 km ensemble by a coarser-resolution ensemble based on the same model. The experimental approach, implemented in the EXP-H2-EPS ensemble, is a direct nesting, where ICs and LBCs are provided by a global ensemble, run at quite coarser resolution. The purpose of this work was to investigate the effect of the direct nesting approach, compared with the one based on the intermediate step, for a specific application, which is ensemble forecasting over Italy with the main focus on the forecast of intense precipitation. Analysis has been performed in terms of geopotential at 500 hPa and precipitation.

Results show that the variability introduced in the geopotential fields by the direct nesting is often smaller than the uncertainty described by the reference ensemble, represented by the ensemble spread. When the forecast error is considered, expressed by the RMSE of the ensemble mean, it appears that the differences between pairs of members having different nesting approaches are generally smaller than this error, therefore hardly detectable from ensemble forecasting evaluation. A case study-based evaluation highlights that few cases show a different behaviour. In four cases the experimental nesting approach leads to a better performance of the ensemble, due to the better performance of one member. Nevertheless, in two other cases, an additional error is introduced on two ensemble members with the direct nesting. These results are confirmed by a similar analysis carried out in terms of mean sea level pressure.

Considering the relation between spread and error over the whole sample, it is found that the direct nesting leads to an improved relation, with spread slightly increased and error significantly reduced.

An analysis of the perturbation spectra indicates that the geopotential perturbations behave similarly in the two approaches, from the point of view of the way the signal propagates at the different scales. Both ensembles are characterised by little signal on the small scales at initial time, as expected since ICs are downscaled from analyses perturbed at coarse resolution. Nevertheless, this feature will not affect the 2.8 km ensemble in its operational set-up, since high-resolution perturbed analyses derived from KENDA will be used as ICs.

In terms of precipitation, it is found that forecasts issued by members with different nesting approaches generally differ on spatial scales which are highly dependent on the case, ranging from 16 up to 180 km. Though these differences are not negligible, it might be difficult to evaluate their impact on the forecast quality, since spatial errors in the precipitation fields may reach scales of 50–100 km even if forecasted by a high-resolution model.

The skill of the ensemble precipitation forecasts has been evaluated over the whole domain against synop data, showing comparable performances of the two ensembles. High-resolution verification has also been performed over a restricted domain, where few intense events took place and a dense observation network was available. Results show that the skill of the two ensembles is similar, with a little outperformance of the direct-nesting ensemble for intense precipitation. This evaluation points out that the overall quality of the 2.8 km ensemble in terms of precipitation forecast is not badly affected by the provision of lower resolution LBCs from the global ensemble.

This work should be regarded as the first step of a deeper analysis, where more cases will be considered, especially including different weather situations. Furthermore, it is intended to study the impact of spatial and temporal resolution separately, considering the usefulness of hourly boundary conditions provided by the global ensemble.

Acknowledgements. This work was partly supported by the Italian Civil Protection Department, in the framework of the MODMDET3 Project. The Italian institutions providing observed precipitation data are also acknowledged. The authors are grateful to Jun Du and to two anonymous referees for their valuable comments, which allowed them to improve the manuscript greatly.

Edited by: R. Buizza

References

Amengual, A., Romero, R., Homar, V., Ramis, C., and Alonso, S.: Impact of the lateral boundary conditions resolution on dynamical downscaling of precipitation in mediterranean Spain, Clim. Dynam., 29, 487–499, doi:10.1007/s00382-007-0242-0, 2007.

Buizza, R., Houtekamer, P. L.,Toth, Z.,Pellerin, G., Wei, M., and Zhu, Y.: A Comparison of the ECMWF, MSC, and NCEP Global Ensemble Prediction Systems, Mon. Weather Rev., 133, 1076–1097, 2005.

Davies, T.: Lateral boundary conditions for limited-area models, Q. J. Roy. Meteor. Soc., 140, 185–196, doi:10.1002/qj.2127, 2014.

de Elia, R., Laprise, R., and Denis, B.: Forecasting skill limits of nested, limited-area models: a perfect-model approach, Mon. Weather Rev., 130, 2006–2023, 2002.

Denis, B., Laprise, R., Côté, J., and Caya, D.: Downscaling ability of one-way nested regional climate models: The big-brother experiment, Clim. Dynam., 18, 627–646, 2001.

Ducrocq, V., Braud, I., Davolio, S., Ferretti, R., Flamant, C., Jansa, A., Kalthoff, N., Richard, E., Taupier-Letage, I., Ayral, P.-A., Belamari, S., Berne, A., Borga, M., Boudevillain, B., Bock, O., Boichard, J.-L., Bouin, M.-N., Bousquet, O., Bouvier, C., Chiggiato, J., Cimini, D., Corsmeier, U., Coppola, L., Cocquerez, P., Defer, E., Delanoë, J., Di Girolamo, P., Doerenbecher, A., Drobinski, P., Dufournet, Y., Fourrié, N., Gourley, J. J., Labatut, L., Lambert, D., Le Coz, J., Marzano, F. S., Molinié, G., Montani, A., Nord, G., Nuret, M., Ramage, K., Rison, B., Roussot, O., Said, F., Schwarzenboeck, A., Testor, P., Van Baelen, J., Vincendon, B., Aran, M., and Tamayo, J.: HyMeX-SOP1, the field campaign dedicated to heavy precipitation and flash flooding in the northwestern Mediterranean, B. Am. Meteorol. Soc., online first, doi:10.1175/BAMS-D-12-00244.1, 2013.

Ferretti, R., Pichelli, E., Gentile, S., Maiello, I., Cimini, D., Davolio, S., Miglietta, M. M., Panegrossi, G., Baldini, L., Pasi, F., Marzano, F. S., Zinzi, A., Mariani, S., Casaioli, M., Bartolini, G., Loglisci, N., Montani, A., Marsigli, C., Manzato, A., Pucillo, A., Ferrario, M. E., Colaiuda, V., and Rotunno, R.: Overview of the first HyMeX Special Observation Period over Italy: observations and model results, Hydrol. Earth Syst. Sci. Discuss., 10, 11643–11710, doi:10.5194/hessd-10-11643-2013, 2013.

Gerard, L., Piriou, J.-M., Brožková, R., Geleyn, J.-F., and Banciu D.: Cloud and Precipitation Parameterization in a Meso-Gamma-Scale Operational Weather Prediction Model, Mon. Weather Rev., 137, 3960–3977, 2009.

HYMEX – HYdrological cycle in the Mediterranean EXperiment, Hymex Project, available at: www.hymex.org, last access: 4 March 2014, 2010–2020.

Marsigli, C., Montani, A., Nerozzi, F., Paccagnella, T., Tibaldi, S., Molteni, F., and Buizza, R.: A strategy for High–Resolution Ensemble Prediction. Part II: Limited–area experiments in four Alpine flood events, Q. J. Roy. Meteor. Soc., 127, 2095–2115, 2001.

Marsigli, C., Montani, A., and Paccagnella, T.: A spatial verification method applied to the evaluation of high–resolution ensemble forecasts, Meteorol. Appl., 15, 125–143, 2008.

Marsigli, C., Montani, A., and Paccagnella, T., 2013. Test of a COSMO-based convection-permitting ensemble in the Hymex framework, COSMO Newsletter No. 13, available at: http://www.cosmo-model.org/content/model/documentation/newsLetters/default.htm, last access: 4 March 2014, 2013

Molteni, F., Buizza, R., Marsigli, C., Montani, A., Nerozzi, F., and Paccagnella, T.: A strategy for High–Resolution Ensemble Prediction. Part I: Definition of Representative Members and Global Model Experiments, Q. J. Roy. Meteor. Soc., 127, 2069–2094, 2001.

Montani, A., Cesari, D., Marsigli, C., and Paccagnella, T.: Seven years of activity in the field of mesoscale ensemble forecasting by the COSMO-LEPS system: main achievements and open challenges, Tellus A, 63, 605–624, 2011.

Nutter, P., Stensrud, D., and Xue, M.: Effects of coarsely resolved and temporally interpolated lateral boundary conditions on the dispersion of limited-area ensemble forecasts, Mon. Weather Rev., 132, 2358–2377, 2004.

Reich, H., Rhodin, A., and Schraff, C.: LETKF for the nonhydrostatic regional model COSMO-DE, COSMO Newsletter No. 11, available at: http://www.cosmo-model.org/content/model/documentation/newsLetters/default.htm, last access: 4 March 2014, 2011.

Steppeler, J., Doms, G., Schattler, U., Bitzer, H. W., Gassmann, A., Damrath, U., and Gregoric, G.: Meso-gamma scale forecasts using the nonhydrostatic model LM, Meteorol. Atmos. Phys., 82, 75–96, 2003.

Termonia, P.: Monitoring and improving the temporal interpolation of lateral-boundary coupling data for limited-area models, Mon. Weather Rev., 131, 2450–2463, 2003.

Termonia, P., Deckmyn, A., and Hamdi, R.: Study of the lateral boundary condition temporal resolution problem and a proposed solution by means of boundary error restarts, Mon. Weather Rev., 137, 3551–3566, 2009.

Warner, T. T., Peterson, R. A., and Treadon, R. E.: A tutorial on lateral boundary conditions as a basic and potentially serious limitation to regional numerical weather prediction, B. Am. Meteorol. Soc., 78, 2599–2617, 1997.

8

Latitudinal variation of stochastic properties of the geomagnetic field

J. A. Wanliss[1], **K. Shiokawa**[2], **and K. Yumoto**[3]

[1]Presbyterian College, 503 South Broad Street, Clinton, SC 29325, USA
[2]Solar-Terrestrial Environment Laboratory, Nagoya University, Furo-cho, Chikusa-ku, Nagoya, Aichi, 464-8601, Japan
[3]Space Environment Research Center, Kyushu University, 53 6-10-1 Hakozaki, Higashi-ku Fukuoka, 812-8581, Japan

Correspondence to: J. A. Wanliss (jawanliss@presby.edu)

Abstract. We explore the stochastic fractal qualities of the geomagnetic field from 210 mm ground-based magnetometers during quiet and active magnetospheric conditions. We search through 10 yr of these data to find events that qualify as quiet intervals, defined by $\mathrm{Kp} \leq 1$ for 1440 consecutive minutes. Similarly, active intervals require $\mathrm{Kp} \geq 4$ for 1440 consecutive minutes. The total for quiet intervals is $\sim 4.3 \times 10^6$ and 2×10^8 min for active data points. With this large number of data we characterize changes in the nonlinear statistics of the geomagnetic field via measurements of a fractal scaling. A clear difference in statistical behavior during quiet and active intervals is implied through analysis of the scaling exponents; active intervals generally have larger values of scaling exponents. This suggests that although 210 mm data appear monofractal on shorter timescales, the scaling changes, with overall variability are more likely described as a multifractional Brownian motion. We also find that low latitudes have scaling exponents that are consistently larger than for high latitudes.

1 Introduction

The transfer of energy, mass, and momentum from the solar wind into the magnetospheric cavity is far from steady state. Rapid transfers of energy cause the magnetosphere to move from a relatively low-energy state into more "exited" states where energy dissipation occurs suddenly in global dynamical processes known as space storms. They are the most dramatic space weather phenomenon that significantly impact modern technology such as satellites, communication and power transmission systems. Space physicists understand their development and morphology, but discussion is still open regarding detailed causes of space storms.

Energy dissipation in the terrestrial magnetosphere often follows intermittent temporal patterns of high activity with abrupt convection increases, particle bursts, impulsive enhancements of $O+ / H+$ density ratio, compressional pulses, intensified fluctuations of field-aligned currents and other bursty processes. These impulsive processes may be separated by periods of quiescence (Korth et al., 2003). Bursty intermittency is evident in both in situ and ground observations (Hori et al., 2005), and are observed under a wide variety of interplanetary conditions (Pulkkinen et al., 2007).

Ground-based geomagnetic indices (Takalo et al., 1999; Wanliss, 2004, 2005; Balasis et al., 2011) and individual magnetometer stations (Vörös, 2000; Wanliss and Reynolds, 2003) provide excellent signals of the fractal nature of space weather energy dissipation. Part of the reason for this is the property of the earth's magnetic field lines to focus and converge when approaching the earth. These field lines extend far into space and since they connect to the earth, nonlinear plasma processes that occur far away map all the way down to the planetary surface. Observation of ground-based magnetometer stations can thus serve as a remote sensing tool of distant magnetospheric features and processes.

Over the years, several indices were developed to monitor geomagnetic activity in a global sense. The most commonly used are the disturbance storm time index (DST), the planetary index (Kp) and the auroral electrojet index (AE, AU and AL). These indices provide global information about magnetospheric activity by combining inputs at different locations around the globe. The Kp index, which we utilize in this paper, is calculated with data from 13 selected, subauroral magnetometer stations at mid- to high latitudes. Because the stations cover a wide range of latitudes it is a relevant reflection of average global magnetospheric activity.

The Kp index is obtained as the average value of the disturbance levels in the two horizontal field components, observed at the reference magnetometer stations (Bartels et al., 1939; Bartels, 1963). Calculation of Kp begins by removing the quiet-day variation from the individual magnetograms (cf. Wanliss and Showalter, 2006). The next step finds the range of the horizontal magnetic field disturbances during 3-hour time intervals, for eight values each day. At each station the range is converted to a quasi-logarithmic scale index with values from 0 to 9. The logarithmic scale normalizes the occurrence frequency of different size disturbances. Finally, an averaging procedure convolves the local indices to form the global geomagnetic index Kp.

If we are interested in the local aspects of geomagnetic activity, i.e., to forecast the geomagnetic conditions for power utilities, we need to develop ways to understand the geomagnetic activity in a more localized way. This is especially important since temporal fluctuations of the geomagnetic field and, more importantly, their nonlinear fractal or multifractal statistical fluctuations depend on geographic location and time (Anh et al., 2007; Yu et al., 2010).

Furthermore, the intermittent character of geomagnetic data suggests that changes in the magnetosphere depend not only on solar wind variability, but also on internal magnetospheric activity (Consolini et al., 1996; Consolini and De Michelis, 1998; Dobias and Wanliss, 2009; Wanliss and Uritsky, 2010). The origin of such an intermittent and multifractal character has been interpreted and modeled in terms of dynamical changes and turbulent dynamics (Wanliss et al., 2005; Anh et al., 2007; Yu et al., 2010; Sharma and Veeramani, 2011).

In this paper we extend previous analyses that considered only global statistics to study the differences between quiet and active magnetospheric times (Wanliss, 2004, 2005). Global studies give only average behavior rather than local information. Our goal is to learn about local behavior of the magnetic field, for differing geomagnetic activity. To do so we will characterize changes in the nonlinear statistics of the earth's magnetic field, by means of the fractal scaling (Hurst) exponent, measured from ground-based magnetometer station.

Previous studies by Wanliss and Reynolds (2003) suggested the possibility that the ubiquitous fractal scaling of geomagnetic data may yet have variability on a latitudinal scale. In addition, while the fractal character of geomagnetic data is ubiquitous, there is strong evidence that the fractal character varies with geomagnetic activity (Wanliss and Dobias, 2007).

The present study explores this further, asking the following questions: (1) Does average fractal scaling vary in a predictable way as a function of latitude? and (2) Are magnetospherically quiet times always, and everywhere, characterized by smaller fractal scaling measures than active times?

Affirmative answers would suggest the possibility that changes in statistics might then be used as local indicators of magnetospheric conditions, which may be useful to develop reliable warning and forecasting systems using information not available in geomagnetic indices.

A second objective is to determine the long-term fractal nature of the geomagnetic field at a local observation site. If the time series can be described as a particular statistical process – fractional Brownian motion for example – then this knowledge can be used for future space weather modeling purposes. Or scaling may be similar to that of multifractional Brownian motion (mBm) (Peltier and Lévy Vehel, 1995; Muniandy and Lim, 2001), where the scaling features depend on time. The statistical structure of the magnetometer time series provides key requirements to guide development of mathematical models. Magnetohydrodynamic models, for example, should correctly reproduce the correct statistical structure seen in the data.

In the following pages we present our examination of these questions. Section 2 describes the analysis techniques employed and important background to statistical methods. In Sect. 3 we describe the criteria used to select and process the data. Section 4 presents results. Finally, in Sect. 5 we form conclusions.

2 Analysis techniques

2.1 Fractional Brownian motions

A signal that displays fractional Brownian motion (fBm) can be expressed as (Mandelbrot and Van Ness, 1968) a stochastic integral of time integrations of fractional Gaussian noise:

$$B_H(t) = \frac{1}{\Gamma(H+1/2)} \left\{ \int_{-\infty}^{0} \left[(t-s)^{H-1/2} - (-s)^{H-1/2} \right] \mathrm{d}W(s) + \int_{0}^{t} (t-s)^{H-1/2} \mathrm{d}W(s) \right\}.$$

Here W is a white noise process defined on $(-\infty, \infty)$, and $H \in (0, 1)$ is known as the Hurst parameter. The Hurst exponent for the signal is its roughness averaged over many length scales. The covariance function is given by

$$\mathrm{cov}\{B_H(s), B_H(t)\} = \frac{1}{2}\{|s|^{2H} + |t|^{2H} - |s-t|^{2H}\},$$

so that $B_H(0) \equiv 0$ and $\mathrm{var}\{B_H(t)\} = t^{2H}$. This means that for the special case $H = 1/2$, fBm reduces to the well-known random walk. Typically, fBm is nonstationary, and thus detection of the presence of memory is a delicate task. It has been observed in a variety fields, including hydrology

(Neuman and Federico, 2003), geophysics (Wanliss and Antoine, 1995; Frisch, 1997), biology (Collins and De Luca, 1994), telecommunication networks (Taqqu et al., 1997), and others.

Traditional estimates of possible fractal scaling employ power spectral density (PSD) analysis. The slope β obtained from the plot of PSD $P(\nu)$ vs. frequency ν determines the correlation level of the signal. If $P(\nu) \propto \nu^{-\beta}$, and $-1 < \beta < 1$ then the signal is fractional Gaussian noise (fGn) and it is stationary, which means that the signal is statistically invariant by translation in time. When the signal $B(t)$ displays fractional Brownian motion (fBm) both the real and imaginary components of the Fourier amplitudes are Gaussian-distributed random variables (Hergarten, 2002). As well, the mean of the Fourier amplitudes can be given by $\overline{\varphi(\nu)} = 0$ and $\varphi(\nu)\varphi(\nu)^* = P(\nu)\delta\left(\nu - \nu'\right)$. The last term is the Dirac delta function.

This means that for $\beta = 2$, fBm reduces to the random walk with power law spectrum varying as an inverse square. Signals with scaling exponents above $\beta = 2$ are called persistent, because if the data at some point have $B(t_{i+1}) > B(t_i)$, for example, then the probability is greater than 0.5 that $B(t_{i+2}) > B(t_{i+1})$. Signals with exponents below 2 are called antipersistent because if $B(t_{i+1}) > B(t_i)$, the probability is greater than 0.5 so that $B(t_{i+2}) < B(t_{i+1})$.

If the signal is fBm, it exhibits power-law scaling with slope $1 < \beta < 3$. In this case the signal is nonstationary but has stationary increments over a range of scales. For fBm

$$\beta = 2H + 1,$$

where H is the scaling exponent, also known as the Hurst exponent. Brownian motion marks the special case where $\beta = 2\ (H = 0.5)$.

Typically, fBm is nonstationary, and detection of memory is a delicate task. Nonstationarity means that statistical properties are not constant through the signal, and one cannot use traditional analysis methods that assume stationarity. The problem with spectral analysis, certainly in the context of the present study, is that it assumes a stationary signal. When implemented with nonstationary signals, spectral analysis is incapable of distinguishing frequency content hidden by the presence of nth-order polynomial trends and requires long time series for reasonable accuracy (Stanley et al., 1999).

2.2 Detrended fluctuation analysis

Although much effort is focused on developing reliable estimators of long-range correlations, there appears to be no clear community consensus on the methods best suited for various types of data. In addition, tests of synthetic time series show that the performance of estimators varies when using different generators of long-range dependent data (Shao et al., 2012).

To determine the self-similarity parameter H we implement detrended fluctuation analysis (DFA) developed by Peng et al. (1995) and recently implemented in space physics research by Wanliss (2004, 2005), and subsequent papers. The technique determines the scaling exponent of nonstationary signals and provides better precision than the power spectral analysis and other more modern techniques. DFA is a modified root mean squared analysis of a random walk designed specifically to be able to deal with nonstationarities in nonlinear data, and is among the most robust of statistical techniques designed to detect long-range correlations in time series (Taqqu et al., 1995; Cannon et al., 1997; Blok, 2000; Shao et al., 2012). DFA has been shown to be robust to the presence of trends (Hu et al., 2001) and nonstationary time series (Kantelhardt et al., 2002; Chen et al., 2002). Bryce and Sprague (2012) argue that DFA has significant problems but they consider only the limited case of fGns with $H = 0.3$, generated using the Davies–Harte algorithm. Shao et al. (2012) consider these and other recent criticisms of DFA and conclude that DFA "remains the method of choice" when the trend in the data is not a priori known.

Briefly, the DFA methodology begins by removing the mean, $\bar{B}$, from the time series, $B(t)$, and then integrating

$$y(k) = \sum_{t=1}^{k} \left[B(t) - \bar{B}\right].$$

The new time-series is then divided into boxes of equal length, n (Fig. 2). The trend, represented by a least-squares fit to the data, is removed from each box; the trend is typically a linear, quadratic, or cubic function. Box n has its abscissa denoted by $y_n(k)$. Next the trend is removed from the integrated time series, $y(k)$, by subtracting the local trend, $y_n(k)$, in each box.

For a given box size n, the characteristic size of the fluctuations, denoted by $F(n)$, is then calculated as the root mean squared deviation between $y(k)$ and its trend in each box

$$F(n) = \sqrt{\frac{1}{N}\sum_{k=1}^{N}\left[y(k) - y_n(k)\right]^2}.$$

The presence of scaling is indicated by a power-law relationship between $F(n)$ and n as follows:

$$F(n) \propto n^{\alpha},$$

where $\alpha = H (0 < \alpha < 1)$ is the scaling exponent. The slope of a log-log curve of $F(n)$ vs. n indicates the value of the scaling exponent. For our data n is in units of s, and $F(n)$ in units of nT^2. If $\alpha = 0.5$ then the signal is white noise. A value of $\alpha < 0.5$ indicates uncorrelated data (antipersistent) and if $0.5 < \alpha < 1$ then there is correlation in the time series (i.e., long-term memory).

3 Data selection

In order to characterize the latitudinal variation of magnetosphere fractal scaling, we use the 190, 210, and 250°

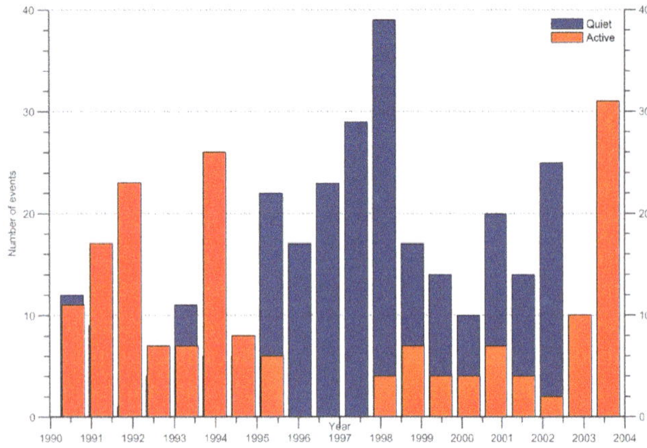

Fig. 1. Histogram of active (red) and quiet (blue) events.

magnetic meridian data (Yumoto et al., 1996). These data feature near-continuous, high-resolution (1 s cadence) data from a magnetometer chain spanning almost 160° in geomagnetic latitude. Magnetometer stations are all close to 210° geomagnetic longitude. Tables 1–3 provide the station locations, given in geographic, magnetic, and L-shell coordinates.

We chose the 3-hour Kp index to discriminate between different levels of magnetospheric activity. We might have adopted other indices, for example DST or AE, but we chose Kp since we considered that it, as a mid- to high-latitude index, would best reflect the mean magnetospheric activity over the wide range of latitudes for the magnetometer data.

Several methods to classify geomagnetic activity using the Kp index have been proposed and used by different authors. Bartels (1963) used $Kp \leq 1$ as an indicator of quiet periods, and $Kp \geq 4$ as an indicator of disturbed periods (Rangarajan and Iyemori, 1997). In this work our interest focuses on two averaged geomagnetic states: active and quiet, so we adopted the Bartels criteria.

Data selected for quiet times (QT) are based on those 24 h periods between 1990 and 2003 where $Kp \leq 1$. Active events (AT) were selected from 24 h periods of time having $Kp \geq 4$. No overlap of events was allowed. In other words, if for instance $Kp \geq 4$ for 48 consecutive hours, then there are two AT events, one beginning at the start of the interval, the other after the elapse of 24 h. But if $Kp \geq 4$ for 40 h we consider this as only one AT event, comprising the first 24 h. Using non-overlapping events helps prevent any single period dominating and biasing the overall statistics.

With these criteria we find a number of 178 active day and 287 quiet day events. Figure 1 shows the yearly distribution of all the events selected, both active and quiet. Most of the active events (red) are close to solar maxima, while most of the quiet events congregate near solar minimum (around 1997).

Local time variations are known to exist in geomagnetic fluctuations, for instance the solar quiet fluctuations near noon (Wanliss and Showalter, 2006) which are absent at local midnight. The extent to which such variations influence fractal scaling is unknown. It is partly because of this unknown variability that we use a 24 h selection criterion. In this case the station samples every local time, reducing the likelihood of unknown local time variability dominating the statistical analysis. In essence, a 24 h period is needed since this allows enough time for any particular magnetometer to sample each different longitudinal region of the magnetosphere. Our analysis thus uses data in a way that is analogous to geomagnetic index data. Geomagnetic indices convolve data from many magnetometers, measured at the same universal time, but at many different local times (and latitudes). Our analysis takes a station at a single geographic location and allows it to sweep through every local time. Thus the possibility that fractal scaling also depends on longitude is negated since the entire longitudinal range is sampled. In addition, this period

Table 1. Location of 190° magnetometer stations used in this study.

Station	IAGA code	Geographic Lat	Geographic Long	CGM Lat	CGM Long	L Val
Tixie	TIK	71.6	129.0	65.7	197.1	5.89
Lunping	LNP	25.0	121.2	13.8	189.5	1.06
Muntinlupa	MUT	14.4	121.0	3.6	191.6	1.00
Pontianak	PTN	−0.1	109.2	−11.4	180.5	1.04
Learmonth	LMT	22.2	114.1	34.1	185.0	1.46
Katanning	KAT	−33.7	117.6	−46.6	188.2	2.12

Table 2. Location of 210° magnetometer stations.

Station	IAGA code	Geographic Lat	Geographic Long	CGM Lat	CGM Long	L Val
Kotel'nyy	KTN	75.9	137.7	69.9	201.0	8.50
Chokhurdakh	CHD	70.6	147.9	64.7	212.1	5.46
Zyryanka	ZYK	65.7	150.8	59.6	216.7	3.91
Magadan	MGD	60.0	150.9	53.6	218.7	2.83
St. Paratunka	PTK	52.9	158.2	46.3	225.9	2.10
Moshiri	MSR	44.4	142.3	37.6	213.2	1.59
Rikubetsu	RIK	43.5	143.8	34.7	210.8	1.48
Onagawa	ONW	38.4	141.5	31.6	212.5	1.38
Kagoshima	KAG	31.5	130.7	25.1	202.2	1.22
Yamakawa	YMK	31.19	130.62	24.85	202.13	1.21
Chichijima	CBI	27.2	142.3	20.6	213.0	1.14
Guam	GUA	13.6	144.9	4.6	214.7	1.01
Yap	YAP	9.3	138.5	−0.3	209.0	1.00
Biak	BIK	−1.1	136.1	−12.2	207.3	1.05
Wewak	WEW	−3.6	143.6	−14.4	215.3	1.06
Darwin	DAW	−12.4	130.9	−23.1	202.7	1.18
Weipa	WEP	−12.7	141.9	−23.0	214.3	1.18
Birdsville	BSV	−25.8	139.3	−36.9	213.1	1.56
Dalby	DLB	−27.2	151.2	−37.1	226.8	1.57
Canberra	CAN	−35.3	149.0	−46.0	226.1	2.07
Adelaide	ADL	−34.7	138.7	−46.5	213.7	2.11

Table 3. Location of 250° magnetometer stations.

Station	IAGA code	Geographic Lat	Geographic Long	CGM Lat	CGM Long	L Val
Kotzebue	KOT	66.9	197.4	64.5	249.7	5.40
Ewa Beach	EWA	21.3	202.0	22.7	269.4	1.17
Macquarie Isl.	MCQ	−54.5	158.9	−64.5	247.8	5.40

allows enough data so that random fluctuations are averaged out.

Studies that use geomagnetic indices implicitly include local time effects due to the nature of the convolution of data to form the index. This produces pitfalls that are usually ignored in the numerous studies of index data. These difficulties do not negate the statistical analyses of index data, as is evident from voluminous publications. Similarly, and though we offer no solution to this issue, our analysis adds to our body of knowledge.

Since not all of the magnetometer stations are operational at the same time, we scanned through the 190, 210, and 250 mm data, selecting data from stations when available, that overlap with the Kp selection criteria.

In the end we used data from 40 magnetometers across almost 160° magnetic latitude. Since the magnetometer data are often noisy, subject to DC shifts, spikes, monochromatic constant amplitude oscillations, and other spurious effects, it is necessary to further cull the magnetometer data. Because of these difficulties with quality we carefully examine magnetometer data by eye to eliminate spurious data. We instituted a laborious routine whereby data were visually examined before analysis. From the Kp data intervals of 24 h were selected which satisfied the appropriate quiet and active criteria. Next, we visually inspected all of the 210 mm data in 24 h blocks falling in the Kp selected intervals. Data having spurious features, mentioned above, were eliminated. Once this was done we still found good quality data for a total of 2.6×10^8 QT data points and 1.2×10^8 AT data points.

Figure 2 gives a sample distribution function of the magnetic field for QTs and ATs. In this plot the solid curve characterizes the X component magnetic field behavior during AT, while the dashed curve represents the QT measured at the MSR station.

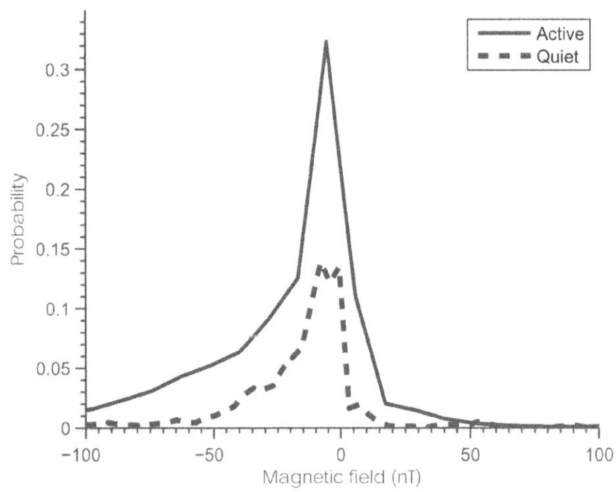

Fig. 2. Distribution functions for the active and quiet events analyzed for the X component magnetic field at the MSR magnetometer station. The dashed line represents the averaged distributions for the quiet events and the solid line represents the averaged distributions for the active events.

4 Results

Once data are selected we apply DFA. Because we analyze 24 h intervals there are $N = 86400$ data points in each series. We use a box size beginning at 8 s and terminating at $N/4 = 21\,600$ s. We use such a cutoff since the low and high box-number edges should be treated with caution. The first few points at the low end must be disregarded since detrending here removes too much of the fluctuation. For larger values of the box size, there are too few boxes for a proper average to be made – a finite size effect – hence we disregard those values. Selecting a box size terminating at less than $N/4$ still allows reasonable estimates of scaling exponents. However, following Hu et al. (2001) and Kantelhardt et al. (2002) we opt for the larger box size since it allows more reliable estimates of the scaling exponent over more than two decades of magnitude.

Figure 3 shows an example of the fluctuation function $F(n)$ against scale size n for the quiet event of 20 January 2000, for three geomagnetic components. The inset shows the magnetic field components for the Biak magnetometer station (BIK). For this event scaling is present for almost four decades of time, with $F(n)$ shown by the circles. Some deviation from the best fit (solid line) is clear near the smallest and largest scales, and is common for all events. Accordingly, to analyze these data we use least-squares fitting over a range from 100 s to 12 h. For this event the scaling exponent is about $\alpha \sim 0.8 - 0.9$. The fits are excellent and uncertainties are on the order of 1–2 % of the value for the scaling exponent. We find that smaller box sizes yield similar scaling exponents, but based on a linear fitting on a smaller scale range, resulting in larger uncertainty.

Figures 4 and 5 summarize the results obtained for the distribution of scaling exponent α for the 3027 quiet and 1422 active events computed for all latitudes. The figures show a Gaussian best fit. Other distributions might be fitted, such as Rayleigh, Maxwell, normal, and log-normal distributions (Wanliss and Showalter, 2006). For quiet events the statistics are fit best by a Gaussian. Active data appear slightly different, having the inkling of stable, heavy-tailed distributions, mainly in the 4-component results. Table 4 summarizes these results, showing that the standard deviations are quite large. The mean values of the scaling exponents α are statistically similar, with $\alpha \sim 0.65 - 0.7$ irrespective of quiet or active times.

As found previously, by Wanliss and Reynolds (2003) and Yu et al. (2010), we find a broad range of scaling exponents, as evidenced by large standard deviations from the mean. This is what one might expect of exponents calculated from magnetometers spanning magnetic latitudes from the equator to auroral regions; if fractal scaling is latitude dependent, a broad α distribution must be the consequence.

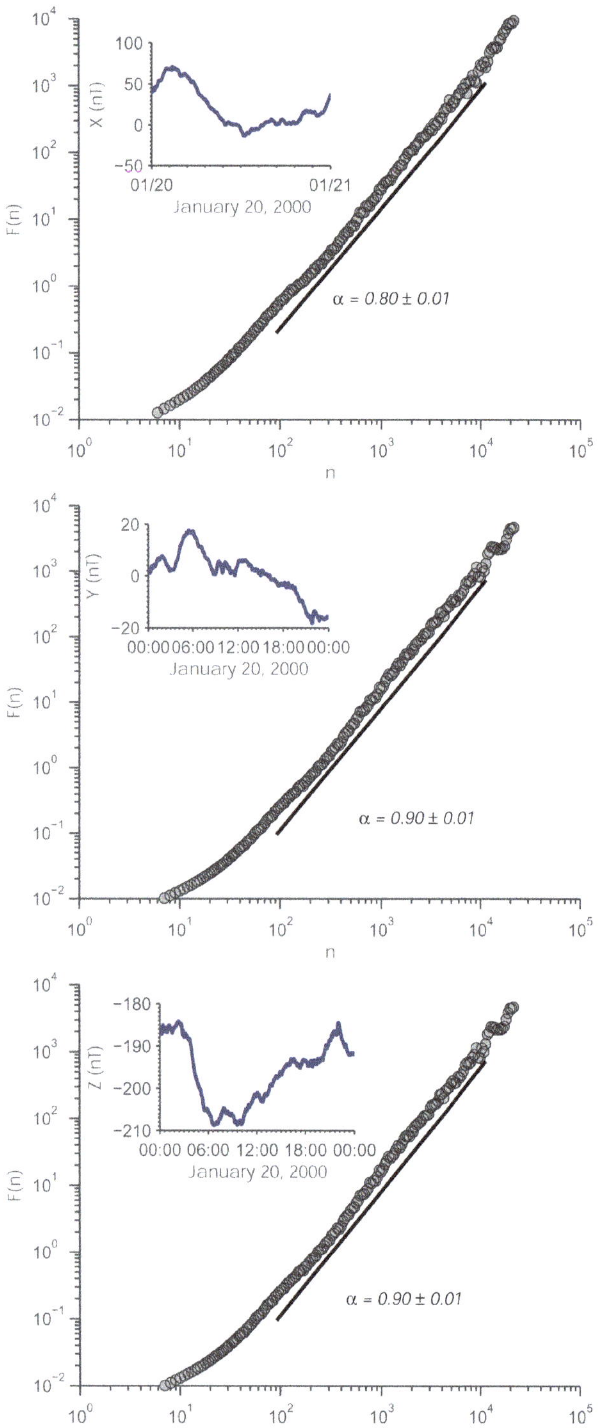

Fig. 3. Circles show an example of the fluctuation versus box size for the quiet event beginning on 20 January 2000. Solid curve below the circles is the best fit linear curve to the data (shifted) and slope α is shown on each graph. The inset in this example shows the component magnetometer data from the BIK station used to compute the fluctuation curves.

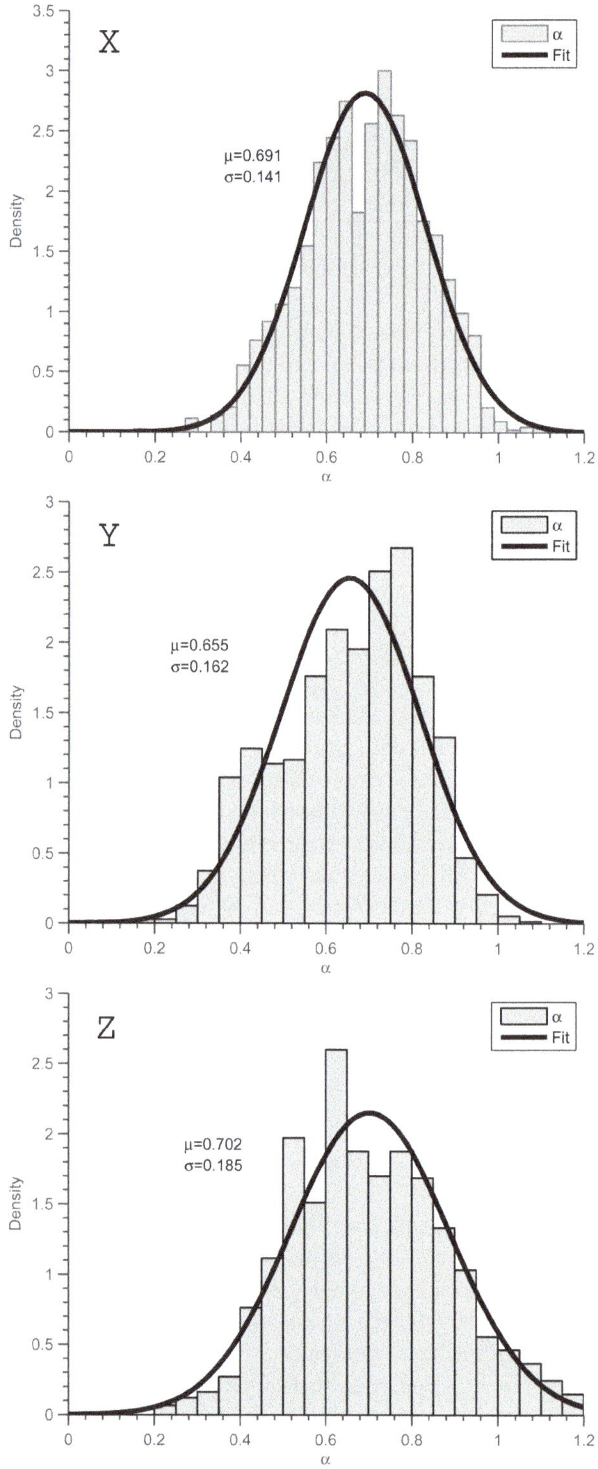

Fig. 4. Distributions of the fractal scaling exponent for quiet events, for all latitudes. These data are fitted to a normal distribution (solid black), with the mean and standard deviation given in the figures.

We find that when all latitudes are included in the analysis there is no statistical difference between fractal exponents measured during quiet and active intervals. However,

Table 4. Scaling exponents for the quiet and active events computed for all latitudes.

	$X\alpha$	$Y\alpha$	$Z\alpha$
Quiet	0.64 ± 0.13	0.65 ± 0.14	0.67 ± 0.16
Active	0.69 ± 0.14	0.66 ± 0.16	0.70 ± 0.19

there is some difference in the overall distribution functions, and the distribution of active component Y component data is slightly asymmetric with a heavier tail for $\alpha < 0.5$.

Skewness is a measure of asymmetry of the data around the sample mean. It is positive(negative) when the data spread out more to the right(left). For a normal distribution, skewness is zero. We find the skewness for quiet events for X, Y, and Z magnetic field components to be, respectively $[-0.40, -0.16, -0.09]$. For active events the corresponding skewness is $[-0.20, -0.40, 0.26]$. None of these distributions are significantly different from a normal distribution for the means are essentially the same as the medians.

Next we examine whether we can distinguish the latitudinal dependency from the rest of the variability. We attempt this by considering the average of many scaling exponents that were computed at a single station, calculating the mean and standard deviation of them all. Figure 6 shows the average scaling exponent against geomagnetic latitude for quiet (magenta circles) and active (red triangles) events. The best-fit polynomial shown is a quadratic function. We fitted various polynomials and found that the second-order polynomial gave the best fit for X and Y. Our preference was to fit the simplest possible model that provides reasonable fits. The order of the best-fit polynomial is determined on the basis of (1) visual examination of the residuals, (2) goodness-of-fit statistics, and (3) confidence bounds. The best-fit is computed with the Levenberg–Marquardt algorithm for nonlinear least squares (Seber and Wild, 2005).

A latitudinal variability emerges from these data. For X and Y geomagnetic components there is a tendency for the scaling exponent to increase as the magnetometer approaches the geomagnetic equator. No such variation is plain in Z. In addition this behavior is consistent across the quiet to active events, thus independent of geomagnetic activity. As well, for each magnetic component the best-fit active events tend to have scaling exponents larger than for quiet times.

To explore whether these quiet and active event fractal scaling exponents are significantly different from the null hypothesis – that the difference is due purely to randomness – we applied the Students paired t test to the distributions. The important output of the paired t test is the value of p, which is the probability the difference in the means of the two distributions is due to random variation. The test focuses on the difference between the paired data and reports the probability that the actual mean difference is consistent with zero.

We found for $[X, Y, Z]$ that $p = [0.01, 0.89, 0.23]$. This shows that statistical differences between the sets are insignificant for the Y and Z components, particularly BY. Though their fluctuations are very different, the overall nonlinear statistics across quiet and active intervals are statistically indistinguishable. However, for X the statistical

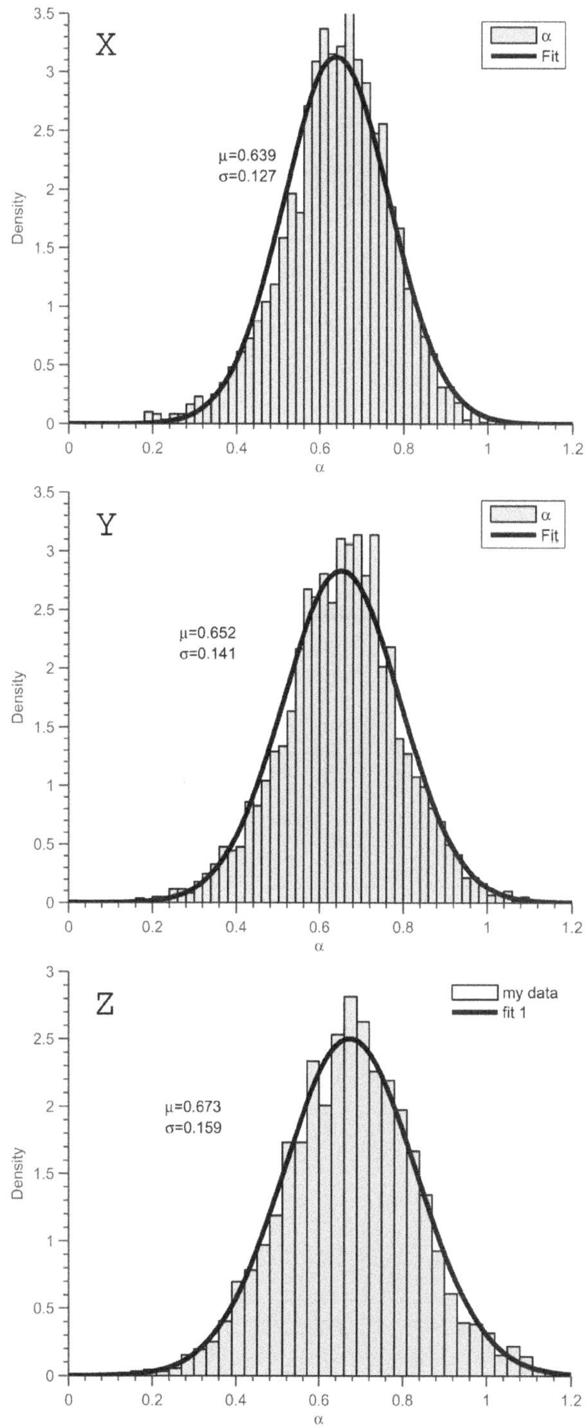

Fig. 5. Distributions of the fractal scaling exponent for active events, for all latitudes. These data are fitted to a normal distribution (solid black), with the mean and standard deviation given in the figures.

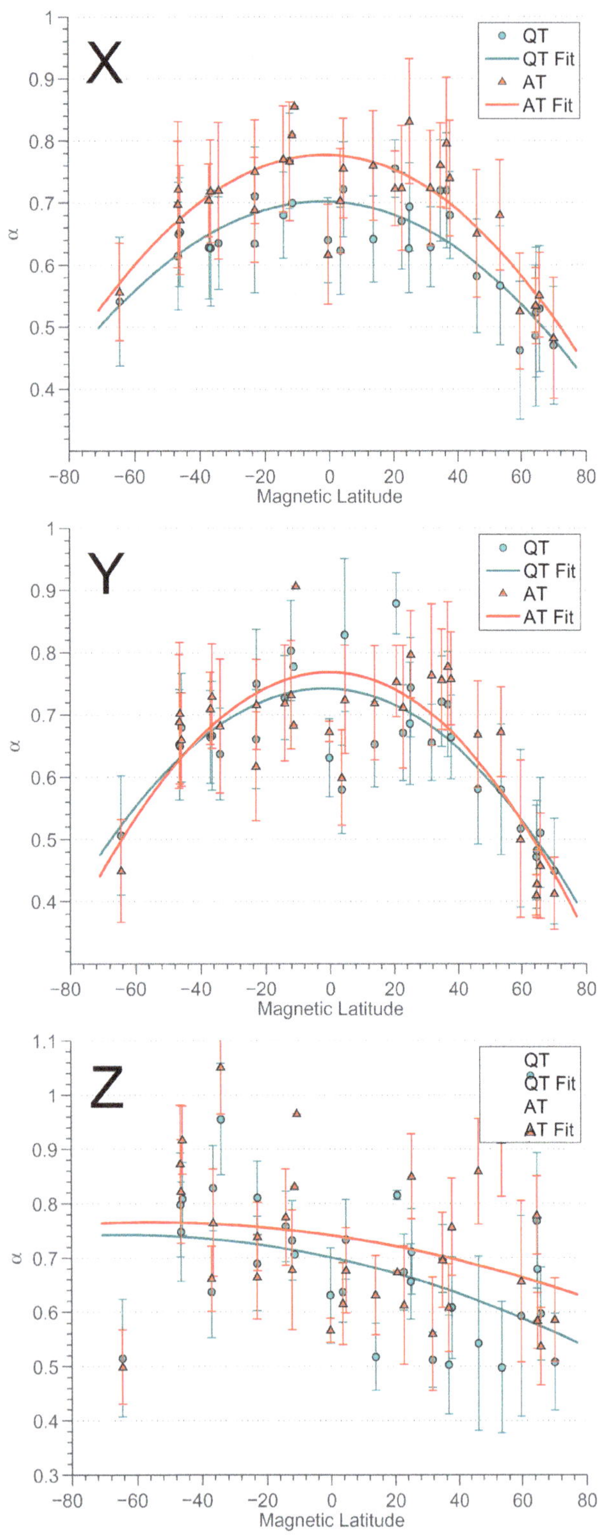

Fig. 6. Fractal scaling exponent for quiet (magenta circles) and active (red triangles) events as a function of latitude for different components of the magnetic field. The best-fit polynomials are shown in red (active) and magenta (quiet).

differences are significantly different from zero – active events have significantly larger scaling exponents than for quiet events.

5 Conclusions

In this paper we characterized fractal behavior of the magnetic field time series obtained from ground-based observatories across a wide range of latitudes. We used the 190, 210, and 250° magnetic meridian data. We classified these data into quiet (QT) and active (AT) periods using the Kp index as the discriminator. For quiet intervals, we required $Kp \leq 1$ for 24 h. Similarly, qualifying as an active interval required $Kp \geq 4$ for 24 h.

We use the DFA technique to compute fractal scaling exponents due to its performance in dealing with nonstationary data. We found evidence that fractional Brownian motion is ubiquitous across all the latitudes encompassed by this study (see e.g., Fig. 2).

Differences presented between QT and AT are less clear for Y and Z, but significant for X (Fig. 6), with events tending to have larger scaling exponents. This implies that active events possess more of a memory effect, while quiet events have data tending to be more random. Scaling exponents at individual stations are variable, irrespective of magnetosphere activity organized in terms of Kp. The latter result suggests that a suitable model of geomagnetic fluctuations is not monofractal but multifractional, i.e., the scaling exponent changes as a function of time. The distributions for quiet and active events encompasses marked differences as the result of different processes dominating the dynamics of the magnetosphere. During quiet times energy is stored and slowly burned, keeping the magnetosphere in a relative low-energy state. But during active times higher energy influx from the solar wind causes the magnetosphere to move to higher energy states, where strong nonlinear processes can dominate the dynamical release of energy.

Both the X and Y geomagnetic components show significant and correlated latitudinal variation, with scaling exponents increasing as the geomagnetic equator is approached. Z shows no such correlation, suggesting that for Z the fingerprints of local magnetic activity are not conserved during the timescale of a particular quiet or active event as determined by our selection criteria.

Our results thus suggest that the magnetic field at a single geomagnetic latitude is better described as a multifractional Brownian motion (mfBm) rather than as a fBm process. The mfBm is a generalized version of fBm in which the scaling exponent α is no longer a constant, but a function of the time index (Peltier and Lévy Vehel, 1995). In this case the increments of mfBm are nonstationary and the process is no longer self-similar.

Another possibility, not explored in this work, is that the well-defined scaling properties of our data are the result of a multifractal process. That is to say, there is no unique scaling exponent, but the presence of a full hierarchy of them (possibly with a stationary distribution, although not required in the generalized scale invariance framework). For instance, Bianchi and Planese (2007) found that in some cases the partition function as well as the scaling function of the mBm, i.e., of a generally non-multifractal process, behave as those of a genuine multifractal process. Indeed, previous work by Wanliss et al. (2005) and Yu et al. (2010) demonstrate the presence of multifractality in magnetospheric data at higher latitudes.

Although this work has only considered the hypothesis of multifractionality, it appears from a synthesis of multiple papers, that the nature of the geomagnetic disturbances (magnetospheric response to solar wind changes) is composed of the two different processes, multifractality (dependence of the increment statistics on the timescale) and multifractionality (corresponding to a unique (across scales), but non-stationary, scaling exponent). The former depends strongly on the solar wind variability and turbulent nature, and the latter is mainly due to the internal magnetospheric dynamics (with a special emphasis to the plasma and current sheet conditions), which could imply a different Markovian nature of the geomagnetic response at certain timescales. The different character between multifractality and multifractionality should manifest in a scale, independent of the fluctuations (increment) statistics. We leave these more detailed explorations to a later work.

Our results can serve as a guide, suggesting the required statistical structure for mathematical models of magnetospheric activity. We also offer a possible explanation relating the physics of QT and AT with their different fractal exponents. Our results are consistent with Consolini and Lui (2000), who examined scaling properties of magnetic fluctuations in the magnetotail. They consistently found a lower scaling exponent before current disruption, followed by higher values afterward. They interpreted the change in scaling exponent as a reorganization during current disruption. Since our ground-based data map the magnetosphere along magnetic field lines, the differences suggest the possibility that, at least for X, the fractal statistics offer another way to monitor transitions from QT to AT.

In conclusion, we have shown evidence that, for ground magnetometer data, average fractal scaling does vary in a consistent way as a function of latitude. And finally, at least for the X component, magnetospheric quiet times are characterized always, and everywhere, by smaller fractal scaling measures than active times.

Acknowledgements. This material is based on work supported by the National Science Foundation under Grant No. 1104364.

Edited by: D. Schertzer

References

Anh, V., Yu, Z.-G., and Wanliss, J. A.: Analysis of global geomagnetic variability, Nonlin. Processes Geophys., 14, 701–708, doi:10.5194/npg-14-701-2007, 2007.

Balasis, G., Daglis, I. A., Papadimitriou, C., Anastasiadis, A., Sandberg, I., and Eftaxias, K.: Quantifying Dynamical Complexity of Magnetic Storms and Solar Flares via Nonextensive Tsallis Entropy, Entropy, 13, 1865–1881, 2011.

Bartels J.: Discussion of time variations of geomagnetic indices Kp and Ap, Ann. Geophys., 19, 19323–19361, 1963.

Bartels, J., Heck, N. H., and Johnston, H. F.: The three-hour-range index measuring geomagnetic activity, J. Geophys. Res., 44, 411–454, 1939.

Bianchi, A. and Planese, A.: Modelling stock price movements: multifractality or multifractionality?, Quant. Financ., 7, 301–319, 2007.

Bryce, R. and Sprague, K.: Forecasting conflict intensity: Afghanistan, Journal of Battlefield Technology, 15, 23–29, 2012.

Chen, Z., Ivanov, P. C., Hu, K., and Stanley, H. E.: Effect of non-stationarities on detrended fluctuation analysis, Phys. Rev. E, 65, 041107, doi:10.1103/PhysRevE.65.041107, 2002.

Collins, J. J. and De Luca, C. J.: Upright, correlated random walks: A statistical-biomechanics approach to human postural control system, Chaos, 5, 57–63, 1994.

Consolini, G. and De Michelis, P.: Non-Gaussian distribution functions of AE-index fluctuations: Evidence for time intermittency, Geophys. Res. Lett., 25, 4087–4090, 1998.

Consolini, G. and Lui, A. T. Y.: Symmetry breaking and nonlinear wave-wave interaction in current disruption: Possible evidence for a phase transition, in: Magnetospheric Current Systems, edited by: Ohtani S.-I., Fujii, R., Hesse, M., and Lysak, R. L., Geophys. Monogr. Ser., 118, p. 395, 2000.

Consolini, G., Marcucci, M. F., and Candidi, M.: Multifractal Structure of Auroral Electrojet Index Data, Phys. Rev. Lett., 76, 4082–4085, 1996.

Dobias, P. and Wanliss, J. A.: Intermittency of storms and substorms: is it related to the critical behaviour?, Ann. Geophys., 27, 2011–2018, doi:10.5194/angeo-27-2011-2009, 2009.

Frisch, U.: Turbulence, Cambridge University Press, New York, 1997.

Hergarten, S.: Self-Organized Criticality in Earth Systems, Springer, Berlin, Heidelberg, New York, 2002.

Hori, T., Lui, A. T. Y., Ohtani, S., C:son Brandt, P., Mauk, B. H., McEntire, R. W., Maezawa, K., Mukai, T., Kasaba, Y., and Hayakawa, H.: Storm-time convection electric field in the near-Earth plasma sheet, J. Geophys. Res., 110, A04213, doi:10.1029/2004JA010449, 2005.

Hu, K., Ivanov, P. C., Chen, Z., Carpena, P., and Stanley, H. E.: Effect of trends on detrended fluctuation analysis, Phys. Rev. E, 64, 011114, doi:10.1103/PhysRevE.64.011114, 2001.

Kantelhardt, J. W., Zschiegner, S. A., Koscielny-Bunde, E., Havlin, S., Bunde, A., and Stanley, H. E.:, Multifractal detrended fluctuation analysis of nonstationary time series, Physica A, 316, 87–114, 2002.

Korth, A., Friedel, R. H. W., Henderson, M. G., Frutos-Alfaro, F., and Mouikis, C. G.: O^+ Transport into the ring current: Storm versus substorms, in: Disturbances in Geospace: The Storm-Substorm Relationship, edited by: Sharma, A. S., Kamide, Y., and Lakhina, G. S., AGU, Washington, DC, Geoph. Monog. Series, 142, 59–73, 2003.

Mandelbrot, B. B. and Van Ness, J. W.: Fractional Brownian Motions, Fractional Noises and Applications, SIAM Rev., 10, 422–437, 1968.

Muniandy, S. V. and Lim, S. C.: Modeling of locally self-similar processes using multifractional Brownian motion of Riemann-Liouville type, Phys. Rev. E, 63, 046104, doi:10.1103/PhysRevE.63.046104, 2001.

Neuman, S. P. and Federico, V. D.: Multifaceted Nature of Hydrogeologic Scaling and Its Interpretation, Rev. Geophys., 41, 1014, doi:10.1029./2003RG000130, 2003.

Peltier, R. and Lévy Vehel, J.: Multifractional Brownian motion: Definition and preliminary results, INRIA, Technical report No. 2645, 39 pp., 1995.

Peng, C. K., Havlin, S., Stanley, H. E., and Goldberger, A. L.: Quantification of scaling exponents and crossover phenomena in nonstationary heartbeat time series, CHAOS, 5, 82–87, 1995.

Pulkkinen, T. I., Partamies, N., Huttunen, K. E. J., Reeves, G. D., and Koskinen, H. E. J.: Differences in geomagnetic storms driven by magnetic clouds and ICME sheath regions, Geophys. Res. Lett., 34, L02105, doi:10.1029/2006GL027775, 2007.

Rangarajan, G. K. and Iyemori, T.: Time variations of geomagnetic activity indices Kp and Ap: an update, Ann. Geophys., 15, 1271–1290, doi:10.1007/s00585-997-1271-z, 1997.

Seber, G. A. F. and Wild, C. J.: Nonlinear Regression, John Wiley & Sons, Inc., Hoboken, NJ, USA, doi:10.1002/0471725315, 2005.

Sharma, A. S. and Veeramani, T.: Extreme events and long-range correlations in space weather, Nonlin. Processes Geophys., 18, 719–725, doi:10.5194/npg-18-719-2011, 2011.

Stanley, H. E., Amaral, L. A. N., Goldberger, A. L., Havlin, S., Ivanov, P. Ch., and Peng, C. K.: Statistical physics and physiology: monofractal and multifractal approaches, Physica A, 270, 309–324, 1999.

Takalo, J., Timonen, J., Klimas, A., Valdivia, J., and Vassiliadis, D.: Nonlinear energy dissipation in a cellular automaton magnetotail field model, Geophys. Res. Lett., 26, 1813–1816, 1999.

Taqqu, M. S., Teverovsky, V., and Willinger, W.: Estimators for long-range dependence: An empirical study, Fractals, 3, 785–798, 1995.

Taqqu, M., Teverovsky, V., and Willinger, W.: Is network traffic self-similar or multifractal?, Fractals, 5, 63–73, 1997.

Vörös, Z.: On multifractality of high-latitude geomagnetic fluctuations, Ann. Geophys., 18, 1273–1282, doi:10.1007/s00585-000-1273-6, 2000.

Wanliss, J.: Fractal properties of SYM-H during quiet and active times, J. Geophys. Res., 110, A03202, doi:10.1029/2004JA010544, 2005.

Wanliss, J. and Uritsky, V.: Understanding bursty behavior in mid-latitude geomagnetic activity, J. Geophys. Res., 115, A03215, doi:10.1029/2009JA014642, 2010.

Wanliss, J. A.: Nonlinear variability of SYM-H over two solar cycles, Earth Planets Space, 56, e13–e16, 2004.

Wanliss, J. A. and Antoine, L. A. G.: Geomagnetic micropulsations: Implications for high resolution aeromagnetic surveys, Explor. Geophys., 26, 535–538, 1995.

Wanliss, J. A. and Dobias, P.: Space Storm as a Phase Transition, J. Atmos. Sol.-Terr. Phy., 69, 675–684, doi:10.1016/j.jastp.2007.01.001, 2007.

Wanliss, J. A. and Reynolds, M. A.: Measurement of the stochasticity of low-latitude geomagnetic temporal variations, Ann. Geophys., 21, 2025–2030, doi:10.5194/angeo-21-2025-2003, 2003.

Wanliss, J. A. and Showalter, K. M.: High-resolution global storm index: Dst versus SYM-H, J. Geophys. Res., 111, A02202, doi:10.1029/2005JA011034, 2006.

Wanliss, J. A., Anh, V. V., Yu, Z.-G., and Watson, S.: Multifractal modeling of magnetic storms via symbolic dynamics analysis, J. Geophys. Res., 110, A08214, doi:10.1029/2004JA010996, 2005.

Yu, Z.-G., Anh, V., Wang, Y., Mao, D., and Wanliss, J.: Modeling and simulation of the horizontal component of the geomagnetic field by fractional stochastic differential equations in conjunction with empirical mode decomposition, J. Geophys. Res., 115, A10219, doi:10.1029/2009JA015206, 2010.

Yumoto, K. and the 210MM Magnetic Observation Group: The STEP 210 magnetic meridian network project, J. Geomagn. Geoelectr., 48, 1297–1310, 1996.

Monte Carlo fixed-lag smoothing in state-space models

A. Cuzol[1] **and E. Mémin**[2]

[1]University of Bretagne-Sud, UMR 6205, LMBA, 56000 Vannes, France
[2]INRIA Rennes-Bretagne Atlantique, Rennes, France

Correspondence to: A. Cuzol (anne.cuzol@univ-ubs.fr)

Abstract. This paper presents an algorithm for Monte Carlo fixed-lag smoothing in state-space models defined by a diffusion process observed through noisy discrete-time measurements. Based on a particle approximation of the filtering and smoothing distributions, the method relies on a simulation technique of conditioned diffusions. The proposed sequential smoother can be applied to general nonlinear and multidimensional models, like the ones used in environmental applications. The smoothing of a turbulent flow in a high-dimensional context is given as a practical example.

1 Introduction

The framework of this paper concerns state-space models described by general diffusions of the form

$$\mathrm{d}\boldsymbol{x}(t) = f(\boldsymbol{x}(t))\mathrm{d}t + \sigma(\boldsymbol{x}(t))\mathrm{d}\boldsymbol{B}(t), \tag{1}$$

which are partially observed through noisy measurements at discrete times. Such models can describe many dynamical phenomena in the environmental sciences and physics, but also in finance or engineering applications. The main motivation of this work concerns environmental applications, where nonlinearity and high-dimensionality arise. Indeed, environmental models and data describe nonlinear phenomena over large domains, with high spatial resolution. The continuous dynamical model (Eq. 1) is defined from a priori physical laws, while observations are supplied by sensors (satellite data for instance) and can appear with very low time frequency. As an example, in the application presented in the last part of this paper, the dimension of the state and observations is of the order of many thousands, and the model is described by the nonlinear Navier–Stokes equation. Filtering and smoothing in such state-space models aim at coupling model and observations, which is called data assimilation. The goal of the filtering is to estimate the system state distribution knowing past and present observations. This allows us for instance to give proper initial conditions to forecast the future state of a system characterizing atmospheric or oceanographic flows. On the other hand, the smoothing aims at estimating the state distribution using past and future observations, and this retrospective state estimation allows us to analyze a spatio–temporal phenomenon over a given time period, for climatology studies for instance. Applications of data assimilation are numerous and interest is growing in environmental sciences with the increase in available data. However, it is still a challenge to develop filtering and smoothing methods that can be used within a general nonlinear and high-dimensional context.

Monte Carlo sequential methods, contrary to standard Kalman filters, are able to deal with the filtering problem in nonlinear state-space models. The particle filtering (Del Moral et al., 2001; Doucet et al., 2000) solves the whole filtering equations through Monte Carlo approximations of the state distribution. On the other hand, ensemble Kalman methods (Evensen, 2003) take into account in some way the nonlinearities in the system, but are based on a Gaussian assumption. For high-dimensional systems, ensemble Kalman methods are preferred in practice to particle filters (Stroud et al., 2010; Van Leeuwen, 2009) since they reach a better performance for a limited number of particles. In order to keep this advantage while alleviating the Gaussian assumption, both methods are combined in Papadakis et al. (2010), leading to a particle filter that can be applied to high-dimensional systems. We will use this technique for the filtering step in the high-dimensional application presented in Sect. 5.

The aim of this paper is to propose a new smoothing method. Within the particle filter framework, the smoothing can be computed backward, reweighting past particles using present observations (Briers et al., 2010; Godsill et al., 2004). There is however one main difficulty for continuous models of type Eq. (1). As a matter of fact, it is necessary to know the transition density of the process between observation times, which is not available for general diffusions. This transition density can be approximated through Monte Carlo simulations, as proposed by Durham and Gallant (2002) to solve inference problems for diffusion processes. However, these approximations are based on Brownian bridge (or modified versions of it) simulations that do not take into account the drift part of the model. For nonlinear and high-dimensional models with a drift term that dominates, such approximations will be inefficient. It is also possible to obtain an unbiased estimate of the transition density (see Beskos et al., 2006), but this approach is not adapted to a multi-dimensional context. As a matter of fact, the use of this technique in a multivariate setting imposes constraints on the diffusion drift (in particular, the drift function has to be of gradient type). In parallel, within the framework of ensemble Kalman methods, Evensen and van Leeuwen (2000) have proposed estimating backward the smoothing distribution in a recursive way, based on existing filtering trajectories; Stroud et al. (2010) presented and applied an ensemble Kalman smoothing method, relying on a linearization of the system dynamics.

All previously mentioned smoothing methods require us to perform specific assumptions or simplifications in order to deal with general nonlinear models of type Eq. (1) in a high-dimensional context. To the best of our knowledge, it remains a challenging problem to develop smoothing methods that can be used in this general setting. In this paper, we deal with this issue sequentially each time a new observation is available, by smoothing the hidden state from this new observation time up to the previous one. This approach, called fixed-lag smoothing, then constitutes a partial answer to the global smoothing problem that would take into account all available observations. Nevertheless, it is reasonable to assume that the distribution of the hidden state depends on future observations through the next observation only, as soon as the time step between measurements is long (which is typically the case in the environmental applications that motivate this work). Under this assumption, a new observation will impact the distribution of the hidden process up to the previous observation only. This point of view justifies the use of a fixed-lag smoothing in our setting as a reasonable approximation of the global smoothing problem.

Such a fixed-lag smoothing may be directly obtained from the particle filtering result, reweighting past trajectories. However, this smoothing will fail in two cases: when the number of particles is too small compared to the size of the system, or when existing trajectories do not correspond to plausible trajectories of the dynamical model. Unfortunately, these two situations have to be faced when smoothing in a high-dimensional state-space model. Firstly, the number of particles has to be reduced for computational reasons. Secondly, particle filters that have been proposed in this high-dimensional context require us to correct trajectories towards the observations (Papadakis et al., 2010; Van Leeuwen and Ades, 2013). This implies that filtering states are consistent at observation times, but also that filtering trajectories may not be plausible realizations of the underlying physical model. In that case, a smoothing based on existing trajectories will fail. Note that these remarks are not only valid for the fixed-lag smoothing, but also for previously mentioned global techniques relying on existing trajectories. In particular, a genealogical smoothing based on ancestral particle lines (Del Moral, 2004) will be deficient in a high-dimensional setting, since many trajectories will share only a few ancestral lines.

In contrast, our method does not rely on existing particles only. It is built on a conditional simulation technique of diffusions proposed by Delyon and Hu (2006) that provides new state trajectories at hidden times between observations. This simulation technique is adapted to a multivariate context where the drift dominates, contrary to techniques based on Brownian bridge sampling (Durham and Gallant, 2002). Moreover, it does not require constraining assumptions for multivariate models, contrary to other techniques based on the exact simulation of diffusions (Beskos and Roberts, 2005; Beskos et al., 2006). The proposed smoothing method can then be applied to high-dimensional systems. Finally, it does not require model linearization nor Gaussian hypotheses, and so is able to deal with general nonlinear models.

The remainder of the paper is organized as follows. Section 2 briefly describes sequential Monte Carlo filtering methods in state-space models, and presents the fixed-lag smoothing problem. Section 3 presents the conditional simulation technique of diffusions of Delyon and Hu (2006), and details the construction of the proposed Monte Carlo estimate of smoothing distributions. The method is then experimented on a one-dimensional example in Sect. 4. Finally, the method is applied in Sect. 5 to a practical nonlinear and high-dimensional case, similar to the problems that have to be faced in environmental applications. A discussion is given in Sect. 6.

2 Monte Carlo filtering and smoothing in state-space models

In this section we recall briefly the particle filtering and smoothing methods for models of type Eq. (1), where the hidden state vector $\boldsymbol{x} \in \mathbb{R}^n$ is observed through the observation vector $\boldsymbol{y} \in \mathbb{R}^m$ at discrete times $\{t_1, t_2, \ldots\}$:

$$\boldsymbol{y}(t_k) = g(\boldsymbol{x}(t_k)) + \boldsymbol{\gamma}_{t_k}. \tag{2}$$

The drift function f and observation operator g can be nonlinear. The dynamical model uncertainty is described by an n-dimensional Brownian motion with covariance $\boldsymbol{\Sigma} =$

$\sigma(\boldsymbol{x}(t))\sigma(\boldsymbol{x}(t))^T$. The functions f, g and σ are assumed to be known, as well as the law of the observation noise $\boldsymbol{\gamma}_{t_k}$.

In particular, we present the standard particle filter and the weighted ensemble Kalman filter, which can be used to face the filtering problem in high-dimensional systems.

2.1 Particle-based filtering methods

Filtering aims at estimating recursively the distribution $p(\boldsymbol{x}_{t_1:t_k}|\boldsymbol{y}_{t_1:t_k})$ (and in particular its marginal distribution $p(\boldsymbol{x}_{t_k}|\boldsymbol{y}_{t_1:t_k})$) at each observation time t_k. This filtering problem can be solved with a Monte Carlo sequential approach, called particle filtering (Del Moral et al., 2001; Doucet et al., 2000). The method relies on a Monte Carlo approximation of the filtering distribution over a set of weighted trajectories $\{\boldsymbol{x}^{(i)}_{t_1:t_k}\}_{i=1:N}$ (called particles):

$$\hat{p}(\boldsymbol{x}_{t_1:t_k}|\boldsymbol{y}_{t_1:t_k}) = \sum_{i=1}^{N} w^{(i)}_{t_k}\delta_{\boldsymbol{x}^{(i)}_{t_1:t_k}}(\boldsymbol{x}_{t_1:t_k}), \tag{3}$$

whose marginal distribution at time t_k is written as

$$\hat{p}(\boldsymbol{x}_{t_k}|\boldsymbol{y}_{t_1:t_k}) = \sum_{i=1}^{N} w^{(i)}_{t_k}\delta_{\boldsymbol{x}^{(i)}_{t_k}}(\boldsymbol{x}_{t_k}). \tag{4}$$

Particle filters rely on a sequential importance sampling scheme that recursively samples particles, and updates their weights at observation times. The weights correspond to the ratio between the target distribution and the importance sampling distribution $\pi(\boldsymbol{x}_{t_k}|\boldsymbol{x}_{t_0:t_{k-1}}, \boldsymbol{y}_{t_1:t_k})$. They are recursively computed as follows:

$$w^{(i)}_{t_k} \propto w^{(i)}_{t_{k-1}} \frac{p(\boldsymbol{y}_{t_k}|\boldsymbol{x}^{(i)}_{t_k})p(\boldsymbol{x}^{(i)}_{t_k}|\boldsymbol{x}^{(i)}_{t_{k-1}})}{\pi(\boldsymbol{x}^{(i)}_{t_k}|\boldsymbol{x}^{(i)}_{t_0:t_{k-1}}, \boldsymbol{y}_{t_1:t_k})}. \tag{5}$$

In practice, a resampling procedure is added in order to avoid degeneracy. This procedure duplicates trajectories with large weights and removes small weighted trajectories.

2.1.1 Standard particle filter

When the proposal distribution π is set to the prior (i.e., $\pi(\boldsymbol{x}_{t_k}|\boldsymbol{x}_{t_0:t_{k-1}}, \boldsymbol{y}_{t_1:t_k}) = p(\boldsymbol{x}_{t_k}|\boldsymbol{x}_{t_{k-1}})$), the weight updating rule (Eq. 5) is simplified to the computation of the data likelihood $p(\boldsymbol{y}_{t_k}|\boldsymbol{x}^{(i)}_{t_k})$. This particular instance of the particle filter is called the *Bootstrap filter* or sequential importance resampling (SIR) filter (Gordon et al., 1993). Due to its simplicity it is the most commonly used particle filter. It is however a very inefficient distribution for high-dimensional space as it does not take into account the current observation and depends only weakly on the past data through the filtering distribution estimated at the previous instant. This distribution requires a great number of particles to explore meaningful areas of the state space.

2.1.2 Weighted ensemble Kalman filter (WEnKF)

One way to incorporate observation within the proposal distribution efficiently consists in relying on the ensemble Kalman filtering mechanism to define this distribution. This is the idea proposed in the WEnKF technique (Papadakis et al., 2010). In the WEnKF approach the importance sampling is taken as a Gaussian approximation of $p(\boldsymbol{x}_{t_k}|\boldsymbol{x}_{t_{k-1}}, \boldsymbol{y}_{t_k})$. This approach is close to the technique proposed in Van Leeuwen (2010). A variation of a similar technique based on a deterministic square-root formulation is also described in Beyou et al. (2013). In order to make the estimation of the filtering distribution exact (up to the sampling), each member of the ensemble must be weighted at each observation instant t_k with appropriate weights $w^{(i)}_{t_k}$, defined from Eq. (5). Therefore, the weighted ensemble Kalman filter (WEnKF) procedure can be simply summarized by Algorithm 1.

Algorithm 1 The WEnKF algorithm

For each $t_k = t_1, t_2, \ldots$:

- Start from particle set $\{\boldsymbol{x}^{(i)}_{t_{k-1}}, i = 1, \ldots, N\}$ and observation $\boldsymbol{y}_{t_k}$
- Obtain particle set $\{\boldsymbol{x}^{(i)}_{t_k}, i = 1, \ldots, N\}$ from:
  - **EnKF step:** Get $\boldsymbol{x}^{(i)}_{t_k}, i = 1, \ldots, N$, from the assimilation of $\boldsymbol{y}_{t_k}$ with an EnKF procedure;
  - **Computation of weights:** $w^{(i)}_{t_k} \propto w^{(i)}_{t_{k-1}} \frac{p(\boldsymbol{y}_{t_k}|\boldsymbol{x}^{(i)}_{t_k})p(\boldsymbol{x}^{(i)}_{t_k}|\boldsymbol{x}^{(i)}_{t_{k-1}})}{p(\boldsymbol{x}^{(i)}_{t_k}|\boldsymbol{x}^{(i)}_{t_{k-1}}, \boldsymbol{y}^{(i)}_{t_k})}$;
  - **Resampling:** For $j = 1, \ldots, N$, sample with replacement index $I(j)$ from discrete probability $\{w^{(i)}_{t_k}, i = 1, \ldots, N\}$ over $\{1, \ldots, N\}$ and set $\boldsymbol{x}^{(j)}_{t_k} = \boldsymbol{x}^{I(j)}_{t_k}$. Set $w^{(i)}_{t_k} = \frac{1}{N}$ $\forall i = 1, \ldots, N$.

Note that particle-based filtering techniques update the filtering distribution at observation times only. However, after the estimate $\hat{p}(\boldsymbol{x}_{t_k}|\boldsymbol{y}_{t_1:t_k})$ has been updated at observation time t_k, the filtering distribution can be predicted in order to have a continuous estimation of $\hat{p}(\boldsymbol{x}_t|\boldsymbol{y}_{t_1:t_k})$ for all $t \in]t_k, t_{k+1}[$ until the next observation time

$$\hat{p}(\boldsymbol{x}_t|\boldsymbol{y}_{t_1:t_k}) = \sum_{i=1}^{N} w^{(i)}_{t_k}\delta_{\boldsymbol{x}^{(i)}_{t}}(\boldsymbol{x}_t), \tag{6}$$

where, for all $i = 1, \ldots, N$, the state $\boldsymbol{x}^{(i)}_t$ is sampled from Eq. (1), starting from $\boldsymbol{x}^{(i)}_{t_k}$.

2.2 Fixed-lag smoothing problem

Contrary to the filtering approach that uses past and present observations, the smoothing in state-space models aims at estimating $p(\boldsymbol{x}_t|\boldsymbol{y}_{t_1:t_{\mathrm{end}}})$ for all $t \in [t_1, t_{\mathrm{end}}]$, using all past and future observations over a given time period. As raised in the introduction, existing smoothing methods do not apply directly to a general nonlinear model of type Eq. (1) in a high-dimensional context, since assumptions have to be made that may not be realistic. Instead of solving the global smoothing, we will concentrate in the rest of the paper on a fixed-lag smoothing, which constitutes a partial answer to the global smoothing problem.

The objective of the fixed-lag smoothing will be to replace the predictive distribution (Eq. 6) by its smoothed version $p(\boldsymbol{x}_t|\boldsymbol{y}_{t_1:t_{k+1}})\ \forall t \in]t_k, t_{k+1}]$ sequentially each time a new observation $\boldsymbol{y}_{t_{k+1}}$ arrives. This will allow us to reduce the temporal discontinuities inherent in the filtering technique that successively predicts the distribution of the state between observations, and updates this distribution at observation times.

To achieve this, by construction of the particle filter that weights entire trajectories (see Eq. 3), it is known (see for instance Doucet et al., 2000) that the fixed-lag smoothing distribution $\hat{p}(\boldsymbol{x}_t|\boldsymbol{y}_{t_1:t_{k+1}})$ can be directly obtained from the marginal at time t of $\hat{p}(\boldsymbol{x}_{t_1:t_{k+1}}|\boldsymbol{y}_{t_1:t_{k+1}})$. The empirical smoothing distribution is then given by

$$\hat{p}(\boldsymbol{x}_t|\boldsymbol{y}_{t_1:t_{k+1}}) = \sum_{i=1}^{N} w_{t_{k+1}}^{(i)} \delta_{\boldsymbol{x}_t^{(i)}}(\boldsymbol{x}_t) \quad \forall t \in]t_k, t_{k+1}]. \tag{7}$$

However, this approximation is simply a reweighting of past existing particle trajectories, and relies on the support of the filtering distribution at time t_k. If the number of particles is too small with respect to the state dimension, the support may be greatly reduced by the correction step (assigning small weights to all particles except a few), leading in practice to a bad estimation of $p(\boldsymbol{x}_t|\boldsymbol{y}_{t_1:t_{k+1}})$. Moreover, if particle trajectories have been forced towards observations during the filtering step (like in the WEnKF procedure), a smoothing based on those particles will fail because it will not be able to correct discontinuities. Consequently, since we are interested in smoothing techniques that are efficient in a high-dimensional context, this direct smoothing technique can not be used in its basic form and has to be improved.

In the following, we propose to use a conditional simulation technique of diffusions that will enable the sampling of new smoothed trajectories between times t_k and t_{k+1}. The approximation of the smoothing distribution (Eq. 7) at each hidden time will then be improved. The conditional simulation technique is presented in the next section, before the resulting smoothing procedure we propose.

3 Fixed-lag smoothing with conditional simulation

The smoothing method we propose is based on a conditional simulation technique that is presented in Sect. 3.1. We then develop in Sect. 3.2 how this technique can be used to improve the estimation of the smoothing distribution (Eq. 7).

3.1 Conditional simulation

Conditional simulation of a diffusion aims at sampling trajectories from a given process

$$\mathrm{d}\boldsymbol{x}(t) = f(\boldsymbol{x}(t))\mathrm{d}t + \sigma(\boldsymbol{x}(t))\mathrm{d}\boldsymbol{B}(t) \tag{8}$$

between two times $t = 0$ and $t = T$, with the constraints $\boldsymbol{x}(0) = \boldsymbol{u}$ and $\boldsymbol{x}(T) = \boldsymbol{v}$. This simulation problem is treated by Delyon and Hu (2006), where the authors show how to obtain the law of the constrained process from a Girsanov theorem. In practice, the proposed algorithms consist in simulating trajectories according to another diffusion process, which is built to respect the constraints and is easy to simulate from. The conditional distribution of the constrained process (Eq. 8) is absolutely continuous with respect to the distribution of the auxiliary process, with explicitly given density. For instance, in the case where the drift is bounded (a similar algorithm is proposed in Delyon and Hu (2006) for the unbounded case) and with σ invertible, the algorithm is based on the simulation of trajectories from the following process:

$$\mathrm{d}\tilde{\boldsymbol{x}}(t) = \left(f(\tilde{\boldsymbol{x}}(t)) - \frac{\tilde{\boldsymbol{x}}(t) - \boldsymbol{v}}{T - t} \right) \mathrm{d}t + \sigma(\tilde{\boldsymbol{x}}(t))\mathrm{d}\boldsymbol{B}(t), \tag{9}$$

with initial condition $\tilde{\boldsymbol{x}}(0) = \boldsymbol{u}$. Note that Lemma 4 in Delyon and Hu (2006) deals with the existence of a unique solution for this equation. This process is a simple modification of Eq. (8), where a deterministic part is added to the drift. It is then easy to simulate unconditional trajectories from this process, and all simulated trajectories will satisfy $\tilde{\boldsymbol{x}}(T) = \boldsymbol{v}$ by construction. For simplicity we will assume in the following that σ is independent of $\boldsymbol{x}(t)$ (note however that this is not an assumption in Delyon and Hu (2006)). The law of the conditioned process is given by

$$\mathbb{E}[h(\boldsymbol{x})|\boldsymbol{x}(0) = \boldsymbol{u}, \boldsymbol{x}(T) = \boldsymbol{v}] = \mathbb{E}\left[h(\tilde{\boldsymbol{x}})\alpha(\tilde{\boldsymbol{x}})\right], \tag{10}$$

for all measurable functions h, where

$$\alpha(\tilde{\boldsymbol{x}}) = \exp\left(-\int_0^T \frac{(\tilde{\boldsymbol{x}}(t) - \boldsymbol{v})^T \boldsymbol{\Sigma}^{-1} f(\tilde{\boldsymbol{x}}(t))}{T - t} \mathrm{d}t \right) \tag{11}$$

is the density coming from the Girsanov theorem (see Delyon and Hu, 2006), with $\boldsymbol{\Sigma} = \sigma(\tilde{\boldsymbol{x}}(t))\sigma(\tilde{\boldsymbol{x}}(t))^T$.

Let us note that the presence of the drift part of model (Eq. 8) in the auxiliary process (Eq. 9) is crucial to making the simulation efficient. The same process had initially been proposed by Clark (1990) to solve the conditional simulation problem. On the other hand, standard Brownian bridges that could be used as auxiliary processes (Durham and Gallant, 2002) lead in practice to poor approximations of the original constrained diffusion in our high-dimensional setting, since Brownian bridge trajectories are too far away from trajectories of Eq. (8).

In the following, the conditional marginal of interest $p(\boldsymbol{x}_t|\boldsymbol{x}(0)=\boldsymbol{u},\boldsymbol{x}(T)=\boldsymbol{v})$ will then be approximated as follows:

$$\hat{p}(\boldsymbol{x}_t|\boldsymbol{x}(0)=\boldsymbol{u},\boldsymbol{x}(T)=\boldsymbol{v})=\sum_{j=1}^{M}\alpha(\tilde{\boldsymbol{x}}^{(j)})\delta_{\tilde{\boldsymbol{x}}_t^{(j)}}(\boldsymbol{x}_t)\quad \forall t\in[0,T], \tag{12}$$

where the M trajectories $\{\tilde{\boldsymbol{x}}_t^{(j)}\}_{j=1:M}$ are simulated from Eq. (9) with $\tilde{\boldsymbol{x}}_0^{(j)}=\boldsymbol{u}$ for all $j=1,\dots,M$.

3.2 Proposed fixed-lag smoothing method

We show in the following how the conditional simulation technique can be used to improve the estimation of the local smoothing distribution $p(\boldsymbol{x}_t|\boldsymbol{y}_{t_1:t_{k+1}})$ for all $t\in]t_k,t_{k+1}]$.

We first note that this distribution can be decomposed as

$$\begin{aligned}p(\boldsymbol{x}_t|\boldsymbol{y}_{t_1:t_{k+1}})&=\int p(\boldsymbol{x}_t,\boldsymbol{x}_{t_k},\boldsymbol{x}_{t_{k+1}}|\boldsymbol{y}_{t_1:t_{k+1}})\mathrm{d}\boldsymbol{x}_{t_k}\mathrm{d}\boldsymbol{x}_{t_{k+1}}\\&=\int p(\boldsymbol{x}_{t_k},\boldsymbol{x}_{t_{k+1}}|\boldsymbol{y}_{t_1:t_{k+1}})p(\boldsymbol{x}_t|\boldsymbol{x}_{t_k},\boldsymbol{x}_{t_{k+1}},\boldsymbol{y}_{t_1:t_{k+1}})\\&\quad\mathrm{d}\boldsymbol{x}_{t_k}\mathrm{d}\boldsymbol{x}_{t_{k+1}}.\end{aligned} \tag{13}$$

Then, from the state-space model properties, we obtain

$$p(\boldsymbol{x}_t|\boldsymbol{y}_{t_1:t_{k+1}})=\int p(\boldsymbol{x}_{t_k},\boldsymbol{x}_{t_{k+1}}|\boldsymbol{y}_{t_1:t_{k+1}})p(\boldsymbol{x}_t|\boldsymbol{x}_{t_k},\boldsymbol{x}_{t_{k+1}})\,\mathrm{d}\boldsymbol{x}_{t_k}\mathrm{d}\boldsymbol{x}_{t_{k+1}}. \tag{14}$$

Moreover, from the particle filter Monte Carlo approximation described by Eq. (3), the joint law $p(\boldsymbol{x}_{t_k},\boldsymbol{x}_{t_{k+1}}|\boldsymbol{y}_{t_1:t_{k+1}})$ can be replaced by

$$\hat{p}(\boldsymbol{x}_{t_k},\boldsymbol{x}_{t_{k+1}}|\boldsymbol{y}_{t_1:t_{k+1}})=\sum_{i=1}^{N}w_{t_{k+1}}^{(i)}\delta_{(\boldsymbol{x}_{t_{k+1}}^{(i)},\boldsymbol{x}_{t_k}^{(i)})}(\boldsymbol{x}_{t_{k+1}},\boldsymbol{x}_{t_k}), \tag{15}$$

where $w_{t_{k+1}}^{(i)}$ are the particle filter importance weights.

Plugging Eq. (15) into Eq. (14) then leads to the following approximation for the fixed-lag smoothing distribution:

$$\hat{p}(\boldsymbol{x}_t|\boldsymbol{y}_{t_1:t_{k+1}})=\sum_{i=1}^{N}w_{t_{k+1}}^{(i)}p(\boldsymbol{x}_t|\boldsymbol{x}_{t_k}^{(i)},\boldsymbol{x}_{t_{k+1}}^{(i)}). \tag{16}$$

The conditional distribution $p(\boldsymbol{x}_t|\boldsymbol{x}_{t_k}^{(i)},\boldsymbol{x}_{t_{k+1}}^{(i)})$ can be estimated using Eq. (12) for each pair of initial and end points $\boldsymbol{x}_{t_k}^{(i)}$ and $\boldsymbol{x}_{t_{k+1}}^{(i)}$:

$$\hat{p}(\boldsymbol{x}_t|\boldsymbol{x}_{t_k}^{(i)},\boldsymbol{x}_{t_{k+1}}^{(i)})=\sum_{j=1}^{M}\alpha(\tilde{\boldsymbol{x}}^{(i)(j)})\delta_{\tilde{\boldsymbol{x}}_t^{(i)(j)}}(\boldsymbol{x}_t), \tag{17}$$

where each $\tilde{\boldsymbol{x}}_t^{(i)(j)}$ is sampled from Eq. (9) with initial constraint $\tilde{\boldsymbol{x}}_{t_k}^{(i)(j)}=\boldsymbol{x}_{t_k}^{(i)}$ and final constraint $\boldsymbol{x}_{t_{k+1}}^{(i)}$.

The estimation of the smoothing distribution of interest is finally written as

$$\hat{p}(\boldsymbol{x}_t|\boldsymbol{y}_{t_1:t_{k+1}})=\sum_{i=1}^{N}w_{t_{k+1}}^{(i)}\sum_{j=1}^{M}\alpha(\tilde{\boldsymbol{x}}^{(i)(j)})\delta_{\tilde{\boldsymbol{x}}_t^{(i)(j)}}(\boldsymbol{x}_t),\quad\forall t\in]t_k,t_{k+1}]. \tag{18}$$

The algorithm we propose to compute the fixed-lag smoothing distribution on a given time interval $[t_k,t_{k+1}]$ is therefore the following:

Algorithm 2 Fixed-lag conditional smoothing

For each $t_k=t_1,t_2,\dots$:

- Store $\{\boldsymbol{x}_{t_k}^{(i)}\}_{i-1:N}$ and compute $\{\boldsymbol{x}_{t_{k+1}}^{(i)}\}_{i-1:N}$ and associated weights $\{w_{t_{k+1}}^{(i)}\}_{i=1:N}$ from a particle filter algorithm;
- For each pair $\{\boldsymbol{x}_{t_k}^{(i)},\boldsymbol{x}_{t_{k+1}}^{(i)}\}$, $i=1,\dots,N$:
  - Simulate M conditional trajectories $\{\tilde{\boldsymbol{x}}_t^{(i)(j)}\}_{j=1:M}$ for $t\in[t_k,t_{k+1}]$ from Eq. (9) with an Euler scheme, with the constraints $\tilde{\boldsymbol{x}}_{t_k}^{(i)(j)}=\boldsymbol{x}_{t_k}^{(i)}$ and $\tilde{\boldsymbol{x}}_{t_{k+1}}^{(i)(j)}=\boldsymbol{x}_{t_{k+1}}^{(i)}$,
  - Compute weights $\alpha(\tilde{\boldsymbol{x}}^{(i)(j)})$ from Eq. (11) for all $j=1,\dots,M$, with final constraint $\boldsymbol{x}_{t_{k+1}}^{(i)}$;
- Compute $\hat{p}(\boldsymbol{x}_t|\boldsymbol{y}_{t_1:t_{k+1}})$ $=\sum_{i=1}^{N}w_{t_{k+1}}^{(i)}\sum_{j=1}^{M}\alpha(\tilde{\boldsymbol{x}}^{(i)(j)})\delta_{\tilde{\boldsymbol{x}}^{(i)(j)}}(\boldsymbol{x}_t)$ for all $t\in]t_k,t_{k+1}]$.

4 One-dimensional simulation study

In this section, the smoothing method is first experimented on a one-dimensional state space model. Since the proposed approach relies on a preliminary particle filtering step, filtering results are first presented in Sect. 4.2 (either considering a standard particle filter or the WEnKF). The results obtained with the standard fixed-lag smoothing method are then shown in Sect. 4.3. Finally, Sect. 4.4 presents the smoothing results obtained with the proposed technique.

4.1 State-space model

The one-dimensional state-space model of interest is a sine diffusion, partially observed with noise (used as an illustration by Fearnhead et al. (2008) for a particle filtering method):

$$\mathrm{d}x(t)=\sin(x(t))\mathrm{d}t+\sigma_x\mathrm{d}B(t), \tag{19}$$

$$y_{t_k}=x_{t_k}+\gamma_{t_k}, \tag{20}$$

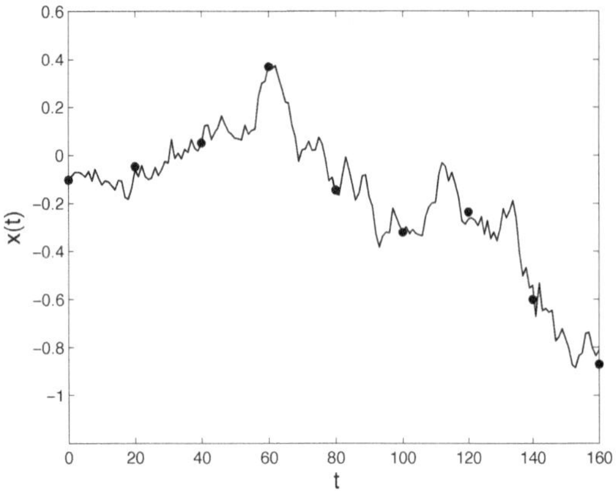

Figure 1. Simulated sine diffusion trajectory $x(t)$ and partial observations $y(t_k)$ (dots) with $t_k - t_{k-1} = 20\Delta t$.

where $\sigma_x^2 = 0.5$ and $\gamma_{t_k} \sim \mathcal{N}(0, \sigma_y)$ with $\sigma_y^2 = 0.01$. One trajectory of the process is first simulated from Eq. (19) with an Euler-type discretization scheme of time step $\Delta t = 0.005$. This trajectory will constitute the hidden process, observed through y_{t_k} generated according to Eq. (20) at every time step t_k, with $t_k - t_{k-1} = 20\Delta t$. The trajectory is plotted in Fig. 1, together with the corresponding discrete observations at times t_k.

4.2 Particle filtering results

The filtering results are presented for the standard particle filter (denoted PF in the following) and the weighted ensemble Kalman filter (WEnKF). Two situations are shown, with a reduced ($N = 20$) and high number ($N = 10\,000$) of particles. The case with a high number of particles is shown as the reference for comparison purpose; note however that this ideal situation is not reachable in a high-dimensional context, since the number of particles has to be reduced for evident computational cost reasons.

The results for the two configurations are presented in Fig. 2, where the dotted lines represent the filtering mean estimates. The filtering distribution $p(x_{t_k}|y_{t_1:t_k})$ is estimated at each observation time t_k using Eq. (4), and predicted between observation times from Eq. (6). The mean is then estimated from weighted particles as $\sum_{i=1}^{N} w_{t_k}^{(i)} x_t^{(i)}$, for all $t \in [t_k, t_{k+1}[$. Figure 2a–b shows that the standard particle filter results diverge from the reference solution between observation times, for low or high number of particles. As a matter of fact, when no observation is available, the state distribution is predicted from the dynamics only, so that particles trajectories are not guided towards the next observation. At observation times t_k, high weights are given to particles that are close to the observation, so that the estimated mean suddenly gets closer to the solution. These discontinuities between measurement times can also be observed in the WEnKF results (Fig. 2c–d), because particle trajectories are

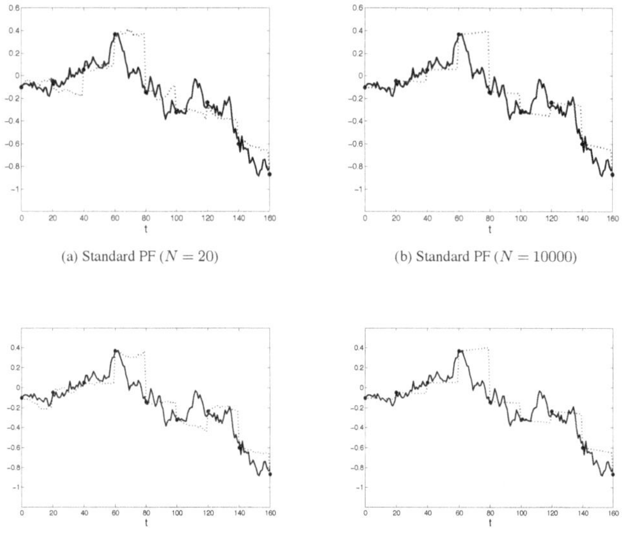

(a) Standard PF ($N = 20$)

(b) Standard PF ($N = 10000$)

(c) WEnKF ($N = 20$)

(d) WEnKF ($N = 10000$)

Figure 2. Standard PF and WEnKF results. Thick line: hidden diffusion; dots: partial observations; dotted line: estimated filtering mean.

brutally corrected with the EnKF step at observation times. A smoothing will aim at reducing these temporal discontinuities, while providing dynamically consistent solutions.

4.3 Standard fixed-lag smoothing results

From the particle filtering results, we now present the results obtained with the direct particle smoothing procedure described in Sect. 2.2. This procedure relies on existing trajectories. The smoothing distribution $\hat{p}(x_t|y_{t_1:t_{k+1}})$ is computed backward for all $t \in]t_k, t_{k+1}]$ using expression Eq. (7) each time a new observation $y_{t_{k+1}}$ becomes available. The smoothing mean is computed as $\sum_{i=1}^{N} w_{t_{k+1}}^{(i)} x_t^{(i)}$ for all $t \in]t_k, t_{k+1}]$, and the standard deviation is computed in the same way from the weighted particles.

It can be observed in Fig. 3a that the smoothing based on the standard particle filter is not efficient when the number of particles N is small: only a few particles are close to the observation at time t_k and have nonzero weights, implying that the smoothing distribution is poorly estimated (see for instance between observation times $t = 100$ and $t = 120$, where the smoothing distribution is artificially peaked but far from the hidden trajectory). The smoothing result obtained from the reference configuration $N = 10\,000$ is plotted in Fig. 3b. In that situation, since many trajectories have high weights at observation times, the estimation of backward smoothing distributions is improved and includes the hidden trajectory.

Moreover, Fig. 3c–d shows that the standard smoothing based on the WEnKF result fails for a low or high number of particles. As a matter of fact, particle trajectories are artificially corrected by the EnKF step at each observation time.

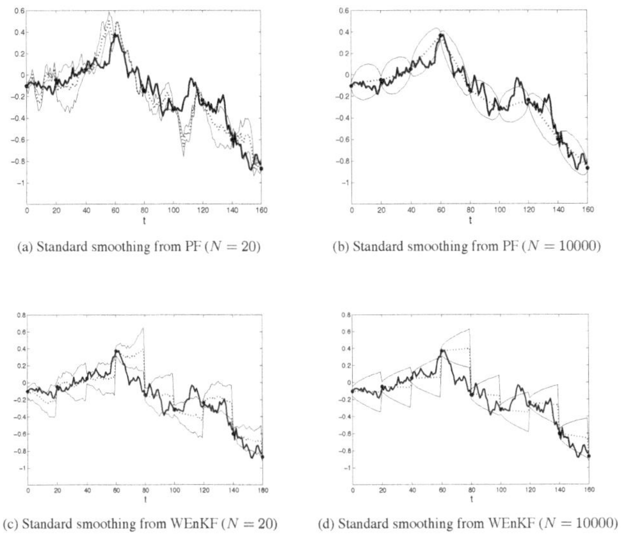

(a) Standard smoothing from PF ($N = 20$) (b) Standard smoothing from PF ($N = 10000$)

(c) Standard smoothing from WEnKF ($N = 20$) (d) Standard smoothing from WEnKF ($N = 10000$)

Figure 3. Standard smoothing from PF and WEnKF results. Thick line: hidden diffusion; dots: partial observations; dotted line: estimated filtering mean.

Resulting trajectories are highly non-plausible. Even for a huge number of particles, a smoothing based on those existing trajectories is not able to reduce the induced time discontinuities.

4.4 Proposed smoothing results

In this section, we show how the proposed method can improve the estimation of backward smoothing distributions when it is not adequate to rely on existing trajectories only. This is the case if the number of particles is too small, as demonstrated from the experiment presented in Fig. 3a, or if the existing trajectories do not correspond to plausible trajectories of the model (as shown for the WEnKF result in Fig. 3c–d).

Our smoothing is first applied using the filtering output of the standard particle filter with $N = 20$ particles. Figure 4a shows the result obtained with a sampling of $M = 50$ conditional trajectories between each pair $\{x_{t_k}^{(i)}, x_{t_{k+1}}^{(i)}\}$, $i = 1, \ldots, N$. The smoothing distribution $\hat{p}(x_t | y_{t_1:t_{k+1}})$ is computed from Eq. (18), so the smoothing mean is computed as $\sum_{i=1}^{N} w_{t_k}^{(i)} \sum_{j=1}^{M} \alpha(\tilde{x}^{(i)(j)}) \tilde{x}_t^{(i)(j)}$ for all $t \in]t_k, t_{k+1}]$, and similarly for the standard deviation. This result highlights the fact that since the proposed method creates new trajectories, it is able to correct the deficiencies of the standard smoothing approach presented in Fig. 3a when the initial number of filtering particles is too small. In Fig. 4b, the same experiment is presented using $M = 500$ conditional trajectories. In that case, the result is very similar to the reference particle smoothing result presented in Fig. 3b, obtained from a particle filter with $N = 10\,000$.

In parallel, the proposed smoothing has been tested using the output of the WEnKF filtering technique with $N = 20$ particles. Again, the smoothing is computed with $M = 50$ and $M = 500$ conditional trajectories, and the corresponding results are presented in Fig. 3c–d. Instead on relying on existing WEnKF trajectories that may not be plausible trajectories of the model (because of the EnKF correction step), the proposed method samples new trajectories between observation times. This leads to a good estimation of the smoothing distributions, contrary to the standard smoothing presented in Fig. 3c. Note that the smoothing results are very similar to the result obtained from the standard particle filter (Fig. 3a–b) because both filters have similar behavior at observation times.

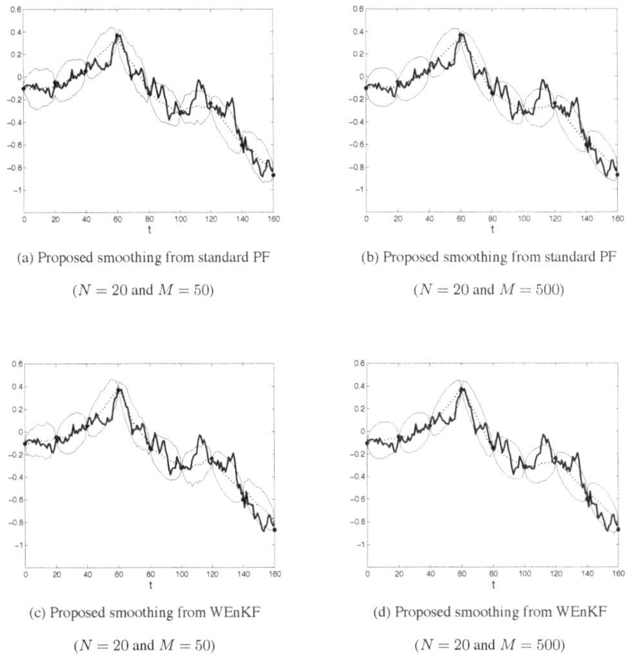

(a) Proposed smoothing from standard PF ($N = 20$ and $M = 50$) (b) Proposed smoothing from standard PF ($N = 20$ and $M = 500$)

(c) Proposed smoothing from WEnKF ($N = 20$ and $M = 50$) (d) Proposed smoothing from WEnKF ($N = 20$ and $M = 500$)

Figure 4. Proposed conditional smoothing result. Thick line: hidden diffusion; dots: partial observations; dotted line: estimated backward smoothing mean; thin line: estimated standard deviation.

5 Application to a high-dimensional assimilation problem

This section aims at illustrating the applicability of our method to a high-dimensional and nonlinear scenario, without extensive study at this stage. The method is applied to a turbulence assimilation problem, where the model of interest is of type Eq. (1). The goal is to recover temporal estimates of velocity/vorticity over a given spatial domain of size $n = 64 * 64$, from a sequence of noisy observations and a continuous a priori dynamical model based on a stochastic version of the Navier–Stokes equation. Within an environmental framework, a direct application would be the estimation of wind fields or sea surface currents from satellite data.

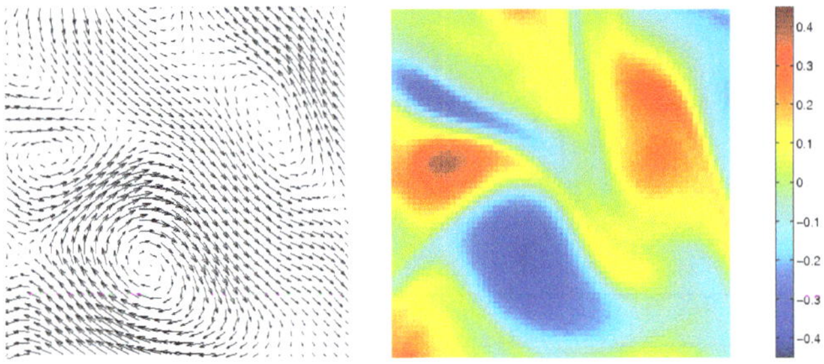

(a) (b)

Figure 5. State example. **(a)** Velocity field $\mathbf{w}_t$; **(b)** associated vorticity map $\boldsymbol{\xi}_t$.

5.1 State-space model

Let $\xi(\boldsymbol{x})$ denote the scalar vorticity at point $\boldsymbol{x} = (x, y)^T$, associated with the 2-D velocity $\boldsymbol{w}(\boldsymbol{x}) = (w_x(\boldsymbol{x}), w_y(\boldsymbol{x}))^T$ through $\xi(\boldsymbol{x}) = \frac{\partial w_y}{\partial x} - \frac{\partial w_x}{\partial y}$. Let $\boldsymbol{\xi} \in \mathbb{R}^n$ be the state vector describing the vorticity over an $n = 64 * 64$ square domain, and $\boldsymbol{w} \in \mathbb{R}^{2n}$ the associated velocity field over the domain. We will focus on incompressible flows such that the divergence of the velocity field is null. A stochastic version of the Navier–Stokes equation in its velocity–vorticity form can then be written as

$$\mathrm{d}\boldsymbol{\xi}_t = -\nabla\boldsymbol{\xi}_t \cdot \boldsymbol{w}_t \mathrm{d}t + \frac{1}{Re}\Delta\boldsymbol{\xi}_t \mathrm{d}t + \sigma \mathrm{d}\boldsymbol{B}_t, \tag{21}$$

where Re denotes the flow Reynolds number ($Re = 3000$). The uncertainty is modeled by a Brownian motion of size n, with covariance $\boldsymbol{\Sigma} = \sigma\sigma^T$, where $\sigma \in \mathbb{R}^n$. A velocity field example, generated from the model Eq. (21), is shown in Fig. 5a, together with the corresponding vorticity map (b).

We assume the hidden vorticity vector $\boldsymbol{\xi}$ is observed through noisy measurements $\boldsymbol{y}_{t_k}$ at discrete times t_k, where $t_k - t_{k-1} = 100\Delta t$, and $\Delta t = 0.1$ is the time step used to discretize Eq. (21). In our experimental setup, measurements correspond to PIV (particle image velocimetry) image sequences used in fluid mechanics applications. Note however that other kind of data can be used similarly within this state space model, like meteorological or oceanographic data for instance. The state and observation are related in our case through $\boldsymbol{y}_{t_k} = g(\boldsymbol{\xi}_{t_k}) + \boldsymbol{\gamma}_{t_k}$, where g is a nonlinear function linking the vorticity to the image data, and $\boldsymbol{\gamma}_{t_k}$ is a Gaussian noise, uncorrelated in time.

5.2 Implementation details

We recall that the smoothing relies first on a particle filter step. Due to the high dimensionality of the state vector, the use of a standard particle filter is not adapted to solve the filtering problem, as discussed by Snyder et al. (2008) or Van Leeuwen (2009). We then make use of the method presented by Papadakis et al. (2010), which combines the benefits of the ensemble Kalman filter, known to perform well in practice for high-dimensional systems (Stroud et al., 2010), and the particle filter (which solves theoretically the true filtering problem, without approximating the filtering distributions with Gaussian distributions). Since the method of Papadakis et al. (2010) is intrinsically a particle filter, it then leads at each observation time t_k to a set of particles and weights $\{\boldsymbol{\xi}^{(i)}_{t_1:t_k}, w_{t_k}\}_{i=1:N}$, as required by the algorithm proposed in Sect. 3.

The particle filter step requires simulations from the dynamical model (Eq. 21), and the conditional simulation step requires us to sample trajectories from its constrained version, which consists in a similar problem with modified drift (see process Eq. 9). The model is discretized in time with the time step $\Delta t = 0.1$; more information about the discretization scheme may be obtained in Papadakis et al. (2010). The random perturbations are assumed to be realizations of Gaussian random fields that are correlated in space with the Gaussian covariance function $\boldsymbol{\Sigma}(\boldsymbol{x}_i, \boldsymbol{x}_j) = \eta\exp(-\frac{||\boldsymbol{x}_i - \boldsymbol{x}_j||^2}{\lambda})$, where $\eta = 0.01$ and $\lambda = 13$. In practice, the simulation of these perturbations is performed in Fourier space, with the method described in Evensen (2003).

Finally, the estimation of the smoothing distributions requires the computation of conditional trajectories weights, corresponding to Girsanov weights given by Eq. (11). After a Riemann sum approximation of the integral, the computation of weights requires the inversion of the matrix $\boldsymbol{\Sigma}$ of size (n, n), where $n = 64 * 64$ is the number of grid points. We choose to compute $\boldsymbol{\Sigma}^{-1}$ empirically using a singular value decomposition computed from the M realizations of the perturbation fields used for the constrained trajectories simulations. Let $\mathbf{Z}$ be the matrix of size (n, M) containing the M-centered fields of size $n = 64 * 64$; the SVD leads to $\mathbf{Z} = \mathbf{UDV}^T$, so that $\mathbf{ZZ}^T = \mathbf{UDD}^T\mathbf{U}^T$. The inverse of the covariance matrix $\boldsymbol{\Sigma}^{-1}$ is finally computed as

$$M(\mathbf{ZZ}^T)^{-1} = M\mathbf{U}(\mathbf{DD}^T)^{-1}\mathbf{U}^T, \tag{22}$$

which only requires the inversion of a diagonal. Note that more efficient procedures could be implemented in our case (homogeneous Gaussian covariance) since the covariance function is separable in the x and y directions. This means that the covariance matrix $\boldsymbol{\Sigma}$ can be written as the Kronecker product of smaller matrices and more easily inverted (Sun et al., 2012). However, the SVD inversion can be applied to any covariance structure, in particular it could deal with a non-homogeneous covariance matrix.

5.3 Results

In this section, we illustrate the capability of the proposed method to reduce the temporal discontinuities inherently introduced by the filtering in continuous–discrete state-space models.

The smoothing result relies on the output of the WEnKF filtering step, computed with $N = 500$ particles. Compared

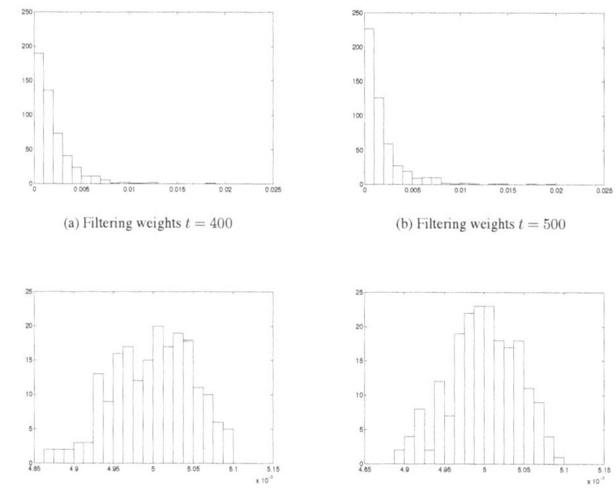

(a) Filtering weights $t = 400$ (b) Filtering weights $t = 500$

(d) Conditional simulation weights $t = 400$ (e) Conditional simulation weights $t = 500$

Figure 6. Filtering and conditional simulation weights

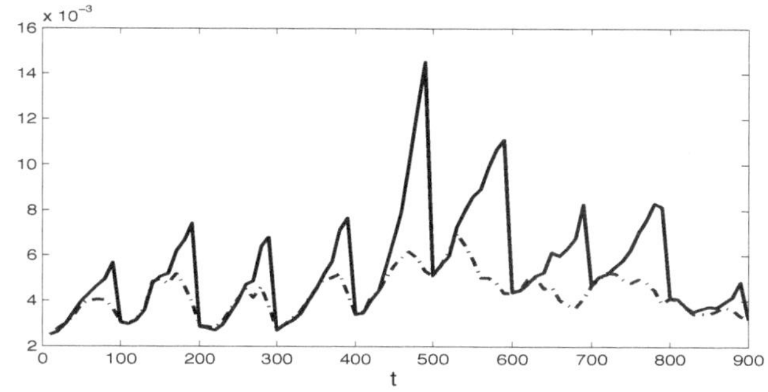

Figure 7. Full line: mean square error between ground truth vorticity and estimated filtering mean; dotted line: mean square error between ground truth vorticity and estimated backward smoothing mean.

to the size of the system, the number of particles is theoretically too small for the filter to be truly efficient. In practice, many filtering trajectories have close to zero weight at observation times. Histograms of filtering weights are given as illustration in Fig. 6a–b at two times $t = 400$ and $t = 500$. Note however that the filter is not degenerate and is able to provide results that get close to the hidden vorticity at measurement times. This can be observed in Fig. 7, where the mean square error is plotted with a full line, averaged at each time over the image domain of size $n = 64 * 64$. Since the ground truth vorticity sequence is known in our experimental setup, the mean square error is computed between the hidden vorticity and the estimated filtering mean, given by $\sum_{i=1}^{N} w_{t_k}^{(i)} \boldsymbol{\xi}_t^{(i)}$ for all $t \in [t_k, t_{k+1}[$. The correction steps lead to successive error decreases at observation times.

The proposed smoothing method has been applied with $M = 200$. Note that we take benefit from the fact that many filterings particles have close to zero weight. Indeed, the smoothing method relies in practice on a reduced number $\tilde{N}M$ of sampled conditional trajectories (with $\tilde{N} << N$), which makes the problem computationally tractable. In this experiment, we have retained around 5 % of initial filtering trajectories. The smoothing distribution $\hat{p}(\boldsymbol{\xi}_t | \boldsymbol{y}_{t_1:t_{k+1}})$ is computed for all $t \in]t_k, t_{k+1}]$ from Eq. (18), and its mean is computed as $\sum_{i=1}^{N} w_{t_{k+1}}^{(i)} \sum_{j=1}^{M} \alpha(\tilde{\boldsymbol{\xi}}^{(i)(j)}) \tilde{\boldsymbol{\xi}}_t^{(i)(j)}$. Histograms of conditional simulation weights $\alpha(\tilde{\boldsymbol{\xi}}^{(i)(j)})$ are given as an illustration in Fig. 6c–d for a given particle (i) at two times, $t = 400$ and $t = 500$.

The mean square error is computed between the true vorticity and the estimated smoothing mean, and plotted in Fig. 7 with a dotted line. As expected, the smoothing method reduces the error at hidden times between observations.

In addition, we present below a qualitative evaluation of the smoothing result for the same experiment, over a specific time interval.

The WEnKF result is first presented in Fig. 8 for the time interval [400, 500] between two observations, where estimated mean vorticity maps are computed as $\sum_{i=1}^{N} w_{400}^{(i)} \boldsymbol{\xi}_t^{(i)}$ for all $t \in [400, 500[$, and as $\sum_{i=1}^{N} w_{500}^{(i)} \boldsymbol{\xi}_t^{(i)}$ for $t = 500$. The temporal discontinuity between estimations can be observed when reaching observation time $t = 500$: the vorticity map is suddenly modified in order to fit the observations, introducing inconsistencies in the vorticity temporal trajectories. Note that the application of the standard particle smoothing (described in Sect. 2.2) will fail here, and not only because the number of particles is too small. As a matter of fact, we recall that the filtering trajectories have been computed from the method presented in Papadakis et al. (2010), which uses the ensemble Kalman filter step as an importance distribution in the particle filter algorithm. The ensemble Kalman filter consists of a prediction step from the dynamical model Eq. (21), and a correction step that shifts particles towards the observation. Because of this correction step, the sampled filtering trajectories between two observation times do not correspond to trajectories of the dynamical model. This implies that from such a particle filter, the standard smoothing based on existing trajectories will not be able to reduce the temporal discontinuities observed in Fig. 8. This can be observed in Fig. 9, where smoothed vorticity maps are computed as $\sum_{i=1}^{N} w_{400}^{(i)} \boldsymbol{\xi}_t^{(i)}$ for $t = 400$, and as $\sum_{i=1}^{N} w_{500}^{(i)} \boldsymbol{\xi}_t^{(i)}$ for all $t \in]400, 500]$. The discontinuity at time $t = 500$ is still present.

The result obtained with the proposed method is plotted in Fig. 10. Estimated mean vorticity maps are computed as $\sum_{i=1}^{N} w_{500}^{(i)} \sum_{j=1}^{M} \alpha(\tilde{\boldsymbol{\xi}}^{(i)(j)}) \tilde{\boldsymbol{\xi}}_t^{(i)(j)}$ for all $t \in [400, 500]$. Spatio–temporal vorticity trajectories are gradually modified until observation time $t = 500$, preserving the fluid flow properties. As a matter of fact, since the proposed method samples new trajectories from the law of the physical process (Eq. 21), the smoothed vorticity trajectories are by construction consistent with the a priori dynamical model. In order to sample the smoothed trajectories, the method relies on the model and on filtering marginals at observation times,

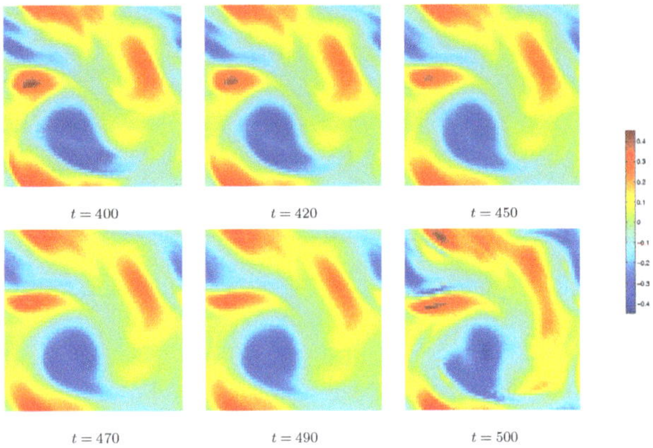

Figure 8. Filtering result with the method of Papadakis et al. (2010). Estimated mean vorticity maps for different times t between observation times $t = 400$ and $t = 500$.

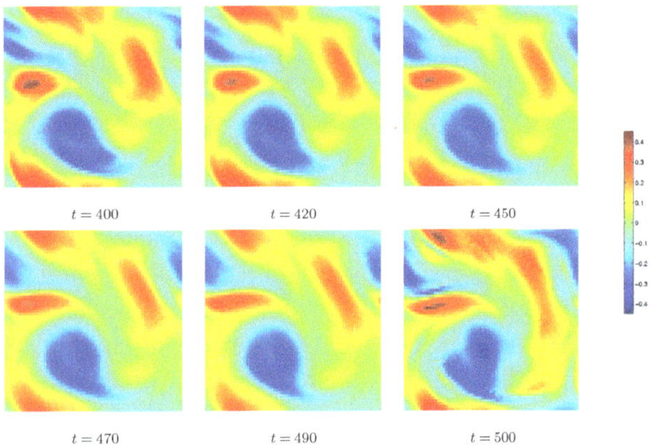

Figure 9. Standard particles smoothing result (see Sect. 2.2). Estimated mean vorticity maps for different times t between observation times $t = 400$ and $t = 500$.

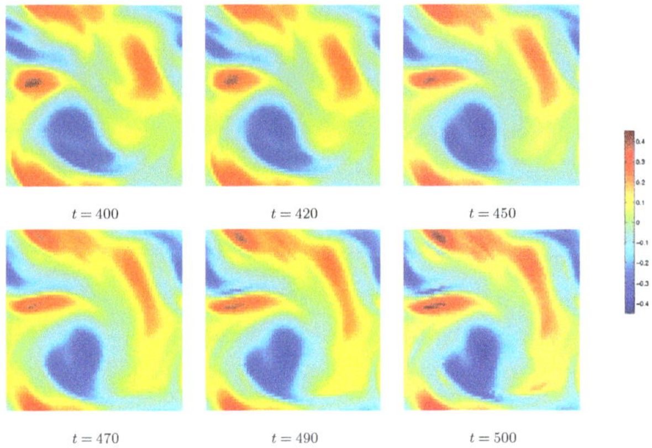

Figure 10. Smoothing result with the proposed method. Estimated mean vorticity maps for different times t between observation times $t = 400$ and $t = 500$.

but not on filtering trajectories at hidden times. It is then able to smooth the discontinuities inherent in the particle filtering technique we have used, contrary to the standard smoothing presented in Fig. 9.

6 Conclusion and discussion

In this paper we introduced a smoothing algorithm based on a conditional simulation technique of diffusions. The proposed smoothing is formulated as fixed-lag, in the sense that it is performed sequentially each time a new observation appears, in order to correct the state at hidden times up to the previous observation. Note that a decomposition similar to Eqs. (13) to (18) can be written from an integration up to a previous time t_{k-h}, with $h > 1$. This implies that the smoother can be formulated with a larger fixed lag, in order to correct the state backward not only up to the previous observation, but up to further measurement times. Yet, due to the successive resampling steps that have been performed in the filtering steps before time t_k, there are in practice only a few distinct filtering trajectories at times t_{k-h} if h is large. Consequently, the estimation of the joint law in Eq. (15) will not be reliable anymore if h is too large.

We have shown the practical applicability of the method to a high-dimensional problem. Nevertheless, the algorithm remains costly since a second Monte Carlo step is added to the Monte Carlo nature of particle filter algorithms. Yet, from an algorithmic point of view, the sequential nature of the proposed technique allows the smoothing to be implemented with a similar structure as filtering methods (sequential sampling and weighting of model trajectories). It is then easy to couple this smoothing to an operational filtering system and benefit from parallelization strategies, for instance.

Since the proposed smoothing uses the filtering result as input, it relies on the success of the underlying particle filter. For high-dimensional systems, a standard particle filter is not adapted and it is necessary to use filtering techniques that guide particles towards observations. In this paper, we use the WEnKF algorithm. In practice, any efficient particle filtering technique with such a guiding can be used within our framework. Note however that the construction of such techniques remains an open area of research.

We plan to work on the application of the smoothing method to a real high-dimensional case (for the estimation of sea surface currents from satellite image sequences). However, such a work will imply numerous difficulties which are not related to the smoothing technique but to the definition of the state-space model: definition of a suitable physical model, good characterization of state noise structure and model parameters. Therefore, this will be part of a future work.

Edited by: P. J. van Leeuwen

References

Beskos, A. and Roberts, G. O.: Exact simulation of diffusions, The Annal. Appl. Probabil., 15, 2422–2444, 2005.

Beskos, A., Papaspiliopoulos, O., Roberts, G. O., and Fearnhead, P.: Exact and computationally efficient likelihood-based estimation for discretely observed diffusion processes (with discussion), J. Roy. Stat. Soc. B, 68, 333–382, 2006.

Beyou, S., Cuzol, A., Gorthi, S., and Mémin, E.: Weighted Ensemble Transform Kalman Filter for Image Assimilation, Tellus A, 65, 18803, doi:10.3402/tellusa.v65i0.18803, 2013.

Briers, M., Doucet, A., and Maskell, S.: Smoothing algorithms for state-space models, Ann. Insti. Stat. Mathe., 62, 61–89, 2010.

Clark, J.: The simulation of pinned diffusions, in: Proceedings of the 29th IEEE Conference on Decision and Control, 1418–1420, 1990.

Del Moral, P.: Feynman-Kac Formulae. Genealogical and Interacting Particle Systems with Applications, Springer, 2004.

Del Moral, P., Jacod, J., and Protter, P.: The Monte Carlo Method for filtering with discrete-time observations, Probabil. Theor. Relat. Fields, 120, 346–368, 2001.

Delyon, B. and Hu, Y.: Simulation of conditioned diffusions and applications to parameter estimation, Stochast. Process. Applic., 116, 1660–1675, 2006.

Doucet, A., Godsill, S., and Andrieu, C.: On sequential Monte Carlo sampling methods for Bayesian filtering, Stat. Comput., 10, 197–208, 2000.

Durham, G. and Gallant, A.: Numerical techniques for maximum likelihood estimation of continuous-time diffusion processes, J. Business Econom. Stat., 20, 297–316, 2002.

Evensen, G.: The ensemble Kalman filter: theoretical formulation and practical implementation, Ocean Dynam., 53, 343–367, 2003.

Evensen, G. and van Leeuwen, P.: An ensemble Kalman Smoother for nonlinear dynamics, Mon. Weather Rev., 128, 1852–1867, 2000.

Fearnhead, P., Papaspiliopoulos, O., and Roberts, G.: Particle filters for partially observed diffusions, J. Roy. Stat. Soc. B, 70, 755–777, 2008.

Godsill, S. J., Doucet, A., and West, M.: Monte Carlo smoothing for nonlinear time series, J. Am. Stat. Assoc., 99, 156–168, 2004.

Gordon, N., Salmond, D., and Smith, A.: Novel approach to nonlinear/non-Gaussian Bayesian state estimation, IEEE Proc. F, 140, 107–113, 1993.

Papadakis, N., Mémin, E., Cuzol, A., and Gengembre, N.: Data assimilation with the weighted ensemble Kalman filter, Tellus A: Dynam. Meteorol. Oceanogr., 62, 673–697, 2010.

Snyder, C., Bengtsson, T., Bickel, P., and Anderson, J.: Obstacles to high-dimensional particle filtering, Mon. Weather Rev., 136, 4629–4640, 2008.

Stroud, J. R., Stein, M. L., Lesht, B. M., Schwab, D. J., and Beletsky, D.: An ensemble Kalman filter and smoother for satellite data assimilation, J. Am. Stat. Assoc., 105, 978–990, 2010.

Sun, Y., Bo, L., and Genton, M. G.: Geostatistics for large datasets, in: Advances and challenges in space-time modelling of natural events, 55–77, Springer, 2012.

Van Leeuwen, P. J.: Particle filtering in Geophysical systems, Mon. Weather Rev., 137, 4089–4114, 2009.

Van Leeuwen, P. J.: Nonlinear data assimilation in geosciences: an extremely efficient particle filter, Q. J. Roy. Meteorol. Soc., 136, 1991–1999, 2010.

Van Leeuwen, P. J. and Ades, M.: Efficient fully nonlinear data assimilation for geophysical fluid dynamics, Comput. Geosci., 55, 16–27, 2013.

10

“Nonlinear waves and chaos in space plasmas”

G. S. Lakhina[1], B. T. Tsurutani[2], A. C.-L. Chian[3], T. Hada[4], G. J. Morales[5], and R. H. J. Grimshaw[6]

[1]Indian Institute of Geomagnetism, New Panvel (W), Navi Mumbai, India
[2]Jet Propulsion Laboratory, California Institute of Technology, Pasadena, CA, USA
[3]Paris Observatory-LESIA, Meudon, France and National Institute for Space Research (INPE), Sao Jose dos Campos, Brazil
[4]Interdisciplinary Graduate School of Engineering Sciences, Kyushu University, Fukuoka, 816-8580, Japan
[5]Physics and Astronomy Department, University of California, Los Angeles, CA 90095, USA
[6]Loughborough University, Department of Mathematical Sciences, Ashby Road, Loughborough, LE11 3TU, UK

Correspondence to: G. S. Lakhina (gslakhina@gmail.com)

Nonlinear waves and chaos (NWC) are believed to play dominant roles in the heating and acceleration of charged particles and generating turbulence in space, astrophysical and laboratory plasmas. A series of workshops on “Nonlinear Waves and Chaos in Space Plasmas” have been held to discuss this subject: in Kyoto, Japan (1994), Köln, Germany (1997), Carlsbad, USA (1999), Tromsø, Norway (2001), Mumbai, India (2003), Fukuoka, Japan (2006), Beaulieu-sur-Mer, France (2008), and La Jolla, USA (2010). The ninth NWC workshop 2013 of this series was held in La Jolla, USA, 3–8 March 2013.

This workshop brought together 31 space plasma, laboratory plasma, and fusion physicists, astrophysicists and atmospheric physicists from various countries. The workshop provided an opportunity to report on the latest results from theory, simulation and data analysis dealing with the nonlinear processes and complex system dynamics in space, laboratory and astrophysical plasmas as well as nonlinear interaction between large-scale waves in the Earth’s atmosphere.

This special issue of *Nonlinear Processes in Geophysics* (NPG) consists of papers drawn from the talks given at NWCW 2013 as well as papers that were submitted in response to the call for the papers for this special issue. All papers had to undergo the usual NPG review process before they were finally accepted for this special issue.

The papers cover a broad range of topics related to nonlinear waves and chaos. These include the multifractal spectrum in the heliosphere and the heliosheath, chaoticity and dynamical complexity in the ionosphere, Alfvén waves in space and astrophysical dusty plasmas, nonlinear electromagnetic waves in magnetized relativistic plasmas, ion acceleration by nonlinear Alfvén wave packets, the relativistic surfatron process for Landau resonant electrons in radiation belts, stochastic electron motion driven by nonlinear waves, turbulence in the interstellar medium, four-dimensional energy spectrum for space–time structure of plasma turbulence, and validity of Taylor’s hypothesis under large-scale flow variations. A brief summary of the special issue papers is given below.

Macek and Wawrzaszek (2014) report on the multifractal scaling of the fluctuations in the interplanetary magnetic field strength as measured onboard Voyager 2 in the heliosphere and the heliosheath. They find that the spectrum is prevalently right-skewed and the degree of multifractality is modulated by the solar activity inside the heliosphere. The results seem to support the idea that the solar wind termination shock is asymmetric.

Ogunsua et al. (2014) analyze total electron content (TEC) time series data during the year 2011 obtained by three global positioning system (GPS) receivers within Nigeria, which lies within the equatorial ionization anomaly region. They find positive Lyapunov exponents (LE) for both quiet and disturbed days, which indicates chaoticity in the system. For different days, the chaoticity of the ionosphere exhibits no definite pattern for either quiet or disturbed days. However, values of the LE were lower for the storm period compared with its nearest relatively quiet periods for all the stations. The monthly averages of LE and entropy also show no definite pattern. The values of the computed correlation dimension range from 2.8 to 3.5, with the lowest values recorded during the storm period of October 2011.

Jatenco-Pereira et al. (2014) present results on the effects of superthermal particles on the dispersive characteristics of Alfvén waves in dusty plasmas in different astrophysical and space regions. It is shown that the presence of dust and superthermal particles significantly modifies the dispersion of Alfvén waves. The authors discuss the competition between different damping processes of kinetic Alfvén and Alfvén cyclotron waves. They briefly review the nonlinear evolution of Alfvén waves to chaos. Some applications of Alfvén waves in the auroral region of space plasmas, as well as stellar winds and star forming regions of astrophysical plasmas are discussed.

Muñoz et al. (2014) have shown that for large-amplitude, circularly polarized waves propagating along a constant magnetic field, an exact solution of the fluid equations can be derived for relativistic temperatures. It is found that relativistic thermal effects produce a decrease in both the effective plasma frequency and the upper frequency cutoff for the Alfvén branch. It is also found that the Alfvén speed decreases with temperature, being zero for infinite temperature. The authors also studied the same system based on the relativistic Vlasov equation, including thermal effects along the direction of propagation. It is concluded that kinetic and fluid results are qualitatively consistent, but have several quantitative differences.

The paper by Nariyuki et al. (2014) reports the nonlinear evolution of parallel propagating Alfvén waves in a radially expanding plasma from numerical simulations of a kinetic-fluid model. It is found that the steepened Alfvén wave packets outwardly accelerate ions, which can be observed as the beam components in the interplanetary space. The energy of imposed Alfvén waves is converted into longitudinal fluctuations by nonlinear steepening and nonlinear Landau damping. Wave shoaling due to the inhomogeneity of the phase velocity is also observed.

Osmane and Hamza (2014) report new results on the relativistic surfatron process for Landau resonant electrons in radiation belts. Their model includes longitudinal wave effects and inhomogeneous magnetic fields. It is found that even though both effects can limit the surfatron acceleration of electrons in the radiation belts, gains in energy of the order of 100 keV can occur in 10^{-4} s. The authors conclude that this is sufficiently strong to be relevant to radiation belt dynamics.

The paper by Khazanov et al. (2014) deals with stochastic electron motion driven by various nonlinear waves. It is found that the interaction between whistler-mode waves and radiation-belt electrons is likely to involve the same mechanism that is responsible for the dynamical balance between the accelerating process and relativistic electron precipitation events. The paper discusses the efficiency of the stochastic surfing acceleration mechanism for cosmic electrons at supernova remnant shock fronts involving Langmuir wave packets. Analytical and numerical methods for studying Hamiltonian chaos and dissipative strange attractors, and for determining the heating extent and energy spectra, are discussed.

Falceta-Gonçalves et al. (2014) give a review of interstellar turbulence focusing on the statistics of its structure and energy cascade. The physics of compressible and incompressible turbulent flows in magnetized plasmas is discussed. The most relevant observational techniques providing quantitative insights of interstellar turbulence are also presented. The main difficulties in developing a three-dimensional view of interstellar turbulence from these observations as well as the main sources of interstellar turbulence are briefly reviewed.

The paper by Narita (2014) discusses a four-dimensional energy spectrum for the space–time structure of plasma turbulence. The model is based on the Eulerian wave number-frequency spectrum developed for describing fluid turbulence. This includes wave vector anisotropies in the three-dimensional wave vector domain by approximating the spectrum to a set of ellipses. The model is applied to the magnetic energy spectrum in the near-Earth solar wind measured by four Cluster spacecraft, and the set of the spectral parameters are determined observationally. In this way, the space–time structure of plasma turbulence can be condensed into a small number of parameters, suitable for evaluating the energy spectra in observational and numerical studies on a quantitative basis.

The paper by Wilczek et al. (2014) discusses the validity of Taylor's hypothesis under large-scale flow variation based on a simple model for the wavenumber-frequency spectrum that has recently been introduced as a generalization of Kraichnan's random sweeping hypothesis. For turbulence with a power-law energy spectrum, it is found that the wavenumber spectrum estimated by Taylor's hypothesis exhibits the same power-law as the true spectrum, but the spectral energy is overestimated due to the large-scale flow variation.

We deeply thank all the referees who donated their time to carefully review the papers and give valuable suggestions/comments which led to improvement of the papers. We hope that the readers of NPG will enjoy this special issue.

References

Falceta-Gonçalves, D., Kowal, G., Falgarone, E. and Chian, A. C.-L.: Turbulence in the Interstellar Medium, Nonlin. Processes Geophys., accepted, 2014.

Jatenco-Pereira, V., Chian, A. C.-L., and Rubab, N.: Alfvén waves in space and astrophysical dusty plasmas, Nonlin. Processes Geophys., 21, 405–416, doi:10.5194/npg-21-405-2014, 2014.

Khazanov, G. V., Tel'nikhin, A. A., and Kronberg, T. K.: Stochastic electron motion driven by space plasma waves, Nonlin. Processes Geophys., 21, 61–85, doi:10.5194/npg-21-61-2014, 2014.

Macek, W. M. and Wawrzaszek, A.: Voyager 2 observation of the multifractal spectrum in the heliosphere and the heliosheath, Nonlin. Processes Geophys., 20, 1061–1070, doi:10.5194/npg-20-1061-2013, 2013.

Muñoz, V., Asenjo, F. A., Domínguez, M., López, R. A., Valdivia, J. A., Viñas, A., and Hada, T.: Large-amplitude electromagnetic waves in magnetized relativistic plasmas with temperature, Nonlin. Processes Geophys., 21, 217–236, doi:10.5194/npg-21-217-2014, 2014.

Narita, Y.: Four-dimensional energy spectrum for space–time structure of plasma turbulence, Nonlin. Processes Geophys., 21, 41–47, doi:10.5194/npg-21-41-2014, 2014.

Nariyuki, Y., Umeda, T., Suzuki, T. K., and Hada, T.: Ion acceleration by parallel propagating nonlinear Alfvén wave packets in a radially expanding plasma, Nonlin. Processes Geophys., 21, 339–346, doi:10.5194/npg-21-339-2014, 2014.

Ogunsua, B. O., Laoye, J. A., Fuwape, I. A., and Rabiu, A. B.: The comparative study of chaoticity and dynamical complexity of the low-latitude ionosphere, over Nigeria, during quiet and disturbed days, Nonlin. Processes Geophys., 21, 127–142, doi:10.5194/npg-21-127-2014, 2014.

Osmane, A. and Hamza, A. M.: Relativistic surfatron process for Landau resonant electrons in radiation belts, Nonlin. Processes Geophys., 21, 115–125, doi:10.5194/npg-21-115-2014, 2014.

Wilczek, M., Xu, H. and Narita, Y.: A note on Taylor's hypothesis under large-scale flow variation, Nonlin. Processes Geophys., accepted, 2014.

11

Barriers to transport in aperiodically time-dependent two-dimensional velocity fields: Nekhoroshev's theorem and "Nearly Invariant" tori

S. Wiggins[1] **and A. M. Mancho**[2]

[1]School of Mathematics, University of Bristol, Bristol BS8 1TW, UK
[2]Instituto de Ciencias Matemáticas, CSIC-UAM-UC3M-UCM, C/Nicolás Cabrera 15, Campus Cantoblanco UAM, 28049 Madrid, Spain

Correspondence to: S. Wiggins (s.wiggins@bristol.ac.uk)

Abstract. In this paper we consider fluid transport in two-dimensional flows from the dynamical systems point of view, with the focus on elliptic behaviour and aperiodic and finite time dependence. We give an overview of previous work on general nonautonomous and finite time vector fields with the purpose of bringing to the attention of those working on fluid transport from the dynamical systems point of view a body of work that is extremely relevant, but appears not to be so well known. We then focus on the Kolmogorov–Arnold–Moser (KAM) theorem and the Nekhoroshev theorem. While there is no finite time or aperiodically time-dependent version of the KAM theorem, the Nekhoroshev theorem, by its very nature, is a finite time result, but for a "very long" (i.e. exponentially long with respect to the size of the perturbation) time interval and provides a rigorous quantification of "nearly invariant tori" over this very long timescale. We discuss an aperiodically time-dependent version of the Nekhoroshev theorem due to Giorgilli and Zehnder (1992) (recently refined by Bounemoura, 2013 and Fortunati and Wiggins, 2013) which is directly relevant to fluid transport problems. We give a detailed discussion of issues associated with the applicability of the KAM and Nekhoroshev theorems in specific flows. Finally, we consider a specific example of an aperiodically time-dependent flow where we show that the results of the Nekhoroshev theorem hold.

1 Introduction

This paper is concerned with "Kolmogorov, Arnold, Moser (KAM) like behavior" in two-dimensional, incompressible, aperiodically time-dependent velocity fields over a finite time interval. We will explain what we mean by this phrase in the course of the introduction, and we will begin by noting that the motivation for this study comes from the dynamical systems approach to Lagrangian transport in fluid flows.

Let $\mathbf{r} \equiv (x, y)$ denote coordinates describing a two-dimensional region. A fluid flow in this region is described by a velocity field, $\mathbf{v}(\mathbf{r}, t) \equiv \big(v_x(x, y, t), v_y(x, y, t)\big)$. The velocity field can be *kinematically defined* (i.e. constructed to describe certain observed features of the flow), dynamically defined (i.e. it is obtained as the solution of a set of partial differential equations that describe the dynamical evolution of the velocity field), or it could be obtained by observation (i.e. through remote sensing of some region of the ocean). In any case, at this point of the discussion, it is not important how one obtains the velocity field, but we assume that by some means we have obtained a velocity field. The equations describing the motion of fluid particles in the velocity field are given by:

$$\dot{\mathbf{r}} = \mathbf{v}(\mathbf{r}, t) \qquad (1)$$

(neglecting molecular diffusion, or possibly the effect of neglected scales of motion, which would be the case if one were considering velocity fields obtained from partial differential equations that describe only certain length and time scales in the ocean). If the flow is incompressible the velocity field can be obtained from the derivatives of a scalar valued function $\psi(x, y, t)$, the stream function, as (see, e.g., Batchelor (1967)):

$$\dot{x} = \frac{\partial \psi}{\partial y}(x, y, t),$$
$$\dot{y} = -\frac{\partial \psi}{\partial x}(x, y, t). \quad (2)$$

Making the connection with the mathematical framework of dynamical systems theory, formally Eq. (2) has the form of Hamilton's canonical equations, where $\psi(x, y, t)$ plays the role of the Hamiltonian function and the corresponding phase space coordinates, (x, y), are actually the physical space coordinates in which the fluid flow takes place. If $\psi(x, y, t)$ is *periodic in the time, t*, then it is standard in dynamical systems theory to study the structure of the trajectories of Eq. (2) by considering the (discrete) orbits of the associated Poincaré map, i.e. to view the continuous time trajectories of Eq. (2) at a sequence of discrete times, where the sequence of times is integer multiples of the (temporal) period of the stream function.

This "Hamiltonian dynamical systems" point of view generated a great deal of interest and further research starting in the early 1980s with the publication of Aref (1984). This occurred at the same time that "applied dynamical systems theory" was flowering as a topic of study across many disciplines in science and engineering. The growing availability of computational resources was giving rise to a large amount of "computational phenomenology" for two-dimensional, area preserving maps (such as the standard map; see e.g. Meiss, 2008). Thus, armed with the point of view described in Aref (1984), one could "see" that phase space structures such as elliptic periodic orbits, hyperbolic periodic orbits and their stable and unstable manifolds, and KAM tori had an immediate interpretation in terms of "structures" in the flow influencing transport and mixing. In this way dynamical systems theory provided an analytical and computational meaning for the notion of "coherent structures" in fluid flows that was becoming a frequent observation in experiments due to advances in flow visualization capabilities (see e.g. Brown and Roshko, 1974). For example, transversely intersecting stable and unstable manifolds of hyperbolic periodic orbits give rise to "chaotic fluid particle trajectories" through the construction on Smale horseshoes, KAM tori trap regions of fluid (therefore preventing them from "mixing" with surrounding fluid), KAM tori are found surrounding elliptic periodic orbits (hence these are a "signature" of regions of unmixed fluid), and the intersecting stable and unstable manifolds give rise to "partial" barriers to transport and "lobe dynamics". This mathematical framework proved to be ideal for realizing the physical picture of mixing put forth earlier by Reynolds (1894), Eckart (1948), Danckwerts (1952), and Danckwerts (1953). Ottino et al. (1994) has described in detail the physical picture of mixing first described by Reynolds and how it had to await the proper mathematical framework, i.e. dynamical systems theory, before it could be analysed and exploited. Reviews of the dynamical systems approach to Lagrangian transport and mixing, mostly for two-dimensional, time-periodic incompressible flows, can be found in Aref and El Naschie (1994), Aref (2002), Acrivos et al. (1991), Babiano et al. (1994), Ottino (1989a, b), Wiggins (1992), Wiggins and Ottino (2004), and Sturman et al. (2006).

The new insights into transport and mixing obtained from the dynamical systems approach for two-dimensional, incompressible, time-periodic flows motivated efforts to extend this approach to more complex flow situations. Two possible extensions would be from two to three space dimensions and from periodic time dependence to a more general time dependence. Some motivation for these extensions came from the desire to use the dynamical systems approach to study Lagrangian transport in the ocean and atmosphere, starting in the early 1990s. Such flows will generally not vary periodically in time. However, the two dimensional approximation does have validity under certain circumstances. Moreover, "realistic" flows are obtained from the solution of a set of partial differential equations derived from the physical situation under consideration. Typically, these partial differential equations are strongly nonlinear and can only be "solved" with a computer. This gives rise to a velocity field defined as a data set over a finite time interval, or a *finite time dynamical system*. Early work on transport in geophysical flows from the point of dynamical systems having aperiodic time dependence and/or defined as a finite time data set can be found in Duan and Wiggins (1996), Miller et al. (1997), Duan and Wiggins (1997), Malhotra and Wiggins (1998), Haller and Poje (1998), Rogerson et al. (1999), and Coulliette and Wiggins (2001). Reviews that describe how these issues in dynamical systems theory arise from the point of view of transport in geophysical flows are Jones and Winkler (2002), Wiggins (2005), Mancho et al. (2006), and Samelson and Wiggins (2006).

This paper is concerned with the extension of the dynamical systems approach to transport in two-dimensional, incompressible flows having more general time dependence than periodic, and their dynamics over a finite time interval. Our focus is on "elliptic behaviour" and invariant tori. However, first we review some of the work on finite time, aperiodically time-dependent dynamical systems. Some of the more recent work in this area has been motivated by the issues raised by some of the work related to transport in geophysical flows noted above. However, there has been a great deal of work in the mathematics community and the control theory community that is relevant that has not been properly recognized. In Sect. 1.1 we describe issues and work on nonautonomous dynamical systems and in Sect. 1.2 we describe issues and work on finite time dynamical systems.

1.1 Nonautonomous dynamical systems

As applied dynamical systems enjoyed an explosion of popularity, starting from the late 1970s and continuing through today, "dynamics" – the study of how the state of a system evolves in time – is typically described by iteration of maps (discrete time) and flows (continuous time). Flows are the group of one-parameter families of transformations of the state space (where the parameter is "time") that are obtained as the solutions of autonomous differential equations (here, we will simplify our discussion by assuming that the solutions of differential equations exist for all positive and negative time). Maps arise naturally from time-periodic differential equations through the so-called "Poincaré map" construction. The success of the geometric approach of dynamical systems theory in applications encouraged efforts to extend the ideas to more complex settings, and time-dependent differential equations whose time dependence is more general than periodic are a natural extension to consider. In this case the Poincaré map construction was no longer possible since this relied on the time periodicity of the differential equation. Moreover, the solutions of nonautonomous differential equations do not define flows in the usual sense of the definition[1]. Therefore, the basic consideration that one must begin with is "how do you describe the dynamics arising from a nonautonomous differential equation"?

In some sense, this problem was solved in the 1960s. Dafermos developed the notion of a *process* and generalized the LaSalle Invariance Principle to this setting (Dafermos (1971)). Miller (1965) and Sell (1967a, b, 1971) developed the notion of *skew product flows* and their associated cocycle property. These ideas are further described from a pedagogical point of view in the recent review article by Balibrea et al. (2010). With descriptions of "time evolution" appropriate to nonautonomous differential equations at hand, the building blocks of a geometrical theory can be developed. This was begun in the works of Dafermos, Miller, and Sell cited above, but there are other important results from this era that do not appear to be well known. Possibly one reason for this is that the particular nature of the time dependence in the ordinary differential equation community was not so important for many lines of investigation. For example, in the classic ordinary differential equation textbook of Coddington and Levinson (1955) it is easy to see that the proof of the stable and unstable manifold theorem for a hyperbolic trajectory does not use any particular form of time dependence (only that the relevant functions are appropriately bounded in time, and that the existence and uniqueness of solutions holds). A stable and unstable manifold theorem for hyperbolic processes is proven by Irwin (1973) and de Blasi and Schinas (1973), and the more recent textbook by Katok and Hasselblatt (1995) develops the framework of dynamics generated by iteration of sequences of maps, which is a possible framework for nonautonomous dynamics. Fenichel (1991) also proves a stable and unstable manifold theorem for hyperbolic trajectories in discrete time aperiodic systems.

By now the framework for the geometrical analysis of nonautonomous dynamical systems is well under development. Notions of attractivity, stability, and asymptotic behaviour have been developed in Kloeden and Schmalfuss (1997), Langa et al. (2002), Meyer and Zhang (1996), and Sell (1967b, 1971). Shadowing lemmas have been developed in Chow et al. (1989) and Meyer and Zhang (1996). Chaos is discussed and analysed in Lerman and Silnikov (1992), Meyer and Sell (1989), Scheurle (1986), Stoffer (1988a, b), Wiggins (1999), and Lu and Wang (2010, 2011). Various aspects of bifurcation theory are developed in Langa et al. (2002), Poetzsche (2011a, b, 2010b), and Rasmussen (2006). A version of normal form theory is developed in Siegmund (2002). Recent work on general discrete nonautonomous systems is described in Kloeden and Poetzsche (2011) and Poetzsche (2010a).

1.2 Finite time dynamical systems

As we mentioned earlier, efforts to use the dynamical systems point of view to analyse transport and mixing in geophysical flows have motivated the study of time-dependent velocity fields that are only defined over a finite time interval, or *finite time dynamical systems*. Initially, one might think that such a notion is completely at odds with the "dynamical systems point of view", since it is often stated that dynamical systems theory is concerned with the "long time behavior" of a system. Indeed, notions such as "stability" and "attraction" describe aspects of the behaviour of trajectories as time approaches infinity. Mathematical proofs of characteristics of collections of trajectories such as "invariance" and "chaos" typically require an appropriate type of control over these collections of trajectories as time approaches infinity. Nevertheless, computer simulations of a wide variety of dynamical systems (necessarily for a finite simulation time) indicate that these infinite time notions provide both a language and structure to describe the results, and this provides some hope that there is a reasonable chance of success for developing analogous "dynamical systems ideas" for nonautonomous dynamical systems that are only defined for a finite time.

There has been a great deal of activity in recent years in developing a "dynamical systems framework" for finite time dynamical systems. However, it should be noted that similar to the situation described above, the differential equations and control theory communities addressed a number of essential issues in this area many years earlier (and it continues to be a topic of interest in control theory). A recent review paper of Dorato (2006) gives an overview of and historical perspective on work on "finite time stability". The paper of Weiss and Infante (1965) also provides a very insightful

[1] Specific examples that illustrate the fact that the solutions of nonautonomous equations do not form flows can be found in Balibrea et al. (2010).

and rigorous discussion on finite time stability. More recently, Duc and Siegmund (2008) developed basic building blocks (i.e. hyperbolic trajectories and their stable and unstable manifolds) for two-dimensional, time-dependent Hamiltonian systems defined only for a finite time interval. Further work along these lines can be found in Berger et al. (2008). The original definition of "hyperbolicity" is intimately connected to a type of infinite time average along trajectories. Restricting such averages to finite time can be problematic. This topic is treated in Duc and Siegmund (2011) and Berger (2011). Another approach to the notion of hyperbolicity for finite times would be through a proper generalization of the idea of the spectrum associated with the linearization about a finite time trajectory. This is discussed in Berger et al. (2009) and Doan et al. (2011).

With an approach for defining and computing hyperbolic trajectories for finite dimensional vector fields in hand, it is natural to consider the computation of the stable and unstable manifolds of the finite time hyperbolic trajectories. This issue had been considered in Haller (2000), Mancho et al. (2004), Mancho et al. (2006), and Branicki and Wiggins (2010). However, it is important to point out a characteristic associated with finite time hyperbolic phenomena – *nonuniqueness*. In general, all methods used to prove the existence of unique invariant manifolds require the use of a type of iterative or recursive technique with a passage to a limit, and a unique invariant manifold is obtained in this limit. Essentially, passage to the limit means taking time to plus or minus infinity (depending on whether or not one is computing unstable or stable manifolds, respectively). Nevertheless, with respect to the notion of barriers to transport, this is not an issue, since the manifolds are constructed (numerically) with trajectories (and therefore uniqueness of solutions implies that trajectories cannot cross manifolds constructed in this way). The nonuniqueness effectively means that the region where the one-dimensional manifolds (in two space dimensions) are numerically constructed has a certain thickness (see estimate in Haller, 2000) which would go to zero if it were possible to allow time to approach infinity.

There is an important point to be made here which will serve to introduce that aspect of aperiodically time-dependent dynamics over finite time intervals that we will be considering in this paper. We emphasize again that we will be considering two-dimensional, time-dependent Hamiltonian systems, i.e. incompressible, two-dimensional velocity fields. Broadly speaking, the stability properties of trajectories and invariant manifolds of Hamiltonian systems are either hyperbolic or elliptic in nature.[2] Very generally, hyperbolic properties of trajectories, or invariant manifolds, are somewhat independent of any "special structure" of the dynamical system, such as Hamiltonian structure. For example, theorems concerning hyperbolic trajectories and their stable and unstable manifolds are generally equally valid in both Hamiltonian and non-Hamiltonian systems. Stability results for "elliptic dynamics", on the other hand, generally rely crucially on the special structure of the dynamical system (e.g. Hamiltonian, reversible) as well as the coordinates in which the Hamiltonian system is expressed (e.g. action-angle coordinates), with the latter being important for specific analytical methods, such as Fourier analysis. One way of understanding this difference is that "hyperbolic phenomena" are generally stable under perturbation of the dynamical system, while "elliptic phenomena" are not. This would seem to indicate that the fate of "elliptic objects" under perturbation requires a more careful analysis of the effect of the perturbation, and this tends to be the case.

The KAM and Nekhoroshev theorems are major results in Hamiltonian dynamics that are concerned with the behaviour of "elliptic objects", i.e. invariant tori, under perturbation. A standard (and the original) setting for these theorems in (canonical) Hamiltonian systems is that of the (Hamiltonian) perturbation of an integrable system expressed in action-angle variables, i.e. the unperturbed Hamiltonian is expressed entirely as a function of the action variables. This is the setting relevant to us, but more general settings can be found in Broer et al. (1996).

The foundations of the KAM theorem were laid in the 1950s and 1960s (Kolmogorov, 1954; Arnold, 1963; Moser, 1962). Succinct overviews of the essential points of KAM theory can be found in Chierchia and Mather (2010), Pöschel (2001), and Sevryuk (2003). It is probably fair to say that KAM theory became known throughout the worldwide dynamics community from the late 1970s onward. However, the Nekhoroshev theorem came much later (Nekhoroshev (1977)), despite the fact that the phenomenon of "stability over exponentially long time scales" was considered earlier than the KAM theorem in Littlewood (1959b, a) and Moser (1955). The Nekoroshev theorem was promoted in the west by the Italian schools associated with Benettin, Gallavotti and Giorgilli. A very accessible proof of the theorem was given in Benettin and Gallavotti (1986), and the website of Prof. Antonio Giorgilli (http://www.mat.unimi.it/users/antonio/) has a wealth of information on both the KAM and Nekhoroshev theorems, as well as a collection of instructive pedagogical articles and applications to fundamental problems in physics.

We now describe the aspects of the KAM and Nekhoroshev theorems that set the context for the purposes of this paper. For more general settings and conditions of applicability, we refer to the references given above.

[2]Of course, this is a bit too simplistic, but it is accurate for our needs. The "boundary" between hyperbolic and elliptic is where bifurcation occurs and requires careful consideration, and "partial hyperbolicity" is also of much current interest (Pesin, 2004).

2 The Kolmogorov–Arnold–Moser (KAM) theorem and Nekhoroshev's theorem

We begin by considering the more familiar autonomous cases for the KAM and Nekhoroshev theorems. The nonautonomous cases, relevant to our work, are discussed in Sects. 2.4 and 3.

We consider a Hamiltonian of the following form:

$$H(I,\theta) = H_0(I) + \epsilon H_1(I,\theta), \quad (I,\theta) \in B \times \mathbb{T}^n, \tag{3}$$

where $B \subset \mathbb{R}^n$ is the ball of radius $\bar{R}$ in $\mathbb{R}^n$, $H_0(I)$ is referred to as the unperturbed part of the Hamiltonian and $H_1(I,\theta)$ is referred to as the perturbation. The Hamiltonian function needs to be "sufficiently differentiable" on a "well-controlled domain", and we will address this issue more precisely when we consider aperiodic time-dependent Hamiltonians in Sect. 3. The coordinates $(I,\theta) \in B \times \mathbb{T}^n$ play a very important role. These are the so-called *action-angle variables* that arise from the structure of the unperturbed, integrable system (Arnold, 1978), and we will have more to say about their role shortly.

The unperturbed Hamiltonian vector field is given by:

$$\begin{aligned} \dot{I} &= -\frac{\partial H_0}{\partial \theta}(I) = 0. \\ \dot{\theta} &= \frac{\partial H_0}{\partial I}(I) \end{aligned} \tag{4}$$

and the trajectories of this vector field are given by:

$$\begin{aligned} I(t) &= I_0 = \text{constant}, \\ \theta(t) &= \frac{\partial H_0}{\partial I}(I_0)t + \theta_0. \end{aligned} \tag{5}$$

Clearly, the n-dimensional action variable $I = I_0$ is constant in time, and the n-dimensional angle variables increase linearly in time at a rate defined by the frequency vector $\frac{\partial H_0}{\partial I}(I_0)$. In this way $I = I_0$ defines an n-dimensional invariant torus and the trajectories on the torus are quasiperiodic, having n frequencies.

2.1 The KAM theorem and sufficient conditions for its application

The KAM theorem is concerned with the preservation of invariant n tori upon perturbation by the term $H_1(I,\theta)$. First, we consider the preservation of a given torus $I = I_0$. This torus will "persist" for the perturbed system with *the same* frequencies $\frac{\partial H_0}{\partial I}(I_0)$ provided the unperturbed Hamiltonian satisfies a nondegeneracy condition, the vector of n frequencies is "strongly nonresonant" and the perturbation is sufficiently small. We discuss the sufficient conditions for its application below.

Action-Angle variables. We assume that the unperturbed system is integrable in a way that action-angle variables exist, i.e. there are n integrals that are independent and in involution (these terms and conditions are defined in Arnold, 1978). The full Hamiltonian, i.e. the unperturbed part and the perturbed part, is then expressed in terms of the action-angle variables of the unperturbed, integrable Hamiltonian (cf. Eq. 3). This is important because a number of the analytical methods used in the proofs of the KAM and Nekhoroshev theorems use characteristics of the action-angle variables.

Dealing with Resonances. The KAM theorem is concerned with the preservation under perturbation of certain nonresonant tori on the unperturbed system. $\frac{\partial H_0}{\partial I}(I_0)$ is the vector of frequencies associated with the n torus $I = I_0$. The standard nonresonant condition for the torus $I = I_0$ that the frequencies must satisfy is:

$$\begin{aligned} &|k \cdot \frac{\partial H_0}{\partial I}(I_0)| > \gamma \parallel k \parallel^{\tau}, \ \gamma > 0, \ \tau > n-1 \quad \text{for all} \\ &\text{nonzero integer vectors} \quad k = (k_1, \ldots, k_n) \in \mathbb{Z}^n - \{0\}, \end{aligned} \tag{6}$$

where $\parallel k \parallel \equiv \sum_{i=1}^{n} |k_i|$. It is a standard result that "almost all" (in the sense of Lebesgue measure) frequencies satisfy such a condition (Broer et al., 1996; Chierchia and Mather, 2010).

Nondegeneracy. Nonresonance is a condition on the first derivative of the unperturbed Hamiltonian. Nondegeneracy is a condition on the second derivative. A standard nondegeneracy condition is the following:

$$\det\left(\frac{\partial^2 H_0}{\partial I^2}(I_0)\right) \neq 0. \tag{7}$$

If the Hamiltonian depends explicitly on time, a different nondegeneracy condition is used, the so-called *isoenergetic nondegenenerацy condition*. This is the standard nondegeneracy condition, but restricted to a level set of the Hamiltonian. For details see Broer et al. (1996) and Chierchia and Mather (2010).

2.2 The Nekhoroshev theorem and sufficient conditions for its application

Now we turn our attention to Nekhoroshev's theorem (Nekhoroshev, 1977). First we state the theorem in the form of a "model statement" (Lochak, 1993), describe what this means, and then give some background and history. Our statement applies to the Hamiltonian Eq. (3) (and the Hamiltonian must be analytic on an appropriate domain; we will comment more on this later).

For an initial condition $I(0) \equiv I_0 \in B$ we have:

$$\parallel I(t) - I_0 \parallel \leq c_1 \varepsilon^b \quad \text{for} \quad |t| \leq \exp\left(c_2/\varepsilon^a\right) \tag{8}$$

for $\varepsilon \leq \varepsilon_0$. Here ε is a parameter that is estimated in the proof of Nekhoroshev's theorem in terms of the defining parameters of the Hamiltonian (to be discussed later on); ε_0 is a "threshold" value for ε. The parameters ε_0, c_1, and c_2 are also estimated in terms of the defining parameters of the Hamiltonian, and the "stability exponents" a and b are estimated as functions of n.

Just as we did for the KAM theorem above, we summarize the sufficient conditions for the applicability of Nekhoroshev's theorem below.

Action-Angle variables. As for the KAM theorem, we assume that the unperturbed system is integrable in a way that action-angle variables exist, exactly as we described for the KAM theorem.

Dealing with Resonances. The Nekhoroshev theorem does not focus on specific values of I corresponding to nonresonant tori. Rather, it provides an estimate of evolution in time of *any* initial action variable over an exponentially long time. The proof of the Nekhoroshev theorem is divided into two parts – an analytic part and a geometric part. The analytic part derives a normal form that is valid in a particular type of nonresonance region. Estimates of the evolution of the action variables can then be obtained for that particular region. The geometric part is the creative element provided by Nekhoroshev. He developed a method that enabled him to show that the entire phase space could be covered by domains in such a way that the normal forms appropriate to these domains, and the associated estimates of the evolution of the action variables, applied to the entire phase space. This construction requires the nondegeneracy condition that we next describe. The geometric argument was improved in Pöschel (1993). A pedagogical discussion of the geometric argument is given in Giorgilli (2002).

Nondegeneracy. The unperturbed Hamiltonian must satisfy a nondegeneracy condition, i.e. a condition on the second derivative of the unperturbed Hamiltonian. However, it is different from the nondegenderacy condition of the KAM theorem. A standard nondegeneracy condition is that the unperturbed Hamiltonian must satisfy a convexity condition of the following form. In particular,

$$\| \frac{\partial^2 H_0}{\partial I^2}(I)v \| \leq M \| v \|, \quad |\frac{\partial^2 H_0}{\partial I^2}(I)v \cdot v| \geq m \| v \|^2$$
$$\text{for all } v \in \mathbb{R}^n,\ m \leq M. \tag{9}$$

Similar to the KAM case, a different nondegeneracy condition may be applied when the perturbation depends explicitly on time. In this case it is assumed that the unperturbed Hamiltonian is *quasiconvex*, i.e. it is convex on a fixed level set of the Hamiltonian. For details see Broer et al. (1996). In his original proof Nekhoroshev used a weaker nondegeneracy condition referred to as "steepness"; see Nekhoroshev (1977).

The idea behind "exponential stability" estimates

"Exponential stability" estimates are not obtained from a "straightforward" application of perturbation theory. Here we give a brief, non-rigorous, discussion of how exponential stability estimates can be obtained from a perturbation expansion. "Non-rigorous" means we do not provide proper estimates of domains and sizes of the remainder in the perturbation expansion. These details are at the heart of the Nekhoroshev theorem. Rather, we show how an exponentially small remainder of a perturbation series can be obtained *if* a perturbation series of a particular type is, somehow, obtained. We discuss the autonomous case since the argument is simpler, and it suffices to convey the main ideas behind exponentially small stability estimates. Our discussion follows from Giorgilli (1995).

We consider the following "near integrable" autonomous Hamiltonian:

$$H(I,\theta) = H_0(I) + \varepsilon \tilde{H}(I,\theta), \quad I \in B \in \mathbb{R}^n, \theta \in \mathbb{T}^n, \tag{10}$$

where B is the open ball centered at the origin of radius R. The associated Hamiltonian vector field is given by:

$$\begin{aligned} \dot{\theta} &= \frac{\partial H}{\partial I}(I,\theta), \\ \dot{I} &= -\frac{\partial H}{\partial \theta}(I,\theta), \end{aligned} \tag{11}$$

and we are interested in the time evolution of the I variables.

We suppose that "r steps" of canonical transformation theory have been performed, which transform Eq. (10) into the normal form:

$$\begin{aligned} H'(I',\theta') = {} & H_0(I') + \varepsilon H_1(I') + \cdots \\ & + \varepsilon^r H_r(I') + \varepsilon^{r+1} R(I',\theta'). \end{aligned} \tag{12}$$

The nature of the domain and the properties of the canonical transformations on this domain that bring Eq. (10) into the form of Eq. (11) are important ingredients in Nekhoroshev's theorem. However, they are not important for the point that we wish to discuss here. Rather, given a normal form of the form Eq. (12) we will describe, roughly, how one obtains an *exponential estimate* (and, in course, what exactly this italicized phrase means)[3]. Hamilton's equations for the Hamiltonian Eq. (12) are given by:

$$\begin{aligned} \dot{\theta}' &= \frac{\partial H'}{\partial I'}(I',\theta') = \frac{\partial H_0}{\partial I'}(I') + \mathcal{O}(\varepsilon), \\ \dot{I}' &= -\frac{\partial H'}{\partial \theta'}(I',\theta') = -\varepsilon^{r+1}\frac{\partial R}{\partial \theta'}(I',\theta'). \end{aligned} \tag{13}$$

We are interested in the time evolution of the action variables, which are given by:

$$I'(t) - I'(0) = -\varepsilon^{r+1} \int_0^t \frac{\partial R}{\partial \theta'}(I'(\tau),\theta'(\tau))d\tau, \tag{14}$$

[3]Of course, the real innovation of Nekhoroshev was showing how the entire action space could be covered with domains on which "appropriate" normal forms could be constructed (i.e. normal forms that were "adapted" to possible resonances on the domains), with associated exponential estimates, and how these estimates could be extended to the entire action space.

and therefore

$$|I'(t) - I'(0)| \leq \varepsilon^{r+1} t \left\| \frac{\partial R}{\partial \theta'} \right\|, \qquad (15)$$

where $\| \cdot \|$ denotes an appropriate norm on functions (a discussion of the particular norm is not important for the present discussion). It follows from these estimates that:

$$|I'(t) - I'(0)| = \mathcal{O}(\varepsilon) \quad \text{if} \quad \epsilon^{r+1} t \left\| \frac{\partial R}{\partial \theta'} \right\| = \mathcal{O}(\varepsilon), \qquad (16)$$

and, therefore, we will have $|I'(t) - I'(0)| = \mathcal{O}(\varepsilon)$ on a time interval $[0, T]$ where

$$T = \frac{\mathcal{O}(\epsilon)}{\varepsilon^{r+1} \left\| \frac{\partial R}{\partial \theta'} \right\|}. \qquad (17)$$

Now if $\| \frac{\partial R}{\partial \theta'} \|$ the result obtained is just the standard perturbation theory estimate after r steps of the normalization process, we have $|I'(t) - I'(0)| = \mathcal{O}(\varepsilon)$ on a time interval of length $\mathcal{O}\left(\frac{1}{\varepsilon^r}\right)$. The problem with this conclusion is that $\| \frac{\partial R}{\partial \theta'} \|$ is not bounded. In general it grows as some power of $r!$ (such an estimate is obtained in the course of the proof of Nekhoroshev's theorem). There is also the fact that an arbitrary parameter r should not play a role in the form of a stability result, but this point will also be addressed in the course of our discussion. Therefore, in general we expect H_r to have the estimate $\mathcal{O}(r!)$. We will ignore constants since they are not essential for understanding the essence of the manner for obtaining exponential stability results. Roughly, in order for Eq. (12) to be of use, the ratio of the order $r+1$ term to the order r term must be smaller than 1, i.e.

$$\frac{(r+1)!\epsilon^{r+1}}{r!\epsilon^r} = (r+1)\varepsilon < 1. \qquad (18)$$

This immediately suggests an "optimal" form for r in terms of ε:

$$r + 1 \approx \frac{1}{\varepsilon}. \qquad (19)$$

Now recall Stirling's formula:

$$r! \approx \sqrt{r}\, r^r\, e^{-r} \qquad (20)$$

Since the remainder term is of the order $(r+1)!\varepsilon^{r+1}$, substituting Eqs. (19) and (20) into this expression will give an expression for the order of the remainder in terms of the "optimal" normalization order as a function of ε:

$$\begin{aligned} (r+1)!\varepsilon^{r+1} &\approx \sqrt{r+1}\,(r+1)^{r+1}\,\varepsilon^{r+1}\,e^{-(r+1)}, \\ &\approx ((r+1)\varepsilon)^{r+1}\,\sqrt{r+1}e^{-(r+1)}, \\ &\approx 1\sqrt{\frac{1}{\varepsilon}}\,e^{-\frac{1}{\varepsilon}}. \end{aligned} \qquad (21)$$

Hence, we see that with this choice of r the remainder term in Eq. (12) is "exponentially small in ε", and using this result with Eq. (17) gives the exponential stability estimate. Of course, we avoided many details that must be dealt with in the course of proving the Nekhoroshev theorem. However, this is the essence of the idea, given a normal form of the form of Eq. (12). A great deal of additional work is required to then show that the entire phase space can be covered with regions on which "appropriate normal forms" having exponential stability estimates are valid, and that these estimates can be used to give a "uniform" estimate valid for the entire phase space (this is the "geometric part" of the Nekhoroshev theorem).

2.3 Verifying that the assumptions of the KAM and Nekhoroshev theorems hold in specific examples

The conditions for the validity of the KAM theorem and the Nekhoroshev theorem in specific applications appear straightforward. However, this situation is somewhat misleading. Most of the work that verifies the applicability of the KAM and Nekhoroshev theorems for specific models has been carried out in the context of models in celestial mechanics; see, e.g., Celletti and Chierchia (2007) and Giorgilli et al. (2009). A specific model problem where detailed calculations of the applicability of Nekhoroshev's theorem are carried out is described in Lochak and Porzio (1989). The issues with applicability start at the very beginning of the consideration of the application. The KAM and Nekhoroshev theorems are stated, and proven, using the action-angle variables of the unperturbed integrable system. Even if one has a model that can be divided into an integrable part plus "a perturbation", it is, in general, highly nontrivial to construct action-angle coordinates for the unperturbed, integrable part. For this reason there have been essentially no applications of the KAM theorem to fluid transport where the conditions for the applicability of the theorem have been verified for a model under consideration. Similarly for the Nekhoroshev theorem, although that result hardly seems known at all by those considering Lagrangian transport issues in the fluids community.

Nevertheless, the KAM theorem provides a "language" to discuss invariant tori, and their manifestation as flow barriers, even though the applicability of the theorem is generally not verified for specific flows. The reason for this is the nature of the KAM theorem itself, and the conditions for its applicability. In particular, we know that, for the unperturbed (two-dimensional, time-independent and incompressible) flow, in a region of closed streamlines action-angle variables exist theoretically, even if we cannot find analytical expressions for the explicit coordinates (Arnold, 1978). Moreover, the nonresonance and nondegeneracy conditions are *generic*. Therefore it would be surprising if they did not hold. Nevertheless, this is no substitute for a quantitative study of the limits of applicability of these theorems in specific examples. For promising recent work on the applicability of KAM

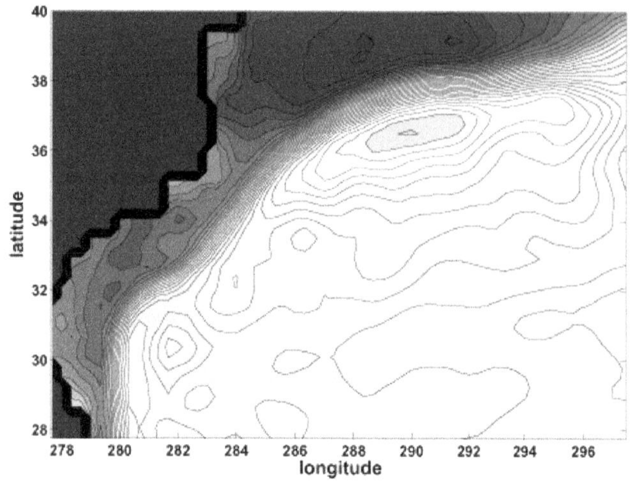

Fig. 1. Average sea-surface height, which is related to streamlines of the velocity field, in the North Atlantic. The "jet" is clearly observable. Image courtesy of Jezabel Curbelo.

theory without action-angle variables see de la Llave et al. (2005).

2.4 Application of the KAM and Nekhoroshev theorems to fluid transport

By now it is "common knowledge" that in the application of the KAM theorem to fluid transport the KAM tori act as complete "barriers to transport". This statement requires much more careful consideration and we want to explore its meaning and full implications in terms of our discussion of the KAM and Nekhoroshev theorems discussed above. First, we motivate our discussion by considering a particular "flow feature" that arises in many geophysical flow studies: a jet. The discussion is also very relevant to our discussion in Sect. 2.3.

A two-dimensional (2-D) "meandering jet" is a common flow feature observed on the surface of the ocean. In particular, specific 2-D jets that are visible on the ocean's surface are, for example, the Gulf Stream and the Kuroshio currents. These circulating patterns are very stable and are often described from the perspective of a stationary reference flow plus a (temporal) variability that acts as a not too large aperiodic perturbation of the reference state (note that we do *not* expect geophysical flows to have time-periodic or quasi periodic variability, even though this has been the subject of investigation for many kinematic jet models). Figure 1 shows the time average of the sea-surface height (SSH) in the North Atlantic. The magnitude of the SSH is locally related to a stream function from which the surface velocity is obtained in the geostrophic approximation. A very similar picture can be obtained for the Kuroshio region.

Consequently, *kinematically* defined 2-D meandering jet models have received much attention over the years. The specific details of those models are not important for our discussion, only the geometry of the streamlines – *and the coordinates*. However, some selected relevant references are Bower (1991), Samelson (1992), Duan and Wiggins (1996), and Samelson and Wiggins (2006), and in these references a specific functional form for the flow field of this particular model of a jet can be found. In a frame of reference moving with the phase velocity of the jet, the streamlines appear as in Fig. 2a. In particular, the flow is steady and spatially periodic. It is important to realise that the horizontal and vertical coordinates in Fig. 2a are the physical coordinates describing the streamlines of the jet, i.e. they are *not* action-angle coordinates. The left and right vertical boundaries of the flow shown in Fig. 2a are identified, i.e. the flow is periodic in the horizontal direction. Consequently, there are five regions of closed trajectories, denoted $R_1, \ldots, R_5$ in the figure. The jet is the central region, denoted R_3 (these trajectories are periodic since the flow is spatially periodic). Immediately above and below the jet are regions of "recirculating trajectories", denoted R_2 and R_4, and at the very top and bottom are two regions of periodic trajectories, denoted R_1 and R_5, that traverse the entire domain and move in the opposite direction as the jet (i.e. trajectories in R_3).

This flow structure describes a steady, incompressible, two-dimensional flow (hence, it is Hamiltonian and integrable) having five regions of qualitatively distinct closed trajectories (we have not considered variability applied to this model – yet). In order to apply the KAM and Nekhoroshev theorems to this flow we must transform the flow to action-angle variables in the regions of closed trajectories. However, the action-angle transformations for the five different regions will generally be different, and action-angle coordinates are not defined on the separatrices that separate the five regions. This is something of a moot point since the transformation to action-angle coordinates, for even one of the regions, has not been carried out for any of the kinematically defined jet models noted above. What is required is that this transformation produces a change of coordinates in which the transformed variables are as follows. The horizontal coordinate (the angle) must be periodic. Additionally, in the new coordinates, the streamlines or contour lines of the Hamiltonian must have a geometry compatible with expression (3). This means that the Hamiltonian must depend, to leading order, on the vertical coordinate (the action) plus a small distortion introduced by the perturbation H_1. This means that for instance the lines in the transformed region of interest of Fig. 2a, before the addition of the perturbation, should be purely horizontal. Following the expression found in Samelson (1992) for the Hamiltonian displayed in Fig. 2a, the unperturbed term H_0 depends both on the horizontal and vertical variables, thus it would not be in the appropriate coordinate system required by the Nekhoroshev theorem. We will consider an example in Sect. 4 that is expressed from the beginning in action-angle variables, that has the geometric features of the jet, and therefore allows us to apply the Nekhoroshev theorem with a variety of time dependencies.

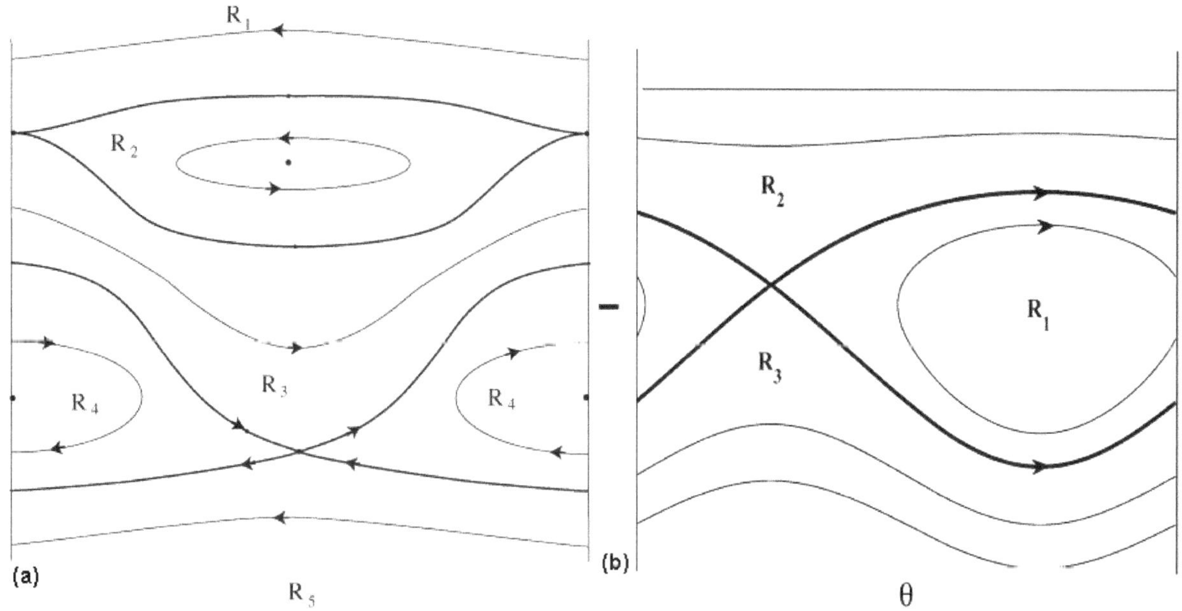

Fig. 2. **(a)** Streamlines associated with a "jet" (figure from Samelson and Wiggins, 2006) shown in the physical coordinates of the flow; **(b)** streamlines associated with the "jet" example from Sect. 4 shown in the action-angle variables. The streamlines are shown for $b(t)=0$ and $\varepsilon=0.1$.

Figure 2b shows the streamlines of the particular example that will be the focus of our study in Sect. 4. This example shows patterns similar to that of Fig. 2a, with a recognizable recirculating region R_1 and two regions R_2 and R_3 with periodic trajectories that transverse the entire domain (the example is periodic in the horizontal direction). This flow is different, though, from the kinematic models illustrated in Fig. 2a in that it is obtained from a Hamiltonian expressed in action-angle variables that correspond to the vertical and horizontal axes, respectively. In particular, the example Hamiltonian in Sect. 4 has the form $H(I,\theta)=H_0(I)+\varepsilon H_1(I,\theta,t)$ so that the Nekhoroshev theorem can be applied immediately. For the streamlines shown in Fig. 2b we have chosen $\varepsilon=0.1$ and $H_1(I,\theta,t)=H_1(I,\theta)$ by setting $b(t)=0$.

Even if we succeed in expressing the stream function in action-angle coordinates in the regions of interest of the flow, we note that the general Hamiltonian given in Eq. (3) will only have relevance as a stream function of a fluid flow for the case $n=1$. In this case it would describe a steady, two-dimensional incompressible flow (in action-angle variables). This is not particularly interesting (from the point of view of mixing, but possibly for transport), since two-dimensional, steady incompressible flows are integrable, and in this case the integral is $H(I,\theta)=H_0(I)+H_1(I,\theta),\quad (I,\theta)\in B\times\mathbb{T}$, where B is an interval in $\mathbb{R}$. Therefore in order for there to be "interesting" mixing and transport in two dimensions, the flows must be time dependent, from which it follows that if the KAM and Nekhoroshev theorems are to play an important role, then the corresponding Hamiltonian must be time dependent:

$$H(I,\theta,t)=H_0(I)+\varepsilon H_1(I,\theta,t),\quad (I,\theta)\in B\times\mathbb{T}. \tag{22}$$

The particular type of time dependence that has been studied in some detail is that of *quasiperiodic* time dependence. This means that the time-dependent perturbation can be written in the following form:

$$H_1(I,\theta,t)\equiv H_1(I,\theta,\phi_1,\dots,\phi_m), \tag{23}$$

where $H_1(I,\theta,\phi_1,\dots,\phi_m)$ is 2π periodic in each ϕ_k, $k=1,\dots,m$ (for each I,θ) where $\phi_k=\omega_k t$ for $k=1,\dots,m$. Of course, the case $m=1$ corresponds to the time-periodic case.

Some early results on periodic and quasiperiodic time dependence of the KAM theorem are discussed in Arnold et al. (1988). More recent results can be found in Broer et al. (1996) and Sevryuk (2007). A detailed discussion, and theorem, for quasiperiodically time-dependent Hamiltonian systems with Hamiltonians of the form of Eq. (23) can be found in Jorba and Simo (1996). Concerning the nature of the tori that persist under the perturbation, the situation is best described by a passage from this paper.

"The frequencies of these tori are those of the unperturbed tori plus those of the perturbation. This can be described by saying that the unperturbed tori are "quasi-periodically dancing" to the "rhythm" of the perturbation. The tori whose frequencies are in resonance with those of the perturbation are destroyed." (Jorba and Simo, 1996).

Hence, the surviving invariant tori, *for one degree-of-freedom, $n=1$*, form "complete" barriers to transport in the sense that these invariant tori are invariant manifolds (i.e. they consist of trajectories) and therefore trajectories not starting on the invariant tori cannot cross the invariant tori. Moreover, by examining the regions of closed trajectories for the "unperturbed" steady flows shown in Fig. 2 one can see that the particular interpretation of these surviving invariant tori in terms of their influence on transport depends on the geometry of the closed streamlines of the unperturbed flow and their relation to the global geometry of the flow. For an application of similar ideas to two-dimensional, quasiperiodically time-dependent flows see Beron-Vera et al. (2010). At present, there is no analogue of the KAM theorem for perturbations having more general time dependence than quasiperiodic, which is particularly notable for geophysical transport applications since one does not expect typical ocean variability to be either periodic or quasiperiodic. However, the observed similarities among the flow structures shown in Figs. 1 and 2 suggest that the transport properties associated with perturbed invariant tori can be related to important transport questions such as, for example, the existence of cross-jet transport of radiative isotopes in the Kuroshio current following the Fukushima accident (Buesseler et al., 2012), or the time of persistence of particles within the jet (this latter problem is directly connected to the results of the Nekhoroshev theorem, as we will discuss later), or similar transport issues.

Next, we turn our attention to the Nekhoroshev theorem. This theorem has not received as much attention from the point of view of perturbations with general time dependence, with the major exception of the remarkable paper of Giorgilli and Zehnder (1992).

"Nearly invariant" tori

Roughly, the estimate in Eq. (8) implies that the action coordinates stay "close" (as measured by some power of a small parameter) to their initial values for a time that is exponentially long (where the exponent is a constant multiplied by the inverse of a (possibly) different power of the same small parameter). The phrases "exponential stability" or 'effective stability" are often used. This is a very special type of finite time stability, as eloquently described by Littlewood (Littlewood, 1959b): "...while not eternity, this is a considerable slice of it."

In this situation the term "nearly invariant tori" is used in the literature (see, e.g., Delshams and Gutierrez, 1996), and this notion is particularly relevant for the notion of "invariant tori" finite time dynamics.

The issue of the existence of invariant tori poses related issues for finite time Hamiltonian vector fields with respect to nonuniqueness and the associated inability to locate precise invariant manifolds as discussed in Sect. 1.2. Invariant tori are barriers to transport – they are invariant manifolds and, therefore, trajectories cannot cross invariant tori. However, they are infinite time objects in the sense that their existence is proved by an iterative or recursive process that requires a passage to a limit. Moreover, there is no existing version of a KAM theorem for general aperiodically time-dependent Hamiltonian systems (other than for quasiperiodic time dependence). However, the Nekhoroshev theorem may be viewed as a type of finite time KAM theorem in the sense that an invariant torus is identified in the unperturbed system and a thickened region is constructed around that invariant torus in which trajectories starting in that region will remain for an exponentially long time. As we have noted, this is similar in spirit to the issue of nonuniqueness for the stable and unstable manifolds of finite time hyperbolic trajectories, and it is probably as good as one might expect for time-dependent vector fields defined on a finite time interval.

Of course, this raises the issue of how useful this is for applications, since one has not identified an exact barrier to transport that is valid for all time. However, practically, this may not be the essential important element. Rather, identifying regions of the flow where trajectories remain for very long times may be more practical since one can only ever observe flow for a finite time.

3 The Nekhoroshev theorem for aperiodic time dependence

Giorgilli and Zehnder (1992) considered time-dependent Hamiltonian systems of the following form:

$$H(\theta, I, t) = \frac{|I|^2}{2} + V(\theta, t), \quad (\theta, I, t) \in \mathbb{T}^n \times \mathbb{R}^n \times \mathbb{R}. \quad (24)$$

This Hamiltonian would appear to have little relevance to the fluid transport settings described earlier, since it would be unusual for a stream function to have the form of "kinetic plus potential energy". Moreover, there is no small parameter in Eq. (24) that would give it the form of the problem of the perturbation of an integrable system. However, a closer examination of Giorgilli and Zehnder (1992) reveals that the techniques used in the paper are much more general than the stated results. The main goal in Giorgilli and Zehnder (1992) was to show that the action variables of Eq. (24) remain bounded over exponentially long time intervals. Casting this problem in the "Nekhoroshev setting" requires $|I|$ to be "large", which will make $V(\theta, t)$ a "small" perturbation of $\frac{|I|^2}{2}$ (and this is why no small parameter appears in the statement of this problem). However, Giorgilli and Zehnder (1992) showed that by rescaling the action variables and time by a small parameter, Eq. (24) could be transformed to the form of a "slow time varying" perturbation of an integrable system in the "standard sense".

There is still the issue of the special form of the Hamiltonian in Eq. (24). Recall from the earlier discussion that the proof of the Nekhoroshev theorem is in two parts – an analytic part and a geometrical part. The analytic part uses standard canonical perturbation theory to derive resonant normal forms on certain regions of phase space. The geometric part shows that the regions on which the normal forms are valid cover the entire phase space. In this way the evolution of trajectories on all of the phase space can be estimated from the dynamics of the normal forms.

It is easy to see that the analytical part in Giorgilli and Zehnder (1992) is very general and does not depend on the special form of the Hamiltonian Eq. (24). The proof of the geometric part given in Giorgilli and Zehnder (1992) is greatly simplified with the special form of Eq. (24). However, this is probably of no consequence for our needs for fluid transport, since in that case we only require $n = 1$. Nevertheless, recently Bounemoura (2013) and Fortunati and Wiggins (2013) have re-visited the work of Giorgilli and Zehnder (1992) and provided a formulation of the Giorgilli and Zehnder (1992) result for a standard perturbation of an integrable Hamiltonian system with arbitrary, slowly varying, time dependence.

It is worth noting the issue of "slow time dependence". The results of Giorgilli and Zehnder (1992), Bounemoura (2013) and Fortunati and Wiggins (2013) all require slow time dependence ("slow" in the sense of the time dependence of the Hamiltonian where the explicit time variable is multiplied by some positive power of the perturbation parameter). Giorgilli and Zehnder (1992) and Fortunati and Wiggins (2013) use a different scheme than Bounemoura (2013) to arrive at the necessary normal form. So, at the moment, it appears that slow time dependence is required to achieve a Nekhoroshev result for aperiodically time-dependent systems and that this should be regarded as the most general form for a Hamiltonian with arbitrary time dependence.

We now state the theorem in a form that is adequate for our needs.

We consider a one degree-of-freedom, aperiodically time-dependent Hamiltonian of the following form:

$$H(\theta, I, t) = H_0(I) + \varepsilon \tilde{H}(\theta, I, \epsilon^c t), \tag{25}$$

$$(\theta, I, t) \in \mathbb{T} \times \mathcal{D}_\rho \times \mathbb{R},$$

where $\mathcal{D}_\rho$ is a ball of radius ρ in $\mathbb{R}$ and $\frac{1}{2} \leq c \leq 1$,

with corresponding Hamiltonian vector field:

$$\dot{\theta} = \frac{\partial H_0}{\partial I}(I) + \varepsilon \frac{\partial \tilde{H}}{\partial I}(\theta, I, \varepsilon^c t),$$

$$\dot{I} = -\varepsilon \frac{\partial \tilde{H}}{\partial \theta}(\theta, I, \varepsilon^c t). \tag{26}$$

It is well known that a time-dependent Hamiltonian can be cast in the form of a time-independent Hamiltonian with an additional degree of freedom. This formulation allows one to treat the problem by an analytic part of the problem by standard canonical perturbation theory, as explained in Giorgilli and Zehnder (1992). We will not be pursuing the proof of the theorem here. However, it is useful to cast the problem in this form in order to understand the role that certain parameters play in the formulation of the result. Towards this end, we re-write the Hamiltonian by redefining a new variable as $\xi = \varepsilon^c t$ and introducing a new extra variable η in an extra term in the following form:

$$H(\theta, I, \xi, \eta) = H_0(I) + \varepsilon^c \eta + \epsilon \tilde{H}(\theta, I, \xi). \tag{27}$$

The corresponding Hamiltonian vector field is:

$$\dot{\theta} = \frac{\partial H_0}{\partial I}(I) + \varepsilon \frac{\partial \tilde{H}}{\partial I}(\theta, I, \xi),$$

$$\dot{I} = -\varepsilon \frac{\partial \tilde{H}}{\partial \theta}(\theta, I, \xi),$$

$$\dot{\xi} = \frac{\partial H}{\partial \eta}(\theta, I, \xi, \eta) = \varepsilon^c,$$

$$\dot{\eta} = -\varepsilon \frac{\partial \tilde{H}}{\partial \xi}(\theta, I, \xi). \tag{28}$$

The variables (I, θ, ξ, η) are now extended to the complex plane $\mathbb{C}$. We define:

$$\mathcal{G}_\delta = \{I \in \mathbb{C} \mid |I - \mathcal{D}_\rho| < \delta\},$$

then the following domain for the Hamiltonian for Eq. (27) is considered:

$$\mathcal{G}_{(\delta,\sigma)} = \mathcal{G}_\delta \times \{\eta \in \mathbb{C}\} \times \{|\mathrm{Im}\,\theta| < \sigma\} \times \{|\mathrm{Im}\,\xi| < \sigma\}. \tag{29}$$

Additionally, we have the following assumptions.

Analyticity The Hamiltonian Eq. (27) is analytic on the domain Eq. (29).

Nondegeneracy of the Integrable Part For $n = 1$ the nondegeneracy condition on the integrable part is particularly simple:

$$M > \left| \frac{\partial^2 H_0}{\partial I^2} \right| > m > 0, \quad \text{for some } M > m > 0, \tag{30}$$

where M and m are upper and lower bounds, respectively, on the magnitude of the frequency.

We can now state the main result due to Giorgilli and Zehnder (1992) (and refined by Bounemoura, 2013 and Fortunati and Wiggins, 2013).

Theorem 1 *Under the assumptions given above, there exists positive constants ε_0, c_1, c_2, c_3 that depend on δ, σ, m, M such that if $\varepsilon \leq \epsilon_0$ such that for all solutions $(\theta(t), I(t))$ of Eq. (26) if $I(0) \in \mathcal{D}_{\frac{\rho}{2}}$ then*

$$|I(t) - I(0)| \leq c_1 \varepsilon^{\frac{1}{2}}, \tag{31}$$

for all

$$|t| \leq c_2 \exp(c_3 \varepsilon^{-\frac{1}{2}}). \tag{32}$$

We make several comments regarding this theorem.

- It is important to understand what is meant by almost invariant tori in the case where the time dependence is not periodic, since by invariant torus typically it is understood that the motion is quasiperiodic. Recall our discussion of almost invariant tori at the end of Sect. 2.4. An almost invariant torus was identified in the unperturbed system and the Nekhoroshev theorem was used to define a thickened region around that invariant torus in which trajectories starting in that region remained for an exponentially long time. Since the torus is identified in the unperturbed problem, this definition still holds when the perturbation is not periodic in time.

- The issue of the choice for the constants c_1, c_2, c_3 is important to note. They are of order 1 and do not depend on the perturbation parameter. Functional forms for them are derived during the course of the proof of the theorem, but general choices are made for various parameters that define the constants and inequalities that arise at different steps in the proof in order to provide a simple and convenient proof of the theorem. Therefore in the proof of the general theorem no effort is made to choose the constants in such a way that is optimal for a specific problem. Moreover, the parameters that go into each constant need to be computed explicitly for each specific example, and for some of the parameters it is not entirely clear how to compute such parameters for a specific example in an optimal manner (e.g. analyticity parameters and ultraviolet cut-off parameters for truncating Fourier series in the normal form). This is why the computations of all of the constants involved in a specific example for application of the KAM theorem (e.g. Celletti and Chierchia, 2007) or the Nekhoroshev theorem (e.g. Lochak and Porzio, 1989) amount to a substantial research project in their own right. It is worth mentioning that it is only in the last 2 years that the nondegeneracy condition for the KAM theorem in the context of the general n body problem of celestial mechanics has been verified (the work of Chierchia and Pinzari, 2011). Nevertheless, even though its rigorous applicability to the n body problem was not established, the KAM theorem provided a valuable theoretical framework for thinking about the problem. We believe that the Nekhoroshev theorem will serve a similar role for finite dimensional fluid transport problems, and this may serve to motivate work on verifying the conditions for the KAM and Nekhoroshev theorems for specific examples that are more directly related to fluid mechanical equations of motion.

  For our example in Sect. 4 we take the constants in Nekhoroshev's theorem to be $c_1 = c_2 = c_3 = 1$. This choice is arbitrary. They are the simplest order 1 constants. Nevertheless, one can see directly from Eqs. (31) and (32) that confinement of trajectories over exponentially long timescales still holds for general order one choices of constants c_1, c_2, c_3 if we take ε sufficiently small. Indeed, for hyperbolic perturbation methods (such as Melnikov's method for determining the existence of transverse intersections of the stable and unstable manifolds) soft analysis is all that is required (i.e. first-order regular perturbation theory, standard implicit function theorems) and, as a result, specific constants are not computed as the results hold for ε sufficiently small. For the KAM/Nekhoroshev theorems one requires perturbation results to all orders and in this case the constants must be estimated at each step, and it is natural to include them in the statement of the theorem (although that may make the theorem extremely difficult to penetrate for a non-specialist). Still, the results hold for ε sufficiently small, but knowing the specific constants enables one to compute how small ε must be.

- The Nekhoroshev estimates hold regardless of whether the unperturbed torus is resonant or not. This has been investigated in detail for the case of n degree of freedom autonomous Hamiltonian systems, and the results are surprising and somewhat counterintuitive. Briefly, the most resonant regions are the most stable in the sense of Nekhoroshev estimates and the least stable regions in the sense of Nekhoroshev estimates are the least resonant regions, e.g. the KAM tori. This phenomena is described in detail in Benettin and Gallavotti (1986) and Lochak (1992).

- The Nekhoroshev estimates do not say anything about the speed at which trajectories move away from an unperturbed invariant torus. For example, it is not ruled out that a trajectory could rather quickly move to the maximum stability radius and then move very little for the rest of the stability time. Whether or not this happens needs to be considered in the context of specific examples.

4 An example

Now we consider an example that "gives us an idea" of the usefulness of Theorem 3. We say "gives us an idea" because we have not computed the threshold value for the perturbations (ε_0) or the constants c_1, c_2, c_3 for this example. Computing these constants would require careful consideration of the proof of the theorem in the context of this particular example. However, for our purposes it is sufficient to know that these are $\mathcal{O}(1)$ constants. With this in mind, the example will illustrate some interesting features of Nekhoroshev's result in the context of aperiodic time dependence. This theorem appears to have been unnoticed by the community dealing with transport in finite time velocity fields, thus the example will allow us to study related issues in a setting where the time dependence can be specified. It is our hope that this will inspire further work on the many issues related to finite time transport associated with elliptic phenomena.

We consider a Hamiltonian of the following form:

$$H(I,\theta,t) = \frac{I^2}{2} + \varepsilon(1-I)(1+b(t))\sin\theta, \quad (\theta, I) \in \mathbb{T}\times\mathbb{R}, \tag{33}$$

where the unperturbed Hamiltonian, H_0, is $H_0 = I^2/2$. Note that, at this point, the time dependence of $b(t)$ is completely arbitrary and can be used to study different aspects of time dependence. For example, $b(t)$ can be chosen to be periodic, quasiperiodic, aperiodic, or to exist only for a finite time. The Hamiltonian vector field corresponding to Eq. (33) is given by:

$$\begin{aligned}\dot\theta &= \frac{\partial H}{\partial I} = I - \varepsilon(1+b(t))\sin\theta,\\ \dot I &= -\frac{\partial H}{\partial \theta} = -\varepsilon(1-I)(1+b(t))\cos\theta.\end{aligned} \tag{34}$$

We describe some of the important features of this example that play a role in our numerical experiments.

An Invariant Torus: $I = 1$. From Eq. (34) it is clear that that for $I = 1$, $\dot I = 0$, which proves that $I = 1$ is an invariant torus. It is important to note that this is independent of both ε and $b(t)$.

The Integrable Case: $\varepsilon = 0$. For the case $\varepsilon = 0$ the vector field is integrable and is given by:

$$\begin{aligned}\dot\theta &= \frac{\partial H}{\partial I} = I,\\ \dot I &= -\frac{\partial H}{\partial \theta} = 0.\end{aligned} \tag{35}$$

This system is clearly integrable. Each value of I corresponds to an invariant circle, and the value of the frequency on the invariant circle is also given by I (hence $I = 1$ has frequency one). Note that the invariant circle corresponding to $I = 0$ is resonant since its frequency is zero.

We remark that we have avoided the issue described in Sect. 2.4 since the unperturbed system is expressed explicitly in terms of action-angle coordinates, and the perturbed system is also expressed in these coordinates.

The Choice of Time Dependence. For the purpose of numerical experiments, we will consider three types of aperiodic time dependence of the following form:

1. Our first choice is a pulse-like time dependence of the form:

$$b(t) = \frac{\operatorname{sech}^2(\varepsilon t)}{\sqrt{2}}\left(\sin\left(\sqrt{3}\varepsilon t\right) + \cos\left(\sqrt{2}\varepsilon t\right)\right), \tag{36}$$

   Figure 3a and b shows the graph of this function for $\varepsilon = 0.01$ and $\varepsilon = 0.1$ respectively. It has a pulse-like structure which is wider for the smaller ε. In both cases, after a time interval, $b(t)$ becomes essentially zero and the system (34) is essentially autonomous. For this reason the system approaches an integrable system, but it is still a perturbed version of H_0.

2. A quasiperiodic time dependence:

$$b(t) = \frac{1}{4\sqrt{2}}\left(\sin\left(\sqrt{3}\varepsilon t\right) + \cos\left(\sqrt{2}\varepsilon t\right)\right), \tag{37}$$

   Fig. 3c and d shows the graph of this function for $\varepsilon = 0.01$ and $\varepsilon = 0.1$. It is observed that due to the time scaling a larger ε produces higher oscillation frequencies.

3. An aperiodic time dependence that is obtained from the chaotic time series of a differential equation. A representation of this time series is shown in Fig. 3e. This series is proportional to the second component of a chaotic trajectory that has been obtained from the integration of a periodically forced Duffing equation (we say "proportional" since we normalise the signal so that it has amplitude one). Since our choice is an arbitrary forcing, in this case we do not rescale the time with ε.

The perturbed system (34) possesses kinematically distinct regions similar to those depicted in Fig. 2b and denoted by R_1, R_2 or R_3, representing, respectively, jets and eddies. In order to obtain a visual representation of the Lagrangian structures of the given examples, similar to what is shown in Fig. 2b, but for the time-dependent case, we will use a recently developed approach based on functions called *Lagrangian descriptors* (see Mendoza and Mancho, 2010, 2012; de la Cámara et al., 2012; Mancho et al., 2013). The Lagrangian descriptor that we use is based on arc length, and is referred to as M_1 in Mancho et al. (2013) or as M in Mendoza and Mancho (2010, 2012) and de la Cámara et al. (2012, 2013). A precise definition and discussion of M_1 is given in the references. Contours of the Lagrangian descriptors highlight singular features (which are related to the lack of regularity of the function) and these have been shown to be directly related to "phase space structures". Significantly for our situation, the method is directly applicable to the aperiodically and finite time-dependent cases (unlike Poincaré maps). Lagrangian descriptors as reported in Mancho et al. (2013) are based on the integration of a positive and bounded intrinsic property of a trajectory along the trajectory itself during a time interval of length 2τ. Revealing the dynamical features requires the use of a long enough τ in order to converge to the singular features.

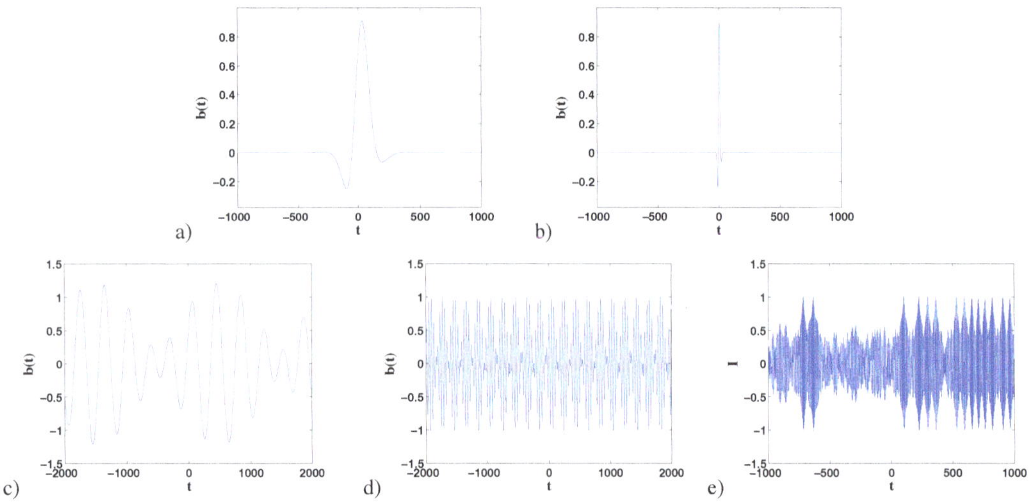

Fig. 3. Representation of different forcing functions $b(t)$ in example (34). **(a)** Forcing as in Eq. (36) for $\varepsilon = 0.01$; **(b)** forcing as in Eq. (36) for $\varepsilon = 0.1$; **(c)** forcing as in Eq. (37) for $\varepsilon = 0.01$; **(d)** forcing as in Eq. (37) for $\varepsilon = 0.1$; **(e)** chaotic forcing.

4.1 Numerical experiments

We will be concerned with the stability properties of three different invariant circles in the unperturbed system:

$I = 0$ For the unperturbed system this is a circle of fixed points (i.e. a resonant invariant circle).

$I = \frac{1}{2}$ For the unperturbed system this is an invariant circle with frequency $\frac{1}{2}$.

$I = 1$ For the unperturbed system this is an invariant circle with frequency 1, which also persists as an invariant circle for any value of ε.

We are interested in the stability properties of these three invariant circles. By stability we mean the intuitive idea of "*if you start close, you stay close*". More precisely, we are interested in the evolution of the action variables of trajectories that "start close" to these invariant circles and whether or not their evolution obeys the Nekhoroshev estimates for "confinement" (Eq. 31) over a given time (Eq. 32).

We consider the following initial conditions for each invariant circle:

$(\theta, I) = (0, 0)$

$(\theta, I) = \left(0, \frac{1}{2}\right)$

$(\theta, I) = (0, 0.99)$ For this case the action value is "slightly offset" since $I = 1$ is invariant for $\varepsilon \neq 0$.

We will consider two values of ε:

$\varepsilon = 0.1, 0.01,$

and for each of these values the "confinement time" denoted by T and defined in Eq. (32) is given by:

$$T \sim 25,\ T \sim 22000, \tag{38}$$

respectively, where we have set $c_2 = c_3 = 1$. For each of these values of ε the "confinement distance", denoted by S and defined in Eq. (31), is given by: for these ϵ values

$$S \sim 0.31623,\ S \sim 0.1, \tag{39}$$

respectively, where we have set $c_1 = 1$.

The numerical experiments consist of the following. For each invariant circle, we integrate the initial condition given above and plot the I value of the resulting trajectory as a function of time. This is done for each time dependence and for the two values of ε. Moreover, in order to understand the nature of trajectories we have provided a "snapshot" of the Lagrangian structure through the use of a *Lagrangian descriptor*. We note that for $\varepsilon = 0.1$ we illustrate the trajectory for a time of *at least* 200 and in some cases up to 400. This is significantly longer than the estimated confinement time of 25. Similarly, for $\varepsilon = 0.01$ for all cases we compute the trajectory for a time of 25 000, which is longer than the estimated confinement time of 22 000. For $\varepsilon = 0.01$ we will see excellent agreement with the Nekhoroshev estimates for all initial conditions. For $\varepsilon = 0.1$ the quality of agreement will vary with the initial condition. This is not unexpected since, a priori, we do not have an estimate of the size of ε, as well as the relevant constants, for each initial condition. The best we can do, in the example under consideration, is to show that "for ε sufficiently small" the Nekhoroshev estimates hold for trajectories with a particular initial condition. This is typical of how most perturbation theories are applied.

Results for the time dependence (Eq. 36) are shown in Fig. 4. The first column shows the results for $\varepsilon = 0.01$ and the second for $\varepsilon = 0.1$. It is easily seen that for the case $\varepsilon = 0.01$ the Nekhoroshev theorem is satisfied, as the trajectories remain within a distance of $S \sim 0.1$ from the initial condition for at least $T \sim 22000$ time units.

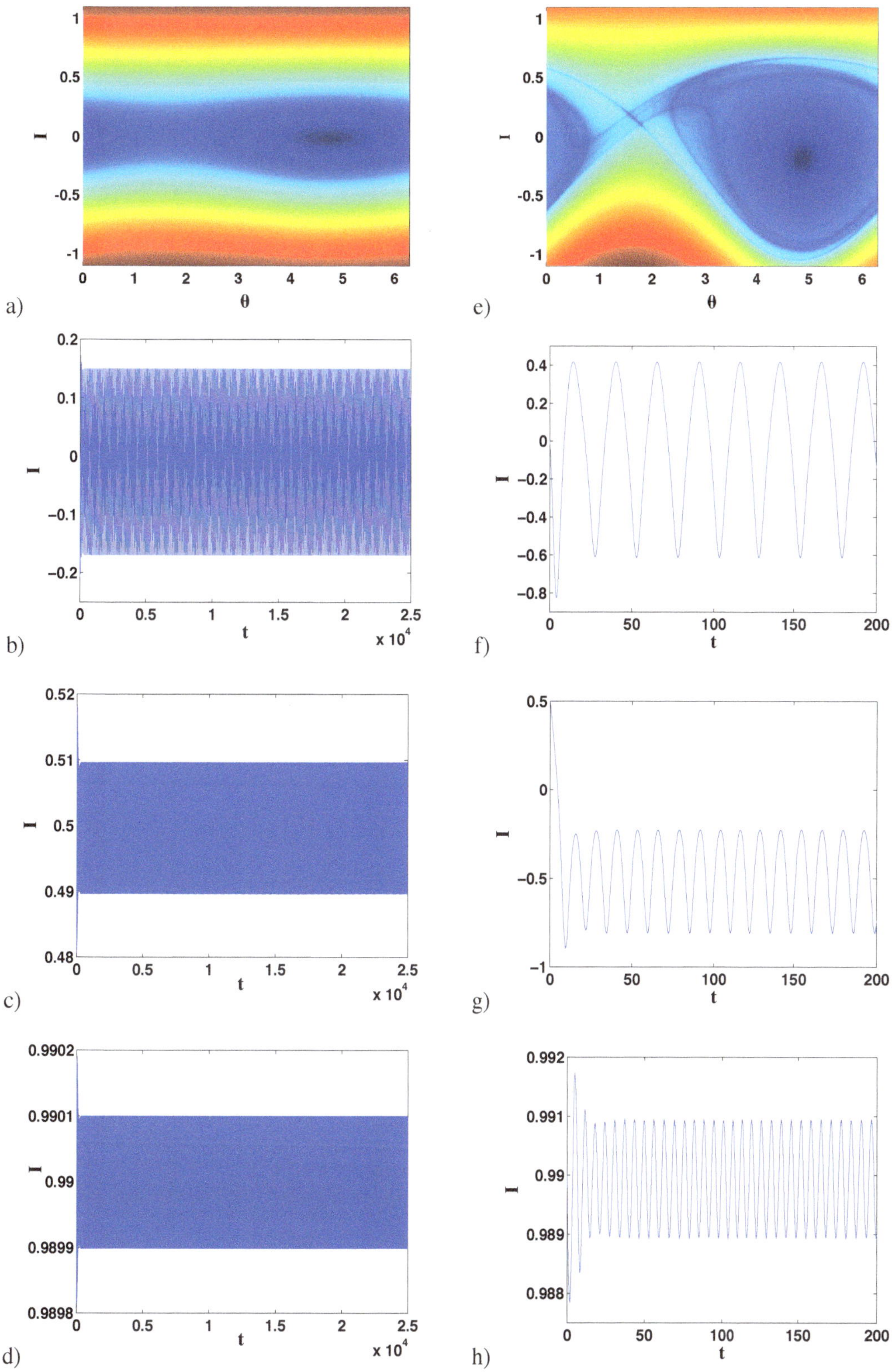

Fig. 4. Results obtained for the time dependence given by Eq. (36). The first column is for $\varepsilon = 0.01$, $\sqrt{\varepsilon} = 0.1$. **(a)** Lagrangian descriptor evaluated in the domain $[0, 2\pi] \times [-1.1, 1.1]$ at $t = 0$ for $\tau = 160$; **(b)**, **(c)**, and **(d)** show the I component of the trajectory as a function of time for the following initial conditions: **(b)** $\theta = 0, I = 0$; **(c)** $\theta = 0, I = 0.5$; **(d)** $\theta = 0, I = 0.99$. The second column is for $\varepsilon = 0.1$, $\sqrt{\varepsilon} = 0.31623$. **(e)** Lagrangian descriptor evaluated in the domain $[0, 2\pi] \times [-1.1, 1.1]$ at $t = 0$ for $\tau = 160$; **(f)**, **(g)**, and **(h)** show the I component of the trajectory as a function of time for the following initial conditions: **(f)** $\theta = 0, I = 0$; **(g)** $\theta = 0, I = 0.5$; **(h)** $\theta = 0, I = 0.99$.

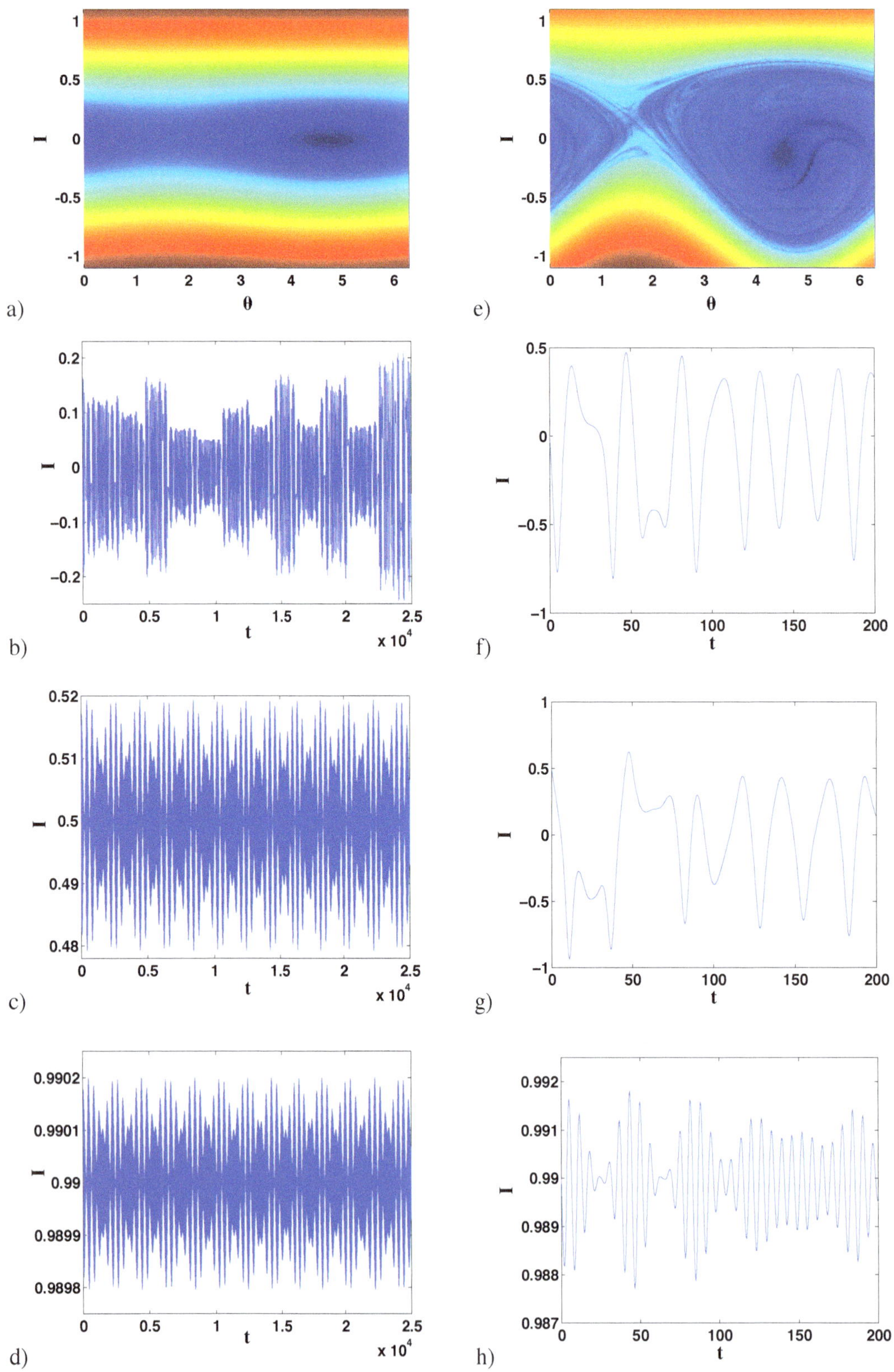

Fig. 5. Results obtained for the time dependence given by Eq. (37). The first column is for $\varepsilon = 0.01$, $\sqrt{\epsilon} = 0.1$. **(a)** Lagrangian descriptor evaluated in the domain $[0, 2\pi] \times [-1.1, 1.1]$ at $t = 0$ for $\tau = 160$; **(b)**, **(c)**, and **(d)** show the I component of the trajectory as a function of time for the following initial conditions: **(b)** $\theta = 0, I = 0$; **(c)** $\theta = 0, I = 0.5$; **(d)** $\theta = 0, I = 0.99$. The second column is for $\varepsilon = 0.1$, $\sqrt{\varepsilon} = 0.31623$. **(e)** Lagrangian descriptor evaluated in the domain $[0, 2\pi] \times [-1.1, 1.1]$ at $t = 0$ for $\tau = 160$; **(f)**, **(g)**, and **(h)** show the I component of the trajectory as a function of time for the following initial conditions: **(f)** $\theta = 0, I = 0$; **(g)** $\theta = 0, I = 0.5$; **(h)** $\theta = 0, I = 0.99$.

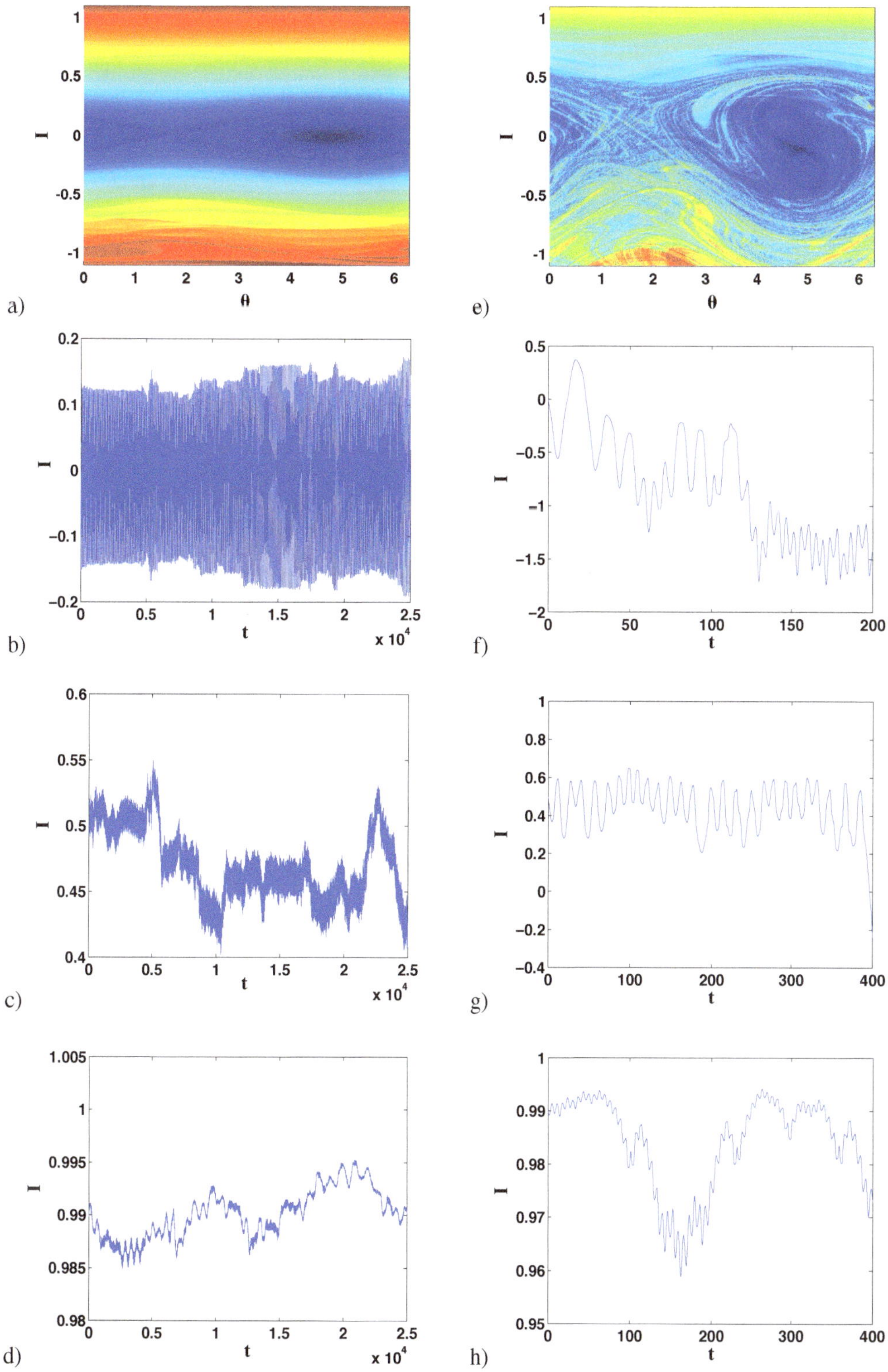

Fig. 6. Results obtained for the time dependence given by the aperiodic forcing obtained from the chaotic output of a differential equation. The first column is for $\varepsilon = 0.01$, $\sqrt{\varepsilon} = 0.1$. **(a)** Lagrangian descriptor evaluated in the domain $[0, 2\pi] \times [-1.1, 1.1]$ at $t = 0$; **(b)**, **(c)**, and **(d)** show the I component of the trajectory as a function of time for the following initial conditions: **(b)** $\theta = 0, I = 0$; **(c)** $\theta = 0, I = 0.5$; **(d)** $\theta = 0, I = 0.99$. The second column is for $\varepsilon = 0.1$, $\sqrt{\varepsilon} = 0.31623$. **(e)** Lagrangian descriptor evaluated in the domain $[0, 2\pi] \times [-1.1, 1.1]$ at $t = 0$; **(f)**, **(g)**, and **(h)** show the I component of the trajectory as a function of time for the following initial conditions: **(f)** $\theta = 0, I = 0$; **(g)** $\theta = 0, I = 0.5$; **(h)** $\theta = 0, I = 0.99$.

In the second column similar agreement with the Nekhoroshev estimates is shown for the trajectories in Fig. 4f and h. In these figures the confinement is of order $S = c\,0.3$ (with c a constant of $O(1)$) for a time of at least $T = 25$. The confinement for Fig. 4g with initial condition ($I = 0.5$ and $\theta = 0$) does not satisfy the Nekhoroshev estimates, since the trajectory rapidly evolves away from the initial condition more than a distance S in time $T < 25$. The perturbation size is too large for this initial condition.

An understanding of how the constants, including perturbation size, affect the estimates of confinement distance and stability time requires detailed estimates of the sizes of these constants, taking into account explicit phase space structures, i.e. resonances. Some results along these lines have been obtained in the time-independent case in Benettin and Gallavotti (1986) and Lochak (1992). Similar results have yet to be obtained for the time-dependent case (where the time dependence is aperiodic), but our simulations indicate that this would be a fruitful direction for further research.

Figure 5 shows results for the time dependence (Eq. 37). As before, the first column shows the results for $\varepsilon = 0.01$ and the second for $\varepsilon = 0.1$. The agreement is as expected in the first column and in the second one the confinement is again not satisfied for the initial condition ($I = 0.5$ and $\theta = 0$), as displayed in Fig. 5g.

For an aperiodically time-dependent forcing obtained from a chaotic trajectory, as illustrated in Fig. 3e, the results are shown in Fig. 6. In the first column the results for $\varepsilon = 0.01$ are in good agreement with what is expected from the Nekhoroshev theorem. In the second column the quality of the results for $\varepsilon = 0.1$ depends on the initial condition. The trajectory shown in Fig. 6f satisfies the confinement estimate for $T < 25$, and shortly after that the trajectory no longer obeys the estimate. Figure 6g confirms that the bound beyond $T \sim 400$ is no longer satisfied by showing a rapid distancing at this point. This is consistent with the estimations obtained from the theorem. Surprisingly for the aperiodic case, the size perturbation $\varepsilon = 0.1$ for the initial condition ($I = 0.5$ and $\theta = 0$) has not been too large as it has been in the previous considered time dependencies. Finally, the initial condition in Fig. 6h displays a confinement that is compatible with the estimates of the theorem. Although not shown, the confinement is maintained up to $T \sim 1500$.

It is notable that for both $\varepsilon = 0.1$ and $\varepsilon = 0.01$ the action variables of trajectories initialised near $I = 1$ change very little during the course of their evolution for the three types of forcing considered. Recall that $I = 1$ is an invariant torus even for $\varepsilon \neq 0$. This result is reminiscent of results on the "stickiness" of invariant tori in time-independent systems; see Perry and Wiggins (1994); Morbidelli and Giorgilli (1995a, b). There are no such "stickiness results" for time-dependent systems (for any time dependence), so this result might serve as a motivation to seek such results.

5 Conclusions and outlook

In this paper we have considered fluid transport in two-dimensional flows from the dynamical systems point of view, with the focus on elliptic behaviour and aperiodic and finite time dependence. We have given an overview of previous work on general nonautonomous and finite time vector fields with the purpose of bringing to the attention of those working on fluid transport from the dynamical systems point of view a body of work that is very relevant, but appears not to be so well known. We then focused on the Kolmogorov–Arnold–Moser (KAM) theorem and the Nekhoroshev theorems. While there is no finite time or aperiodically time-dependent version of the KAM theorem, the Nekhoroshev theorem, by its very nature, is a finite time result, but for a "very long" (i.e. exponentially long with respect to the size of the perturbation) time interval. In this sense it provides a rigorous description of the notion of "nearly invariant tori" over a finite time interval. Moreover, there is an aperiodically time-dependent version of the Nekhoroshev theorem due to Giorgilli and Zehnder (1992) (recently refined by Bounemoura (2013) and Fortunati and Wiggins (2013)) which is directly relevant to fluid transport problems. We give a detailed discussion of issues associated with the applicability of the KAM and Nekhoroshev theorems in specific flows. Finally, we consider a specific example where we show that the results of the Nekhoroshev theorem hold. We note that the general aperiodic time dependence in this particular version of Nekhoroshev's theorem is "slow" (in a sense that is made clear in the statement of the theorem). Work is underway to generalise this to more general aperiodic time dependence.

Acknowledgement. S. Wiggins would like to acknowledge useful e-mail exchanges with Antonio Giorgilli. The research of S. Wiggins was supported by ONR grant no. N00014-01-1-0769. The research of A. M. Mancho is supported by MINECO under grant MTM2011-26696. S. Wiggins and A. M. Mancho acknowledge support from MINECO: ICMAT Severo Ochoa project SEV-2011-0087.

Edited by: J. Duan

References

Acrivos, A., Aref, H., and Ottino, J. M. (Eds.): Symposium on Fluid Mechanics of Stirring and Mixing, Phys. Fluids A, Part 2, 723–1469, 1991.

Aref, H.: Stirring by chaotic advection, J. Fluid Mech., 143, 1–21, 1984.

Aref, H.: The development of chaotic advection, Phys. Fluids, 14, 1315–1325, 2002.

Aref, H. and El Naschie, M. S. (Eds.): Chaos Applied to Fluid Mixing, Chaos, Solitons, and Fractals, 4, 745–1116 , 1994.
Arnold, V. I.: Proof of A. N. Kolmogorov's theorem on the preservation of quasiperodic motions under small perturbations of the Hamiltonian, Russ. Math. Surveys, 18, 9–36, 1963.
Arnold, V. I.: Mathematical Methods of Classical Mechanics, Springer-Verlag, New York, Heidelberg, Berlin, 1978.
Arnold, V. I., Kozlov, V. V., and Neishtadt, A. I.: Mathematical Aspects of Classical and Celestial Mechanics, Dynamical Systems, Vol. III, Springer-Verlag, New York, Heidelberg, Berlin, 1988.
Babiano, A., Provenzale, A., and Vulpiani, A. (Eds.): Chaotic Advection, Tracer Dynamics, and Turbulent Dispersion, Proceedings of the NATO Advanced Research Workshop and EGS Topical Workshop on Chaotic Advection, Conference Centre Sereno di Gavo, Italy, 24–28 May 1993, Physica D, Vol. 76, 1994.
Balibrea, F., Caraballo, T., Kloeden, P. E., and Valero, J.: Recent develpments in dynamical systems: Three perspectives, Int. J. Bif. Chaos, 20, 2591–2636, 2010.
Batchelor, G. K.: An Introduction to Fluid Dynamics, Cambridge University Press, Cambridge, 1967.
Benettin, G. and Gallavotti, G.: Stability of motions near resonances in quasi-integrable Hamiltonian systems, J. Stat. Phys., 44, 293–338, 1986.
Berger, A.: On finite time hyperbolicity, Comm. Pure App. Anal., 10, 963–981, 2011.
Berger, A., Son, D. T., and Siegmund, S.: Nonautonomous finite-time dynamics, Discrete Cont. Dyn.-B, 9, 463–492, 2008.
Berger, A., Doan, T. S., and Siegmund, S.: A definition of spectrum for differential equations on finite time, J. Diff. Eq., 246, 1098–1118, 2009.
Beron-Vera, F. J., Olascoaga, M. J., Brown, M. G., Kocak, H., and Rypina, I.: Invariant-tori-like Lagrangian coherent structures in geophysical flows, Chaos, 20, 017514, doi:10.1063/1.3271342, 2010.
Bounemoura, A.: Effective stability for slow time-dependent near-integrable Hamiltonians and application, C. R. Math. Acad. Sci. Paris, 351, 673–676, 2013.
Bower, A. S.: A simple kinematic mechanism for mixing fluid parcels across a meandering jet, J. Phys. Oceanogr., 21, 173–180, 1991.
Branicki, M. and Wiggins, S.: Finite-time Lagrangian transport analysis: stable and unstable manifolds of hyperbolic trajectories and finite-time Lyapunov exponents, Nonlin. Processes Geophys., 17, 1–36, doi:10.5194/npg-17-1-2010, 2010.
Broer, H. W., Huitema, G. B., and Sevryuk, M. B.: Quasi-Periodic Motions in Families of Dynamical Systems, Lecture Notes in Mathematics, Vol. 1645, Springer-Verlag, New York, Heidelberg, Berlin, 1996.
Brown, G. L. and Roshko, A.: Density effect and large structure in turbulent mixing layers, J. Fluid Mech., 64, 775–816, 1974.
Buesseler, K., Jayne, S. R., Fisher, N. S., Rypina, I. I., Baumann, H., Baumann, Z., Breier, C. F., Douglass, E. M., George, J., Macdonald, A. M., Miyamoto, H., Nishikawa, J., Pike, S. M., and Yoshida, S.: Fukushima-derived radionuclides in the ocean and biota off japan, Proc. Natl. Acad. Sci., 109, 5984–5988, 2012.
Celletti, A. and Chierchia, L.: KAM stability and celestial mechanics, Mem. Am. Mathe. Soc., 187, no. 878, 2007.
Chierchia, L. and Mather, J. N.: Kolmogorov-Arnold-Moser theory, Scholarpedia, 5, 2123, doi:10.4249/scholarpedia.2123, 2010.
Chierchia, L. and Pinzari, G.: The planetary N-body problem: symplectic foliation, reductions, and invariant tori, Invent. Math., 186, 1–77, 2011.
Chow, S. N., Lin, X. B., and Palmer, K.: A shadowing lemma with applications to semilinear parabolic equations, SIAM J. Math. Anal., 20, 547–557, 1989.
Coddington, E. A. and Levinson, N.: Theory of Ordinary Differential Equations, McGraw-Hill, New York, 1955.
Coulliette, C. and Wiggins, S.: Intergyre transport in a wind-driven, quasigeostrophic double gyre: An application of lobe dynamics, Nonlin. Processes Geophys., 8, 69–94, doi:10.5194/npg-8-69-2001, 2001.
Dafermos, C. M.: An invariance principle for compact processes, J. Diff. Eq., 9, 239–252, 1971.
Danckwerts, P. V.: The definition and measurement of some characteristics of mixtures, Appl. Sci. Res., A3, 279–296, 1952.
Danckwerts, P. V.: Theory of mixtures and mixing, Research, 6, 355–361, 1953.
de Blasi, F. S. and Schinas, J.: On the stable manifold theorem for discrete time dependent processes in banach spaces, Bull. London Math. Soc., 5, 275–282, 1973.
de la Cámara, A., Mancho, A. M., Ide, K., Serrano, E., and Mechoso, C.: Routes of transport across the Antarctic polar vortex in the southern spring, J. Atmos. Sci., 69, 753–767, 2012.
de la Cámara, A., Mechoso, R., Mancho, A. M., Serrano, E., and Ide, K.: Quasi-horizontal transport within the antarctic polar night vortex: Rossby wave breaking evidence and lagrangian structures, J. Atmos. Sci., 70, 2982–3001, 2013.
de la Llave, R., González, A., Jorba, A., and Villanueva, J.: KAM theory without action-angle variables, Nonlinearity, 18, 855–895, 2005.
Delshams, A. and Gutierrez, P.: Effective stability and KAM theory, J. Diff. Eq., 128, 415–490, 1996.
Doan, T. S., Palmer, K., and Siegmund, S.: Transient spectral theory, stable and unstable cones and Gershgorin's theorem for finite-time differential equations, J. Diff. Eq., 250, 4177–4199, 2011.
Dorato, P.: An overview of finite-time stability, in: Current Trends in Nonlinear Systems and Control: In Honor of Petar Kokotovic and Turi Nicosia, edited by: Menini, L., Zaccarian, L., and Abdallah, C. T., Systems and Control-Foundations and Applications, 185–194, Birkhauser, Boston, 2006.
Duan, J. Q. and Wiggins, S.: Fluid exchange across a meandering jet with quasi-periodic time variability, J. Phys. Oceanogr., 26, 1176–1188, 1996.
Duan, J. and Wiggins, S.: Lagrangian transport and chaos in the near wake of the flow around an obstacle: a numerical implementation of lobe dynamics, Nonlin. Processes Geophys., 4, 125–136, doi:10.5194/npg-4-125-1997, 1997.
Duc, L. H. and Siegmund, S.: Hyperbolicity and invariant manifolds for planar nonautonomous systems on finite time intervals, Int. J. Bif. Chaos, 18, 641–674, 2008.
Duc, L. H. and Siegmund, S.: Existence of finite-time hyperbolic trajectories for planar hamiltonian flows, J. Dyn. Diff. Eq., 23, 475–494, 2011.
Eckart, C.: An analysis of the stirring and mixing processes in incompressible fluids, J. Mar. Res., 7, 265–275, 1948.
Fenichel, N.: Hyperbolicity and exponential dichotomy for dynamical systems, in: Dynamics Reported, edited by: Jones, C. K.

R. T., Kirchgraber, U., and Walther, H. O., Vol. 5, 1–25, Springer-Verlag, New York, Heidelberg, Berlin, 1991.

Fortunati, A. and Wiggins, S.: Normal form and Nekhoroshev stability for nearly-integrable Hamiltonian systems with unconditionally slow aperiodic time dependence., available at: http://arxiv.org/abs/1311.6290, 2013.

Giorgilli, A.: Quantitative methods in classical perturbation theory, in: From Newton to chaos: modern techniques for understanding and coping with chaos in N-body dynamical systems, edited by: Roy, A. E. and Steves, B. D., 21–38, New York, Plenum Press, 1995.

Giorgilli, A.: Notes on exponential stability of Hamiltonian systems, in: Dynamical Systems. Part I. Hamiltonian Systems and Celestial Mechanics, Pisa, Centro di Recerca Matematica Ennio De Giorgi, Scuola Normale Superiore, 2002.

Giorgilli, A. and Zehnder, E.: Exponential stability for time dependent potentials, Z. angew. Math. Phys. (ZAMP), 43, 827–855, 1992.

Giorgilli, A., Locatelli, U., and Sansottera, M.: Kolmogorov and Nekhoroshev theory for the problem of three bodies, Cel. Mech. Dyn. Astron., 104, 159–173, 2009.

Haller, G.: Finding finite-time invariant manifolds in two-dimensional velocity fields, Chaos, 10, 99–108, 2000.

Haller, G. and Poje, A.: Finite time transport in aperiodic flows, Physica D, 119, 352–380, 1998.

Irwin, M. C.: Hyperbolic time dependent processes, Bull. London Math. Soc., 5, 209–217, 1973.

Jones, C. K. R. T. and Winkler, S.: Invariant manifolds and Lagrangian dynamics in the ocean and atmosphere, in: Handbook of dynamical systems, 55–92, North-Holland, Amsterdam, 2002.

Jorba, A. and Simo, C.: On quasiperiodic perturbations of elliptic equilibrium points, SIAM J. Math. Anal., 27, 1704–1737, 1996.

Katok, A. and Hasselblatt, B.: Introduction to the Modern Theory of Dynamical Systems, Cambridge University Press, Cambridge, 1995.

Kloeden, P. E. and Poetzsche, C.: Non-autonomous difference equations and discrete dynamical systems. J. Difference Eq. Appl., 17, 129–130, 2011.

Kloeden, P. and Schmalfuss, B.: Nonautonomous systems, cocycle attractors, and variable time-step discretization, Num. Algor., 14, 141–152, 1997.

Kolmogorov, A. N.: On conservation of conditionally periodic motions under small perturbations of the Hamiltonian, Dokl. Akad. Nauk. USSR, 98, 527–530, 1954.

Langa, J. A., Robinson, J. C., and Suàrez, A.: Stability, instability, and bifurcation phenomena in non-autonomous differential equations, Nonlinearity, 15, 887–903, 2002.

Lerman, L. and Silnikov, L.: Homoclinical structures in nonautonomous systems: Nonautonomous chaos, Chaos, 2, 447–454, 1992.

Littlewood, J. E.: The Lagrange configuration in celestial mechanics, Proc. London Math. Soc., 9, 525–543, 1959a.

Littlewood, J. E.: On the equilateral configuration of the restricted problem of three bodies, Proc. London Math. Soc., 9, 343–372, 1959b.

Lochak, P.: Canonical perturbation theory via simultaneous approximation, Russ. Math. Surveys, 47, 57–133, 1992.

Lochak, P.: Hamiltonian perturbation theory: periodic orbits, resonances, and intermittency, Nonlinearity, 6, 885–904, 1993.

Lochak, P. and Porzio, A.: A realistic exponential estimate for a paradigm Hamiltonian, Annales de l'I. H. P, section A, 51, 199–219, 1989.

Lu, K. and Wang, Q.: Chaos in differential equations driven by a nonautonomous force, Nonlinearity, 23, 2935–2973, 2010.

Lu, K. and Wang, Q.: Chaotic behavior in differential equations driven by a Brownian motion, J. Diff. Eq., 251, 2853–2895, 2011.

Malhotra, N. and Wiggins, S.: Geometric structures, lobe dynamics, and Lagrangian transport in flows with aperiodic time-dependence, with applications to Rossby wave flow, J. Nonlinear Science, 8, 401–456, 1998.

Mancho, A. M., Small, D., and Wiggins, S.: Computation of hyperbolic trajectories and their stable and unstable manifolds for oceanographic flows represented as data sets, Nonlin. Processes Geophys., 11, 17–33, doi:10.5194/npg-11-17-2004, 2004.

Mancho, A. M., Small, D., and Wiggins, S.: A tutorial on dynamical systems concepts applied to Lagrangian transport in oceanic flows defined as finite time data sets: Theoretical and computational issues, Phys. Reports, 437, 55–124, 2006.

Mancho, A. M., Wiggins, S., Curbelo, J., and Mendoza, C.: Lagrangian descriptors: A method for revealing phase space structures of general time dependent dynamical systems, Communic. Nonlin. Sci. Num. Simul., 18, 3530–3557, 2013.

Meiss, J. D.: Visual explorations of dynamics: The standard map, PRAMANA-J. Phys., 70, 965–988, 2008.

Mendoza, C. and Mancho, A. M.: The hidden geometry of ocean flows, Phys. Rev. Lett., 105, 038501, doi:10.1103/PhysRevLett.105.038501, 2010.

Mendoza, C. and Mancho, A. M.: Review Article: "The Lagrangian description of aperiodic flows: a case study of the Kuroshio Current", Nonlin. Processes Geophys., 19, 449–472, doi:10.5194/npg-19-449-2012, 2012.

Meyer, K. R. and Sell, G. R.: Melnikov transforms, bernoulli bundles, and almost periodic perturbations, Trans. Amer. Math. Soc., 314, 63–105, 1989.

Meyer, K. R. and Zhang, X.: Stability of skew dynamical systems, J. Diff. Eq., 132, 66–86, 1996.

Miller, P. D., Jones, C. K. R. T., Rogerson, A. M., and Pratt, L. J.: Quantifying transport in numerically generated velocity fields, Physica D, 110, 105–122, 1997.

Miller, R. K.: Almost periodic differential equations as dynamical systems with applications to the existence of almost periodic solutions, J. Diff. Eq., 1, 337–395, 1965.

Morbidelli, A. and Giorgilli, A.: On an connection between KAM and Nekhorshev's theorems. Physica D, 86, 514–516, 1995a.

Morbidelli, A. and Giorgilli, A.: Superexponential stability of KAM tori, J. Stat. Phys., 78, 1607–1617, 1995b.

Moser, J.: Stabilitätsverhalten kanonischer Differentialgleichungssysteme, Nachrichten der Akademie der Wissenschaften in Göttingen. II. Mathematisch-Physikalische Klasse, 1955, 87–120, 1955.

Moser, J.: On invariant curves of an area preserving mappings of an annulus, Nachr. Akad. Wiss. Gött., II. Math.-Phys. Kl., 1–20, 1962.

Nekhoroshev, N. N.: An exponential estimate on the time of stabilty of nearly-integrable Hamiltonian systems, Russ. Math. Surveys, 32, 1–65, 1977.

Ottino, J.: The Kinematics of Mixing: Stretching, Chaos, and Transport, Cambridge University Press, Cambridge, 1989a.

Ottino, J. M.: The mixing of fluids, Scientif. Am., 260, 56–67, 1989b.

Ottino, J. M., Jana, S. C., and Chakravarthy, V. S.: From Reynolds stretching and folding to mixing studies using horseshoe maps, Phys. Fluids, 6, 685–699, 1994.

Perry, A. D. and Wiggins, S.: KAM tori are very sticky: rigorous lower bounds on the time to move away from an invariant Lagrangian torus with linear flow tori are very, Physica D, 71, 102–121, 1994.

Pesin, Y.: Lectures on Partial Hyperbolicity and Stable Ergodicity, Zurich Lectures in Advanced Mathematics, European Mathematical Society, 2004.

Poetzsche, C.: Geometric Theory of Discrete Nonautonomous Dynamical Systems, Lecture Notes in Mathematics, Vol. 2002, Springer-Verlag, Berlin, 2010a.

Poetzsche, C.: Nonautonomous bifurcation of bounded solutions I, A Lyapunov-Schmidt approach, Discrete and continuous dynamical systems-series B, 14, 739–776, 2010b.

Poetzsche, C.: Nonautonomous bifurcation of bounded solutions II, A shovel bifurcation pattern, Discrete Contin. Dyn. Syst., 31, 941–973, 2011a.

Poetzsche, C.: Persistence and imperfection of nonautonomous bifurcation patterns, J. Diff. Eq., 250, 3874–3906, 2011b.

Pöschel, J.: Nekhoroshev estimates for quasi-convex Hamiltonian systems, Math. Z., 213, 187–216, 1993.

Pöschel, J.: A lecture on the classical KAM theorem, Proc. Symp. Pure Math., 69, 707–732, 2001.

Rasmussen, M.: Towards a bifurcation theory for nonautonomous difference equations. J. Difference Eq. Appl., 12, 297–312, 2006.

Reynolds, O.: Study of fluid motion by means of coloured bands, Nature, 50, 161–164, 1894.

Rogerson, A. M., Miller, P. D., Pratt, L. J., and Jones, C. K. R. T.: Lagrangian motion and fluid exchange in a barotropic meandering jet, J. Phys. Oceanogr., 29, 2635–2655, 1999.

Samelson, R. M.: Fluid exchange across a meandering jet, J. Phys. Oceanogr., 22, 431–440, 1992.

Samelson, R. and Wiggins, S.: Lagrangian Transport in Geophysical Jets and Waves: The Dynamical Systems Approach, Springer-Verlag, New York, 2006.

Scheurle, J.: Chaotic solutions of systems with almost periodic forcing, Z. Angew. Math. Phys., 37, 12–26, 1986.

Sell, G. R.: Nonautonomous differential equationa and topological dynamics I, The basic theory, Trans. Amer. Math. Soc., 127, 241–262, 1967a.

Sell, G. R.: Nonautonomous differential equationa and topological dynamics II, Limiting equations, Trans. Amer. Math. Soc., 127, 263–283, 1967b.

Sell, G. R.: Topological Dynamics and Differential Equations. Van Nostrand-Reinhold, London, 1971.

Sevryuk, M. B.: The classical KAM theory at the dawn of the twenty-first century, Moscow Math. J., 3, 1113–1144, 2003.

Sevryuk, M. B.: Invariant tori in quasiperiodic non-autonomous dynamical systems via Herman's method, Discrete Contin. Dyn. Syst., 18, 569–595, 2007.

Siegmund, S.: Normal forms for nonautonomous differential equations, J. Diff. Eq., 178, 541–573, 2002.

Stoffer, D.: Transversal homoclinic points and hyperbolic sets for non-autonomous maps i, J. Appl. Math. and Phys. (ZAMP), 39, 518–549, 1988a.

Stoffer, D.: Transversal homoclinic points and hyperbolic sets for non-autonomous maps ii, J. Appl. Math. and Phys. (ZAMP), 39, 783–812, 1988b.

Sturman, R., Ottino, J. M., and Wiggins, S.: The Mathematical Foundations of Mixing, Cambridge Monographs on Applied and Computational Mathematics, Vol. 22, Cambridge University Press, 2006.

Weiss, L. and Infante, E. F.: On the stability of systems defined over a finite time interval, Proc. Nat. Acad. Sci., 54, 44–48, 1965.

Wiggins, S.: Chaotic Transport in Dynamical Systems, Springer-Verlag, New York, 1992.

Wiggins, S.: Chaos in the dynamics generated by sequences of maps, with applications to chaotic advection in flows with aperiodic time dependence, Z. Angew. Math. Phys., 50, 585–616, 1999.

Wiggins, S.: The dynamical systems approach to Lagrangian transport in oceanic flows, Annu. Rev. Fluid Mech., 37, 295–328, 2005.

Wiggins, S. and Ottino, J.: Foundations of chaotic mixing, Phil. Trans. Roy. Soc., 362, 937–970, 2004.

12

Analysis of stochastic model for nonlinear volcanic dynamics

D. V. Alexandrov, I. A. Bashkirtseva, and L. B. Ryashko

Department of Mathematical Physics, Ural Federal University, Lenin ave. 51, 620000 Ekaterinburg, Russia

Correspondence to: L. Ryashko (lev.ryashko@urfu.ru)

Abstract. Motivated by important geophysical applications we consider a dynamic model of the magma-plug system previously derived by Iverson et al. (2006) under the influence of stochastic forcing. Due to strong nonlinearity of the friction force for a solid plug along its margins, the initial deterministic system exhibits impulsive oscillations. Two types of dynamic behavior of the system under the influence of the parametric stochastic forcing have been found: random trajectories are scattered on both sides of the deterministic cycle or grouped on its internal side only. It is shown that dispersions are highly inhomogeneous along cycles in the presence of noises. The effects of noise-induced shifts, pressure stabilization and localization of random trajectories have been revealed by increasing the noise intensity. The plug velocity, pressure and displacement are highly dependent of noise intensity as well. These new stochastic phenomena are related to the nonlinear peculiarities of the deterministic phase portrait. It is demonstrated that the repetitive stick–slip motions of the magma-plug system in the case of stochastic forcing can be connected with drumbeat earthquakes.

1 Introduction

It is well-known that the behavior of volcanic systems is enormously complex so that a lot of nonlinear feedbacks lead to multiple states even during a single eruption (Tanaka et al., 2014). Without better modeling forecasts of these dynamic processes, the highly important questions of where, when and how volcanic eruptions occur will remain substantially empirical. Nowadays, an elaboration of the adequate mathematical models for volcanic dynamics is a challenging problem (Melnik and Sparks, 1999; Barmin et al., 2002; Nakanishi and Koyaguchi, 2008; Costa et al., 2012).

Many uncertainties in physical parameters of volcanic dynamics (Woo, 2000) lead to a conclusion that like the climate systems (see, among others, Saltzman, 2002; Alexandrov et al., 2014), volcanoes, representing stochastic and chaotic systems, need to be described in terms of probabilities (Sparks, 2003; Bebbington and Marzocchi, 2011). Stochastic approaches and mathematical formalisms can be found in Gardiner (2009).

It is well-known, that an interplay between nonlinearity and noise can generate various probabilistic phenomena such as noise-induced transitions (Horsthemke and Lefever, 1984), stochastic resonance (McDonnell et al., 2008; Pikovsky and Kurths, 1997; Arathi, 2013), and noise-induced chaos (Lai and Tél, 2011; Bashkirtseva et al., 2012). Stochastic effects in nonlinear models are the subjects of intensive investigations in various research domains (Horsthemke and Lefever, 1984; Lindner et al., 2004; Bashkirtseva et al., 2013; Alexandrov et al., 2013).

Some of the silicic volcanoes analyzed in detail over the last few decades represent complex periodic systems (Denlinger and Hoblitt, 1999; Michaut et al., 2013). The dome-building eruption of Mount St. Helen (MSH) during 2004 and 2005 has represented a near-equilibrium cyclic system with the solid plug uplift caused by magma ascent from below with a nearly steady rate of roughly 1–2 $m^3 s^{-1}$. This eruption was accompanied by drumbeat earthquakes that recurred every 1–2 min with magnitudes <2 and focal depth <1 km (Iverson et al., 2006; Moore et al., 2008; Matoza and Chouet, 2010). A stick–slip mechanism explains a cyclic behavior of such earthquakes as a consequence of stick–slip motion of a plug pushed by compressible magma (Denlinger and Hoblitt, 1999). A new dynamic approach based on this mechanism (connecting the MSH behavior with a damped oscillator) was suggested by Iverson et al. (2006). We use this

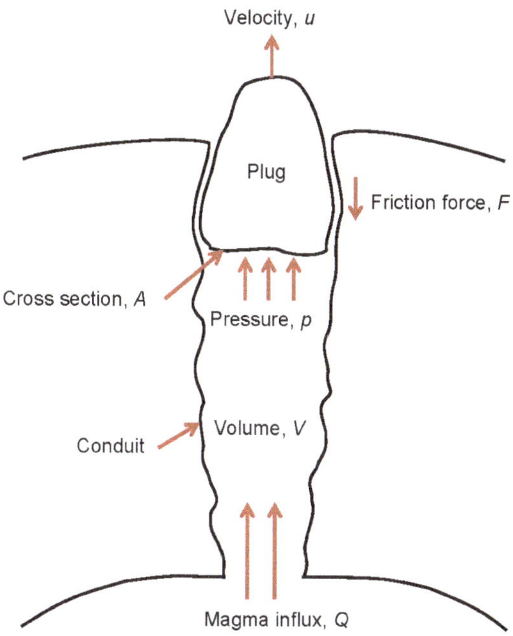

Figure 1. A scheme of the plug dynamics.

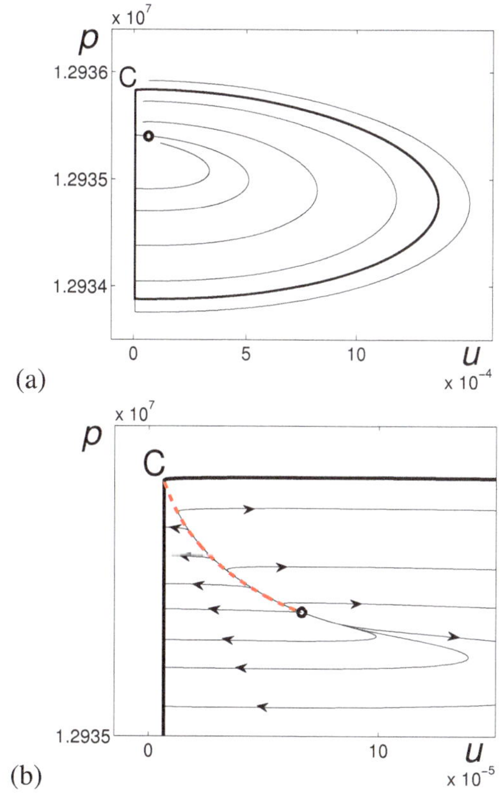

Figure 2. **(a)** projection of the phase portrait of the deterministic system, p (Pa) vs. u ($\mathrm{m\,s^{-1}}$). **(b)** illustrates an enlargement near unstable equilibrium (open circle). The dashed red line is a pseudo-separatrix. Physical parameters of the system under consideration are (Iverson et al., 2006) $B = Q = 2\,\mathrm{m^3 s^{-1}}$, $m_0 = 3.6 \times 10^{10}$ kg, $A = 30\,000\,\mathrm{m}^2$, $p_0 = 12936\times 10^3$ Pa, $\alpha_1 = 10^{-7}\,\mathrm{Pa}^{-1}$, $\alpha_2 = 10^{-9}\,\mathrm{Pa}^{-1}$, $F_0 = 3528 \times 10^7\,\mathrm{kg\,m\,s^{-2}}$, $u_{\mathrm{ref}} = 0.1 Q/A = 6.67 \times 10^{-6}\,\mathrm{m\,s^{-1}}$, $V_0 = 6.32 \times 10^5\,\mathrm{m}^3$, $c = 1.7 \times 10^{-4}$, $g = 9.8\,\mathrm{m\,s^{-2}}$, $\rho_0 = \rho_{\mathrm{r}} = 2000\,\mathrm{kg\,m^{-3}}$, and $\kappa = 0$.

model to demonstrate unusual dynamic behavior of similar volcanic systems under the influence of parametric noises.

In order to explain interactions between solid-state extrusion and persistent drumbeat earthquakes at MSH, Iverson et al. (2006) developed a model based on recurrent stick–slip motions of the solid plug along its margins with the friction force F (Fig. 1). Let us briefly discuss the main principles of this dynamic model. The magma influx comes to the base of an eruptive conduit from below with a nearly steady rate Q. A solid dacite plug of solidified magma blocks the conduit from above so that its lower boundary is mobile due to the effects of pressure and basal accretion with mass rate ρB from below (ρ and B stand for the magma bulk density and the volumetric rate of magma crystallization). The total plug mass m changes with time because the difference in mass rates ρB and $\rho_{\mathrm{r}} E$ as $m = m_0 + \kappa t$ (here ρ_{r} is the plug bulk density, E is the volumetric rate of surface erosion, m_0 is the initial plug mass, t is the process time, and $\kappa = \rho B - \rho_{\mathrm{r}} E$ is assumed constant). The horizontal cross-sectional area A, the magma compressibility α_1 and the conduit wall compliance α_2 are estimated by Iverson et al. (2006). The dynamic process of plug extrusion is controlled by the plug weight mg and the friction force F dependent of the plug velocity u (g is the acceleration due to gravity) whereas the conduit volume V is governed by the law of mass conservation. A three-parametric differential model connecting independent variables u, p and V (p is the pressure) was derived by Iverson et al. (2006). Below we use this model to demonstrate some new special aspects of nonlinear dynamics of volcanic systems under the influence of stochastic noises.

2 The model and its deterministic behavior

The following system of reduced governing equations based on the laws of conservation of the solid plug linear momentum, solid plug mass and conduit fluid mass was derived and discussed in detail by Iverson et al. (2006). These equations can be written in the form of

$$\frac{\mathrm{d}u}{\mathrm{d}t} = -g + \frac{1}{m_0 + \kappa t}(pA - \kappa u - F), \tag{1}$$

$$\frac{\mathrm{d}p}{\mathrm{d}t} = -\frac{1}{(\alpha_1 + \alpha_2)V}(Au + RB - Q), \tag{2}$$

$$\frac{\mathrm{d}V}{\mathrm{d}t} = \frac{\alpha_1}{\alpha_1 + \alpha_2}(Au + RB - Q) + Q - B, \tag{3}$$

where $R = 1 - \rho/\rho_{\mathrm{r}} = 1 - (\rho_0/\rho_{\mathrm{r}})\exp[\alpha_1(p - p_0)]$ is found from the isothermal equation of state (Iverson et al., 2006). Here, p_0 and ρ_0 represent the static equilibrium pressure and magma density. The key aspects of MSH friction force measured in experiments (Moore et al., 2008) can be described

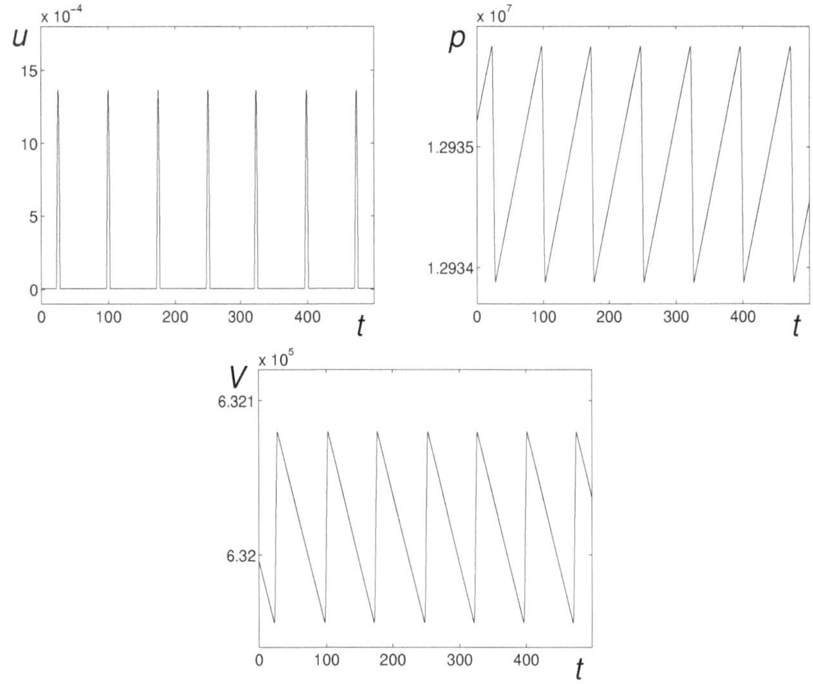

Figure 3. Time series of the cycle shown in Fig. 2: u ($\mathrm{m\,s^{-1}}$), p (Pa) and V ($\mathrm{m^3}$) as functions of time t (s).

by a function (Iverson et al., 2006):

$$F(u) = \mathrm{sgn}(u) f_0(u), \;\; f_0(u) = F_0\left(1 - c\sinh^{-1}|u/u_{\mathrm{ref}}|\right), \quad (4)$$

where $\mathrm{sgn}(u)$ is the sign of u, F_0 is the friction force at static equilibrium, $c \ll 1$ is a rate-weakening parameter and u_{ref} is a reference value of u (Iverson et al., 2006). Equation (4) includes the main physical aspects of the process: the friction force at $u = 0$ abruptly changes its sign due to the fact that the gravity force, which shifts the plug in downward direction, is opposite to the friction force. However, an abrupt behavior of Eq. (4) is not a good physical approximation of the friction force. Therefore, let us model this force by the close continuous function:

$$F(u) = \mathrm{sgn}(u) f_1(u), \quad (5)$$

where

$$f_1(u) = \begin{cases} f_0(u), & |u| \geq u_{\mathrm{ref}} \\ F_0 |\frac{u}{u_{\mathrm{ref}}}|, & 0 < |u| < u_{\mathrm{ref}} \end{cases}.$$

In present paper we focus on the autonomous case, when $\kappa = 0$. The model (Eqs. 1–3) demonstrates the stick–slip oscillations (see Figs. 2 and 3). An important point is that this system has only one unstable equilibrium $(\overline{u}, \overline{p}, \overline{V})$ for any V_0, where $\overline{u} = \dfrac{Q}{A} = \dfrac{2}{3} \times 10^{-4}\,\mathrm{m\,s^{-1}}$, $\overline{p} = p_0 = 12\,936 \times 10^3$ Pa, and $\overline{V} = V_0$. This equilibrium is plotted by an open circle in Fig. 2.

In Fig. 2, u and p projections of phase trajectories of the system (Eqs. 1–3) for the fixed initial value $V_0 = 6.32 \times 10^5\,\mathrm{m^3}$ are plotted by the thin black lines. These trajectories tend to the closed curve (thick black line) of the cycle. Time series of this cycle are presented in Fig. 3. A vertical left part of this cycle in Fig. 2 corresponds to the slow movement, and the other arc part of the cycle reflects fast movement. The slow dynamics become to fast at the corner point C in the case of movement in a clockwise direction along the thick black curve. The stability of this cycle is highly nonuniform: the vertical part is extremely stable, but the arc curve possesses a neutral stability.

Essential details of the phase portrait are shown in Fig. 2b by an enlarged fragment of Fig. 2a. As one can see, there exists a pseudo-separatrix (dashed red line) which divides two types of dynamics. If the initial state lies to the left of this red curve, then the trajectory quickly verges towards the vertical part of the cycle (arrows pointing to the left). If the initial state lies to the right of this pseudo-separatrix, then the phase trajectory goes away from the cycle (arrows pointing to the right), and only after a long excursion, the trajectory approaches to the vertical part of the cycle. Physically it means that small deviations in u at sufficiently large p may redeploy the dynamic system through its pseudo-separatrix. This feature of the deterministic phase portrait playing an important role in understanding of stochastic phenomena will be discussed below.

Note, that this cycle is not a limit cycle in the classical mathematical sense. Indeed, for different values of V_0, the

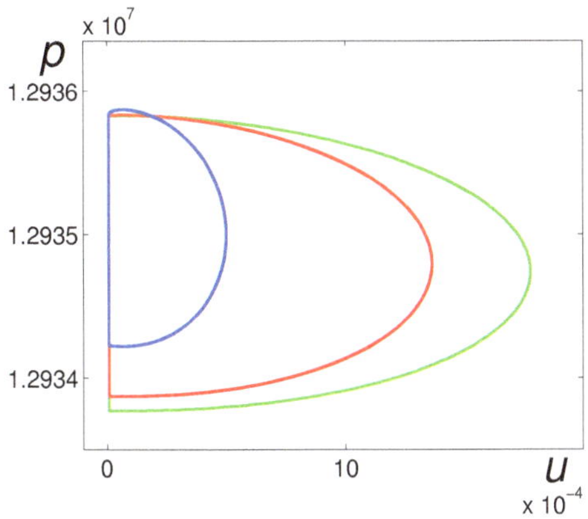

(a)

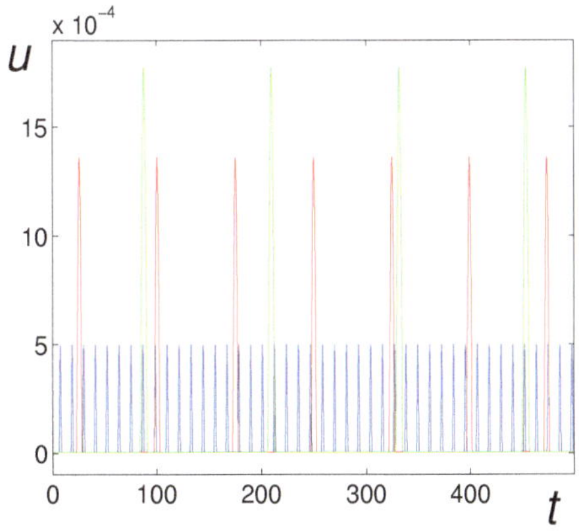

(b)

Figure 4. (a) illustrates a projection of the phase portrait p (Pa) vs. u ($\mathrm{m\,s^{-1}}$) of the deterministic system. **(b)** shows u ($\mathrm{m\,s^{-1}}$) as a function of t (s). These dependencies are plotted for different values V_0: $V_0 = 10^5\,\mathrm{m}^3$ (blue), $V_0 = 6.32 \times 10^5\,\mathrm{m}^3$ (red), and $V_0 = 10^6\,\mathrm{m}^3$ (green).

nonlinear system (Eqs. 1–3) exhibits different closed curves. The cycles and time series for various values of V_0 are compared in Fig. 4. Note that an increase of V_0 implies an increase of both amplitude and period of oscillations.

3 The role of stochastic forcing

In order to study possible deviations of the friction force from expression (Eq. 5) let us consider the parametric random disturbances. Such disturbances simulate the influence of different physical processes and phenomena leading to variations in the friction force behavior (e.g., the effects of frictional melting, and temperature-dependent friction).

At first we analyze the following F noise: $F_0 \to F_0(1 + \varepsilon\xi(t))$, where $\xi(t)$ is a standard Gaussian white noise with parameters $\langle\xi(t)\rangle = 0$, $\langle\xi(t)\xi(\tau)\rangle = \delta(t-\tau)$, and ε is a noise intensity. The corresponding stochastic system includes Eqs. (2) and (3) whereas Eq. (1) should be replaced by

$$\frac{\mathrm{d}u}{\mathrm{d}t} = -g + \frac{1}{m_0}[pA - F(u)] - \frac{\varepsilon}{m_0}F(u)\xi(t). \tag{6}$$

Note that under stochastic disturbances, random trajectories leave the deterministic cycle and form a bundle of stochastic trajectories.

If the noise intensity is small enough, such bundle has a small dispersion and is localized near the deterministic cycle (green lines in Fig. 5a). As the noise intensity increases, along with the natural increase of dispersion, the following unexpected phenomenon is observed: the bundle's right side of stochastic trajectories is shifted inside the deterministic cycle (blue and red lines in Fig. 5a). Some details of the corresponding probabilistic distributions are presented in Figs. 5a, b. The probability density functions of u coordinates of intersection points of the random trajectories with the line $p = 1.2935 \times 10^7$ Pa are plotted for three values of the F-noise intensity in Fig. 5b whereas the probability density functions of time intervals between successive intersections are shown in Fig. 5c. As one can see, with increasing noise, both the amplitude and period of stochastic oscillations decrease.

This stochastic phenomenon can be explained by the phase portrait peculiarities of an initial deterministic system (see Fig. 2b) near the upper part of vertical fragment of the cycle. In the deterministic case, the phase trajectory slowly moves along the vertical part of the cycle up to point C. At point C, this trajectory abruptly changes the direction, and begins to move along the arc part quickly. Under the stochastic disturbances, random trajectories deviate from this vertical part of the deterministic cycle. As a result of this deviation, the random trajectory can cross the red pseudo-separatrix, and then it falls within the region of large arc-form excursions. In this case, the random trajectory turns right before point C. The more noise, the earlier this turn. Such stochastic deformation of the random flow results in a decrease of u- and p-oscillation amplitudes and of the period.

Under the further increase of noise intensity, random states of the system (Eqs. 2, 3, 6) are localized and leave the interior of the deterministic cycle. This noise-induced shift is demonstrated in Fig. 6. Here, an essential decrease of the dispersion of the p coordinate is observed. In other words, p stabilizes near its certain value with increase in the noise intensity.

The dynamics of plug displacement is shown in Fig. 7. If the noise intensity is large enough so that the system leaves its cycle, the plug displacement increases with noise. If the system is within its cycle, the displacement is also within the corresponding deterministic stepwise curve (black line in Fig. 7).

In order to study an influence of possible changes in magma influx, let us consider a role of Q noise: $Q \to Q(1 + \delta\xi(t))$, where $\xi(t)$ is a Gaussian white noise, and δ is a Q-noise intensity. In this case, a nonlinear dynamic system consists of Eq. (1) as well as of the following stochastic equa-

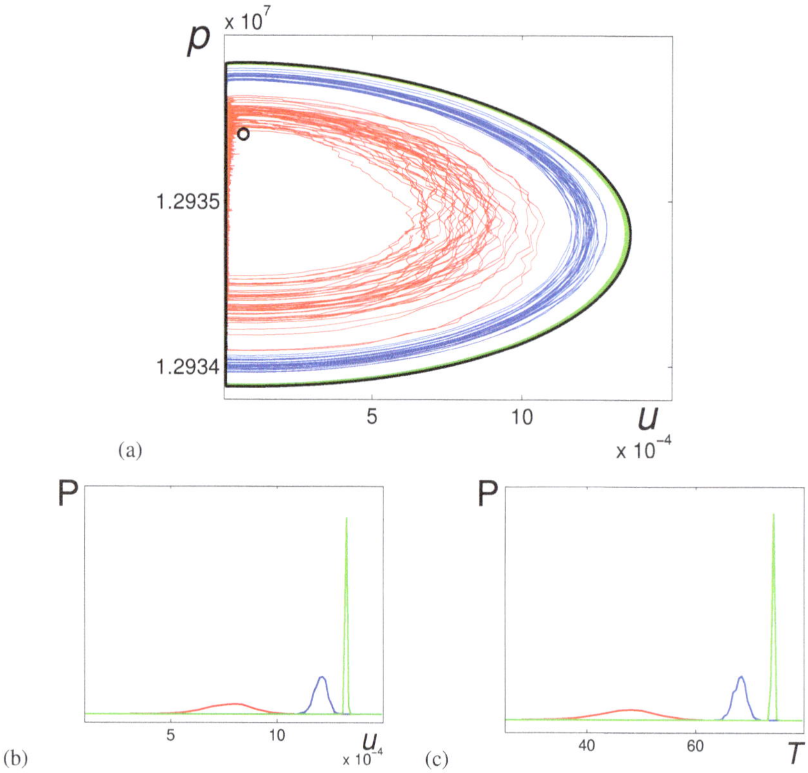

Figure 5. Stochastic cycles for $\varepsilon = 10^{-6}$ (green), $\varepsilon = 10^{-5}$ (blue), and $\varepsilon = 5 \times 10^{-5}$ (red): **(a)** random trajectories p (Pa) vs. u ($\mathrm{m\,s^{-1}}$), **(b)** probability density functions of u coordinates ($\mathrm{m\,s^{-1}}$), and **(c)** probability density functions of the period T (s).

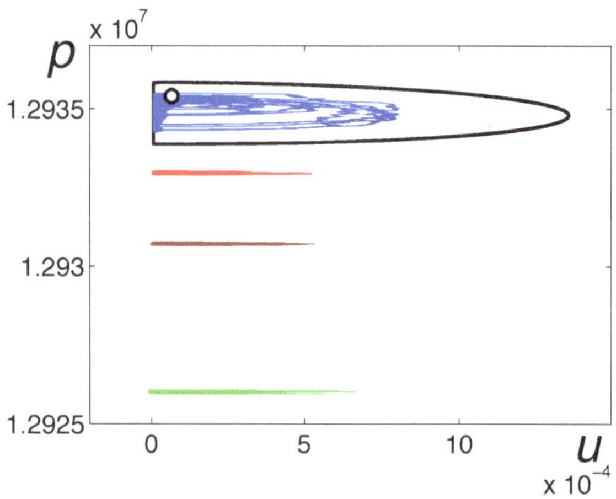

Figure 6. Random trajectories p (Pa) vs. u ($\mathrm{m\,s^{-1}}$) for $\varepsilon = 10^{-4}$ (blue), $\varepsilon = 5 \times 10^{-4}$ (red), $\varepsilon = 7 \times 10^{-4}$ (brown), and $\varepsilon = 10^{-3}$ (green). The open circle designates the point of unstable equilibrium.

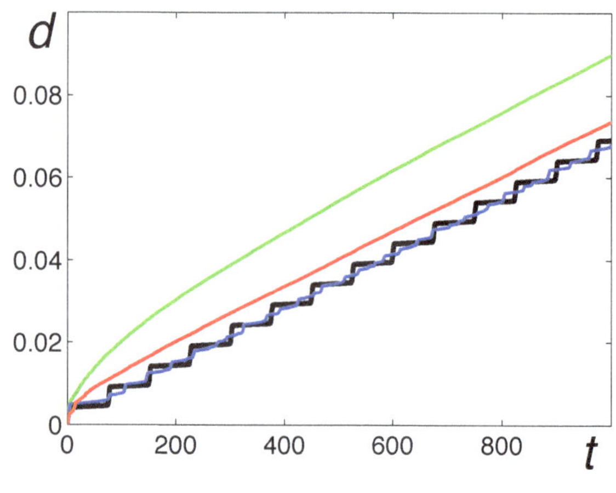

Figure 7. Displacement d (m) as a function of time t (s) under the influence of F noise: deterministic case (black), $\varepsilon = 10^{-4}$ (blue), $\varepsilon = 5 \times 10^{-4}$ (red), and $\varepsilon = 10^{-3}$ (green).

tions:

$$\frac{\mathrm{d}p}{\mathrm{d}t} = -\frac{1}{V(\alpha_1 + \alpha_2)}[Au - Qe^{\alpha_1(p-p_0)}] + \frac{\delta}{V(\alpha_1 + \alpha_2)} Qe^{\alpha_1(p-p_0)}\xi(t), \tag{7}$$

$$\frac{\mathrm{d}V}{\mathrm{d}t} = \frac{\alpha_1}{\alpha_1 + \alpha_2}[Au - Qe^{\alpha_1(p-p_0)}] - \frac{\delta\alpha_1}{\alpha_1 + \alpha_2} Qe^{\alpha_1(p-p_0)}\xi(t). \tag{8}$$

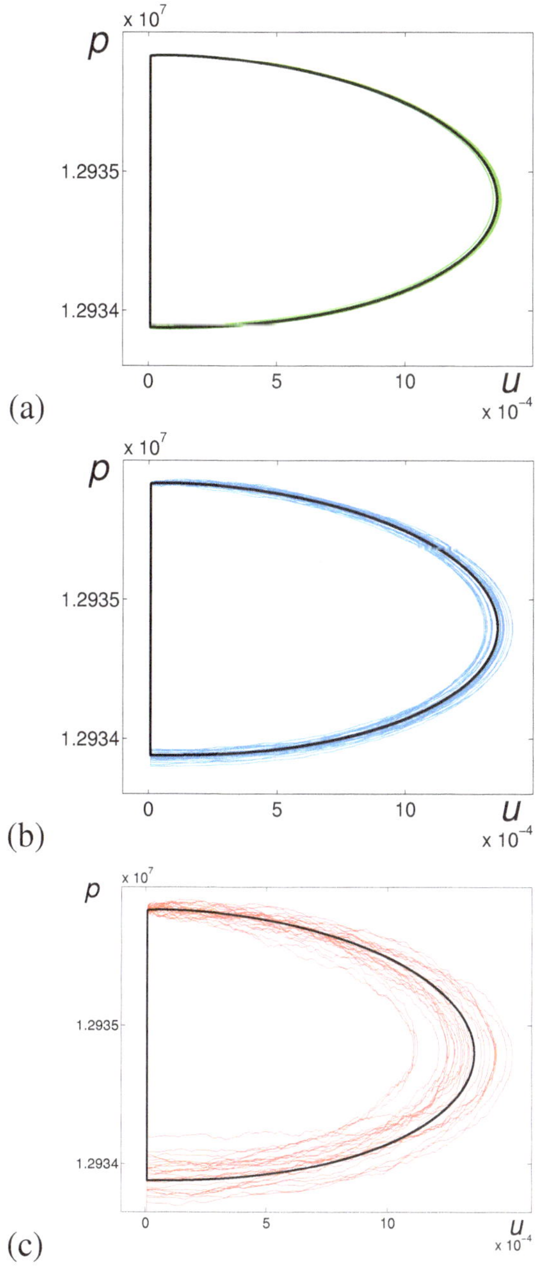

Figure 8. Stochastic cycles p (Pa) vs. u ($\mathrm{m\,s^{-1}}$) for $\delta = 0.1$ **(a)**, $\delta = 0.5$ **(b)**, and $\delta = 2$ **(c)**.

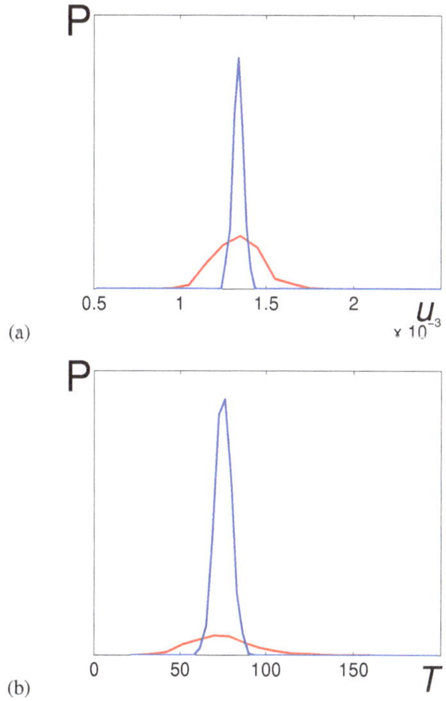

Figure 9. Influence of Q noise for $\delta = 0.5$ (blue), and $\delta = 2$ (red): **(a)** probability density functions of u coordinates ($\mathrm{m\,s^{-1}}$), **(b)** probability density functions of the period T (s).

For weak noise, stochastic trajectories are localized near the deterministic cycle (Fig. 8a). As noise intensity increases, the dispersion of random trajectories increases as well (Figs. 8b, c). It can be seen that the dispersion is extremely nonuniform along the cycle. For the vertical part, a dispersion is small even for large noise, and the random trajectories do not differ from the deterministic cycle. Along the arc part, the dispersion of the random trajectories increases. The probability density functions of u coordinates of intersection points of the random trajectories with the line $p = 1.2935 \times 10^7$ Pa are plotted for three values of the Q-noise intensity in Fig. 9a. Panel b of this figure shows the probability density functions of time intervals between successive intersections. As one can see, the dispersions grow and the mean values are practically unchangeable with increasing noise.

As one can see, the volcanic model under consideration demonstrates quite different qualitative and quantitative responses to the random perturbations of different parameters. The system is extremely sensitive to F noise so that even a weak F noise implies a crucial deformation of the oscillatory behavior. An increase of F noise leads to a decrease of the period and amplitude of oscillations. Note that the system is also sensitive to Q noise so that a large noise intensity implies a dispersion increase of the arc part of stochastic oscillations.

4 Conclusions

The phase portrait of a deterministic system contains a point of unstable equilibrium and a pseudo-separatrix, which subdivides the system into different dynamic areas (point C and dashed red line in Fig. 2). In addition, if point (u, p) of the phase plane lies below this pseudo-separatrix (Fig. 2b), the system quickly reaches its equilibrium state (thick vertical line in Fig. 2). If, however, the phase point (u, p) is above this pseudo-separatrix, the system tends to its equilibrium state in course of a long time interval. An important point of the deterministic behavior is that a certain constant value of the plug velocity u establishes at different pressures p in the state of equilibrium (thick vertical line in Fig. 2b). This equilibrium state also persists at different conduit volumes V_0 (Fig. 4). A time of system transition to its equilibrium state (vertical line in Fig. 4a) therewith increases with increasing V_0. In addition, more broad conduits might have a rather big variation in u and p than more narrow ones during the deterministic process of the volcanic plug evolution. By this is meant that a time required to attain the equilibrium state increases by increasing the conduit's cross-sectional area.

In order to analyze the role of variations of two main parameters of the plug motion (friction force F and magma influx Q), two types of noises have been introduced in the model equations: F noise and Q noise. It was shown that these noises lead to different evolutionary types of the dynamic system. Let us summarize the main aspects of this behavior. In the first place, random trajectories are scattered either on both sides of the deterministic cycle (Q noise) or on its internal side (F noise). Dispersions corresponding to random trajectories of both noises upon that grow by increasing the noise intensities. As this takes place, one can see that both dispersions are highly inhomogeneous along cycles (Figs. 5, 8). Note that these dispersions are small enough in the vertical parts of corresponding cycles for F and Q noises.

An important point is that an increase in dispersion occurs in the vicinity of the pseudo-separatrix under the influence of F noise. This is due to the fact that phase trajectories intersect the pseudo-separatrix and the phase points undergo transitions across it under the action of F noise. Let us especially emphasize that F-noise-phase trajectories leave the corresponding deterministic cycle and form a stochastic bundle shifted into the cycle's interior. As this takes place, the bundle's dispersion increases while the period and amplitude of oscillations decrease by increasing the F-noise intensity (Fig. 5). By this is meant that the presence of F noise reduces possible variations in the plug velocity u and pressure p and decreases a time required to attain the equilibrium state (thick vertical line in Fig. 5a). It is significant that the effect of pressure stabilization near a certain value (dependent of the noise intensity) occurs with a rise in the F-noise intensity. The random trajectories therewith leave the corresponding cycle and are localized in the vicinity of this value (Fig. 6). A dynamic behavior of the plug displacement is dependent of whether the dynamic system is within or beyond its phase cycle. In the former case, the plug displacement oscillates within the bounds of the corresponding deterministic stepwise curve (Fig. 7). In the latter case, when the noise intensity is sufficiently large, the plug displacement increases drastically.

It is known that the eruption of Mount St. Helens was accompanied by rather regular repetitive long-period (or drumbeat) earthquakes over a long time. Moreover, such drumbeat events were more random from time to time. In addition, subevents in the form of randomly occurring series of smaller seismic events (produced by a separate random process) have been imposed upon these long-period events (Matoza and Chouet, 2010). The present study demonstrates that repetitive stick–slip motions of the plug representing stochastic oscillations can be connected with these drumbeat earthquakes. The calculated period between drumbeats (see Figs. 5 and 9) is in agreement with experimental data (30–300 s, Iverson et al., 2006). The physical reason is that such earthquakes observed at shallow depths (<1 km at MSH) can be caused by the stick–slip motions of the magma-plug system under the influence of noises where the driving force acting on a compliant crustal body is large enough (the force drop responsible for this kind of seismicity can be estimated from our calculations as $\Delta F = \Delta p A \sim 6 \times 10^7\ \mathrm{kg\,m\,s^{-2}}$, where $\Delta p \sim 2 \times 10^3$ Pa).

Acknowledgements. This work was supported by the Ministry of Education and Science of the Russian Federation under the project no. 315.

Edited by: R. Gloaguen

References

Alexandrov, D. V., Bashkirtseva, I. A., Malygin, A. P., and Ryashko, L. B.: Sea ice dynamics induced by external stochastic fluctuations, Pure Appl. Geophys., 170, 2273–2282, 2013.

Alexandrov, D. V., Bashkirtseva, I. A., and Ryashko, L. B.: Stochastically driven transitions between climate attractors, Tellus A, 66, 23454, doi:10.3402/tellusa.v66.23454, 2014.

Arathi, S., Rajasekar, S., and Kurths, J.: Stochastic and coherence resonances in a modified chua's circuit system with multi-scroll orbits, Int. J. Bifurcat. Chaos, 23, 1350132, doi:10.1142/S0218127413501320, 2013.

Barmin, A., Melnik, O., and Sparks, R. S. J.: Periodic behavior in lava dome eruptions, Earth Planet. Sc. Lett., 199, 173–184, 2002.

Bashkirtseva, I., Chen, G., and Ryashko, L.: Analysis of noise-induced transitions from regular to chaotic oscillations in the Chen system, Chaos, 22, 033104, doi:10.1063/1.4732543, 2012.

Bashkirtseva, I., Neiman, A. B., and Ryashko, L.: Stochastic sensitivity analysis of the noise-induced excitability in a model of a hair bundle, Phys. Rev. E, 87, 052711, doi:10.1103/PhysRevE.87.052711, 2013.

Bebbington, M. S. and Marzocchi, W.: Stochastic models for earthquake triggering of volcanic eruptions, J. Geophys. Res., 116, B05204, doi:10.1029/2010JB008114, 2011.

Costa, A., Wadge, G., and Melnik, O.: Cyclic extrusion of a lava dome based on a stick-slip mechanism, Earth Planet. Sci. Lett., 337–338, 39–46, 2012.

Denlinger, R. P. and Hoblitt, R. P.: Cyclic behavior of cilicic volcanoes, Geology, 27, 459–462, 1999.

Gardiner, C.: Stochastic Methods: A Handbook for the Natural and Social Sciences, Springer, Berlin, 2009.

Horsthemke, W. and Lefever, R.: Noise-Induced Transitions: Theory and Applications in Physics, Chemistry, and Biology, Springer, Berlin, 1984.

Iverson, R. M., Dzurisin, D., Gardner, C. A., Gerlach, T. M., LaHusen, R. G., Lisowski, M., Major, J. J., Malone, S. D., Messerich, J. A., Moran, S. C., Pallister, J. S., Qamar, A. I., Schilling, S. P., and Vallance, J. W.: Dynamics of seismogenic volcanic extrusion at Mount St Helens in 2004–05, Nature, 444, 439–443, 2006.

Lai, Y. C. and Tél, T.: Transient Chaos: Complex Dynamics on Finite Time Scales, Springer, Berlin, 2011.

Lindner, B., Garcia-Ojalvo, J., Neiman, A., and Schimansky-Geier, L.: Effects of noise in excitable systems, Phys. Rep., 392, 321–424, 2004.

Matoza, R. S. and Chouet, B. A.: Subevents of long-period seismicity: Implications for hydrothermal dynamics during the 2004–2008 eruption of Mount St. Helens, J. Geophys. Res., 115, B12206, doi:10.1029/2010JB007839, 2010.

McDonnell, M. D., Stocks, N. G., Pearce, C. E. M., and Abbott, D.: Stochastic Resonance: from Suprathreshold Stochastic Resonance to Stochastic Signal Quantization, Cambridge University Press, Cambridge, 2008.

Melnik, O. E. and Sparks, R. S. J.: Nonlinear dynamics of lava dome extrusion, Nature, 402, 37–41, 1999.

Michaut, C., Ricard, Y., Bercovici, D., and Sparks, R. S. J.: Eruption cyclicity at silicic volcanoes potentially caused by magmatic gas waves, Nat. Geosci., 6, 856–860, 2013.

Moore, P. L., Iverson, N. R., and Iverson, R. M.: Frictional properties of the Mount St Helens gouge, US Geological Survey Professional Paper 1750, US Geological Survey, Reston, Virginia, 2008.

Nakanishi, M. and Koyaguchi, T.: A stability analysis of a conduit flow model for lava dome eruptions, J. Volcanol. Geoth. Res., 178, 46–57, 2008.

Pikovsky, A. S. and Kurths, J.: Coherence resonance in a noise-driven excitable system, Phys. Rev. Lett., 78, 775–778, 1997.

Saltzman, B.: Dynamical Paleoclimatology: Generalised Theory of Global Climate Change, Academic Press, San Diego, 2002.

Sparks, R. S. J.: Forecasting volcanic eruptions, Earth Planet. Sci. Lett., 210, 1–15, 2003.

Tanaka, H. K. M., Kusagaya, T., and Shinohara, H.: Radiographic visualization of magma dynamics in an erupting volcano, Nat. Commun., 5, 3381, doi:10.1038/ncomms4381, 2014.

Woo, G.: The Mathematics of Natural Catastrophes, Imperial College Press, London, 2000.

13

Multiple-scale error growth in a convection-resolving model

F. Uboldi[1] **and A. Trevisan**[2]

[1]Independent researcher, Milan, Italy
[2]CNR-ISAC, Bologna, Italy

Correspondence to: F. Uboldi (uboldi@magritte.it)

Abstract. The properties of the multiple-scale instabilities present in a non-hydrostatic forecast model are investigated. The model simulates intense convection episodes occurring in northern Italy. A breeding technique is used to construct ensembles of perturbations of the model trajectories aimed at representing the instabilities that are responsible for error growth on various timescales and space scales. By means of perfect model twin experiments it is found that, for initial errors of the order of present-day analysis error, a non-negligible fraction of the forecast error can be explained by a bred vector ensemble of reasonable size representing the growth of errors on intermediate scales. In contrast, when the initial error is much smaller, the spectrum of bred vectors representing the fast convective-scale instabilities becomes flat, and the number of ensemble members needed to explain even a small fraction of the forecast error becomes extremely large. The conclusion is that as the analysis error is decreased, it becomes more and more computationally demanding to construct an ensemble that can describe the high-dimensional subspace of convective instabilities and that can thus be potentially useful for controlling the error growth.

1 Introduction

In the last 10–15 years there has been operational interest in non-hydrostatic, convection-resolving models (e.g. Dixon et al., 2009; Seity et al., 2011; Baldauf et al., 2011) and in the possibility of making use of data assimilation (e.g. Zhang et al., 2003; Kain et al., 2010; Schenkman et al., 2011; Claussnitzer et al., 2011) to improve short-range forecasting and nowcasting of weather fields, or the accuracy of a past event trajectory reconstruction. Some attention has been devoted to evaluating whether a higher resolution and a more realistic description of convective events are in fact associated with a quantitative improvement in forecasting their location, timing and intensity (e.g. Weisman et al., 2008; Schwartz et al., 2009).

Prediction and data assimilation systems for the ocean and the atmosphere are largely founded upon the theory, developed over the last decades, of predictability and state estimation in chaotic dynamical systems. Atmospheric NWP (numerical weather prediction) models that predict the synoptic-scale evolution of mid-latitude weather systems are based on the hydrostatic assumption and include parameterisation of convective processes. Ensemble forecasting and data assimilation for such models have reached a mature stage, whereas non-hydrostatic models face the hard task of dealing with the fast instabilities typical of convection. Predictability in the presence of multiple scales has been investigated in several studies starting with the pioneering work of Lorenz (1969). Errors in the small-scale weather features such as thunderstorms grow very rapidly and reach saturation within a timescale comparable to their brief lifetime, while errors in larger scales may still be at a linear stage of growth. A comparison of the timescales involved in prediction of hydrostatic versus non-hydrostatic models has been performed by Hohenegger and Schär (2007a), casting some doubts on the feasibility of borrowing algorithms developed in the context of the former for the application to the latter models.

In the present study we will try to look more deeply into this question, using as a "laboratory" a realistic model (MOLOCH, Sect. 3) and applying to a real case study the classical breeding method. The relation between bred vectors and Lyapunov vectors, developed in a rigorous theoretical framework, states that as the rescaling amplitude goes to zero, Lyapunov vectors are recovered from bred vectors (BVs hereafter). This argument can be exploited when there fundamentally is one typical instability scale, as in synoptic-scale models. In a convection-resolving model, we expect to find a very large number of positive, competing Lyapunov exponents with very large values, making it unfeasible to compute all or even a reasonable number of their corresponding modes.

The flow-dependent character of forecast uncertainty and its estimation has become an important issue in NWP in the last decades, starting from the early works of Kalnay and Dalcher (1987) and Lorenz (1996), among others. In the forecasting practice, before Lorenz clarified the dynamical mechanism of selection of instabilities in his 1996 model (Lorenz, 1996), researchers at the National Meteorological Centre (NMC, now NCEP) were well aware of this problem, and the breeding method (Toth and Kalnay, 1993, 1997) was devised for its solution. When instabilities in a wide range of scales is present, breeding provides a useful tool for selecting the instability scale one wants to investigate by tuning the breeding parameters, i.e. the rescaling amplitude and time interval. For instance, if the rescaling amplitude is larger than the typical saturation amplitude of a given scale, unstable structures characteristic of that scale will not appear in the bred vectors.

While the importance for prediction of a rigorous description of flow instabilities in terms of the Lyapunov vectors with positive exponents that define the unstable subspace has long been recognised, the introduction of these concepts and the use of Lyapunov vectors in the state estimation problem and data assimilation algorithms is more recent. Starting with the paper by Trevisan and Uboldi (2004), a number of studies (Uboldi and Trevisan, 2006; Carrassi et al., 2007, 2008a, b; Palatella et al., 2013) have shown that, by confining the Assimilation of observations in the Unstable Subspace (AUS) of the system, one can construct data assimilation algorithms that are computationally more efficient (EKF-AUS, Trevisan and Palatella, 2011b) and in some cases more accurate (4DVar-AUS, Trevisan et al., 2010) than their classical counterparts: the extended Kalman filter (EKF) and four-dimensional variational assimilation (4DVar). A discussion of the role of instabilities in weather forecasting and data assimilation is found in Trevisan and Palatella (2011a).

For geophysical models, the huge number of degrees of freedom constitutes the main difficulty whenever a description of the error probability distribution is attempted, in particular in the application of data assimilation algorithms. During the forecast stage, errors grow in the unstable subspace, which can be of a much lower dimension than the full phase-space dimension of the model: the algorithms that perform the assimilation in the unstable subspace take full advantage of the reduced dimension in which they operate. In high-resolution non-hydrostatic models, the presence of a large number of fast instabilities and spatial scales as small as convective cells makes the problem of high dimensionality particularly critical.

One of the goals of the present paper is to investigate, in a realistic high-resolution, convection-resolving model, what the structure of the forecast error is in relation to the unstable modes on various scales that can be estimated by breeding. One question we try to address is the following: in line of principle, can an assimilation algorithm keep under control a reasonable fraction of the errors that grow on such a wide spectrum of space scales and timescales?

2 Preliminary remarks

Limited-area models are forced by lateral boundary conditions that represent an error source in addition to initial state errors and to errors intrinsic to the model dynamics (model error), which are present in all operational systems. Moreover, initial and model errors evolve in a convoluted manner (a relevant subject in recent studies, e.g. Nicolis et al., 2009; Carrassi and Vannitsem, 2011).

The goal of this work is, however, to characterise the error growth on various dynamical scales. This is obtained by studying the evolution and growth of errors that are only introduced at initial time: both model and boundary conditions are then "perfect", i.e. they are the same in the "true" trajectory and in all, control and perturbed, trajectories. The reason for this choice is that the introduction of model and/or boundary error would obscure the interpretation of results: it is important not to attribute to newly introduced errors any behaviour which is only and specifically due to error dynamical evolution. At the same time, we feel the need to caution the reader against too optimistic an interpretation of any results obtained in the absence of model and boundary errors.

A reader not particularly familiar with the perturbation breeding technique, besides the original papers (Toth and Kalnay, 1993, 1997), may refer to the book by Kalnay (2003), where the relation between bred and Lyapunov vectors is discussed; see also Trevisan and Pancotti (1998), Legras and Vautard (1996), and Wolfe and Samelson (2007). A classical method to compute Lyapunov exponents and vectors is to make use of frequent Gram–Schmidt re-orthogonalisation of small perturbations (Benettin et al., 1980). The review by Trevisan and Palatella (2011a) discusses the role of unstable Lyapunov vectors in atmospheric predictability and data assimilation. Finally, the perturbative equations relevant for breeding in both an unforced dynamical system and in a system forced by the assimilation of observations are presented in Sect. 2.1 of the paper by Uboldi and Trevisan (2006).

Although throughout this work we broadly associate fast growth with small scales and slow growth with large scales, this should not be intended as a very rigid spatial characterisation of error and perturbation fields, especially in the non-linear regime of growth.

3 Model and case study

The model is MOLOCH, developed at CNR-ISAC, Bologna, Italy (Buzzi et al., 2004; Zampieri et al., 2005; Malguzzi et al., 2006; Davolio et al., 2006, 2009a)[1]. MOLOCH is a convection-resolving, non-hydrostatic atmospheric model, running at about 2.2 km resolution with 50 hybrid vertical levels. State variables include temperature, pressure, three velocity components, specific humidity, and concentrations of cloud water and ice and of three precipitation phases: rain, snow, and hail (plus temperature and water content at five ground levels). The domain includes the Alps, northern Italy and Corsica, with portions of the Ligurian, Tyrrhenian and Adriatic seas. The reference trajectory is a simulation of the real case of 26 September 2006 (Davolio et al., 2009b), with initial and boundary conditions from GFS (Global Forecasting System)[2] and BOLAM (Buzzi et al., 1998, 2004; Zampieri et al., 2005; Malguzzi et al., 2006). In this case study, the circulation at 500 hPa is characterised by a deep trough west of the Alps. Near the surface, a pressure low is located in the Genoa Gulf, and strong winds from the southeast over the Adriatic Sea. The Venice area is hit by intense convective precipitation, mostly during the morning. Four main episodes can be identified in the model trajectories used in this work. (1) 00:00–05:00 UTC: scattered convection over the Po Plain and the Alps; (2) 05:00–09:00 UTC: no convection in the plain, orographic convection only; (3) 09:00–12:00 UTC: organised convection associated with a mesoscale front with intense precipitation, mainly concentrated on the Alps, the eastern Po Plain (Venice area) and the northern Adriatic Sea; (4) 12:00–18:00 UTC: various convergence areas with intense winds and precipitation progressively extending to the whole domain.

A BOLAM trajectory provides boundary conditions every hour from 12:00 UTC of 25 September 2006 to 18:00 UTC on 26 September 2006. Initial conditions for MOLOCH can be chosen among these BOLAM fields.

4 Characterising the fastest instabilities: an extension of previous results

The results reported in this section regard the fastest instabilities associated with the smallest dynamical scales resolved by the model. Results presented in this section were obtained by simulations performed in a smaller domain, not including Corsica and the Western Alps (Uboldi, 2010; Uboldi et al., 2010), and the integrations only covered the time interval between 00:00 and 12:00 UTC on 26 September. The timescales that characterise the growth rate and saturation of instabilities associated with convection in the 26 September case are investigated in the line of Hohenegger and Schär (2007a).

[1] http://www.isac.cnr.it/dinamica/projects/forecasts/index.html
[2] http://www.emc.ncep.noaa.gov/index.php?branch=GFS

First, a perturbation breeding technique has been used: two initial random, independent perturbations, with very small initial amplitude, are added to the non-linear model reference trajectory at 00:00 UTC. Their rms (root mean square) value is $0.05\,\mathrm{m\,s^{-1}}$ for horizontal velocity at model level 5 (about 925 hPa over sea), with each variable scaled by its estimated variability (estimated values, 5 °C for temperature and $5\,\mathrm{m\,s^{-1}}$ for horizontal velocity, are obtained by performing an average over the control trajectory). Perturbed states evolve following the model non-linear dynamics, and perturbations are rescaled to the initial amplitude every 5 min. Bred vectors quickly get organised in spatially coherent structures and, after about 90 min, the two bred vectors (not shown here) show very similar spatial structures, localised in dynamically active areas, where intense winds and convective precipitation are located, differences only appearing in small-scale structures. The emergence of a larger-scale unstable pattern shows that the linear regime of growth is not disrupted by the strongly non-linear processes present in moist convection.

In order to estimate the linearity time, two perturbed states are obtained by adding, to the state of the reference trajectory, two perturbations with the same structure (an organised bred vector), but with opposite signs. These perturbed states are non-linearly evolved, and the spatial correlation of the perturbations (i.e. the cosine of the angle between the perturbation vectors) is used as a linearity indicator (Hohenegger and Schär, 2007a). The two perturbations, which initially (01:30 UTC) have the same direction (the cosine is -1), progressively lose their alignment as non-linearity becomes important. Conventionally, linearity is considered lost when the spatial correlation exceeds the threshold -0.25 (for random states it would be $+0.5$, Hohenegger and Schär, 2007b). The scalar product used to compute spatial correlation is extended to the whole domain, and to variables T, U and V after scaling each of them with its own variability (estimated as described above). A minor dependence on variables has been found when the spatial correlation is computed for each variable separately: global linearity appears to be lost when it is lost in the horizontal velocity field. Estimates obtained at different times along the same control trajectory, i.e. starting the perturbed trajectories, in two different experiments, at 01:30 and at 06:00 UTC (26 September 2006), indicate that the time required to lose linearity is shorter during the night ($T_{\mathrm{lin}} \simeq 2.5\,\mathrm{h}$), when the main activity is scattered convection, and longer ($T_{\mathrm{lin}} \simeq 4.0\,\mathrm{h}$) when organised convection, forced by a mesoscale front, takes place during the morning.

Recalling that the perturbed states evolve according to the full non-linear dynamics, the growth rate has been estimated by averaging during the linear growth regime, when the logarithm of the amplification factor is approximately a linear function of time (and the growth rate is approximately constant). The estimates of the doubling time are $T_{\mathrm{D}} \simeq 2.5\,\mathrm{h}$, approximately the same value as T_{lin}, during the night, and $T_{\mathrm{D}} \simeq 2.0\,\mathrm{h}$, smaller than T_{lin}, during the morning.

Comparing T_D with T_{lin} is important, because when T_{lin} is shorter than T_D, errors may reach non-linear saturation so quickly that one can hardly see a linear growth phase. In principle, a reduction of the initial error cannot guarantee a reduction of the forecast error after a time when non-linear error saturation has been reached (Hohenegger and Schär, 2007a).

By comparing T_D with T_{lin} in the present case, the morning episode of organised convection, though more intense, appears more stable and predictable than the scattered convection episode occurring during the night.

The values obtained here (for this system and case study) for the ratio T_{lin}/T_D appear a bit more promising than those obtained by Hohenegger and Schär (2007a), at least in the perspective of very short-range forecasting or nowcasting.

5 Experiments

In Sect. 4, we described perturbation growth and saturation properties with regard to the fastest-growth instabilities associated with the smallest dynamical scales. This was obtained by breeding perturbations with a very high rescaling frequency and a very small rescaling amplitude. Such an amplitude is much smaller, in fact, than the order of magnitude presently achievable in operational practice for analysis error.

With a rescaling amplitude of the same order of magnitude as the analysis error, and an appropriately lower rescaling frequency, the bred vectors do not describe the fast instabilities, which then have sufficient time to saturate between successive rescaling times.

As an example of a saturated small-scale error, the position of a particular, localised and isolated convective feature can be off by some distance in a "true" and in a "control" trajectory. As long as this offset is shorter than the spatial width of the disturbance itself, the error may grow in a linear fashion. When the offset is larger than the disturbance, though, the error is determined by the presence, in the two state fields, of two separate, non-overlapping signals of approximately the same size. As they drift away from each other, the rms state difference (any variable) does not change steadily, so that this error component does not effectively grow anymore. Saturated small-scale instabilities are not directly responsible for further error growth, but are expected to affect, by non-linear interaction, larger scales that may still be in a regime of linear growth. These large scales are mainly responsible for error growth at an error level comparable to the analysis error.

Under these premises, we perform an experiment where the error, i.e. the state difference between a trajectory representing the truth and a control run representing the forecast, has an initial amplitude (norm) compatible with the present-day analysis error. We then examine the ability of bred vectors to capture the forecast error. In order to do so and in view of the different scales that may be involved in the growth of forecast errors, a range of values for rescaling amplitude and frequency is taken into consideration. In a second parallel experiment, a hypothetical analysis error 1 order of magnitude smaller is considered.

The experimental framework is that of a twin experiment, where a model trajectory is considered to represent the "truth" and is compared with one or more control trajectories. Here, the true trajectory is a free MOLOCH evolution initiated by an external model state (from the same BOLAM trajectory providing boundary conditions) at 21:00 UTC on 25 September 2006 and integrated until 18:00 UTC on 26 September. Two different control trajectories have been considered, both computed using the same model and the same boundary conditions as the true trajectory. The first one is obtained by simply initiating a MOLOCH trajectory from the BOLAM state at 18:00 UTC on 25 September (3 h before the initial condition of the true trajectory) and integrating for 24 h, with the same boundary conditions as the "truth". At 00:00 UTC, its rms error (calculated as described in Sect. 5.1) is about 0.3, corresponding to about 0.4 °C for temperature and 1.9 m s^{-1} for horizontal velocity at level 8 (about 1500 hPa over sea), and 0.25 °C, 1.8 m s^{-1} at level 25 (about 500 hPa). This first control trajectory, labelled 18H, is used for breeding vectors meant to estimate instabilities on a larger scale. The second control trajectory has been obtained from the first one, by rescaling by 0.1 all components of its error vector at 21:00 UTC on 25 September in order to have a smaller initial error and again using the same boundary conditions. This second control trajectory, labelled R21, is used to construct bred vectors intended to estimate small-scale, fast instabilities (still slower, though, than those studied in Sect. 4).

In any case, no comparison is done before 00:00 UTC on 26 September, so that all trajectories have time enough to develop the dynamical scales typical of a MOLOCH non-hydrostatic, convection-resolving simulation. The experiments described in Sect. 4 suggest waiting an initial 3 h period for both control and "true" states to approach the system attractor and for the bred vectors to acquire structures representative of the unstable directions of the system. Note that the bred vectors are initiated as random noise (Sect. 5.2) at 21:00 UTC on 25 September 2006 and that, at this very initial time, the error is due to features present in the external model field. Instead, after 00:00 UTC on 26 September, forecast error and bred vector structures are those typical of the convection-resolving model dynamics, making it possible and significant to compare them.

The procedure used in the experiments is very similar to what is often done in operational practice, when the state used as the initial condition for a non-hydrostatic model is taken from a hydrostatic model that also provides boundary conditions. This procedure implies the absence, in the initial control state, of small and fast scales typical of a convection-resolving model: these dynamical scales, present in reality, are also present in our experiments, as we gave enough integration time for them to develop.

5.1 Norm and scalar product definition

For both control trajectories, 18H and R21, the error vector is orthogonally projected onto orthonormalised bred vectors, and the square error component, which is a function of the number of bred vectors used, is expressed as a percentage of the square norm of the whole error vector (Sects. 5.2 and 6.2). In order to orthogonalise the bred vectors and to compute the error orthogonal projection, the definition of a scalar product is needed. Our choice, appearing as the most natural, is the sum of component products extended to the whole three-dimensional domain (actually vertical levels 5–45 only) for temperature and horizontal velocity, each variable normalised with its own variability (Sect. 4). This is called the TUV scalar product hereafter. The norm of a vector is by definition the square root of the scalar product of the vector with itself. The TUV norm is then dimensionless; it can be interpreted as a fraction of the overall state variability. Since the state vector is composed of many other variables besides T, U, and V (pressure, specific humidity, vertical velocity, concentration of condensed water phases, etc.), this is more properly a seminorm: it is considered sufficient, though, to evaluate errors and state differences, because the chosen variables have the most direct impact on the dynamics. The error size is defined as the TUV norm of the vector difference between the control and true states.

5.2 Breeding and orthonormalisation procedure

A breeding scheme is characterised by a rescaling period and amplitude. The rescaling period is set to 30 min for trajectory 18H (larger initial error), and, to estimate faster instabilities, to 15 min for trajectory R21 (smaller initial error). Several different values of the rescaling amplitude have been tested for each trajectory. Boundary conditions are the same for all perturbed and unperturbed integrations.

Bred vectors are initiated ("seeded") by adding perturbations to the control state (either that of trajectory 18H or that of trajectory R21) at 21:00 UTC on 25 September. Note that all initial perturbations are randomly generated point by point in the three-dimensional domain on variables T, U, and V, so that they do not have a particular spatial structure at this time. Only their amplitude (TUV norm) is imposed to be the same used for rescaling (see below). The spatial structure that bred vectors have at later times (when comparisons are performed) is the result of the breeding technique, combining model evolution and rescaling.

Every three hours, all bred vectors are Gram–Schmidt orthonormalised with the TUV scalar product; the error, i.e. the vector difference between control and true state, is then projected onto the orthonormal bred vectors. The ratio between the sum of square error projections and the square norm of

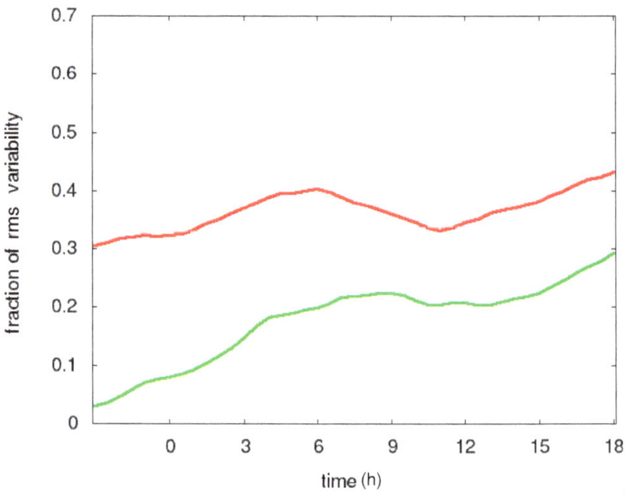

Figure 1. The rms forecast error of control trajectories with different initial errors. Times indicated on the x axis correspond to UTC times of 26 September 2006. Values on the y axis are non-dimensional (TUV norm defined in the text) and are expressed as fractions of natural variability. Red: trajectory labelled 18H (actually starting at 18:00 UTC on 25 September), which at 21:00 UTC has error 0.3, of the order of the present-day analysis error. Green: trajectory labelled R21, obtained from the previous one by rescaling, at 21:00 UTC on 25 September, its error to 0.03, i.e. 1 order of magnitude smaller than the present-day analysis error

the whole error is calculated and expressed as a percentage: this is an increasing function of the number of bred vectors used (it would be 100 % for a complete orthonormal basis). A similar technique was used by Wei and Toth (2003) to compare forecast error estimates with member subspaces in an ensemble prediction system.

In each experiment, 12 bred vectors are computed (this was the maximum allowed by computer memory). A second set of 12 bred vectors was computed in some cases, to assess the dependence on increasing subspace dimension of the projected error component.

For trajectory 18H, having fixed the rescaling period to 30 min, five different rescaling amplitudes have been tested, each in a different run: 0.20, 0.30, 0.36, 0.40, and 0.50 (rms TUV norm values). For R21, with rescaling period 15 min, six rescaling amplitudes have been tested: 0.033, 0.067, 0.100, 0.133, 0.167, and 0.200 (these values are still 1 order of magnitude larger than those used for the fastest instabilities of Sect. 4). In both cases, the chosen amplitude values span the range of variability of the corresponding control trajectory error, as can be seen in Fig. 1.

6 Results

6.1 Multiple-scale instabilities in the forecast error and in bred vector subspaces

As stated in the introduction, one of the aims of the present work is to extract information about the instabilities at play from the inspection of bred vectors and their associated growth rates. A second related question is whether the forecast error can be described by BVs in a model such as the present one, which includes the fast small scales typical of convection. In the case of a positive answer, the algorithms developed for ensemble weather forecasting and ensemble data assimilation could be successfully used also in the context of non-hydrostatic models.

We start discussing Fig. 1, which shows the forecast error as a function of time for the two control trajectories starting from initial errors with the same spatial structure, but with different amplitudes: one, R21, 1 order of magnitude smaller than the other, 18H. For both trajectories, the growth rate is positive in the initial phase, until about 05:00 UTC for trajectory 18H and 09:00 UTC for trajectory R21, approximately corresponding to episode (1) of Sect. 3, and in the last part of the simulation, from 11:00 UTC to the end, when the intense phenomena progressively extend from north-eastern Italy to almost the whole domain.

We observe that, whereas the ratio of the initial errors in the two control trajectories was initially as large as 10 (in view of the hypothetical assumption that the analysis error could be reduced by 1 order of magnitude), after 21 h, the ratio of the forecast errors has become smaller than 1.5. The error during the first 10 h in fact grows much faster in the small initial error trajectory (R21), whereas, during the second episode, the error growth of the two trajectories is very similar. This is because the growth rate is dominated by (small-scale) fast instabilities when the initial error is small (regardless of its initial spatial structure, which is the same in both cases), but, as the forecast error grows in time, these components saturate and slower instabilities associated with larger dynamical scales become dominant[3]. This is the mechanism suggested in the Lorenz (1996) paper, as well as in Toth and Kalnay (1993), that we discuss here relative to a real case study. We now will see how bred vectors can be useful for understanding this behaviour and for quantifying the characteristics of instabilities on various scales.

Figure 2 shows the "spectrum" of BV (3 h averaged) growth exponents for the two control trajectories of Fig. 1. Here, for trajectory 18H (large initial error, top panel), BVs are constructed with rescaling period 30 min and TUV amplitude 0.36; for trajectory R21 (small initial error, bottom panel), BVs are constructed with period 15 min and TUV

[3] In fact, they dominate the growth of the whole vector norm, even if small-scale signals corresponding to saturated unstable structures may still be present and, locally, some of them may even still be growing.

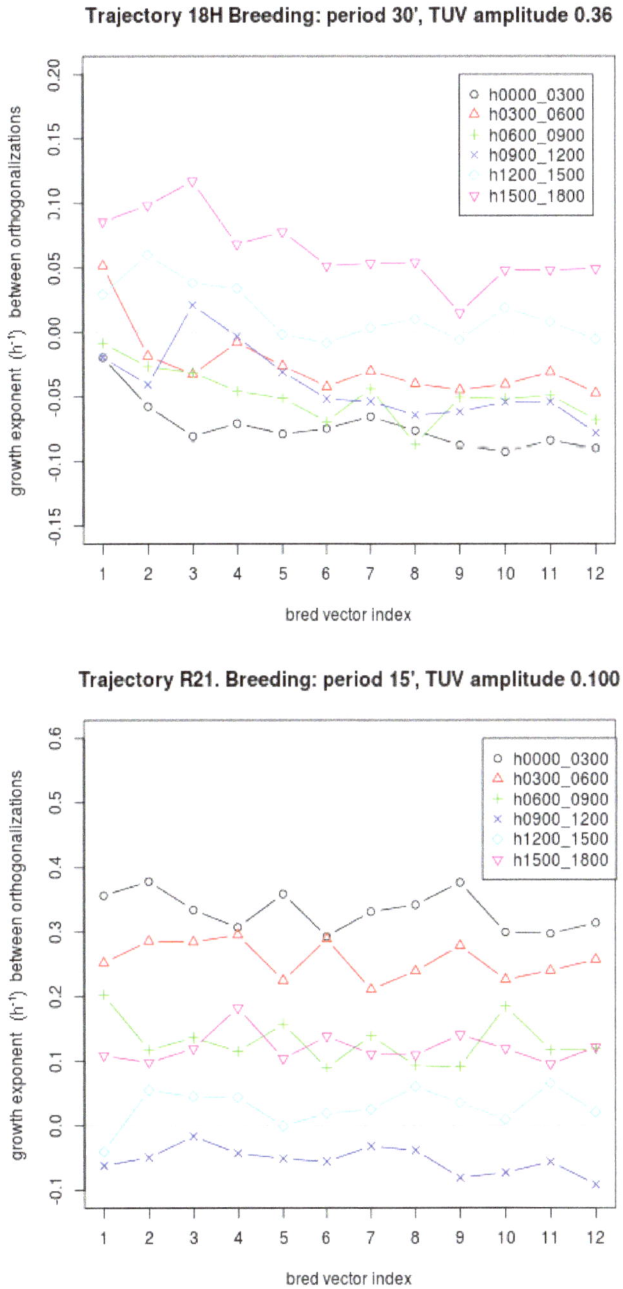

Figure 2. Growth rates of the 12 bred vectors at different forecast times, as a function of the bred vector index (the growth rate has been averaged over 3 h intervals, as indicated in the legend box). Different colours refer to different forecast times. As indicated in the panel titles, the BVs are constructed differently for trajectory 18H (large initial error, top panel) and for trajectory R21 (small initial error, bottom panel).

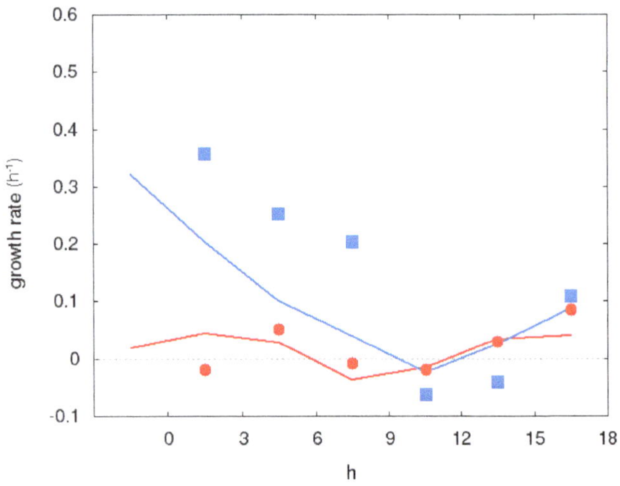

Figure 3. Growth rates of forecast errors for the two control trajectories of Fig. 1 and the corresponding leading bred vector. Times indicated on the x axis correspond to UTC times on 26 September 2006. Breeding parameters: period 30 min, amplitude 0.36 (red dots) for trajectory 18H (large initial error, red line); period 15 min, amplitude 0.100 (blue squares) for trajectory R21 (small initial error, blue line).

amplitude 0.100. In Fig. 2, different colours refer to different forecast times.

The time variability of the exponents reflects the time variability of the forecast errors: BV growth rates are larger when the forecast error increases, during the two periods of intense convection activity. The number of BVs with a positive growth rate at each time provides some indication of how complex the structure of the error is, i.e. how many directions are unstable and possibly useful for describing errors on the various scales present in the forecast error. We observe that in the set of BVs that has been constructed for the control trajectory with larger initial error (18H) and meant to describe larger-scale errors present in the forecast, the growth rate generally decreases with the BV index. In general, these BVs are more active (have larger growth rates) at the end of the integration period, when slower scales become dominant, and less active at the beginning. There are not many BVs with a positive growth rate in this set, except during the second episode of error growth.

In contrast, all the BVs of the set for the control trajectory with small initial error (R21) have large growth rates during the entire period, except for a time interval between the error growth episodes; furthermore, the growth rate in this set of BVs does not decrease at all with the BV index, indicating that a large number of directions are present in the small-scale unstable subspace. In Sect. 6.2, it is shown that this number is much larger than 12.

Figure 3 shows the growth rate of the leading bred vector of each set together with the growth rate of the forecast error shown in Fig. 1. The close correspondence in the time

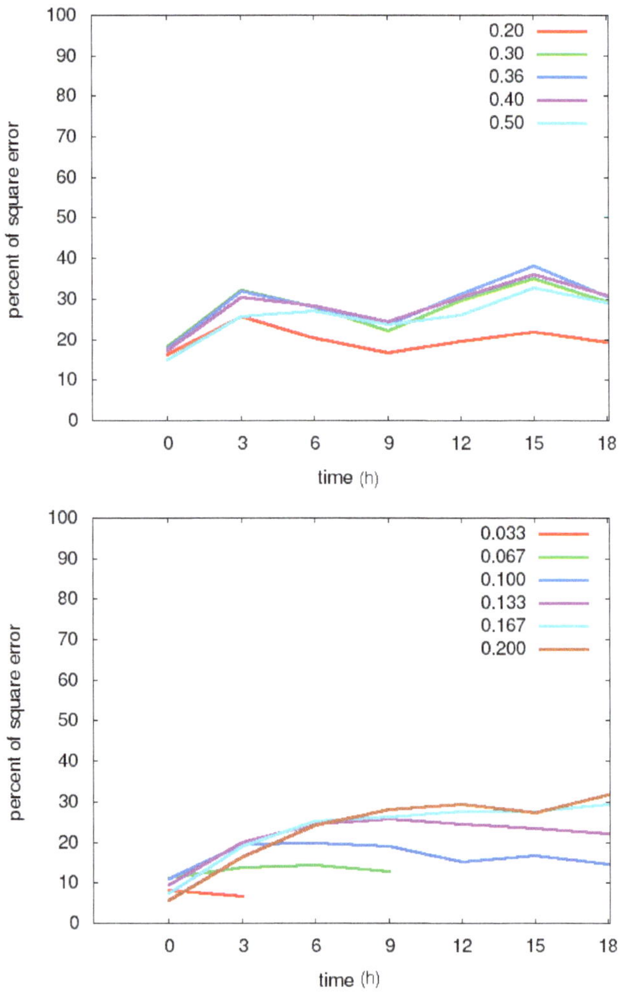

Figure 4. Square norm of forecast error orthogonal projection onto the subspace of the 12 leading bred vectors (percentage of error square norm) as a function of time. Different curves in each panel correspond to different rescaling norms. Top panel: control trajectory 18H (large initial error), rescaling period 30 min, rescaling rms TUV norms from 0.20 to 0.50. Bottom panel: control trajectory R21 (small initial error), rescaling period 15 min, rescaling rms TUV norms from 0.033 to 0.200. Times indicated on the x axis correspond to UTC times on 26 September 2006. The rescaling norm range in each panel should be compared with the corresponding control trajectory error in Fig. 1. The same data are plotted as a function of rescaling amplitude in Fig. 5.

evolution of the growth rate for the forecast error and the associated BV suggests that the BVs embed the same instabilities that are active in the same time interval in the forecast error. Only in the first stages of the trajectory with small initial error (R21) do the fast instabilities dominate, whereas, towards the end of the forecast period, when they have saturated, only those associated with larger scales, with smaller growth rates, contribute to the error in both experiments; the values of the growth rates at this stage are all very close to each other.

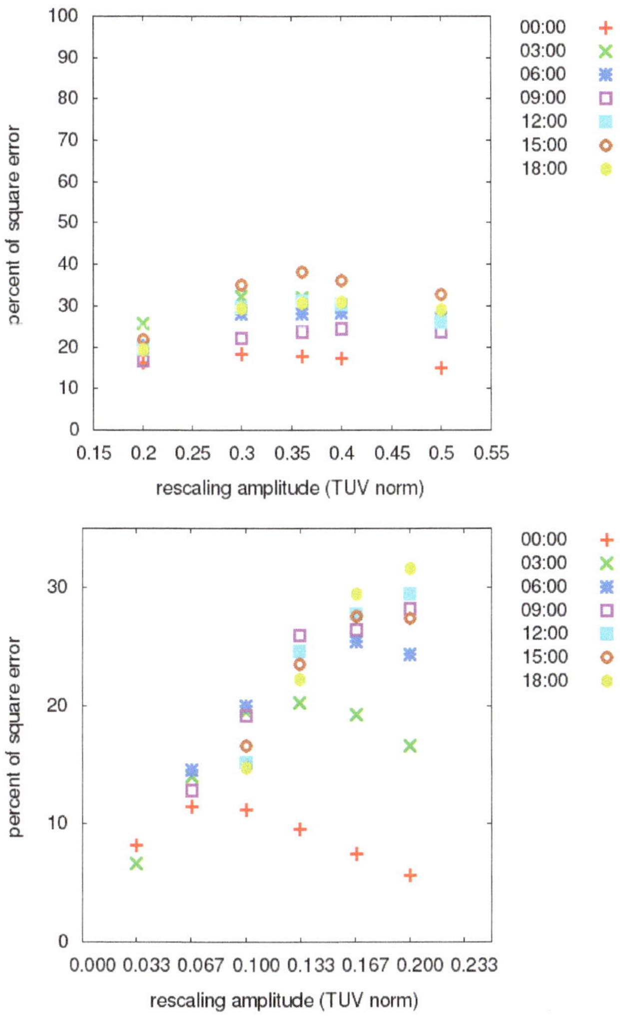

Figure 5. Square norm of error orthogonal projection onto the subspace of the first 12 bred vectors (percentage of error square norm), as in Fig. 4, but plotted here as a function of rescaling amplitude. Different symbols are used in each panel to mark different times (UTC, 26 September 2006). Top panel: control trajectory 18H (large initial error), longer rescaling period 30 min. Bottom panel: control trajectory R21 (small initial error), shorter rescaling period 15 min. The same data are plotted as a function of lead time in Fig. 4.

In Sect. 6.2 we examine in more detail whether the forecast error of the two trajectories can be described in terms of the two sets of BVs.

6.2 Forecast error component in orthonormalised bred vector subspaces

As described in Sect. 5, we performed the orthogonal projection of the forecast error on the subsets of BVs that were constructed for each of the two control trajectories. The breeding parameters were varied and we explored a wide range of values. Results are shown for several values of the amplitude and the appropriate renormalisation frequency.

Figures 4–6 show these results. In all these figures, the top panel refers to BVs constructed with rescaling period 30 min for control trajectory 18H, while the bottom panel refers to BVs constructed with rescaling period 15 min for control trajectory R21. Notice that, here, percentages of the square error are shown, rather than rms values.

The square norm of the error orthogonal projection onto the 12-bred-vector subset is shown as a function of simulation time in Fig. 4, where different colours correspond to different rescaling amplitudes, and as a function of rescaling amplitude in Fig. 5, where different symbols indicate different simulation times (rms rescaling amplitude values on the x axis may be compared with the range of values spanned by the error evolution for the two trajectories in Fig. 1).

For the large initial error trajectory 18H, as is evident in the top panel of Fig. 5, the largest error component at each time (i.e. for each set of symbols) is almost always obtained for the "optimal" rescaling amplitude 0.36. This value corresponds to the average of TUV error values of trajectory 18H at orthonormalisation times.

On the contrary, as can be seen in the bottom panels of both Fig. 4 and Fig. 5, for the small initial error trajectory R21, the maximum error component is obtained as time progresses for an increasing rescaling amplitude. As a consequence, no "optimal value" is found in this case. The maximum error component is found for smaller rescaling amplitudes at the beginning of the simulation, and for larger rescaling amplitudes at later times. In other words, bred vectors corresponding to larger and slower scales progressively become dominant for increasing time. This result is consistent both with the results of Sect. 5.1 and with the theory: small-scale unstable structures grow fast in the initial phases and, when they begin to saturate, they trigger growth at larger and slower scales, which become more important later on.

An important result is that, when the initial error is small, and as long as it remains small, as in the initial part (00:00–03:00 UTC) of trajectory R21, because of the many fast instabilities present in the first stages of error growth, only a small fraction of the forecast error is captured by the bred modes. This is evident in the bottom panels of Figs. 4 and 5.

Twelve additional bred vectors have been computed for two selected cases[4]: for control trajectory 18H with rescaling amplitudes 0.36 and period 30 min; and for control trajectory R21 with 0.100 amplitude and 15 min period. A total of 24 orthonormal bred vectors is then available in these two cases, and the square norm of the error orthogonal projection onto each set is shown as a function of the subspace dimension in Fig. 6. The top panel shows that, for trajec-

[4]Since 12 BV was the maximum allowed by computer memory in one run, it has been necessary to save on disk all states (i.e. all T, U, and V components on the whole three-dimensional grid) of the previous 12-vector set at all orthogonalisation times, and to read them from disk during the new run to perform the Gram–Schmidt orthogonalisation on the 24 vectors.

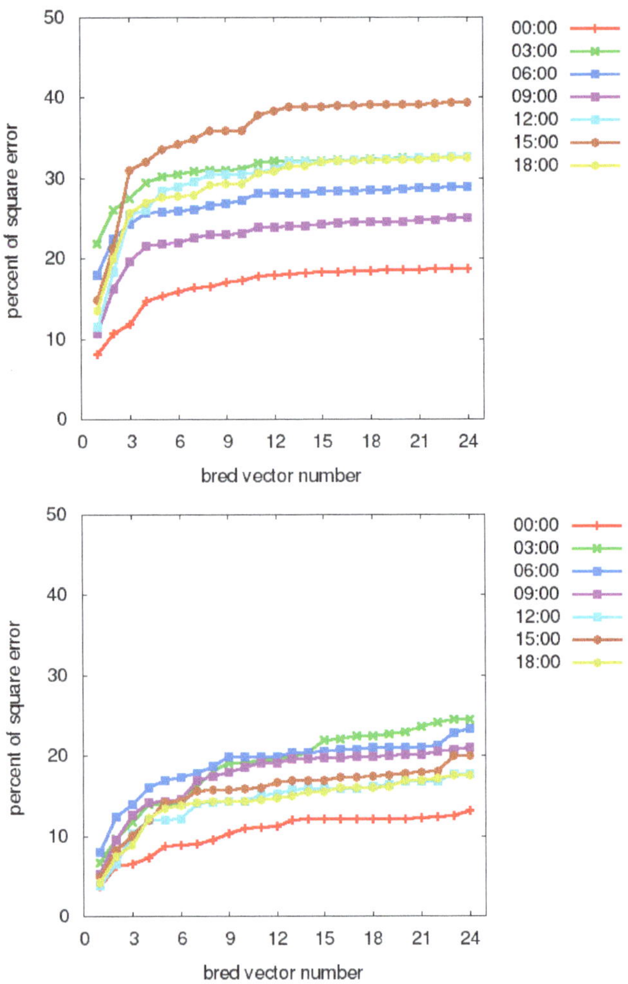

Figure 6. Square norm of error orthogonal projection onto subsets of bred vectors (percentage of error square norm) as a function of subspace dimension. Different curves in each panel correspond to different times (UTC, 26 September 2006). Top panel: control trajectory 18H (large initial error), rescaling period 30 min, rescaling rms TUV norm 0.36. Bottom panel: control trajectory R21 (small initial error), rescaling period 15 min, rescaling rms TUV norm 0.10.

tory 18H (large initial error), though very different error percentages are reached at different simulation times, increasing the number of bred vectors above 12 does not increase the projected error component in an important way. Assuming that slower growth at very large scales (that could hypothetically be found by further increasing the subspace dimension) is suppressed by boundary forcing, this means that most of the error components are not, in fact, growing when the initial error is so large. Non-growing error components can be composed of localised unstable structures which have already reached growth saturation, and, also, of wider signals, originally present in the large initial error, that are still present in areas that are dynamically rather inactive (the 18H initial state has in fact been constructed from a forecast state of the BOLAM hydrostatic model, Sect. 5). Such larger non-growing signals may appear, in the error field, as locally organised structures, which persist or drift in the domain, accounting for an important portion of the error.

Finally, in the bottom panel of Fig. 6, corresponding to trajectory R21 (small initial error), the error components appear to continue increasing, more or less steadily, with increasing subspace dimension. This means that the number of small-scale fast instabilities is large (larger than 24, at least).

7 Discussion and further developments

In a preliminary study, it was possible to evaluate growth and saturation properties of the fastest instabilities associated with the smallest (convective) dynamical scales (Sect. 4). Their doubling time, characterising linear error growth, and their tangent-linear time, characterising the duration of growth in the linear regime, were found to be comparable in magnitude, leaving, on the one hand, some theoretical possibilities to make use of some characterisation of the linear regime to control error growth. On the other hand, both these resulting characteristic times are quite short (about 2 h) with respect to lead-time lengths of interest for weather forecasting, even if, in principle, still of some interest for nowcasting (6–12 h).

A possibility that would remain open in that context would be the implementation of a sequence of dynamically consistent analysis steps, with appropriate frequency, say every hour, with the perspective of attempting control of the assimilation system trajectory.

The forecast error, built in a twin-experiment framework, has been compared with sets of orthogonalised bred vectors, and its orthogonal projection onto bred vector subspaces has been computed (Sect. 6.2).

When the initial error is comparable with the present-day analysis error, the largest forecast error component on a subspace of 12 bred vectors has been found to have a square norm of about 30 % of the whole square error norm. This is a remarkable result, considering the huge linear dimension of the state vector and the limited number of bred vectors used. Another remarkable point is, though, that when the number of orthogonal bred vectors is increased, the error component in the BVs subspace does not grow much further. Slower mesoscale instabilities, not mainly convective, which remain longer in the linear regime of growth, appear then to be controllable using a small number of independent directions. The "unexplained" 70 % of (square) forecast error appears to be distributed on saturated small-scale unstable structures and on non-growing larger-scale signals present in the error field, directly inherited from the initial state: a field of the larger-scale hydrostatic model providing boundary conditions. These last error components could be reduced or eliminated by assimilating observations, but in fact they are re-introduced each time the model is restarted from an external model field.

Starting with a smaller initial error (using a higher rescaling frequency), the "explained" square percentage of forecast error grows very slowly with the number of bred vectors. The fastest growing, small-scale instabilities (purely convective), which are also known to saturate quickly, are important when the error is initially small and appear to be active in a very large number, impossible to control within a reasonably small dimension subspace.

7.1 Error evolution after subtraction of the forecast error projection onto the bred vector subspace

In a successive experiment, a new state at 03:00 UTC is obtained by taking the vector difference between the 18H control state and its error projection onto the twelve-dimensional BV subspace. In this way, the square error of the new state is 30 % smaller than before.

Figure 7 shows the maps of the 18H control state error (top panel) at 03:00 UTC, of its orthogonal projection onto the 12 BV subspace (middle panel), and of their difference, the residual error of the new state (bottom panel). Comparing the top and bottom panels, it is evident that many structures disappear. The field in the middle panel shows which error components are detected by the 12 BV together. Undetected error components include many very localised signals, but also a few larger structures (for example, one located in the northern Apennines).

The (rms) error evolution of the new model trajectory restarted from this new state is shown in Fig. 8 together with, for comparison, the error in the 18H control forecast of Fig. 1. The error appears to grow faster than before: the new trajectory error reaches, in about 3 h, the level of the unperturbed control trajectory, even slightly exceeding it before settling on a similar behaviour.

After restart, the rms error goes in 3 h from about 0.3 to about 0.4: this corresponds to a doubling time of about 7 h, much longer than that estimated in Sect. 4 for the fastest-growing instabilities, but similar to that of the initial part of control trajectory R21 started from a small initial error (Fig. 1). The error reduction obtained by subtracting the error projection onto the BV subspace brings some small-scale error features below their saturation level: in this condition, fast error growth can be fueled by strong instabilities.

We think that we can summarise all this by drawing the following picture.

Perturbations that grow fast in a non-hydrostatic, convection-resolving system are associated with small dynamical scales, typical of convection. These fast-growing instabilities may be present in a large number and quickly reach linear growth saturation, and they do this at an error level smaller than the present-day analysis error. An important portion of larger-scale instabilities is effectively controlled by boundary forcing. Between these "extremes", there exists

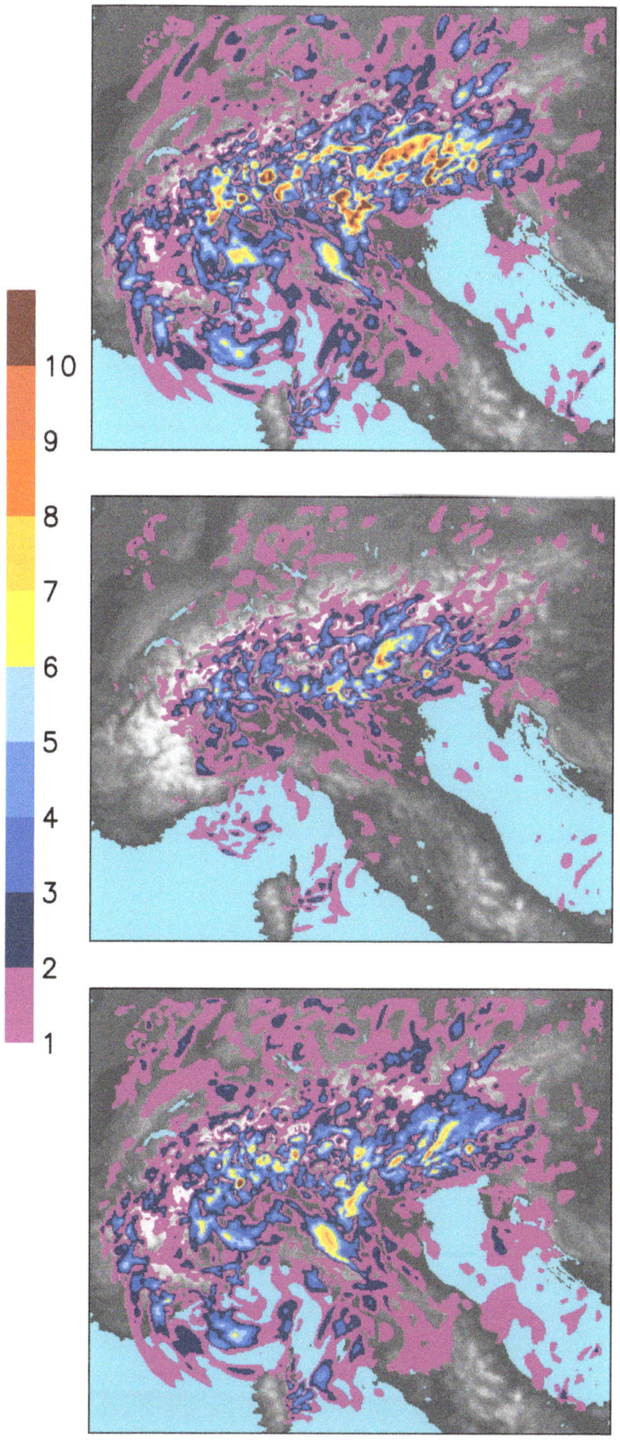

Figure 7. Maps of error at 03:00 UTC on horizontal velocity (norm of two-dimensional vector differences, m s^{-1}). Top: error of control state 18H (trajectory started with large initial error). Middle: projection of the error onto the 12 BV subspace. Bottom: difference between the two above fields, i.e. error of the restart state of Fig. 8.

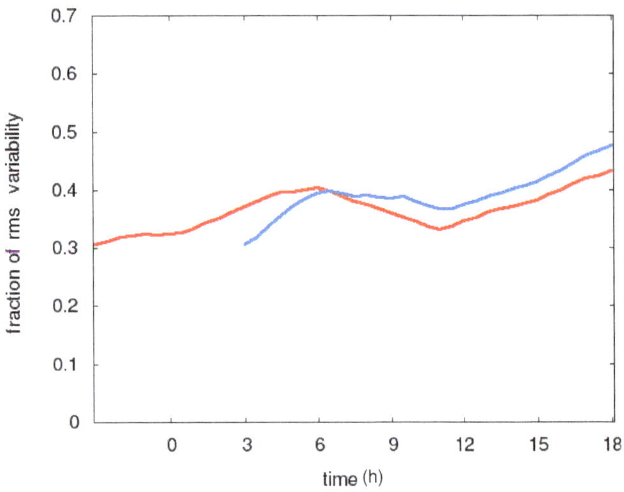

Figure 8. Red: rms error evolution of the 18H control trajectory, same as in Fig. 1. Blue: error of the trajectory restarted after subtraction of the orthogonal error component on the 12 bred vectors' subspace. Times indicated on the x axis correspond to UTC times on 26 September 2006.

some linear growth, so that part of the error can be estimated by breeding (30 % of its square norm in this case) and, in principle, could be eliminated by assimilating observations. However, after subtracting the BV-estimated component, the growth of fast instabilities is reactivated and rapidly brings the error to its previous level in the subsequent free evolution. This means that it would be necessary to assimilate observations very frequently, with an interval of the order of 1 h in the forecast-analysis cycle process, in order at least to control the error growth of slower instabilities. Moreover, setting up such a forecast-analysis cycle process might in principle make unnecessary the periodical restart from an external model field (as is customary in the operational practice), thus avoiding the periodical re-introduction of errors due to the different nature of the external model solution.

8 Conclusions

In this paper, we have investigated the properties of error evolution in the forecast of a real case study with a convection-resolving model. We use the model as a laboratory where we assume that a model trajectory represents the truth and we can vary the amplitude and structure of the initial condition error. We produce forecasts that start from two hypothetical analyses, one with an error comparable to and one with an error ten times smaller than the present-day analysis error. In this way, it has been possible to quantify the properties of instabilities with different space scales and timescales that are representative of the evolution of the forecast error. According to the theory, when the analysis and the ensuing forecast are so accurate as to reproduce the details of the single convective cells, the error growth rate is that typical of this phenomenon, while, as the error increases and convective-scale instabilities saturate, the error growth is due to slower instabilities. This result is based on the theory of Lyapunov exponents and vectors that characterise the unstable, neutral and stable subspace of the system. It is impossible to compute the entire set of Lyapunov vectors with a positive growth rate that would give us all the information on the instabilities characterising the linear growth regime of the non-hydrostatic model. To quantify the instabilities, we then used the breeding technique, which, moreover, enables selection of those that are relevant for forecast errors of a given typical amplitude. We tuned the breeding parameters and used different rescaling amplitudes and frequencies to adapt them to the assumed initial condition error. The results can be summarised as follows. When the initial error is comparable to the present-day analysis error, the unstable subspace, as estimated with the breeding technique, has the following properties: the growth rate always decreases with the BV index, and the number of BVs that are active during the convective episodes is not extremely large: just one or two in this case, during the night and the morning; only in the afternoon does their number increase, to 8 after 12:00 UTC and to 12 (at least) after 15:00 UTC (Fig. 2), in agreement with the flow-dependent instabilities evident in the forecast error evolution (Fig. 1). By projecting the forecast error onto this subspace, we find that most of the forecast error projection is onto the leading BVs, and it does not significantly increase when the number of BVs is raised from 12 to 24. Typical values of growth rates for the leading BVs are about 0.05–0.07 h^{-1}, corresponding to doubling times of 10–14 h.

When we decrease the analysis error by 1 order of magnitude, we find a completely different scenario. The spectrum of BVs is flat: there is a large number of BVs with competitive growth rates, three or four times larger than those found in the previous case (large initial error), corresponding to doubling times of 2–6 h. In addition, the projection of the forecast error onto the leading 24 BVs is very small and continues to increase very slowly and steadily, suggesting the need for a very large number of BVs to account for even a small fraction of the forecast error. This makes us believe that the unstable subspace on the convective scale really has a very large dimension. The conclusion is that as we attempt to decrease the analysis error significantly (e.g. by improving on the quality of observations and assimilation schemes), the need to increase the number of members in an ensemble forecast or an ensemble-based assimilation scheme may very well turn out to be so computationally demanding as to become an insurmountable obstacle.

The situation appears to be different for errors typical of present-day analysis errors, when the instabilities at play are not growing so fast, because the fast growth of errors on the small convective scales is saturated. In this case, we have seen that a non-negligible fraction of the forecast error is captured by the BVs.

We performed an experiment which can be interpreted as a preliminary test of the expected performance of an assimilation in the present situation. We assumed a hypothetical assimilation scheme, able to subtract exactly from the forecast error its projection onto the BV subspace (once assigned the subspace where the analysis increment is confined, this would be an upper limit to scheme performance and observation quality), and looked at the forecast error of the subsequent free evolution (i.e. after this single "assimilation" step). A relevant error reduction is obtained (in this perfect-model, perfect-boundary framework): subtracting 30 % of square norm means reducing the rms "analysis" error to about 84 % of the rms forecast error. However, after restart, the fast instabilities again take over and the forecast error recovers its previous levels in a matter of 3 h. This means that a frequent assimilation, every hour or so, would be required to control the large-scale, slower instabilities, but rapid growth should still be expected after each analysis, due to the action of small-scale fast instabilities. Their large number makes the possibility of controlling them by data assimilation (by means of a hypothetical "multiple-scale" breeding or ensemble) particularly challenging.

Acknowledgements. We wish to thank Silvio Davolio, Piero Malguzzi, and Oxana Drofa for their support in using the MOLOCH model and Andrea Buzzi for the partial financial support of this work under project PROSA of the Italian Space Agency (ASI).

Edited by: Z. Toth

References

Baldauf, M., Seifert, A., Förstner, J., Majewski, D., Raschendorfer, M., and Reinhardt, T.: Operational convective-scale numerical weather prediction with the COSMO model: description and sensitivities, Mon. Weather Rev., 139, 3887–3905, doi:10.1175/MWR-D-10-05013.1, 2011.

Benettin, G., Galgani, L., Giorgilli, A., and Strelcyn, J.-M.: Lyapunov characteristic exponents for smooth dynamical systems and for Hamiltonian systems; a method for computing all of them. Part 1: Theory, Meccanica, 15, 9–20, doi:10.1007/BF02128236, 1980.

Buzzi, A., Tartaglione, N., and Malguzzi, P.: Numerical simulations of the 1994 Piedmont Flood: role of orography and moist processes, Mon. Weather Rev., 126, 2369–2383, doi:10.1175/1520-0493(1998)126<2369:NSOTPF>2.0.CO;2, 1998.

Buzzi, A., Davolio, S., D'Isidoro, M., and Malguzzi, P.: The impact of resolution and of MAP reanalysis on the simulations of heavy precipitation during MAP cases, Meteorol. Z., 13, 91–97, doi:10.1127/0941-2948/2004/0013-0091, 2004.

Carrassi, A. and Vannitsem, S.: Treatment of the error due to unresolved scales in sequential data assimilation, Int. J. of Bifurcation and Chaos, 21, 3619–3626, doi:10.1142/S0218127411030775, 2011.

Carrassi, A., Trevisan, A., and Uboldi, F.: Adaptive observations and assimilation in the unstable subspace by breeding on the data-assimilation system, Tellus A, 59, 101–113, doi:10.1111/j.1600-0870.2006.00210.x, 2007.

Carrassi, A., Ghil, M., Trevisan, A., and Uboldi, F.: Data assimilation as a nonlinear dynamical systems problem: stability and convergence of the prediction-assimilation system, Chaos, 18, 023112, doi:10.1063/1.2909862, 2008a.

Carrassi, A., Trevisan, A., Descamps, L., Talagrand, O., and Uboldi, F.: Controlling instabilities along a 3DVar analysis cycle by assimilating in the unstable subspace: a comparison with the EnKF, Nonlin. Processes Geophys., 15, 503–521, doi:10.5194/npg-15-503-2008, 2008b.

Claussnitzer, A., Schartner, T., Névir, P., Stephan, K., and Cubasch, U.: The data assimilation method "Latent Heat Nudging" assessed with the Dynamic State Index, Meteorol. Z., 20, 165–172, doi:10.1127/0941-2948/2011/0220, 2011.

Davolio, S., Buzzi, A., and Malguzzi, P.: Orographic influence on deep convection: case study and sensitivity experiments, Meteorol. Z., 15, 215–223, doi:10.1127/0941-2948/2006/0118, 2006.

Davolio, S., Buzzi, A., and Malguzzi, P.: Orographic triggering of long lived convection in three dimensions, Meteorol. Atmos. Phys., 103, 35–44, doi:10.1007/s00703-008-0332-5, 2009a.

Davolio, S., Mastrangelo, D., Miglietta, M. M., Drofa, O., Buzzi, A., and Malguzzi, P.: High resolution simulations of a flash flood near Venice, Nat. Hazards Earth Syst. Sci., 9, 1671–1678, doi:10.5194/nhess-9-1671-2009, 2009b.

Dixon, M., Li, Z., Lean, H., Roberts, N., and Ballard, S.: Impact of data assimilation on forecasting convection over the United Kingdom using a high-resolution version of the Met Office Unified Model, Mon. Weather Rev., 137, 1562–1584, doi:10.1175/2008MWR2561.1, 2009.

Hohenegger, C. and Schär, C.: Atmospheric predictability at synoptic versus cloud-resolving scales, B. Am. Meteorol. Soc., 88, 1783–1793, doi:10.1175/BAMS-88-11-1783, 2007a.

Hohenegger, C. and Schär, C.: Predictability and error growth dynamics in cloud-resolving models, J. Atmos. Sci., 64, 4467–4478, doi:10.1175/2007JAS2143.1, 2007b.

Kain, J. S., Xue, M., Coniglio, M. C., Weiss, S. J., Kong, F., Jensen, T. L., Brown, B. G., Gao, J., Brewster, K., Thomas, K. W., Wang, Y., Schwartz, C. S., and Levit, J. J.: Assessing advances in the assimilation of radar data and other mesoscale observations within a collaborative forecasting-research environment, Weather Forecast., 25, 1510–1521, doi:10.1175/2010WAF2222405.1, 2010.

Kalnay, E.: Atmospheric Modeling, Data Assimilation and Predictability, Cambridge University Press, 2003.

Kalnay, E. and Dalcher, A.: Forecasting forecast skill, Mon. Weather Rev., 115, 349–356, doi:10.1175/1520-0493(1987)115<0349:FFS>2.0.CO;2, 1987.

Legras, B. and Vautard, R.: A guide to Liapunov vectors, in: Proceedings 1995 ECMWF Seminar on Predictability, vol. 1, pp. 143–156, 1996.

Lorenz, E. N.: The predictability of a flow which possesses many scales of motion, Tellus, 21, 289–307, doi:10.1111/j.2153-3490.1969.tb00444.x, 1969.

Lorenz, E. N.: Predictability: a problem partly solved, in: Proceedings of ECMWF Seminar on Predictability, 1–18, 4–8 September 1995, Reading, UK, 1996.

Malguzzi, P., Grossi, G., Buzzi, A., Ranzi, R., and Buizza, R.: The 1966 "century" flood in Italy: a meteorological and hydrological revisitation, J. Geophys. Res., 111, D24106, doi:10.1029/2006JD007111, 2006.

Nicolis, C., Perdigao, R. A. P., and Vannitsem, S.: Dynamics of prediction errors under the combined effect of initial condition and model errors, J. Atmos. Sci., 66, 766–778, doi:10.1175/2008JAS2781.1, 2009.

Palatella, L., Carrassi, A., and Trevisan, A.: Lyapunov vectors and assimilation in the unstable subspace: theory and applications, J. Phys. A Math. Theor., 46, 254020, doi:10.1088/1751-8113/46/25/254020, 2013.

Schenkman, A. D., Xue, M., Shapiro, A., Brewster, K., and Gao, J.: The analysis and prediction of the 8–9 May 2007 Oklahoma tornadic mesoscale convective system by assimilating WSR-88D and CASA radar data using 3DVAR, Mon. Weather Rev., 139, 224–226, doi:10.1175/2010MWR3336.1, 2011.

Schwartz, C. S., Kain, J. S., Weiss, S. J., Xue, M., Bright, D. R., Kong, F., Thomas, K. W., Levit, J. J., and Coniglio, M. C.: Next-day convection-allowing WRF model guidance: a second look at 2-km versus 4-km grid spacing, Mon. Weather Rev., 137, 3351–3372, doi:10.1175/2009MWR2924.1, 2009.

Seity, Y., Brousseau, P., Malardel, S., Hello, G., Bénard, P., Bouttier, F., Lac, C., and Masson, V.: The AROME-France convective-scale operational model, Mon. Weather Rev., 139, 976–991, doi:10.1175/2010MWR3425.1, 2011.

Toth, Z. and Kalnay, E.: Ensemble Forecasting at NMC: the generation of perturbations, B. Am. Meteorol. Soc., 74, 2317–2330, doi:10.1175/1520-0477(1993)074<2317:EFANTG>2.0.CO;2, 1993.

Toth, Z. and Kalnay, E.: Ensemble forecasting at NCEP and the Breeding Method, Mon. Weather Rev., 125, 3297–3319, doi:10.1175/1520-0493(1997)125<3297:EFANAT>2.0.CO;2, 1997.

Trevisan, A. and Palatella, L.: Chaos and weather forecasting: the role of the unstable subspace in predictability and state estimation problems, Int. J. Bifurcat. Chaos, 21, 3389–3415, doi:10.1142/S0218127411030635, 2011a.

Trevisan, A. and Palatella, L.: On the Kalman Filter error covariance collapse into the unstable subspace, Nonlin. Processes Geophys., 18, 243–250, doi:10.5194/npg-18-243-2011, 2011b.

Trevisan, A. and Pancotti, F.: Periodic Orbits, Lyapunov Vectors, and Singular Vectors in the Lorenz System, J. Atmos. Sci., 55, 390–398, doi:10.1175/1520-0469(1998)055<0390:POLVAS>2.0.CO;2, 1998.

Trevisan, A. and Uboldi, F.: Assimilation of standard and targeted observations in the unstable subspace of the observation-analysis-forecast cycle system, J. Atmos. Sci., 61, 103–113, doi:10.1175/1520-0469(2004)061<0103:AOSATO>2.0.CO;2, 2004.

Trevisan, A., D'Isidoro, M., and Talagrand, O.: Four-dimensional variational assimilation in the unstable subspace and the optimal subspace dimension, Q. J. Roy. Meteor. Soc., 136, 487–496, doi:10.1002/qj.571, 2010.

Uboldi, F.: Inspecting the trajectory instabilities of a convection-resolving model, in: Poster presentation at the ECODYC workshop (Exploring COmplex DYnamics in high-dimensional Chaotic systems): from weather forecasting to oceanic flows, 25–29 January 2010, Dresden, Germany, available at: http://www.pks.mpg.de/~ecodyc10/Contributions/Uboldi.pdf (last access: 5 January 2015), 2010.

Uboldi, F. and Trevisan, A.: Detecting unstable structures and controlling error growth by assimilation of standard and adaptive observations in a primitive equation ocean model, Nonlin. Processes Geophys., 13, 67–81, doi:10.5194/npg-13-67-2006, 2006.

Uboldi, F., Trevisan, A., and Davolio, S.: Evolution and growth of perturbations in a convection-resolving model, in: Poster presentation at the HyMeX workshop, (HYdrological cycle in Mediterranean EXperiment), 8–10 June 2010, Bologna, Italy, available at: http://www.hymex.org/public/workshops/4/posters/poster_HPF38.pdf (last access: 5 January 2015), 2010.

Wei, M. and Toth, Z.: A new measure of ensemble performance: perturbation versus error correlation analysis (PECA), Mon. Weather Rev., 131, 1549–1565, doi:10.1175//1520-0493(2003)131<1549:ANMOEP>2.0.CO;2, 2003.

Weisman, M. L., Davis, C., Wang, W., Manning, K. W., and Klemp, J. B.: Experiences with 0–36-h explicit convective forecasts with the WRF-ARW model, Weather Forecast., 23, 407–437, doi:10.1175/2007WAF2007005.1, 2008.

Wolfe, C. L. and Samelson, R. M.: An efficient method for recovering Lyapunov vectors from singular vectors, Tellus A, 59, 355–366, doi:10.1111/j.1600-0870.2007.00234.x, 2007.

Zampieri, M., Malguzzi, P., and Buzzi, A.: Sensitivity of quantitative precipitation forecasts to boundary layer parameterization: a flash flood case study in the Western Mediterranean, Nat. Hazards Earth Syst. Sci., 5, 603–612, doi:10.5194/nhess-5-603-2005, 2005.

Zhang, F., Snyder, C., and Rotunno, R.: Effects of moist convection on mesoscale predictability, J. Atmos. Sci., 60, 1173–1185, doi:10.1175/1520-0469(2003)060<1173:EOMCOM>2.0.CO;2, 2003.

14

Can irregularities of solar proxies help understand quasi-biennial solar variations?

A. Shapoval[1,3,4], J. L. Le Mouël[2], M. Shnirman[1,2], and V. Courtillot[2]

[1]IEPT RAS, Profsoyuznaya str. 84/32, 117 997 Moscow, Russia
[2]IPGP, 1 rue Jussieu, 75005 Paris, France
[3]Financial University, Leningradsky pr. 49, 125 167 Moscow, Russia
[4]National Research University Higher School of Economics, 20 Myasnitskaya Ulitsa, 101 000 Moscow, Russia

Correspondence to: V. Courtillot (courtil@ipgp.fr)

Abstract. We define, calculate and analyze irregularity indices λ_{ISSN} of daily series of the International Sunspot Number ISSN as a function of increasing smoothing from $N = 162$ to 648 days. The irregularity indices λ are computed within 4-year sliding windows, with embedding dimensions $m = 1$ and 2. λ_{ISSN} displays Schwabe cycles with $\sim$ 5.5-year variations ("half Schwabe variations" HSV). The mean of λ_{ISSN} undergoes a downward step and the amplitude of its variations strongly decreases around 1930. We observe changes in the ratio R of the mean amplitude of λ peaks at solar cycle minima with respect to peaks at solar maxima as a function of date, embedding dimension and, importantly, smoothing parameter N. We identify two distinct regimes, called Q1 and Q2, defined mainly by the evolution of R as a function of N: Q1, with increasing HSV behavior and R value as N is increased, occurs before 1915–1930; and Q2, with decreasing HSV behavior and R value as N is increased, occurs after $\sim$ 1975. We attempt to account for these observations with an autoregressive (order 1) model with Poissonian noise and a mean modulated by two sine waves of periods T_1 and T_2 ($T_1 = 11$ years, and intermediate T_2 is tuned to mimic quasi-biennial oscillations QBO). The model can generate both Q1 and Q2 regimes. When $m = 1$, HSV appears in the absence of T_2 variations. When $m = 2$, Q1 occurs when T_2 variations are present, whereas Q2 occurs when T_2 variations are suppressed. We propose that the HSV behavior of the irregularity index of ISSN may be linked to the presence of strong QBO before 1915–1930, a transition and their disappearance around 1975, corresponding to a change in regime of solar activity.

1 Introduction

Regular and irregular features of solar activity reflect the behavior of the solar dynamo. Their spectrum contains low-frequency "cycles", from decadal to centennial scales, whose durations and amplitudes vary with time, and a higher frequency spectrum with much stronger irregularities, notably in the 1–3-year pseudoperiod range. The case of quasi-biennial oscillations (QBO) has been widely discussed in the recent literature (e.g., McIntosh et al., 1992; Lawrence et al., 2008; Mursula et al., 2003; Rouillard and Lockwood, 2004). The range of 1–3-year quasi-periodicities has been studied in a number of time series, using different techniques such as power spectral analysis (Rouillard and Lockwood, 2004; Valdes-Galicia et al., 1996), wavelet analysis (Kudela et al., 2002; Mursula et al. , 2003), empirical mode decomposition (Vecchio et al., 2010), or the successive approximation technique (Mavromichalaki et al., 2003). All techniques confirm the reality of these quasi-periodicities, with time-varying amplitude and "frequency".

Several papers discuss variations with periods close to 27 days (related to the Sun's rotation as seen from Earth). For instance, in an earlier paper (Le Mouël et al., 2007), we considered the series of the International Sunspot Number (ISSN; SIDC team, 2005) and magnetic *aa* index: we computed their energy for periods of around 27 days and found that this energy roughly followed the initial time series it was computed from. More detailed analysis revealed a significant increase of energy approximately two decades prior to the increase in solar activity that occurred in the 1930s. Other papers deal with the long-term evolution of short-term variations of different time series by standard wavelet analysis (Lawrence et al., 2008) or using some modification of Kolmogorov entropy (Blanter et al., 2005, 2006). These papers reveal the existence of different regimes in the long-term evolution of the high-frequency part of the spectrum (estimated locally in time).

In a previous paper (Shapoval et al., 2013), we introduced the irregularity index of a given time series as the convergence (or divergence) rate of nearby points in a certain phase space, under a "one-step" translation. In the case of low-dimensional dynamical systems, the irregularity index corresponds to the maximal Lyapunov exponent (e.g., Bergé et al., 1984). Lyapunov exponents characterize the convergence (resp. divergence) rate of infinitesimally close trajectories of a dynamical system to (resp. from) its attractor in phase space. There is a link between the magnitude of the Lyapunov exponent and the regularity of the process: the larger the exponent, the stronger the irregularities. In contrast to the maximal Lyapunov exponent, the irregularity index can be computed for shorter time series with a significant random component.

In Shapoval et al. (2013), we explored variations of the irregularity index $\lambda_m(t)$ of the daily ISSN series as a function of time for intermediate values (4–6) of the embedding dimension m: $\lambda_m(t)$ generally attains strong main maxima at ISSN minima, has secondary maxima at ISSN maxima and minima at the time of the descending and ascending phases of the Schwabe cycles. Such a pattern of "half-Schwabe cycles", with a large amplitude of λ main maxima, remained stable between 1850 and 1915, then changed to a new pattern (with significantly smaller maxima) that remained stable from 1935 to 2005. We interpreted this pattern change as an indication of a "hidden" change in the regime of solar activity, the years 1915–1935 being a transitional interval. We could reproduce the observed behavior of λ with a synthetic signal, consisting of an autoregressive process of order 1 with Poisson noise, modulated by an 11-year sine. The switch between the two regimes was obtained by a change in autocorrelation, itself linked to the lifetime of sunspots. In a second paper (Shapoval et al., 2014), we found additional evidence of the two regimes of the irregularity index using embedding dimensions from 3 up to 32. During the first regime R1, from 1850 to 1915, λ values were larger than during the second regime R2. The difference is most remarkably seen at the minima of the Schwabe cycles. The value of λ at the recent minimum between cycles 23 and 24 was found to be as large as the largest value of λ prior to 1915, and much larger than values between 1915 and 2000. This could signal a return of solar activity to regime R1. In Shapoval et al. (2014), we established that the two regimes of λ were stable with respect to the parameters used in the computation and to detrending ("decycling") of the Schwabe cycles.

In Shapoval et al. (2013, 2014), we studied the two regimes of the irregularity index with embedding dimension m between 3 and 16. In the present paper, we concentrate on the smallest values of m (1 and 2). However our analysis cannot be performed at many solar minima because the distances between nearby points in the phase space contain too many zeros. Therefore, we first preprocess the data by smoothing them over $N = 162, 324$ or 648 successive values (these numbers are chosen as multiples of 27 to suppress the influence of solar rotation on the times series).

Several authors have suggested that observed solar (magnetic) time series are generated by an (as yet) unknown low-dimensional dynamical system (see Zhang, 1996, and Sello, 2001, for a review and original results). Attempts to reconstruct the dynamical system and to use it to predict the future behavior of the time series have led, according to the reports of their authors, to reasonable medium-term predictions of the Schwabe cycle. The efficiency of different predictions is out of the scope of this paper. Pesnell's review (2012) of the prediction of ongoing cycle 24 together with Love and Rigler's (2012) and Choudhuri and Karak's (2012) finding of random walk properties exhibited by some cycle-to-cycle characteristics constitute a useful introduction to the subject. The horizon of the predictions based on chaotic models is linked to the estimates of Lyapunov exponents (Bershadskii, 2009; Zhang, 1996; Sello, 2001). In these studies, Lyapunov exponents are focused on the low-frequency part of the data spectrum, and the dynamical system is reconstructed based on at least decades of observation. In the present paper, we use the irregularity index with embedding dimensions $m = 1$ and 2 to characterize higher-frequency variations of ISSN in the period range of the QBO.

Section 2 recalls the definition of the irregularity index and previous attempts to use them in trying to characterize the solar dynamo. Section 3 illustrates further applications of the irregularity index to the Wolf number ISSN and also to the geomagnetic index aa, with results on the evolution of its higher frequency content. A simple autoregressive model is next constructed in Sect. 4, in order to try and reproduce some of the observed properties of the irregularity index and, in particular the appearance (depending on the fundamental parameters of the irregularity index and of model parameters) of half-Schwabe cycle peaks. The discussion and conclusion are given in Sect. 5.

2 Basic tools

This section recalls the definition of classical Lyapunov exponents and of the irregularity index first introduced in Shapoval et al. (2013). Further remarks that may be useful to better appreciate the characteristics of the method are given in the Appendix.

2.1 Theoretical background

2.1.1 Definition

Lyapunov exponents are well defined for dynamical systems. Let F map a m dimensional Euclidian space Ω into itself. The Lyapunov exponent λ measures the rate of exponential convergence or divergence of initially close points in a phase space under the map F:

$$\|\mathbf{J}\varepsilon\| \sim \|\varepsilon\| e^{\lambda}, \quad \varepsilon \in \Omega,$$

where $\mathbf{J}$ is the linear part (Jacobian matrix) of F, $\|\cdot\|$ is the norm in the phase space, and $\|\varepsilon\|$ is small.

Formally, we define the trajectory $U_0, U_1, U_2, \ldots,$ as

$$U_1 = F(U_0), \quad U_2 = F(U_1), \ldots,$$

for an arbitrary point U_0 of the phase space. The small distance ε_n in the neighborhood of U_n becomes $\varepsilon_{n+1} = J(U_n)\,\varepsilon_n$ under the map F. Thus,

$$\varepsilon_{n+1} = J_n\,\varepsilon_0, \quad J_n = J(U_n)\,J(U_{n-1}) \ldots J(U_0).$$

The limit,

$$\lambda = \lim_{n\to\infty} \lim_{\varepsilon_0\to 0} \frac{1}{n} \log\left(\frac{\|J_n\,\varepsilon_0\|}{\|\varepsilon_0\|}\right), \tag{1}$$

is the Lyapunov exponent. For so-called ergodic systems, the limit (Eq. 1) is the same for almost any initial point U_0 (Oseledets, 1968; Eckmann and Ruelle, 1985).

2.1.2 Reconstruction of a dynamical system

Sometimes, when an underlying dynamical system does exist but is not known, a Lyapunov exponent can still be computed for a time series $u_1, u_2, \ldots, u_L$, allowing one to reconstruct the key features of the dynamical system, i.e., the embedding dimension m and the map F (Wolf et al., 1985; Rosenstein et al., 1993; Kantz, 1994). The vectors of the phase space are supposed to be

$$U_1 = \left(u_1, u_{T+1}, \ldots, u_{(m-1)T+1}\right),$$
$$U_2 = \left(u_2, u_{T+2}, \ldots, u_{(m-1)T+2}\right), \tag{2}$$

and so on, where T is a delay. The Lyapunov exponent is computed for the map F defined on the set $\{U_i\}$ by

$$F(U_n) = U_{n+1}. \tag{3}$$

A corollary of the fundamental Takens theorem (Takens, 1981) underlies this computation: let the time series be a projection of the orbit of a dynamical system that lies on its attractor A and be dense on it. Then the Lyapunov exponents of the attractor A and of the set $\{U_i\}$ are the same for an arbitrary delay T.

2.1.3 Standard computational technique

In practice, the time series under study are always finite and noisy. Values of the delay T and of the embedding dimension m must first be selected in order to estimate the Lyapunov exponent. The delay is frequently taken to be the time of the first minimum of the autocorrelation function of the series, or that of its mutual information (Fraser and Swinney, 1986). The embedding dimension is chosen to be the minimal value m such that the map F transforms a neighborhood of each point U_i defined in Eq. (2) into a neighborhood of $F(U_i)$.

Given T and m, the Lyapunov exponent is to be inferred from the quantities

$$\log\left(\frac{\|J(U-V)\|}{\|U-V\|}\right) \tag{4}$$

for sufficiently close points U, V in the phase space Ω. Algorithms introduced by Rosenstein et al. (1993) and Kantz (1994) have been used with success in recent analyses of solar time series (Macek et al., 2006; Li and Li, 2007). In order to circumvent the rather slow computation of the Jacobian, Ding and Li (2007) use the initial nonlinear map F rather than its linearization J when computing the ratio (Eq. 4).

2.2 Definition of the irregularity index

Based on the standard technique described in Sects. 2.1.2 and 2.1.3, this section introduces a straightforward definition of the quantity computed in the paper. In order to determine the irregularity index, we relax the requirement that close points in the phase space must be remote along the time axis, contrary to what is done for the Lyapunov exponent.

2.2.1 Phase space

We consider a sliding window of L values $u_1, u_2, \ldots, u_L$, where u_i is the ith daily value of a given index, counted within the window. Given the embedding dimension m and delay T, define the vectors $\boldsymbol{U}_i$ in the phase space by Eq. (2).

2.2.2 The map

Let F be the displacement along the orbits given by Eq. (3).

2.2.3 Nearest neighbors

For each U_i, find the nearest point U_j which does not coincide with U_i. Specifically, take $j = \Psi(i)$ such that dist(U_i, U_j) = $\min_{U_i \neq U_l}$ dist(U_i, U_l), and $l = 1, 2, \ldots, L$; the distance between two vectors $\boldsymbol{U}_i$ and $\boldsymbol{U}_j$ is the square root of the sum of the squares of the differences between each vector coordinate: $\left(\Sigma_{k=1}^{m}\left(u_{(k-1)T+j} - u_{(k-1)T+i}\right)^2\right)^{1/2}$.

2.2.4 Space-close points

We next build the sequence Θ of the distances corresponding to the different[1] pairs $(U_i, U_{\Psi(i)})$, where i goes from 1 to L. Let $\tilde{L} = |\Theta|$ be the number of these pairs and d^* be the left α quantile ($\alpha \in [0, 1]$) of Θ; in other words, the pairs (U_i, $U_{\Psi(i)}$), and $i \in \{1 \ldots, L\}$ are ordered according to the distance between the two elements of each pair, so that the ordered sequence is $\{U_{i_k}, U_{\Psi(i_k)}\}$ $k = 1, \ldots, \tilde{L}$, where dist(U_{i_k}, $U_{\Psi(i_k)}$) $\leq$ dist(U_{i_k+1}, $U_{\Psi(i_k)+1}$). P is defined as the first α fraction of the ordered pairs, i.e., $P = \{(U_{i_k}, U_{\Psi(i_k)}) : k \leq \tilde{L}\alpha\}$.

2.2.5 Small distances for Eq. (2)

We enlarge P to the set $\tilde{P}$ by adding the pairs displaced along the orbit of each $(U_{i_k}, U_{\Psi(i_k)}) \in P$ until the distance between the elements in each pair becomes large enough (see below) or the end of the window is reached. Formally,

$$\tilde{P} = \bigcup_{k=1}^{\tilde{L}} \{(U_{i_k}, U_{\Psi(i_k)}), (F\,U_{i_k}, F\,U_{\Psi(i_k)}), \ldots, (F^l\,U_{i_k}, F^l\,U_{\Psi(i_k)})\}$$

such that (i) dist($F^{l'}\,U_{i_k}$, $F^{l'}\,U_{\Psi(i_k)}$) $\leq d^*, l' = 0, 1, \ldots, l$, and (ii) either $l + i_k = L$ or $l + \Psi(i_k) = L$ (the end of the window is reached) or dist ($F^{l+1}U_{i_k}$, $F^{l+1}\,U_{\Psi(i_k)}$) $> d^*$ (the distance is large enough).

2.2.6 Irregularity index

For each pair $(U, V) \in \tilde{P}$, we compute log [dist(FU, FV)/dist(U, V)] and define the irregularity index λ as the median of these numbers. The computed irregularity index is assigned to the middle of the sliding window of length L. Let g be the lag between two successive sliding windows (we use $g = L/8$; i.e., 6 months); we construct the new time series:

$$\lambda_{L/2}, \lambda_{L/2+g}, \lambda_{L/2+2g}, \ldots,$$

consisting of the irregularity indices found for sliding windows $[1, L]$, $[g+1, g+L]$, $[2g+1, 2g+L]$, and so forth. This new time series is considered as an additional solar index series.

2.3 Some specifics of the irregularity index

Many papers have aimed at reconstructing the dynamical system underlying long time series, such as the daily Wolf numbers (Spiegel and Wolf, 1987; Lawrence et al., 1995), the monthly Wolf numbers (Ruzmaikin et al., 1992; Price et al., 1992), and also some yearly series (Ostryakov and Usoskin, 1990). These series are either sufficiently long (tens of thousands of points for daily Wolf numbers) or smooth (since at least the 27-day variations are averaged). The embedding dimension for these systems is generally taken to be at least 7, and the delay T is of the order of months (see e.g., Greenkorn, 2009, for a summary table). However, smaller embedding dimensions ($4 < m < 9$) have been used by Greenkorn (2009) for daily data over one Schwabe cycle. The latter paper shows that the Lyapunov exponent is only weakly sensitive to the value of delay T as long as it remains small (a few days). The orbit corresponding to the time series mentioned above now and then returns to the same regions in the phase space. Usually, only points that are close in the phase space but far from each other on the time axis are used to estimate the Lyapunov exponent (Rosenstein et al., 1993).

In this paper, on the contrary, we do not set any limit to the distance in time of points that are close in the phase space, and the exact definition of points being "close" is adapted to the data being studied, using the α quantile of the smallest distances, as explained above. When the embedding dimension $m = 1$, the smallest positive distance between two points in the phase space is 1 because of the integer nature of ISSN. We find that at the minima of the ISSN series many points lying at distance 1 are mapped along the corresponding trajectories to points lying at exactly the same distance, so that the values of the irregularity index computed at the signal minima can be inadequate (many values of the ratio – Eq. 4 – are zero). The transition from $m = 1$ to $m = 2$ changes the properties of the exponent. Whereas we studied embedding dimensions m from 4 to 6 in Shapoval et al. (2013) and up to 32 in Shapoval et al. (2014), we concentrate here on the cases $m = 1$ and 2 that are the simplest, and because they shed light on the occurrence of the quasi-biennial variations, at the focus of the present paper.

We first smooth the ISSN daily series and investigate the properties of the irregularity index computed for different smoothing, with delay $T = 1$. We compute the irregularity index within a 4-year sliding window; the choice of this 4-year length is a compromise between two opposite requirements: first, the window must be sufficiently large to obtain a stable determination of the irregularity index; second, it should be shorter than the Schwabe cycles. We have checked that the values of the irregularity index calculated as explained are only weakly sensitive to changes of window length inside a 3–5-year interval.

3 Data analysis

Solar activity is estimated in the paper with the Wolf (ISSN, sunspot) numbers, involving the number of groups and the number of spots in each particular group. The number of groups reflects the emerging magnetic field and is an indicator of activity. The number of spots within a group depends on the magnetic field as such and also on the interaction

[1] If U_i is the nearest neighbor of U_j and U_j is the nearest neighbor of U_i then the distance dist (U_i, U_j) is considered only once.

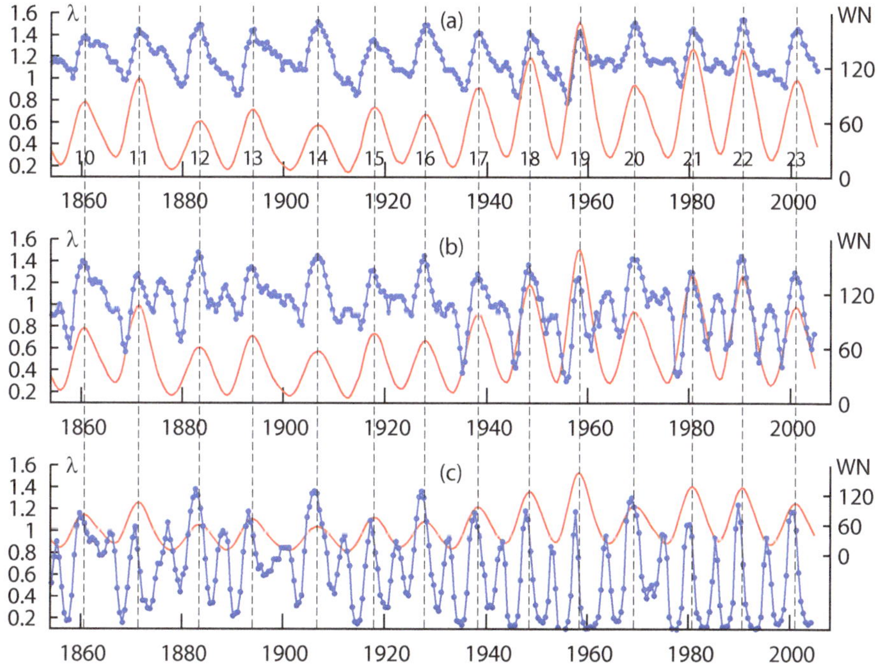

Figure 1. The irregularity index λ (blue) computed in a 4-year sliding window for the ISSN averaged over 162 (top panel), 324 (middle panel), and 648 (bottom panel) days; embedding dimension $m = 1$. The Wolf numbers averaged first over $N = 162$, 324, and 648 days, and then over 4 years are shown in red, together with solar cycle number. Dashed black vertical lines are located at the maxima of ISSN.

between the magnetic and velocity fields. In this paper, we mainly study sunspot numbers, but we also present some preliminary results on group sunspot numbers (Hoyt and Schatten, 1998) in order to check that results are not affected by the way in which ISSN is determined (including possible data heterogeneities).

The Wolf (sunspot) numbers (ISSN) are defined as $K(10G+s)$, where G is the number of sunspot groups, s is the number of individual spots, and K is a factor that is relative to the observer. Daily data series of ISSN are available from 1849 onwards (SIDC-team, 2005). We now apply our algorithm to the ISSN series $w(t)$. The time series is strongly affected by the ~ 27-day signal connected to solar rotation and reflected axis asymmetry of solar activity (Bartels, 1934; Kitchatinov and Olemskoy, 2005; Howe 2009; Le Mouël et al., 2007). Therefore the series is first smoothed in order to reduce the influence of the Sun's rotation. Namely, it is averaged over multiples of 27 (days), i.e., $N = 162$ (27×6), 324 (27×12), and 648 (27×24) days. This results in new series $\mathrm{ISSN}_{(t)} = \sum_{k=t-[N/2]}^{t-[N/2]+N} w(k)$, where $[x]$ is the integer part of x.

3.1 Case $m = 1$

The evolution of the irregularity index λ computed with $m = 1$ and the three values of N given above is shown (in blue) in Fig. 1, together with the original *ISSN* series smoothed over the same 4-year window (in red). With 162-day averaging (Fig. 1a), there is a clear one-to-one correspondence between Schwabe cycles of ISSN and λ. The maxima of the cycles coincide precisely with each other in time. The λ "Schwabe cycles" exhibit asymmetry: the rising segments are shorter and steeper than the decreasing ones. There is some structure in the decreasing segments, sometimes in the form of a secondary maximum; minima in the irregularity index cycles occur later than minima in the ISSN series.

When the data are smoothed over larger windows, oscillations with a period close to 5.5 years, i.e., half the period of the Schwabe cycle, appear (Fig. 1b, c). We call these "half-Schwabe variations" (HSV; see Shapoval et al., 2013, 2014). In the following, we use HSV to refer to the presence of irregularity maxima at solar minima (thus generating a 5.5-year quasi-periodicity), since the irregularity maxima at solar maxima are almost always present. We also sometimes refer to HSV to refer to the amplitude or amplitude changes of the irregularity peaks and their ratios (see below). In Fig. 1c, both maxima and minima of the ISSN Schwabe cycles correspond to maxima of HSV. Averaging over 324 days leads to an intermediate behavior of the irregularity index (Fig. 1b): secondary peaks at solar cycle minima appear clearly in the 1870s and 1880s and after 1950, but some are not or hardly visible at 1865, 1900 or 1975.

In order to provide a more quantitative measure of HSV behavior, we determine the ratio R of the amplitude of λ oscillations near maxima of ISSN ($\Delta_{S_{\max}}$) to that near minima of ISSN ($\delta_{S_{\min}}$; $R = \delta_{S_{\min}}/\Delta_{S_{\max}}$ in Fig. 2; the method is introduced in Shapoval et al. (2013) and further explained in the present paper). Figure 2 presents a schematic "Schwabe cycle" (smoothed artificial signal in red; actually somewhat more than one full period) and its irregularity index λ in blue. λ attains main maxima $\lambda_{S_{\max}}$ at the maxima of the (smoothed) original signal, secondary maxima $\lambda_{S_{\min}}$ at the minima of the signal, and its minima λ_{mid} on the descending and ascending phases of the signal (subscripts in this notation correspond to the *signal* not to λ itself). Three local minima occur in Fig. 2 (because somewhat more than one cycle is represented) λ^1_{mid}, λ^2_{mid}, and λ^3_{mid}. λ_{mid} is defined as their mean. Let $\Delta_{S_{\max}} = \lambda_{S_{\max}} - \lambda_{\text{mid}}$ and $\delta_{S_{\min}} = \lambda_{S_{\min}} - \lambda_{\text{mid}}$. $\Delta_{S_{\max}}$ and $\delta_{S_{\min}}$ measure the amplitude of λ peaks at signal maxima (minima) when they exist. Finally, $R = \delta_{S_{\min}}/\Delta_{S_{\max}}$ measures "HSV performance", such that a decrease of R accompanies a clearer appearance of HSV.

The quantities $\lambda_{S_{\max}}$, $\lambda_{S_{\min}}$, and λ_{mid} can be determined even if the HSV structure is subdued. $\lambda_{S_{\min}}$ is taken to be the *maximal* value of λ in a neighborhood of the *minimum* of the smoothed original signal. We extend the construction of R to several solar cycles. In such a case the quantities $\lambda_{S_{\max}}$, $\lambda_{S_{\min}}$, and λ_{mid} are obtained by averaging the corresponding quantities for all cycles included in the time window of interest.

We see in Fig. 1 that $\delta_{S_{\min}}$ increases significantly (from ~ 0.1 to ~ 0.7) when N is increased from 162 to 648 (HSV is actually hardly visible when $N = 162$ and almost as strong as peaks at solar maxima when $N = 648$). This is mainly due to the large drop of λ on descending and ascending phases as smoothing is increased. In Fig. 1b, we note differences in behavior between the period before and after $\sim$ 1930 (for instance lower overall mean value of λ and larger amplitude of HSV and Schwabe cycles after 1930).

3.2 Case $m = 2$

The above computations are repeated for $m = 2$ (Fig. 3). Different epochs can readily be distinguished. First, both the amplitude of variations and actual values of λ change around 1930, in a much more visible way than in the case $m = 1$, confirming that 1930 is a time of first-order regime change. This is particularly clear in Fig. 3b where λ drops from a mean value of about 0.3 to 0.2. This implies that the ISSN series becomes less irregular after 1930 (see discussion). In Fig. 3a (162-day averaging), HSV is clearly visible in cycles 21–23, i.e., from 1975 onwards; it is still visible from about 1915 to 1975 but is barely recognizable prior to 1915. 1915 and 1975 therefore appear as possible second-order regime changes or at least singularities. Increased smoothing strengthens HSV behavior. In Fig. 3b, with 324-day averaging, HSV is seen with varying shapes and amplitudes from 1867 to 1930 and from 1945 to 2005, with a gap at the times of cycle minima 16–17 and 17–18. In Fig. 3c, with 648-day averaging, HSV is quite clearly present from 1867 to 1915, but it is subdued from 1975 to 2005. We further note that some λ maxima at

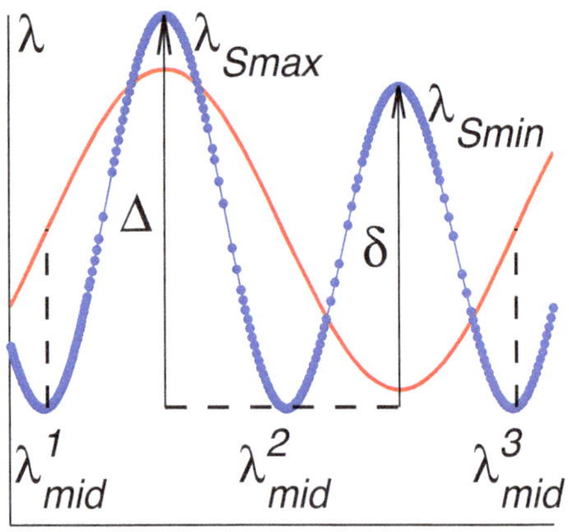

Figure 2. Construction of $R = \delta_{S_{\min}}/\Delta_{S_{\max}}$ (see text); smoothed artificial signal (red) and its irregularity index (blue); main maxima $\lambda_{S_{\max}}$, secondary maxima $\lambda_{S_{\min}}$, and local minima λ^i_{mid}. Black lines show construction of $\Delta_{S_{\max}}$ and $\delta_{S_{\min}}$.

solar minima are enhanced by increasing smoothing prior to 1915 (cycles 12–13 and 13–14), whereas others remain similar or are reduced after 1975 (cycles 21–22 and 22–23).

Using the same notations as introduced for the case $m = 1$, we see (Table 1a) that indeed $\lambda_{S_{\min}}$ is larger for the period 1867–1915 than for 1975–2008, both decreasing with increasing N, the former from 0.37 to 0.22 and the latter from 0.30 to 0.10, when N increases from 162 to 648. We see in Fig. 3 that HSV is better marked when N is increased for the period 1867–1915 and when N is decreased for the period 1975–2008. This can be expressed by the evolution of the ratio $R = \delta_{S_{\min}}/\Delta_{S_{\max}}$ of the mean amplitudes of HSV peaks at solar minima vs. solar maxima as a function of N. When N increases from 162 to 648, this ratio increases from 0.67 to 0.79 for the period 1867–1915 but decreases from 0.71 to 0.34 for 1975–2008 (Table 1a).

In summary, the λ curves shown in Figs. 1 and 3 allow us to distinguish different epochs. In Fig. 1, we see the very strong appearance of HSV with an amplification of λ peaks at solar cycle minima, and an indication of a change in behavior of $\delta_{S_{\min}}$ values as a function of N before and after $\sim$ 1930. In Fig. 3, we also see that λ decreases in both mean value and amplitude of variations in the 1930s. The R ratio increases with N for the period 1867–1915, but it decreases in 1975–2008. So, evolution of the irregularity index reveals a first-order singular date $\sim$ 1930. Finer analysis of HSV properties (R ratio evolution as a function of smoothing) reveals second-order singular dates around 1915 and 1975.

4 A model

We have already made a first attempt at constructing a model that would embody HSV behavior of irregularity in solar activity as observed in real data in Shapoval et al. (2013). We extend here the method, with slightly different choices of

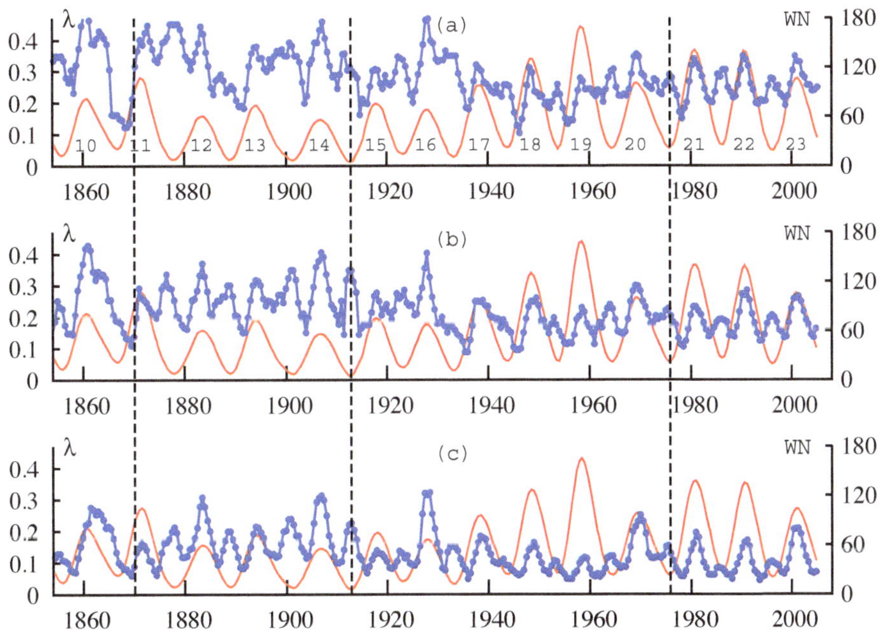

Figure 3. Blue curves : the irregularity index λ computed in 4-year sliding windows for the Wolf numbers averaged over N days; red curves : the Wolf numbers averaged over N days and then over 1461 days (4 years), where N is 162 (top panel), 324 (center panel), and 648 (bottom panel) days; $m = 2$. Dashed black vertical lines are located at times of possible regime change of λ.

relevant parameters, to concentrate on what could be related to the QBO.

4.1 Definition of the model

Consider a first-order autoregressive AR(1) process $x(t)$:

$$x(t) = a\,x(t-1) + \eta(t),$$

where the random variable $\eta(t)$ is Poissonian with mean $\mu(t)$, $P\{\eta = n\} = e^{-\mu}\mu^n/n!$. The mean $\mu(t)$ is modulated by the sum of two periodic functions (as opposed to only one in Shapoval et al., 2013) with periods T_1 and T_2, $T_1 > T_2$. The longer period T_1 is set to 11 years (11 × 365 days), corresponding to the Schwabe cycle. We choose the shorter (which we call intermediate) period T_2 in the range from 1 to 3 years, such that it includes the QBO. Therefore,

$$\mu(t) = h\left(-\cos\frac{2\pi t}{T_1} - k\cos\frac{2\pi t}{T_2} + c\right), \qquad (5)$$

where $h > 0, 0 < k < 1; c > 1 + k$ is a vertical shift. The synthetic signal $w(t)$ is defined by

$$w(t) = [M\,x(t)]. \qquad (6)$$

M is set to 10 to mimic the factor in the definition of the group sunspot number. This yields the same order of magnitude for synthetic and observed ISSN values.

The model is a function of variable t (time, in days) and depends on five adjustable parameters: a, h, T_2, k and c. The value of parameter a in the auto-regressive process determines the correlation of the data. Modeling the sunspot series by an autoregressive AR(1) model connects a to the lifetime of sunspots (Blanter et al., 2005). h controls the smoothness of the signal. The factor $k < 1$ controls the relative amplitude of the T_2 vs. T_1 modulations. The vertical shift c controls the ratio of the maximum to the minimum of λ. Figure 4 shows an example of a realization of the model with parameters given in the legend.

4.2 Modeling results

4.2.1 Appearance of HSV with data smoothing (case $m = 1$)

Already with $m = 1$ and without intermediate period modulation T_2 ($k = 0$), HSV appears in response to increasing data smoothing (Fig. 5). For $N = 162$ days, λ shows no HSV maxima at solar minima. When N is increased to 648, all λ values decrease, but their overall structure changes markedly. λ remains approximately the same at solar minima, decreases slightly at solar maxima and falls dramatically in intermediate intervals (corresponding to the ascending and descending phases of the "solar" cycle). As a result, λ now peaks sharply not only at solar maxima but also at solar minima: this behavior is indeed reminiscent of that observed for ISSN (compare Figs. 5 and 1a, c), as looked for when building the model.

Table 1. Calculation of R ratio values (see text). N is the number of days over which the ISSN data are smoothed to remove in particular the effect of solar rotation; $\lambda_{S_{min}}$ is the mean value of the maxima of the irregularity index at the times of solar minima over the period range indicated under the heading "years" (by default 1867–2008 when not indicated); $\lambda_{S_{max}}$ is the mean value of the maxima of the irregularity index at the times of solar maxima over the same period range; λ_{mid} is the mean value of the minima of the irregularity index as indicated in Fig. 2 in the text; $\delta_{S_{min}} = \lambda_{S_{min}} - \lambda_{mid}$; $\Delta_{S_{max}} = \lambda_{S_{max}} - \lambda_{mid}$; and $R = \delta_{S_{min}} / \Delta_{S_{max}}$. Each part of the table corresponds to a figure as indicated.

N	$\lambda_{S_{min}}$	$\lambda_{S_{max}}$	λ_{mid}	$\delta_{S_{min}}$	$\delta_{S_{max}}$	R	years
(a) (corresponding to Fig. 3)							
162	0.37	0.43	0.23	0.14	0.20	0.67	1867–1910
324	0.31	0.35	0.15	0.16	0.19	0.83	
648	0.22	0.25	0.10	0.12	0.15	0.79	
162	0.30	0.34	0.18	0.12	0.17	0.71	1970–2008
324	0.21	0.26	0.12	0.09	0.14	0.64	
648	0.10	0.18	0.06	0.04	0.12	0.34	
(b) (corresponding to Fig. 6)							
Fig. 6a ($T = 450$, $a = 0.8$, $h = 0.4$, $c = 1.7$, $k = 0.35$)							
162	0.34	0.48	0.30	0.03	0.17	0.19	
648	0.21	0.32	0.11	0.10	0.21	0.47	
Fig. 6b ($T = 610$, $a = 0.8$, $h = 0.4$, $c = 1.7$, $k = 0.35$)							
162	0.37	0.47	0.31	0.06	0.16	0.39	
648	0.24	0.35	0.10	0.14	0.25	0.56	
Fig. 6c ($T = 700$, $a = 0.8$, $h = 0.4$, $c = 1.7$, $k = 0.35$)							
162	0.35	0.46	0.30	0.05	0.15	0.31	
648	0.24	0.31	0.10	0.14	0.21	0.70	
Fig. 6d ($T = 800$, $a = 0.8$, $h = 0.4$, $c = 1.7$, $k = 0.35$)							
162	0.36	0.51	0.32	0.05	0.19	0.25	
648	0.18	0.34	0.10	0.08	0.23	0.35	
(c) (corresponding to Fig. 7)							
Fig. 7b ($a = 0.8$, $h = 0.4$, $k = 0$, $c = 1.2$)							
162	0.14	0.54	0.09	0.05	0.45	0.12	
648	0.09	0.30	0.05	0.04	0.25	0.16	
Fig. 7c ($a = 0.8$, $h = 0.4$, $k = 0$, $c = 1.6$)							
162	0.43	0.54	0.32	0.12	0.22	0.53	
648	0.20	0.33	0.09	0.11	0.24	0.46	
Fig. 7d ($a = 0.8$, $h = 0.4$, $k = 0$, $c = 1.7$)							
162	0.48	0.53	0.33	0.15	0.20	0.77	
648	0.27	0.30	0.10	0.17	0.20	0.83	
(d) (corresponding to Fig. 8)							
($a = 0.8$, $h = 0.8$, $k = 0$, $c = 1.7$)							
162	0.40	0.46	0.28	0.12	0.17	0.69	
648	0.15	0.26	0.07	0.08	0.18	0.43	

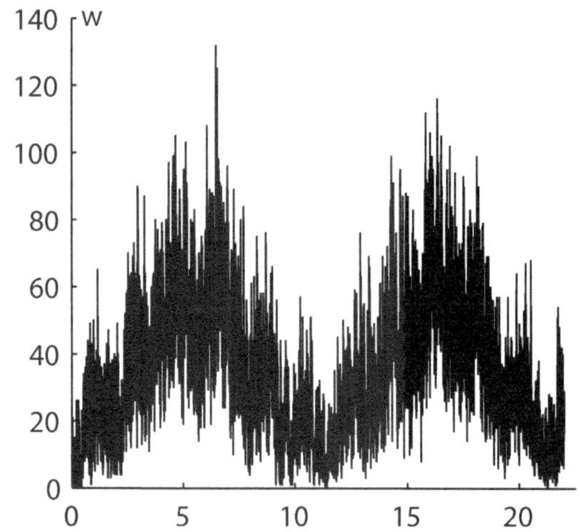

Figure 4. A realization of the AR(1) process introduced in Sect. 3.1, shown prior to smoothing by N days and over the 4-year interval over which it will next be averaged (see text). Model parameters: $a = 0.8$, $h = 0.4$, $k = 0.35$, $c = 1.7$, $T_1 = 11$ years, and $T_2 = 700$ days.

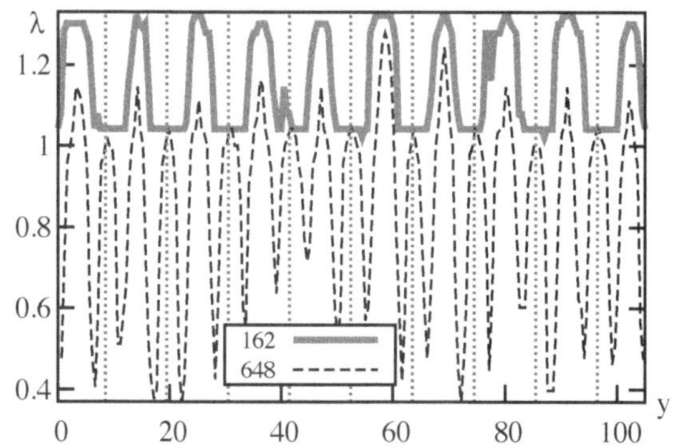

Figure 5. The irregularity index computed in 4-year sliding windows for synthetic data averaged over 162 (solid gray) and 648 (dashed black) days ($m = 1$). Model parameters: $a = 0.8$, $h = 0.4$, $k = 0.0$, and $c = 1.7$. Red dashed vertical lines at solar sunspot minima.

4.2.2 Increase in HSV behavior as a function of data smoothing (case $m = 2$)

The behavior of the irregularity index for $m = 2$ is significantly richer. We set intermediate period variations at a relatively strong level ($k = 0.35$; Fig. 6): the irregularity index exhibits HSV that increases as smoothing is increased (see Fig. 6a–c when averaging interval goes from 162 days – solid gray lines – to 648 days – dashed black lines). In the case when $T_2 = 610$, $N = 648$, strong HSV peaks are always present except in 1 out of 10 possible occurrences (at $y = 11$; Fig. 6b). On the contrary, for the same T_2 but with $N = 162$, HSV peaks are quite subdued (yet generally visible), though again only 1 out of 10 is missing (at $y = 41$; Fig. 6b). In that case, increasing data smoothing in the model results in amplifying HSV behavior: R increases from 0.39 to 0.56

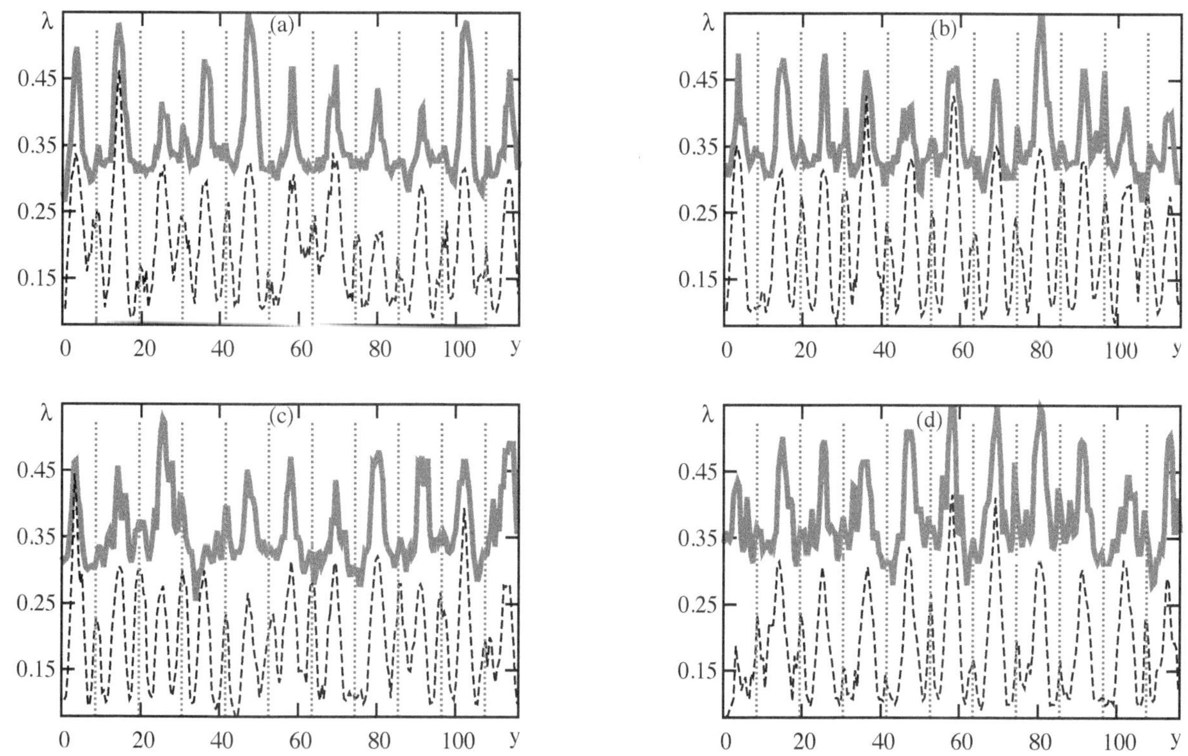

Figure 6. The irregularity index computed in 4-year sliding windows for synthetic data averaged over 162 (solid gray) and 648 (dashed black) days ($m = 2$). Model parameters: $a = 0.8$, $h = 0.4$, $c = 1.7$, $k = 0.35$, and intermediate period variation T_2 set at 450 **(a)**, 610 **(b)**, 700 **(c)**, 800 **(d)** days. Red dashed vertical lines at solar sunspot minima.

(Table 1b). HSV increases (that is R increases) with smoothing when T_2 is in the interval [450, 700] (Fig. 6a–c). The effect slowly disappears when T_2 reaches 800 (Fig. 6d).

Figure 7 illustrates the effect of changing the value of the "vertical shift" c when there is no intermediate period modulation ($k = 0$). HSV behavior becomes increasingly significant as c is increased: R grows from ~ 0.1 to 0.8 as c increases from 1.2 to 1.7 (Table 1c).

Comparing Fig. 6c (where $k = 0.35$) to Fig. 7d (where $k = 0$, $a = 0.8$, $h = 0.4$, and $c = 1.7$ in both cases), we see that HSV behavior is more visible in the case of a smaller N (162) when k is smaller (R is then respectively 0.31 vs. 0.77; Table 1c). When $k = 0$, there is little or no HSV increase (R increases from 0.77 to 0.83; Table 1c; Fig. 7b–d) whereas with $k = 0.35$ it grows significantly (from 0.31 to 0.70; Table 1b). Therefore, k is an important factor controlling HSV behavior.

4.2.3 Decrease of HSV as a function of data smoothing (case $m = 2$)

When intermediate period (T_2) variations are suppressed and parameter h (that controls the smoothness of the signal) is increased (Fig. 8), we find another regime in which HSV decreases with increasing smoothing. When N increases from 162 to 648, the R ratio decreases from 0.69 to 0.43 (Table 1d).

4.2.4 A direct comparison of model with observations

In Fig. 9, we model the data of cycles 21–23 (which have similar durations, to allow a comparison with a model where $T_1 = 11$ years). Parameter k is set to zero and a to 0.9. This choice of a reflects the increase of the lifetime of sunspots found by Blanter et al. (2005). We can now directly compare the irregularity index computed for the synthetic and actual signals.

Since the model contains a random ingredient and the computation of the irregularity index is sensitive to particular realizations, two of them are shown in Fig. 9. The irregularity index for the model follows rather precisely that for ISSN for $N = 162$ (Fig. 9, middle row panels, except at the minimum between cycles 22 and 23 in the realization on the right side). The quality of the fit is somewhat less for $N = 648$ (Fig. 9, bottom row panels). Nevertheless, the model realizations globally reproduce the λ pattern of the real data quite faithfully.

An explicit comparison (Fig. 9) of the irregularity index for the model and ISSN time series is possible for cycles 21–23 (1975–2005) because λ constructed with ISSN exhibits a smooth, quasi-cyclic and regular behavior, as is the case for

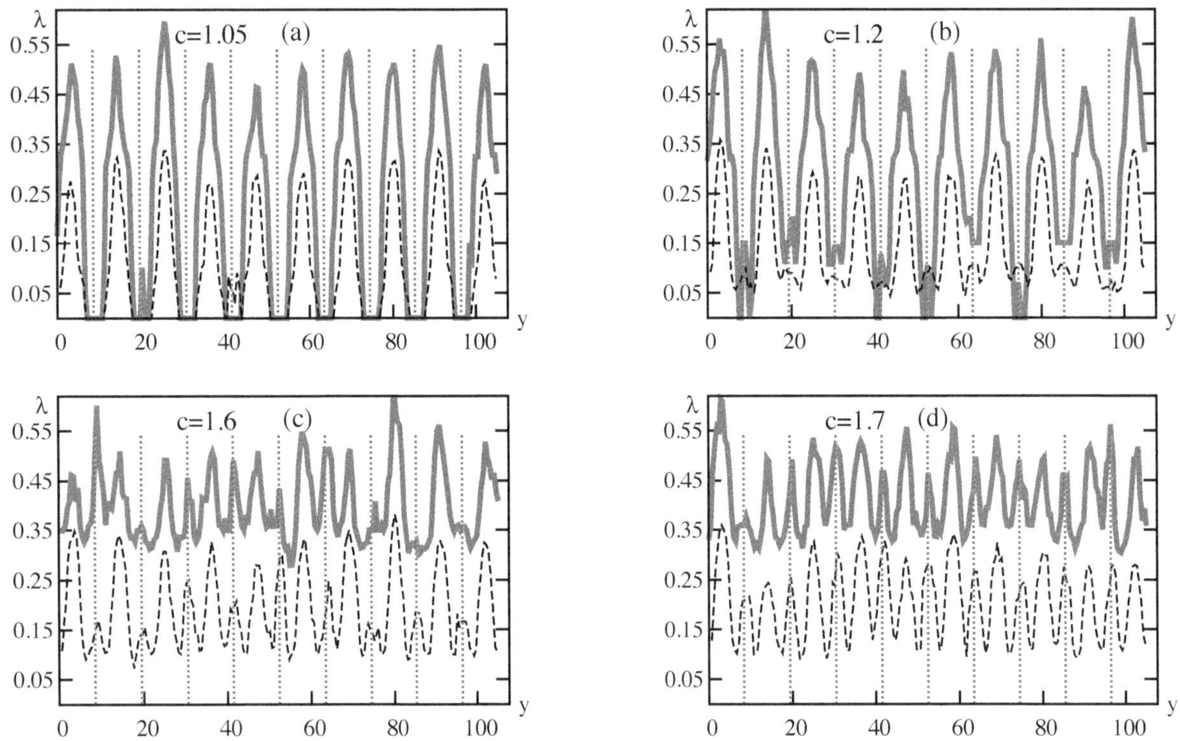

Figure 7. The irregularity index computed in 4-year sliding windows for synthetic data averaged over 162 (solid gray) and 648 (dashed black) days ($m = 2$). Model parameters are: $a = 0.8$, $h = 0.4$, $k = 0$, and $c = 1.05$ **(a)**, 1.2 **(b)**, 1.6 **(c)** and 1.7 **(d)**. Red dashed vertical lines at solar sunspot minima.

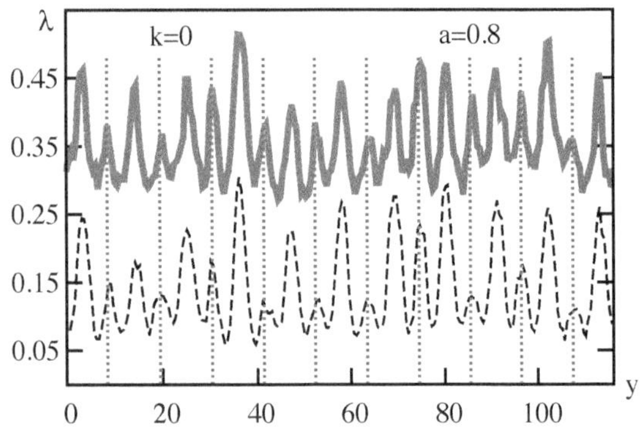

Figure 8. The irregularity index computed in 4-year sliding windows for synthetic data averaged over 162 (solid gray) and 648 (dashed black) days ($m = 2$). Model parameters: $a = 0.8$, $h = 0.8$, $k = 0$, and $c = 1.7$. Red dashed vertical lines at solar sunspot minima.

model realizations. The regime observed in 1867–1915 does not display such regular cycles of λ and therefore does not allow such an easy comparison.

5 Summary, discussion and conclusion

The evolution of the daily values of sunspot number ISSN from 1850 to 2005 has been studied in this paper, using tools from dynamical systems. Some interesting results are obtained with the irregularity index, a new method introduced in Shapoval et al. (2013, 2014). The method computes the rate of divergence of close trajectories in the phase space under a one-step translation mapping. This index is akin to the maximal Lyapunov exponent for time series calculated for low-dimensional dynamical systems, but is applicable to short time series with a random component. We have computed the irregularity index λ of ISSN for embedding dimensions $m = 1$ and 2 within a 4-year sliding window, after first averaging the data over $N = 162$, 324 and 648 days (multiples of the solar rotation period). The irregularity index for $N = 162$ follows the Schwabe cycle (Fig. 1), with sharp, high peaks at solar cycle maxima. But when N becomes large enough, it also exhibits sharp maxima at solar cycle minima (see also Shapoval et al., 2013), resulting in 5.5-year time variations, i.e., half the period of the Schwabe cycles (Fig. 1 middle and bottom panels; Fig. 3); we call them *half-Schwabe variations* (HSV).

The mean level of the irregularity index for ISSN undergoes a downward step around 1930 (particularly clear with embedding dimension $m = 2$, as seen in Fig. 3). This can be linked to the observation by Bershadskii (2008) that a change in the fractal properties of ISSN took place at that time. For a given time period, HSV can be characterized by the mean differences $\delta_{S_{\min}}$ of λ values at the times of Schwabe

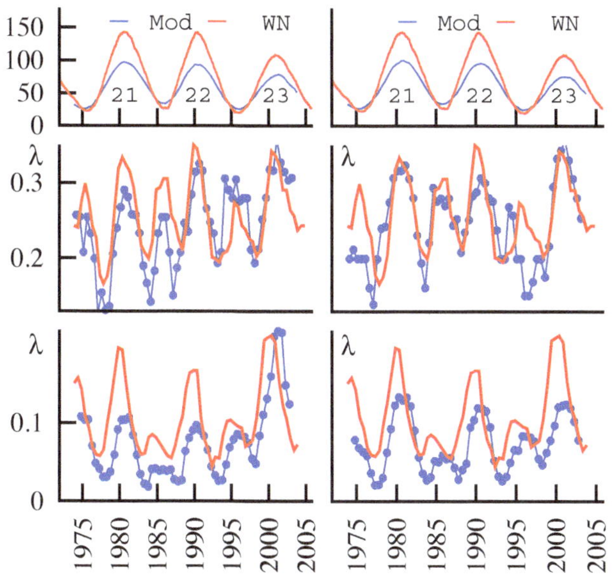

Figure 9. Top row panels: two synthetic signals (blue; see text) and ISSN (red) averaged over 4 years. The two columns of the figure are relative to different synthetic signals. The irregularity index ($m=2$) for model (blue) and ISSN (red) series are shown averaged over 162 (middle row) and 648 days (bottom row). $a=0.9$, $h=0.4$, and $k=0$.

cycle minima ($\lambda_{S_{\min}}$) and the mean of λ minima (λ_{mid}) at the middle times of the descending and ascending phases of the Schwabe cycles; we use the ratio R of $\delta_{S_{\min}}$ over its equivalent $\Delta_{S_{\max}}$ taken at the times of Schwabe cycle maxima. A first regime (denoted by Q1) is characterized by R increasing with N, and a second one (Q2) has R decreasing with N. For ISSN, with $m=2$, Q1 is observed until ~ 1915, whereas Q2 appears after ~ 1975; the main transition may be around 1930 (Fig. 3).

HSV as such may not be regarded as a result with great importance. Our functional λ can attain its extrema on both ascending and descending phases. If such is the case, HSV appears because of a certain similarity between these ascending and descending phases. That is why we cannot yet discuss the physics underlying the essence of HSV. However, we are entitled to look for simple time series that would display the properties observed for ISSN.

A synthetic signal, generated by a simple autoregressive model of order 1 exhibits many of the above-mentioned properties of the irregularity index of ISSN. The random part of this synthetic signal is taken to be Poissonian. Its mean is modulated by the sum of two periodic functions with periods $T_1=11$ years and $T_2<T_1$, the latter being tunable in an interval that can range from months to years. The introduction of intermediate oscillations (T_2) allows one to reproduce both the Q1 and Q2 regimes (Table 1). When the embedding dimension m is 1, HSV (5.5-year pseudoperiod) oscillations appear even if T_2 variations are absent ($k=0$; Fig. 5). When m is equal to 2, the behavior of the irregularity index series becomes richer: regime Q1, in which HSV behavior increases with smoothing N, is observed for larger values of k (Fig. 6), whereas regime Q2, in which HSV decreases with N, is obtained when intermediate period (T_2) variations are absent ($k=0$) and parameter h (which controls the smoothness of the signal) is increased (Fig. 8). We conclude that high frequency components of ISSN have much in common with an AR(1) process. The presence and then disappearance of ~ 1–2-year (T_2) oscillations seem to be required to produce a transition between regimes Q1 and Q2 (when $m=2$). We propose that these oscillations may be linked to the QBO, the second most powerful solar variation after the 11-year cycle (e.g., Ivanov et al., 2002).

At first order, the observed change in the mean level of λ around 1930 found in this paper could mark a shift of solar activity to a new regime (a transition of the solar dynamical system to a new state). This regime change is also marked by a (second order) change in the way HSV amplitude varies as the data is increasingly smoothed. These observed features can be reproduced by the model: the R ratio increases with increasing N (Q1) prior to 1915 and decreases after 1975 (Q2). Although several model parameters interact to promote one or the other regime, the most important one appears to be parameter k, which reflects the presence or absence of intermediate T_2 variations in the process. The shift of the irregularity index of ISSN from regime Q1 to Q2 may be due to the decrease or even disappearance of QBO.

In contrast to a standard statistical analysis, one cannot introduce a reasonable null hypothesis in the present study. Instead, we have checked the stability of the observed phenomena with respect to the parameters that control the irregularity index and we have tested the significance of our conclusions with the auto-regressive model.

The homogeneity of the ISSN-series is a long-debated question. Svalgaard (2010, 2012) points to an abrupt increase of ISSN in ~1945 and argues that this increase is caused by changes in the measurement rules. The NASA website (http://solarscience.msfc.nasa.gov/greenwch.shtml) also notes that the sunspot series is not uniform; abrupt changes occurred in 1941–1942 (sunspot numbers) and 1976–1977 (sunspot areas, not used in our paper). However, our conclusions about regime changes are not seriously affected by such events, because we use ratios (Eq. 4). Moreover, the date of the ~ 1930 singularity is remote from 1941 to 1942 (or 1945).

In order to see whether the observed behavior of the irregularity index of ISSN could be affected by such data problems, we have computed the irregularity index for another proxy influenced by solar activity but derived completely independently, namely the geomagnetic index *aa* (available at http://isgi.latmos.ipsl.fr/source/indices/aa/). With $m=1$, we computed the irregularity index for *aa* as such (without any prior averaging over multiples of 27 days). The values of $\lambda(t)$ (Fig. 10) exhibit both a noticeable increase in mean level (from about 1.75 to 2) and a decrease in range (from 0.5 to 0.2) in the 1930s. The sign of the change in mean value

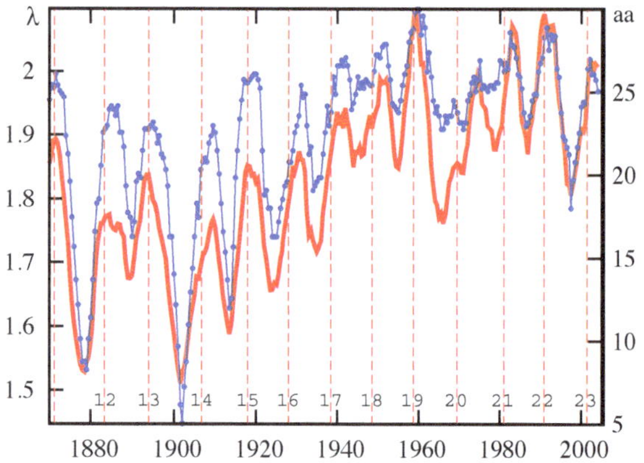

Figure 10. Red curve: the daily *aa* averaged over 4 years. Blue curve: the irregularity index computed in a 4-year sliding window with $m = 1$. Vertical dashed lines are at the maxima of Wolf numbers (averaged over 4 years). The number of each Schwabe cycle is indicated.

is opposite to that found for the irregularity index of ISSN (Fig. 10 vs. Fig. 3), but the same singularity in solar behavior could be at the origin of both.

Although this paper focuses mainly on changes of the irregularity index with smoothing, we also describe briefly changes of λ with time, in order to provide further evidence of the robustness of the technique. Our previous paper (Shapoval et al., 2013) examines the time evolution of λ computed for ISSN with a 4-year sliding window and different embedding dimensions and finds a change of regime in approximately 1915–1940. The same computation has been repeated for the Hoyt and Schatten group sunspot numbers (GSN, Hoyt and Schatten, 1998). Despite differences in inhomogeneities and potential problems with the two series, the main results are quite similar, excluding the possibility of an artefact due to the choice of an imperfect time series. The irregularity index of GSN exhibits two different regimes with a clear transition in the period 1915–1940 (details and figures in Appendix B). This strengthens the result obtained for ISSN and published in Shapoval et al. (2013) and further supports our approach, as used in the present paper.

It would be good to find some physical evidence to support our hypothesis that the change of regime can be linked to QBO rarefactions. There has been a significant amount of research on oscillations in the 1–2-year period range in cosmic rays (Valdes-Galicia et al., 1996; Kudela et al., 2002; Rouillard and Lockwood, 2004) and in midlatitude coronal holes area (McIntosh et al., 1992). Obridko and Shelting (2007) give a brief review of these works, together with new results (see also e.g., Ivanov et al., 2002). Intermediate variations do not seem to have been reported up to now for ISSN, but further interesting observations have been made for *aa* (Lockwood, 2001; Mursula et al., 2003). Using an extended *aa* index over 160 years, Mursula et al. (2003) have found that the power of "mid-term quasi-periodicities" (identical to QBO) is larger at periods alternating between 1.3 and 1.6 years; maxima of 1.3-year oscillations occur at the maxima of Schwabe cycles 18 and 22, while the 1.6-year oscillations peak at the maxima of cycles 16 and 21. The spectral power of *aa* is high during periods of high solar activity and would reflect the strength of the solar dynamo. Sudden disappearance of power is considered as a precursor for long-term decreases in solar activity. Our observations of a post-1975 decrease of R ratio with smoothing and our modeling of this regime by using an AR(1) process without T_2 variations are in line with the work of Mursula et al. (2003).

At least two different mechanisms could generate QBO. On one hand, Ivanov et al. (2002) show that the QBO of solar magnetic fields are mainly revealed in their large-scale component; they argue that QBO actually reflect variations in the equatorial dipole (and to a lesser extent quadrupole); for these authors, QBO sources are located near the base of the convection zone and remain invariable. Vecchio et al. (2012), using magnetic synoptic maps from 1976 to 2003, propose that QBO are fundamental modes associated with poleward magnetic flux migration from low to high latitudes (part of meridional circulation) during the maximum and descending phases of the solar cycle. A strong link between QBO and the solar dynamo is inferred from these and other works. Time variations of QBO might therefore provide information on changes in meridional flow. On the other hand, nonlinearity of the solar dynamo itself could be the source of QBO. Using a nonlinear Babcock–Leighton model, Charbonneau et al. (2007) support the hypothesis that the nonlinear component of the solar dynamo prevails over the stochastic one. Mayr and Schatten (2012) argue that the strong nonlinearity in the Charbonneau et al. (2007) equations could generate QBO without any time-dependent solar excitation.

The irregularity index of ISSN computed in this paper may provide a measure of the irregular behavior of the solar dynamo. Duhau and de Jager (2008) propose that ISSN may be used as a proxy of the toroidal component of the Sun's magnetic field and *aa* of the poloidal component. The irregularity index of *aa* as such presents a change in the 1930s, with a sign opposite to that for ISSN. We could therefore interpret our observations of changes in regime of the irregularity indices of ISSN and *aa* as indicating respectively a decrease in the irregular character of the toroidal field and an increase in the irregularity of the poloidal field in the 1930s, date of the advent of a grand maximum period in solar activity. Our analysis also suggests that another change may have started around 1975, as witnessed by decreasing HSV as a function of smoothing (see Fig. 3a–c in that order). This may have heralded the 2005 change found in our complementary studies of the irregularity index (Shapoval et al., 2013, Figs. 2–5; Shapoval et al., 2014, Figs. 1–3).

The irregularity index method is promising but still not a fully understood tool. It appears to be able to uncover singular phenomena and solar activity changes that cannot easily be seen by other means, but the tool depends on a number of parameters, particularly the embedding dimension and changes of behavior as the embedding dimension is changed. We note that the regime changes R1 and R2 uncovered by Shapoval et al. (2013, 2014) are not identical to the regime changes Q1 and Q2 found in the present paper. The R1/R2 regimes are marked by different levels of the irregularity index (computed with embedding dimensions from 4 to 32). In the present paper, when embedding dimension m is 2, we also find some evidence of the R1/R2 regimes (Fig. 3a, b). But we introduce an additional tool, the analysis of the irregularity index as a function of data smoothing (N), and this is what reveals the Q1/Q2 regimes. Although R1 and Q1, ending around 1915–1930, could correspond to the same regime, R2 (starting after 1930 and possibly ending in 2008) and Q2 (emerging clearly only after 1975) do not coincide. The physical nature of these singularities and the differences in their timing and behavior remain to be deciphered.

Appendix A: Examples of computation of the irregularity index

In order to illustrate some aspects of the computations involved in this paper, we generate synthetic data with the formula

$$u(t) = \sin\frac{2\pi t}{20} + \eta(t), \quad t = 1, 2, \ldots, 19,$$

where $\eta(t)$ is a random variable uniformly distributed over $[-0.05, 0.05]$. The embedding space is one-dimensional ($m = 1$).

Figure A1 exhibits 19 points of the sample. For each point, we first find the nearest neighbor. In the present case, only 13 pairs of nearest neighbors are different, so that the set of distances consists of 13 values. Let $\alpha = 0.25$. Since the integer part of 0.25×13 is 3, the α quantile d^* is the third distance from the lowest one. Distances less than or equal to d^* are "small", according to our definition. For each pair of nearest neighbors, the map F defined in Eq. (3) that moves points along their trajectory is applied, the distance between the corresponding images being small. In the example above, the pair at times 6 and 19 (red points in Fig. A1) possesses the smallest distance. A move to the right is impossible because 19 is the largest time in the window. Thus, this pair does not contribute to the computation of the exponent.

The points at times 4 and 11 form the next pair (blue filled circles). The distance between $u(5)$ and $u(12)$ (blue empty circles), which is $|u(12) - u(5)|$, is larger than d^*; therefore, this pair generates a single quantity log $[|u(12) - u(5)|/|u(11) - u(4)|]$ as a candidate irregularity index.

The last pair under consideration is $[u(1), u(11)]$. Since the value $u(11)$ appears twice in the pairs we consider, the corresponding point in the graph is marked first by a blue circle and second by a green circle (the green circle is smaller). Although the values $u(2)$ and $u(12)$ are very close, the distance between $F(u(1)) = u(2)$ and $F(u(11)) = u(12)$ is larger than the critical distance. Thus, the quantity log $[|u(12) - u(2)|/|u(11) - u(1)|]$ becomes the second candidate to the irregularity index. The irregularity index is chosen as the median of the candidate values.

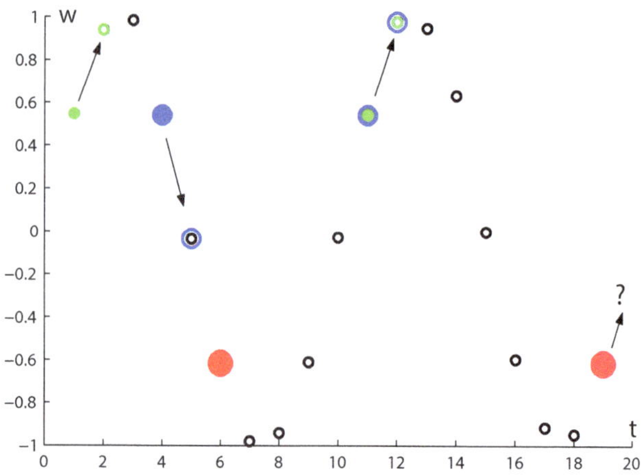

Figure A1. Computation of λ. Synthetic signal vs. time (see text).

Appendix B: Regime change of λ of GSN and ISSN

The irregularity index λ of both ISSN and GSN is computed with a 4-year sliding window, 8-day delay, and embedding dimensions 2–32. According to Figs. B1 and B2, λ of both the series exhibits two different patterns before ~1915 and after ~1940 with a transition during 1915–1940. The patterns for ISSN differ by the values of the irregularity index. High values of the irregularity index, most markedly seen at the minimum of cycles 14 and 15, underlie the pattern prior to 1915. The second pattern continues with the minimum

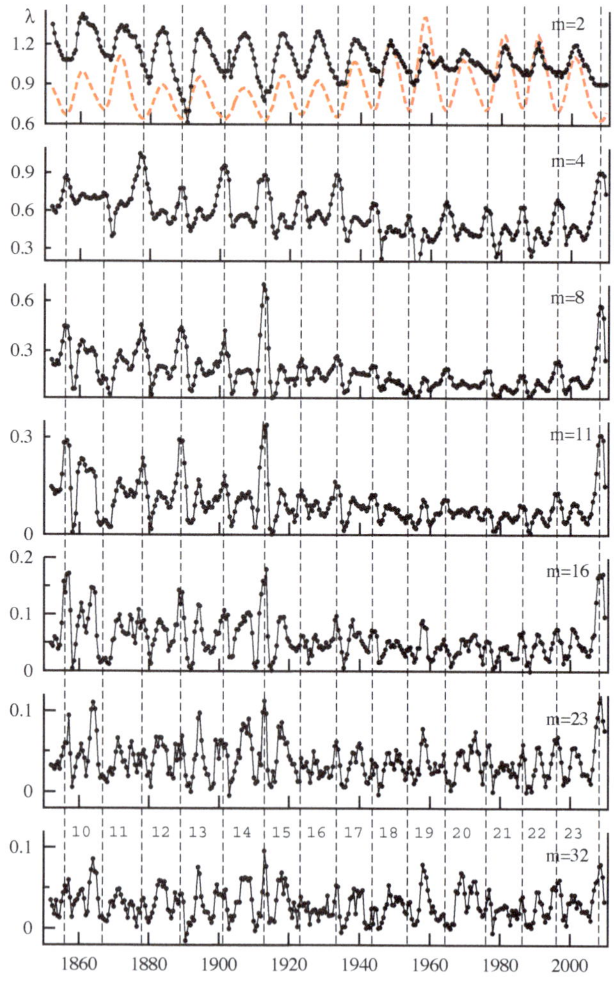

Figure B1. The irregularity index λ computed for ISSN within 4-year sliding windows; the embedding dimension m is indicated; vertical lines are at solar cycle minima.

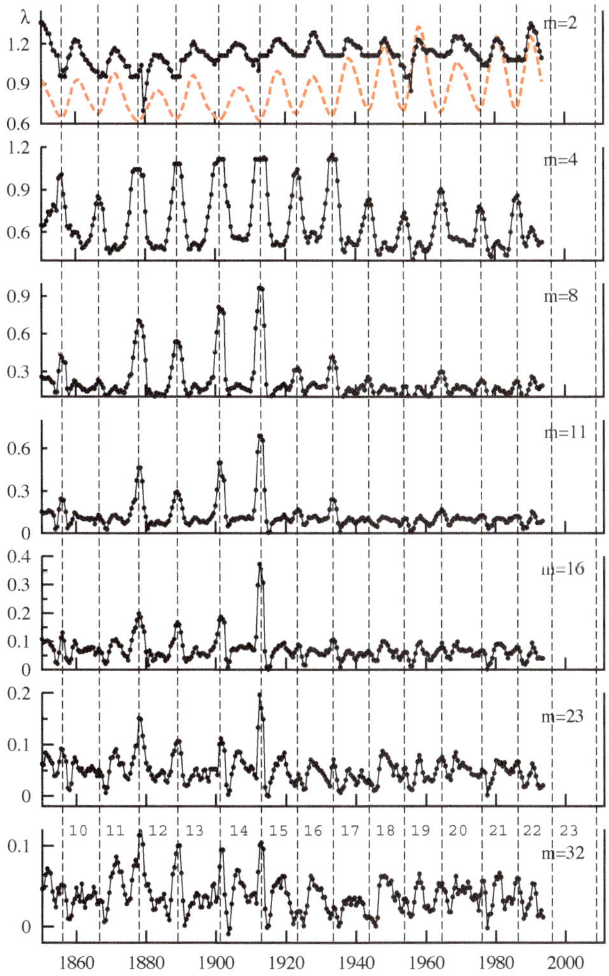

Figure B2. The irregularity index λ computed for GSN (Hoyt and Schatten, 1998) within 4-year sliding windows; the embedding dimension *m* is indicated; vertical lines are at solar cycle minima.

of cycles 23 and 24 when λ achieves another remarkable maximum.

The irregularity index of GSN also exhibits four high maxima at the minima of cycles 11–15. We cannot check the existence of the cycle 23–24 peak since recent data, following the Hoyt and Schatten technique, are not available as open sources. This confirmation of the regime change of solar activity between 1915 and 1940 vindicates our approach and results.

Acknowledgements. IPGP provided support to A. Shapoval and M. Shnirman during their visit to the institute. A. Shapoval was partially supported by RFBR grants 14-01-00346 and 14-01-00773. IPGP contribution NS 3503.

Edited by: G. Lapenta

References

Bartels, J.: Twenty-Seven Day Recurrences in Terrestrial-Magnetic and Solar Activity, 1923–1933, Terr. Magnet. Atmos. Elect., 39, 201–202, 1934.

Bergé, P., Pomeau, Y., and Vidal, C.: L'Ordre dans le Chaos, Hermann, Paris, France, 353 pp., 1984.

Bershadskii, A.: New dynamics of the Sun convection zone and global warming, arXiv:0805.2108v1 [astro-ph.SR], 2008.

Bershadskii, A.: Chaotic mean wind in turbulent thermal convection and long-term correlations in solar activity, arXiv:0908.4008v4 [astro-ph.SR], 2009.

Blanter, E. M., Shnirman, M. G., and Le Mouël, J.-L.: Solar variability: Evolution of correlation properties, J. Atmos. Solar-Terr. Phys. 67, 521–534, 2005.

Blanter, E. M., Le Mouël, J.-L., Perrier, F., and Shnirman, M. G.: Short-term correlation of solar activity and sunspot: evidence of lifetime increase, Solar Phys. 237, 329–350, 2006.

Charbonneau, P., Beaubien, G., and St-Jean, C.: Fluctuations in Babcok-Leighton dynamos: II. Revisiting the Gnevyshev-Ohl rule, Astrophys. J., 658, 657–662, 2007.

Choudhuri, A. R. and Karak, B. B.: Origin of Grand Minima in Sunspot Cycles, Phys. Rev. Lett., 109, 171103–171106, 2012.

Ding, R. and Li, J.: Nonlinear finite-time Lyapunov exponent and predictability, Phys. Lett. A 364, 396–400, 2007.

Duhau, S. and de Jager, C.: The Solar Dynamo and its Phase Transitions during the Last Millenium, Solar Phys. 250, 1–15, 2008.

Eckmann, J.-P. and Ruelle, D.: Ergodic theory of chaos and strange attractors, Rev. Modern Phys., 57, 617–656, 1985.

Fraser, A. M. and Swinney, H. L.: Independent coordinates for strange attractors from mutual information, Phys. Rev. A, 33, 1134–1140, 1986.

Greenkorn, R. A.: 2009, Analysis of Sunspot Activity Cycles, Solar Phys., 255, 301–323, doi:10.1007/s11207-009-9331-z, 2009.

Howe, R.: Solar interior rotation and its variation, Living Rev. Solar Phys., 6, 1–91, 2009.

Hoyt, D. V. and Schatten, K. H.: Group Sunspot Numbers: A new Solar Activity Reconstruction, Solar Phys., 181, 491–512, 1998.

Ivanov, E. V., Obridko, V. N., and Shelting, B. D.: Quasi-biennial oscillations of the solar magnetic fields, in: Solar variability: from core to outer frontiers, The 10th European Solar Physics Meeting, 9–14 September 2002, Prague, Czech Republic, edited by: Wilson, A., ESA SP-506 v. 2, ESA Publications Division, Noordwijk, 847–850, 2002.

Kantz, H.: A robust method to estimate the maximal Lyapunov exponent of a time series, Phys. Lett. A, 185, 77–87, 1994.

Kitchatinov, L. L. and Olemskoi, S. V.: Active longitudes of the sun: The rotation period and statistical significance, Astron. Lett., 31, 280–284, 2005.

Kudela, K., Rybak, J., Antalova, A., and Storini, M.: Time evolution of low-frequency periodicities in cosmic ray intensity, Solar Phys. 205, 165–175, 2002.

Lawrence, J. K., Cadavid, A. C., and Ruzmaikin, A. A.: Turbulent and chaotic dynamics underlying solar magnetic variability, Astron. Astrophys., 455, 366–375, 1995.

Lawrence, J. K., Cadavid, A. C., and Ruzmaikin, A. A.: Rotational Quasi-Periodicities and the Sun-Heliosphere Connection, Solar Phys., 252, 179–193, 2008.

Le Mouël, J.-L., Shnirman, M. G., and Blanter, E. M.: The 27-Day Signal in Sunspot Number Series and the Solar Dynamo, Solar Phys., 246, 295–307, 2007.

Li, Q.-X. and Li, K.-J.: 2007, Low dimensional chaos from the group sunspot numbers, Chinese J. Astron. Astrophys., 7, 435–440, 2007.

Lockwood, M.: Long-term variations in the magnetic fields of the sun and the heliosphere: their origin, effects and implications, J. Geophys. Res., 106, 16021–16038, 2001.

Love, J. J. and Rigler, E. J.: Sunspot random walk and 22-year variation, Geophys. Res. Lett., 39, L10103–L10108, 2012.

Macek, W. M., Bruno, R., and Consolini, G.: Testing for multifractality of the slow solar wind, Adv. Space Res. 37, 461–466, 2006.

Mavromichalaki, H., Preka-Papadema, P., Petropoulos, B., Tsagouri, I., Georgakopoulos, S., and Polygiannakis, J.: Low- and high-frequency spectral behavior of cosmic-ray intensity for the period 1953–1996, Ann. Geophys., 21, 1681–1689, doi:10.5194/angeo-21-1681-2003, 2003.

Mayr, H. G. and Schatten, K. H.: Nonlinear oscillators in space physics, J. Atmos. Sol.-Terr. Phys., 74, 44–50, 2012.

McIntosh, P. S., Thompson, R. J., and Willock, E. C.: A 600-day periodicity in solar coronal holes, Nature, 360, 322–324, 1992.

Mursula, K., Zieger, B., and Vilppola, J. H.: Mid-term quasi-periodicities in geomagnetic activity during the last 15 solar cycles: connection to solar dynamo strength, Solar Phys., 212, 201–207, 2003.

Obridko, V. N. and Shelting, B. D.: Occurrence of the 1.3-year periodicity in the large-scale solar magnetic field for 8 solar cycles, Adv. Space Res., 40, 1006–1014, 2007.

Oseledets, V. I.: A multiplicative ergodic theorem. Lyapunov characteristic numbers for dynamical systems, Trans. Moscow Math. Soc., 19, 197–231, 1968.

Ostryakov, V. N. and Usoskin, I. G.: On the dimension of solar attractor, Solar Phys., 127, 405–412, 1990.

Pesnell, W. D.: Solar Cycle Predictions (Invited Review), Solar Phys., 281, 507–532, 2012.

Price, C. P., Prichard, D., and Hogenson, E. A.: Do the sunspot numbers form a chaotic set?, J. Geophys. Res., 97, 19113–19120, 1992.

Rosenstein, M. T., Collings, J. J., and De Luca, C. J.: A practical method for calculating largest Lyapunov exponents from small data sets, Physica D, 65, 117–134, 1993.

Rouillard, A. and Lockwood, M.: Oscillations in the open solar magnetic flux with a period of 1.68 years: imprint on galactic cosmic rays and implications for heliospheric shielding, Ann. Geophys., 22, 4381–4395, doi:10.5194/angeo-22-4381-2004, 2004.

Ruzmaikin, A., Feynman, J., Kosacheva, V.: On Long-Term Dynamics of the Solar Cycle, in: The solar cycle; Proceedings of the National Solar Observatory/Sacramento Peak 12th Summer Workshop 27, edited by: Harvey, K. L., ASP Conference Series, San Francisco, 547–556, 1992.

Sello, S.: Solar cycle forecasting: A nonlinear dynamics approach, Astron. Astrophys., 377, 312–320, 2001.

Shapoval, A., Le Mouël, J.-L., Courtillot, V., and Shnirman, M.: Two regimes in the regularity of sunspot numbers, Astrophys. J., 779, 108–116, 2013.

Shapoval, A., Le Mouël, J.-L., Courtillot, V., and Shnirman, M.: submitted, Is a sudden increase of irregularity of sunspot numbers a precursor of a return to low solar activity?, J. Geophys. Res., in review, 2014.

SIDC-team: World Data Center for the Sunspot Index, Royal Observatory of Belgium, Monthly Report on the International Sunspot Number, online catalogue of the sunspot index: http://www.sidc.be/sunspot-data/, 1850–2005, 2005.

Spiegel, E. A. and Wolf, A.: Chaos and the Solar Cycle, in: Chaotic Phenomena in Astrophysics, vol. 497, edited by: Buchier, J.-R. and Eichhorn, H., Ann. N.Y. Acad. Sci., New York, 55–60, 1987.

Svalgaard, L., Updating the Historical Sunspot Record, arXiv:1003.4666 [astro-ph.SR], 2010.

Svalgaard, L.: How well do we know the sunspot number? Comparative Magnetic Minima: Characterizing quiet times in the Sun and Stars, Proceedings of the International Astronomical Union, IAU Symposium 286, Mendoza, Argentina, 27–33, 2012.

Takens, F.: Detecting strange attractors in turbulence, edited by: Rand, D. A., and Young, L. S., Springer, Berlin, 366 – 381, 1981.

Valdes-Galicia, J. F., Perez-Enrizuez, R., and Otaola, J. A.: The cosmic ray 1.68 year variation: a clue to understand the nature of the solar cycle?, Solar Phys., 167, 409–417, 1996.

Vecchio, A., Laurenza, M., Carbone, V., and Storini, M.: Quasi-biennial modulation of solar neutrino flux and solar and galactic comsic rays by solar cyclic activity, Astrophys. J. Lett., 709, L1–L5, 2010.

Vecchio, A., Laurenza, M., Meduri, D., Carbone, V., and Storini, M.: The dynamics of the solar magnetic field: polarity reversals, butterfly diagrams, and quasi-biennial oscillations, Astrophys. J., 749, 27–36, doi:10.1088/0004-637X/749/1/27, 2012.

Wolf, A., Swift, J. B., Swinney, H. L., and Vastano, J. A.: Determining Lyapunov Exponent from a Time Series, Physica D, 16, 285–317, 1985.

Zhang, Q.: A nonlinear prediction of the smoothed monthly sunspot numbers, Astron. Astrophys. 310, 646–650, 1996.

Brief Communication: The double layer in the separatrix region during magnetic reconnection

J. Guo[1,2] **and B. Yu**[1]

[1]School of Mathematics and Physics, Qingdao University of Science and Technology, Qingdao, Shandong 266061, China
[2]Sate Key Laboratory of Space Weather, Chinese Academy of Sciences, Beijing 100190, China

Correspondence to: J. Guo (guojun@qust.edu.cn, guoj2005@163.com)

Abstract. With two-dimensional (2-D) particle-in-cell (PIC) simulations we investigate the evolution of the double layer (DL) driven by magnetic reconnection. Our results show that an electron beam can be generated in the separatrix region as magnetic reconnection proceeds. This electron beam could trigger the ion-acoustic instability; as a result, a DL accompanied with electron holes (EHs) can be found during the nonlinear evolution stage of this instability. The spatial size of the DL is about 10 Debye lengths. This DL propagates along the magnetic field at a velocity of about the ion-acoustic speed, which is consistent with the observation results.

1 Introduction

Magnetic reconnection is a very important process in space plasma, which can convert the magnetic field energy into kinetic energy of electrons and ions (Pritchett, 2006; Lu et al., 2010). In previous work, more attention is paid to the whistler waves (Fujimoto and Sydora, 2008; Deng and Matsumoto, 2001; Guo, 2011; Xiao et al., 2007, 2008, 2010) and electron accelerations (Fu et al., 2006; Huang et al., 2010; Pritchett, 2006). Recently, the electrostatic waves have been found in the simulations and observations. With the particle-in-cell (PIC) simulation, Drake et al. (2003) reported that the electrostatic waves and electron holes (EHs) are produced by the Buneman instability. In their simulation, the initial electron drift speed is above the threshold to trigger the Buneman instability, except for a strong guide field. However, Fujimoto and Machida (2006) suggested that the electrostatic waves are excited by the electron two-stream instability which has a relationship with magnetic reconnection. Particularly, no guide field is used in their initial condition. The observation (Li et al., 2009) proved that the guide field has no direct effect on the generation of the electrostatic wave in or around the reconnection diffusion region. These reported electrostatic waves appear with bipolar electric field and EHs, while the signature of the double layer (DL) is a unipolar electric field. The electrons and ions will be accelerated, decelerated, or reflected by the electric field when they enter the DL. The current-driven DLs have been carefully studied by using a series of one-dimensional (1-D) Vlasov simulations (Singh, 1980, 1982, 2000). It is found that EHs are a common feature of a strong DL. The evolution of the electron beam accelerated by a DL would contribute to continuously creating moving EHs. Additionally, the DL can accelerate the ion beam which can generate fluctuations above the ion cyclotron frequency in the 2.5-D PIC simulations (Singh and Khazanov, 2003).

Electron holes are frequently observed by the spacecraft observation during magnetic reconnection. Double layers are frequently observed in the auroral ionosphere (Boström, 1992; Ergun et al., 2001). Only recently, Ergun et al. (2009) reported DLs in the plasma sheet in the magnetotail, but they could not confirm the relation between the DL and magnetic reconnection in their work. Wang et al. (2014) presented the first evidence of DLs during magnetic reconnection, and found that DLs are moving away from X line at a velocity of about ion-acoustic speed in the separatrix region. Meanwhile, a series of EHs moving toward the X line are observed in the wake of the DL. In addition, the multiple DLs are also observed in the simulations (Singh et al., 2011).

The electrostatic waves have a close relationship with electron beams. Previous simulations show that electron beams

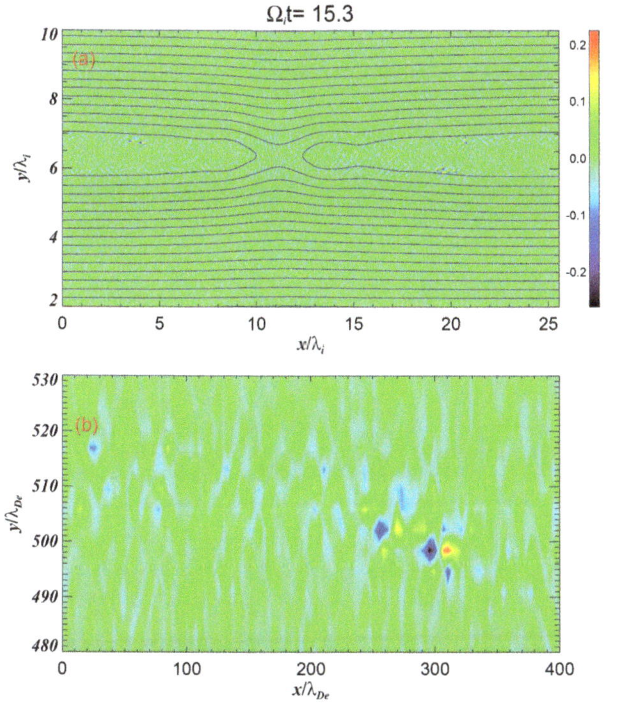

Figure 1. The electric field parallel to the local magnetic filed $E_{//}$ and magnetic field lines at $t\Omega_\mathrm{i} = 15.3$ **(a)**, the magnification of the left-hand plane **(b)**.

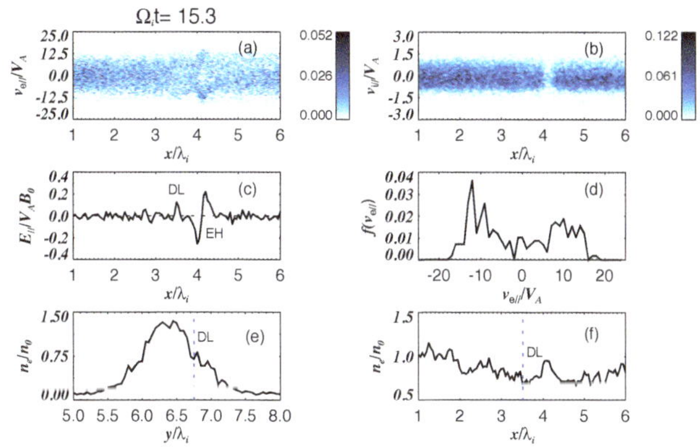

Figure 2. The $v_{//} - x$ phase-space distributions of electron **(a)** and ion **(b)**, and the electric field $E_{//}$ at $y/\lambda_\mathrm{i} = 6.75$ **(c)**, the parallel velocity distribution of electron $f(v_{\mathrm{e}//})$ at about $x/\lambda_\mathrm{i} = 4.1$ and $y/\lambda_\mathrm{i} = 6.75$ **(d)**, the density distribution of electron along y axis at $y/\lambda_\mathrm{i} = 3.5$ where the double layer (DL) appear **(e)** and the density distribution of electron along x axis at $y/\lambda_\mathrm{i} = 6.75$ **(f)**.

form mainly close to the separatrix (Hoshino et al., 2001). Using 3-D simulation in the presence of a guide field, Pritchett and Coroniti (2004) found the enhanced parallel electric field and electron velocity are confined to one pair of separatrix. In this case strong DLs can be generated at the reconnection site. The purpose of our study is to investigate the generation mechanism of the DL during the magnetic reconnection without a guide field. Our results suggest that both the DL and EH can be formed in our simulation. These polarized structures should be excited by the ion-acoustic instabilities.

2 Simulation model

The initial magnetic field distribution is set to be $B_x(y) = B_0 \tanh[(y - L_y/2)/\delta_0]$, where δ_0 is the half width of the initial current sheet. The particle densities read $n(y) = n_0 \mathrm{sech}^2[(y - L_y/2)/\delta_0] + n_\mathrm{b}$, where n_0 and n_b represent the current sheet and background densities. δ_0 is equal to $0.5\lambda_\mathrm{i}$, where λ_i is the ion inertial length given by n_0. All of the particles have the initial Maxwellian velocity distributions. In the present simulation, the simulation size is $L_x \times L_y = 25.6$ $\lambda_\mathrm{i} \times 12.8\lambda_\mathrm{i}$. The other parameters are $T_\mathrm{i}/T_\mathrm{e} = 5, n_\mathrm{b} = 0.1n_0$, $m_\mathrm{i}/m_\mathrm{e} = 256$, and $c/V_\mathrm{A} = 30$, where V_A is the Alfvén speed based on B_0 and n_0; the ratio of ion gyrofrequency to plasma frequency is $\omega_\mathrm{pi} / \Omega_\mathrm{i} = c/V_\mathrm{A} = 30$. The spatial resolution is $\Delta x = \Delta y = 0.05\lambda_\mathrm{i}$. The time step is $\Delta t\Omega_\mathrm{i} = 0.001$, where Ω_i is the proton cyclotron frequency. Along the x axis, the periodic boundary conditions are used, and the ideal conducting boundary conditions are used in the y axis. The particles will be reflected when they reach the boundary in the y axis. The magnetic and electric fields are calculated with a full explicit algorithm. About 6×10^6 particles per species are employed in the simulation. A small initial magnetic perturbation is superposed in the form of which is the same as in the references (Lu et al., 2013; Guo, 2014).

3 Simulation results

Figure 1a shows the electric field parallel to the local magnetic field $E_{//}$ and magnetic field lines at $t\Omega_\mathrm{i} = 15.3$. Figure 1b shows the magnification of $E_{//}$ in left-hand plane. We plot Fig. 1b as a function of the electron Debye length λ_De instead of the ion inertial length; $\lambda_\mathrm{De} = v_\mathrm{the}/\omega_\mathrm{pe}$ is the initial electron Debye length. At this time, the electron thermal speed is about $v_\mathrm{the} \sim v_\mathrm{th0} = (2T_\mathrm{e}/m_\mathrm{e})^{1/2} \sim 6.5V_\mathrm{A}$. It is obvious that electric fields with polarized structures are mainly located in the separatrix region. These structures are almost along the magnetic field lines. As the magnetic reconnection proceeds, the polarized electric fields will also appear in the diffusion region, which have been discussed in our previous work (Guo, 2014).

Shown in Fig. 2 are the $v_{//} - x$ phase-space distributions of electron (a) and ion (b), and the electric field $E_{//}$ at $y/\lambda_\mathrm{i} = 6.75$ (c). Figure 2a and c show clearly that there is an electron phase-space hole at $x/\lambda_\mathrm{i} \sim 4.1$. Besides that, a DL with a width about $10\lambda_\mathrm{De}$ lies at $x/\lambda_\mathrm{i} \sim 3.5$, which is in the wake of a EH. Figure 2d is the parallel velocity distribution of electron $f(v_{\mathrm{e}//})$ at about $x/\lambda_\mathrm{i} = 4.1$ and $y/\lambda_\mathrm{i} = 6.75$, which corresponds to the EH shown in Fig. 2a. This distribution appears an electron beam with $v_{\mathrm{e}//}/V_\mathrm{A} \sim -12.0$ and a

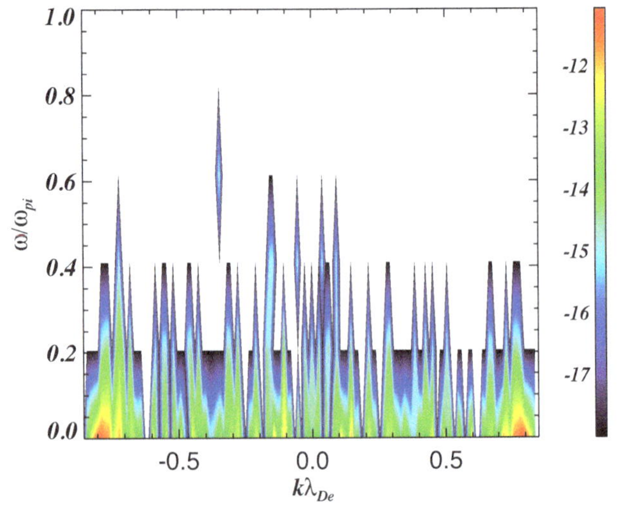

Figure 3. The wave spectrum of parallel electric field. The sampling region is $0 < x/\lambda_i < 6.4$ and the time interval is $15.0 < t\Omega_i < 15.5$.

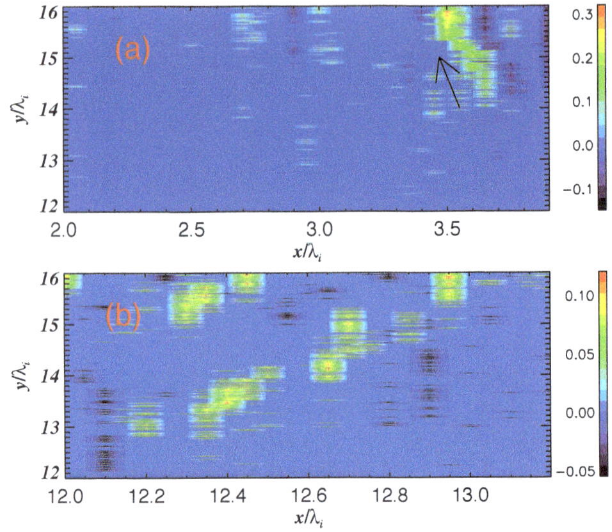

Figure 4. The time evolution of the DL shown in Fig. 1b at $y/\lambda_i = 6.75$ (**a**) and the DL found in the separatrix near the X line (**b**).

broad peak centered around $v_{e//}/V_A \sim 10.0$ (the flat-top distribution). On the other hand, the low-velocity portion of the distribution is also dominated by a plateau formation. Figure 2e is the density distribution of electron along y axis at $y/\lambda_i = 3.5$ where the DL appears. Figure 2f is the density distribution of electron along x axis at $y/\lambda_i = 6.75$. By coupling Fig. 2e and f, it is found that the density of electrons where the DL appears is lower than the surrounding, along both x and y axes. Based on the observations (Wang et al., 2013, 2014), the DL and EH are observed together. Moreover, the structure of DLs in our simulation is basically consistent with the observation results.

The magnetic field line passes through $y/\lambda_i = 6.75$, where the magnetic field is nearly parallel to the x axis in the region $x/\lambda_i < 7$. The plateau function means that the Buneman or ion-acoustic modes probably appear. Figure 3 is the wave spectrum of a parallel electric field. The sampling region is $0 < x/\lambda_i < 6.4$ and the time interval is $15.0 < t\Omega_i < 15.5$. It is found that the normalized wave number associated with the highest wave energy density is $k\lambda_{De} \sim \pm 0.8$. These forward and backward modes correspond to the distribution shown in Fig. 2d. During this period, the frequency of the dominant mode is far below the ion plasma frequency. All of the above suggests that the DL located in the separatrix region shown in Fig. 2c should be produced by the ion-acoustic instabilities.

Both numerical simulations and observations have been performed during the past decades which show that DLs are highly variable structures moving with time (Singh et al., 2005; Wang et al., 2014). Figure 4a is the time evolution of the polarized electric field shown in Fig. 1b at $y/\lambda_i = 6.75$. The DL in our simulation seems to only propagate outward along the magnetic field for a short distance. The propagation velocity can be estimated to be $0.2\ V_A$, while the DL found in the separatrix near the X line shown in Fig. 4b can propagate

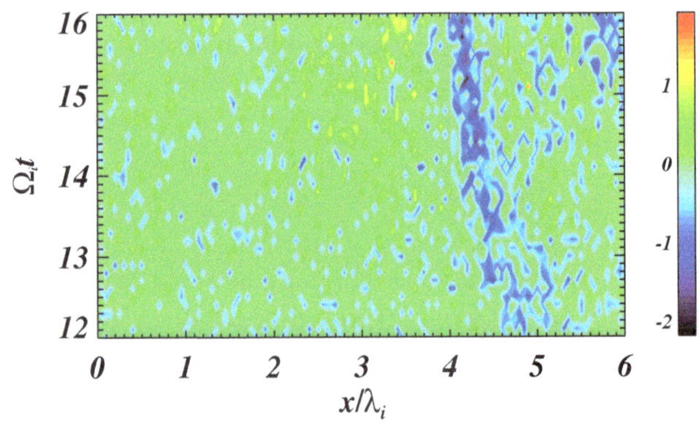

Figure 5. The time evolution of the electron flow velocity parallel to the magnetic field at $y/\lambda_i = 6.75$.

for a long time and distance. The propagation velocity can be estimated to be $0.4\ V_A$. In fact, these two velocities are on the same order of the ion-acoustic speed, although the former is a little small. The observation (Wang et al., 2014) also suggests that the DL propagates away from the X line at a velocity of about ion-acoustic speed; that is, the results shown in Fig. 4 are very consistent with those observations. However, the DL found in the separatrix at a distance of several ion inertial lengths away from the X line shown in Fig. 1b cannot propagate for a long distance. A reasonable explanation for Fig. 4a is due to the periodic boundary condition.

Previous simulation results about magnetic reconnection (Hoshino et al., 2001; Pritchett and Coroniti, 2004) show that electron beams form mainly close to the separatrix. In this condition, strong DLs can be generated at the reconnection site. These simulation results have been confirmed by the cluster observations published by Vaivads et al. (2004).

Figure 5 shows the time evolution of the electron flow velocity parallel to the magnetic field at $y/\lambda_i = 6.75$. The localized electron beam with negative drift velocity propagates along the magnetic field toward the left. At time $t\Omega_i \sim 12$, this electron beam lies at $x/\lambda_i \sim 4.8$, while the propagation stops at $x/\lambda_i \sim 4.1$ due to the periodic boundary condition at time $t\Omega_i \sim 15.5$. This evolution is consistent with that of the parallel electric field shown in Fig. 4a. Of course, the DL appears after the electron beam formed for a period of time. On the other hand, as the time goes on, the width of electron flow becomes smaller, while the beam velocity becomes larger. The figure shows that only when the velocity of electron beam reaches a certain value the electrostatic waves can be excited. Compared with Fig. 2, the electron beam is located in the region where the EH appears. Our previous simulation results (Guo and Li, 2013) suggest that the ion-acoustic waves are always accompanied by electron beams. Together with Fig. 2d, it can be thus concluded that the localized electron beam excites the ion-acoustic instabilities, which results in the formation of DL. However, the electron beam cannot drift for a long time due to the periodic boundary, which leads the DL to not propagate over a long distance.

4 Conclusions

In this paper, the evolution of DLs during magnetic reconnection is studied by 2-D electromagnetic PIC simulation. The simulation results show that the DLs can be found in the separatrix region, which are generated by ion-acoustic instability. The DL found in our simulation is about 10 Debye lengths and propagates away from the diffusion region with a velocity of about the ion-acoustic speed. These results are consistent with the observation results (Wang et al., 2014). The physical evolution process can be inferred from the following processes. At first, the electrons are accelerated near the X line and then flow out of the diffusion region along the magnetic field. When the velocity of the electron beam is large enough, the localized electron beam excites the ion-acoustic instabilities, which produced the DL. But the DLs at a few ion inertial lengths away from the X line can only propagate for a short distance, because the propagation of electron beam tends to be stable due to periodic boundary conditions. However, even though we get some interesting results, there are still many questions that remain to be addressed. For instance, a sufficiently long system will be used to study the propagation of DLs in future work.

Acknowledgements. This research was supported by the National Natural Science Foundation of China under grant no. 41204115 and the Specialized Research Fund for State Key Laboratories.

Edited by: B. Tsurutani

References

Boström, R.: Obdervations of weak double-layers on auroral field lines, IEEE Trans. Plasma Sci., 20, 756–763, 1992.

Deng, X. H. and Matsumoto, H.: Rapid magnetic reconnection in the Earth's magnetosphere mediated by whistler waves, Nature, 410, 557–560, 2001.

Drake, J. F., Swisdak, M., Cattell, C., Shay, M. A., Rogers, B. N., and Zeiler, A.: Formation of electron holes and particle energization during magnetic reconnection, Science, 299, 873 877, 2003.

Ergun, R. E., Su, Y. -J., Andersson, L., Carlson, C. W., McFadden, J. P., Mozer, F. S., Newman, D. L., and Goldman, M. V.: Direct observation of localized parallel electric fields in a space plasma, Phys. Rev. Lett., 87, 045003, doi:10.1103/PhysRevLett.87.045003, 2001.

Ergun, R. E., Andersson, L., Tao, J., Angelopoulos, V., Bonnell, J., McFadden, J. P., Larson, D. E., Eriksson, S., Johansson, T., Cully, C. M., Newman, D. N., Goldman, M .V., Roux, A., LeContel, O., Glassmeier, K.-H., and Baumjohann, W.: Observation of double layers in Earth's plasma sheet, Phys. Rev. Lett., 102, 155002, doi:10.1103/PhysRevLett.102.155002, 2009.

Fu, X. R., Lu, Q. M., and Wang, S.: The process of electron acceleration during collisionless magnetic reconnection, Phys. Plasma, 13, 012309, doi:10.1063/1.2164808, 2006.

Fujimoto, K. and Machida, S.: A generation mechanism of electrostatic waves and subsequent electron heating in the plasma sheet-lobe boundary region during magnetic reconnection, J. Geophys. Res., 111, A09216, doi:10.1029/2005JA011452, 2006.

Fujimoto, K. and Sydora, R. D.: Whistler waves associated with magnetic reconnection, Geophys. Res. Lett., 35, L19112, doi:10.1029/2008GL035201, 2008.

Guo, J.: Excitation of whistler mode waves in collisionless magnetic reconnection with a particle-in-cell simulation, Plasma Phys. Control. Fusion, 53, 022001, doi:10.1088/0741-3335/53/2/022001, 2011.

Guo, J. and Li, B.: The electron-ion streaming instabilities driven by drift velocities of the order of electron thermal velocity in a nonmagnetized plasma, Astrophys. Space Sci., 347, 99–107, 2013.

Guo, J.: Electric field with bipolar structure during magnetic reconnection without a guide field, Astrophys. Space Sci., 351, 159–163, 2014.

Huang, C., Lu, Q. M., and Wang S.: The mechanisms of electron acceleration in anti-parallel and guide filed magnetic reconnection, Phys. Plasma, 17, 072306, doi:10.1063/1.3457930, 2010.

Hoshino, M., Mukai. T., Terasawa, T., and Shinohara, I.: Suprathermal electron acceleration in magnetic reconnection, J. Geophys. Res., 106, 25979–25997, 2001.

Li, S. Y., Deng, X. H., Zhou, M., Tang, R. X., Liu, K., Kojima, H., and Matsumoto, H.: Statistical study of electrostatic solitary waves associated with reconnection: Geotail observations, Adv. Space Res., 43, 394–400, 2009.

Lu, Q. M., Huang, C., Xie, J. L., Wang, R. S., Wu, M. Y., Vaivadds, A., and Wang, S.: Features of separatrix regions in magnetic reconnection: Comparison of 2D particle-in-cell simulations and Cluster observations, J. Geophys. Res., 115, A11208, doi:10.1029/2010JA015713, 2010.

Lu, Q. M., Lu, S., Huang, C., Wu, M. Y., and Wang S.: Self-reinforcing process of the reconnection electric field in the electron diffusion region and onset of collisionless magnetic reconnection, Plasma Phys. Control. Fusion, 55, 085019, doi:10.1088/0741-3335/55/8/085019, 2013.

Pritchett, P. L.: Relativistic electron production during guide field magnetic reconnection, J. Geophys. Res., 111, A10212, doi:10.1029/2006JA011793, 2006.

Pritchett, P. L. and Coroniti, F. V.: Three-dimensional collisionless magnetic reconnection in the presence of a guide field, J. Geophys. Res., 109, A01220, doi:10.1029/2003JA009999, 2004.

Singh, N.: Computer experiments on the formation and dynamics of electric double layers, Plasma Phys., 22, 1–24, 1980.

Singh, N.: Double-layer formation, Plasma Phys., 24, 639–660, 1982.

Singh, N.: Electron holes as a common feature of double-layer-driven plasma waves, Geophys. Res. Lett., 27, 927–930, 2000.

Singh, N. and Khazanov, I.: Dynamic planar double layers: Filamentary substructures, electron holes and ion heating, Geophys. Res. Lett., 30, 2262, doi:10.1029/2003GL018480, 2003.

Singh, N., Deverapalli, C., and Khazanov, I.: Comparison of electric fields and density structures seen in simulations and satellite observations, Geophys. Res. Lett., 32, L13108, doi:10.1029/2005GL022920, 2005.

Singh, N., Araveti, S., and Wells, E. B.: Mesoscale PIC simulation of double layers and electron holes affecting parallel and transverse accelerations of electrons and ions, J. Geophys. Res., 116, A00K09, doi:10.1029/2010JA016323, 2011.

Vaivads, A., Khotyaintsev, Y., Andre, M., Retinò, A., Buchert, S. C., Rogers, B. N., Decreau, P., Paschmann, G., and Phan, T. D.: Structure of the magnetic reconnection diffusion region from four-spacecraft observations, Phys. Rev. Lett., 93, 105001, doi:10.1103/PhysRevLett.93.105001, 2004.

Wang, R. S., Du, A. M., Nakamura, R., Lu, Q. M., Khotyaintsev, Y. V., Volwerk, M., Zhang, T. L., Kronberg, E. A., Daly, P. W., and Fazakerley, A. N.: Observation of multiple sub-cavities adjacent to single separatrix, Geophys. Res. Lett., 40, 2511–2517, doi:10.1002/grl.50537, 2013.

Wang, R. S., Lu, Q. M., Khotyaintsev, Y. V., Volwerk, M., Du, A. M.,Nakamura, R., Gonzalez, W. D., Sun, X., Baumjohann, W., Li, X., Zhang, T. L., Fazakerley, A. N., Huang, C., and Wu, M. Y.: Observation of double layer in the separatrix region during magnetic reconnection, Geophys. Res. Lett., 41, 4851–4858, doi:10.1002/2014GL061157, 2014.

Xiao, F. L., Chen, L. X., Zhou, Q. H., He, H. Y., and Wen, Y. J.: Bounce-averaged acceleration of energetic electrons by whistler mode chorus in the magnetosphere, Chin. Phys. Lett., 24, 294–297, 2007.

Xiao, F. L., Chen, L. X., He, H. Y., and Zhou, Q. H.: Second-order resonant interaction of ring current protons with whister-mode waves, Chin. Phys. Lett., 25, 336–339, 2008.

Xiao, F. L., Zong, Q. G., Pu, Z. Y., Su, Z. P., Cao, J. B., He, J. S., Wang, Y. F., and Zheng, H. N.: Electron acceleration by whistler-mode waves around the magnetic null during 3D reconnection, Plasma Phys. Control. Fusion, 52, 052001, doi:10.1088/0741-3335/52/5/052001, 2010.

16

Relativistic surfatron process for Landau resonant electrons in radiation belts

A. Osmane[1,2] and A. M. Hamza[1]

[1]University of New Brunswick, Physics Department, Fredericton, New Brunswick, Canada
[2]Aalto University, Radio Science and Engineering, Espoo, Finland

Correspondence to: A. Osmane (a.osmane@unb.ca)

Abstract. Recent theoretical studies of the nonlinear wave-particle interactions for relativistic particles have shown that Landau resonant orbits could be efficiently accelerated along the mean background magnetic field for propagation angles θ in close proximity to a critical propagation θ_c associated with a Hopf–Hopf bifurcation condition. In this report, we extend previous studies to reach greater modeling capacities for the study of electrons in radiation belts by including longitudinal wave effects and inhomogeneous magnetic fields. We find that even though both effects can limit the surfatron acceleration of electrons in radiation belts, gains in energy of the order of 100 keV, taking place on one tenth of a millisecond, are sufficiently strong for the mechanism to be relevant to radiation belt dynamics.

1 Introduction

The following report aims at extending a theoretical model for wave-particle coherent interaction (Osmane and Hamza, 2012a; Osmane and Hamza, 2012b) to characterize electron dynamics in the Earth's radiation belts. Electrons with energies in the MeV range have been frequently measured in the inner and outer component of the radiation belt. An increase in relativistic electron flux observed for short time scales (from few hours to few days) and in correlation with an increase of magnetic activity during the recovery phase of geomagnetic substorms (Friedel et al., 2002; O'brien et al., 2003). Not only relativistic electrons constitute a threat to satellites and spacecrafts in orbit, but their production has to be understood in order to account for the magnetospheric energy budget. Thus far, numerous theoretical models have been proposed. They can be grouped into two categories: mechanisms relying on radial transport alone as well as those that rely on internal mechanisms such as wave-particle interactions and recirculation models (see, e.g., Friedel et al., 2002 and reference therein, Albert, 2002; Shprits et al., 2006; Summers and Omura, 2007). Radial transport is often described as a diffusion mechanism driven by the fluctuations in the large scale magnetosphere electric and magnetic fields. As the particles spread from the outer magnetosphere to smaller equatorial radial distances, L, the first two adiabatic invariants are conserved while the third one is violated, resulting in an increase of energy (Kulsrud, 2005). On the other hand, local wave-particle interactions, and other local mechanisms accelerate particles already present in the inner magnetosphere. Even though it is accepted that radial diffusion is an important transport mechanism, particle energization on time scales shorter than the drift period predicted by radial diffusion has lead to hold wave-particle interaction responsible for a number of observables. Events occurring on time scale of days are believed to originate from electromagnetic cyclotron waves and whistler waves through pitch angle scattering, whereas the more intense and monochromatic chorus waves are believed to be the source of the strong energy increase occurring on the shortest time scales (Miyoshi et al., 2002; Thorne et al., 2005; Albert, 2002; Summers and Omura, 2007). More recently, observations of peaks of phase space density, in contradiction with inward radial diffusion, have shown that wave-particle interactions are dominant mechanisms (Green and Kivelson, 2004; Horne et al., 2005; Chen et al., 2007).

Aside from the outstanding problem of flux enhancement of relativistic electrons in the radiation belt, wave-particle interaction could also be proven to hold a decisive role in a number of other magnetospheric problems. Among them lies the generation of relativistic electron microbursts observed in association with VLF chorus waves (Lorentzen et al., 2001; Summers and Omura, 2007; Omura Hikishima et al., 2010) as well as precipitation rates of electrons entering the loss cone in pitch angle.

Perhaps more importantly, the most recent waveforms measured in the radiation belts have revealed additional motivating reasons to consider the wave-particle interaction as a dominating energy-momentum exchange mechanism in radiation belt problems. Large-amplitude, monochromatic, obliquely propagating, and bursty waveforms were not only repeatedly measured in the radiation belt (Catell et al., 2008; Kellog et al., 2010; Kersten et al., 2011; Wilson III et al., 2011), but appeared correlated with electron energization (Wilson III et al., 2011) as well as relativistic microbursts events (Kersten et al., 2011). The correlation between chorus waves and electron energization in the radiation belts is not recent, but it is suspected that if such waveforms were more commonly present in the radiation belts they could be the dominant trigger responsible for the energization of electrons on short timescales. A study by Yoon (2011) has shown that if one solves the plasma equations self-consistently, such waveforms were indeed capable of accelerating electrons on kinetic time scales consistent with the observations. Even though our study lacks the levels of self-consistency provided by the numerical method developed by Yoon (2011), we will show hereafter that we arrive at similar conclusions if we choose parameters consistent with the radiation belt-measured waveforms.

The large-amplitude wave forms are observed with a longitudinal component and the analysis above needs to be conducted with the addition of this compressive electric component. Whereas the addition of the electrostatic field with the same phase as the electromagnetic components of the fields would result in the same condition for the surfatron process, a difference in phase would shift the Hopf bifurcation and have non-trivial effects that need to be scrutinized. Moreover, radiation belt electrons are confined in the magnetic field of the Earth, and one must take into account the effect of field inhomogeneities. We therefore proceed in this in this report by modifying a previously derived dynamical system (Osmane and Hamza, 2012a) to allow a study of relativistic electrons in radiation belts.

2 Longitudinal effects

We follow the procedure described in Osmane and Hamza (2012a); Osmane and Hamza (2012b) for the derivation of a dynamical system to study the interaction of an ion with an obliquely propagating wave composed of a transverse and longitudinal component. As shown in Fig. 1, the electromagnetic wave is composed of a transverse component along the $(\hat{x}, \hat{y})$ plane and a longitudinal component along the $\hat{z}$

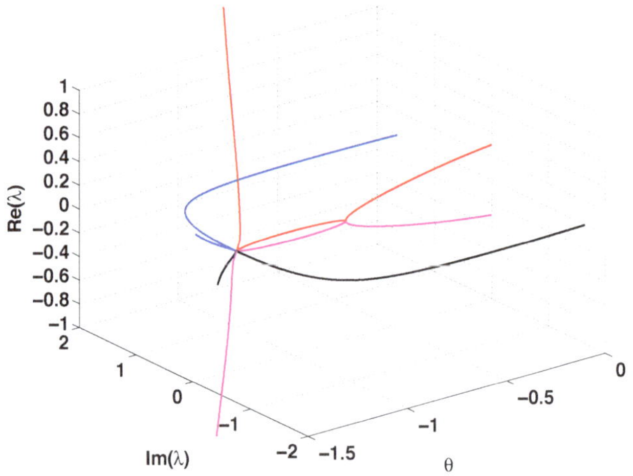

Fig. 1. Electromagnetic field configuration. A circularly polarized wave propagating obliquely to a background magnetic field $\mathbf{B}_0$ with a longitudinal component.

direction superposed onto a background magnetic field B_0 in the $(\hat{y}, \hat{z})$ plane:

$$\begin{aligned} \boldsymbol{E}(\boldsymbol{x}, t) &= \boldsymbol{\delta} \boldsymbol{E}_{\mathrm{em}}(\boldsymbol{x}, t) + \boldsymbol{\delta} \boldsymbol{E}_k(\boldsymbol{x}, t) \\ \boldsymbol{B}(\boldsymbol{x}, t) &= \boldsymbol{B}_0 + \delta \mathbf{B}(\boldsymbol{x}, t). \end{aligned} \tag{1}$$

We write the longitudinal component as a function of parameters η and Ψ as follows:

$$\boldsymbol{\delta} \boldsymbol{E}_k(\boldsymbol{x}, t) = \eta \delta E \sin(kz - \omega t + \Psi)\hat{z}. \tag{2}$$

The parameter η quantify the amplitude of the longitudinal component of the wave with respect to the electromagnetic component. Setting $\eta = 1$ would therefore result in having equal electric field amplitudes parallel and perpendicular to the wave vector $\boldsymbol{k}$. Setting $\eta = 0$ recovers the case treated in Osmane and Hamza (2012b). The parameter Ψ is a phase difference between the longitudinal component and the transverse component. It is added for the sake of completeness. The dynamical system equation is therefore not fundamentally modified. The difference resides in the addition of an electric field component along the z component of the Lorentz force. It is easy to show that the dynamical system takes the following form (Osmane and Hamza, 2012b) (see Appendix for detailed derivation of the purely transverse case):

$$\begin{cases} \dot{p}'_x = \Omega_0 p'_y \cos(\theta) - \Omega_1 p'_z \cos(kz') + \Omega_0 \left(p'_z + p_\phi\right) \sin(\theta) \\ \dot{p}'_y = -\Omega_0 p'_x \cos(\theta) + \Omega_1 p'_z \sin(kz') \\ \dot{p}'_z = -p'_x \Omega_0 \sin(\theta) + \Omega_1 \frac{n^2-1}{n^2} \left(p_x \cos(kz') - p'_y \sin(kz')\right) \\ \qquad - \frac{1}{n^2} \Omega_1 \eta p'_z \sin(kz' + \Psi) + \Omega_1 \frac{n^2-1}{n^2} \eta p_\phi \sin(k' + \Psi) \\ \dot{z}' = p'_z v_\Phi / p_\Phi \end{cases} \tag{3}$$

for the dynamical gyrofrequencies $\Omega_0 = \frac{e B_0}{m \gamma c}$ and $\Omega_1 = \frac{e \delta B}{m \gamma c}$, relativistic momentum $p_i = m \gamma v'_i$, refractive index $n = \frac{c}{v_\phi}$, phase speed $v_\Phi = \omega / k$ and dots indicating time derivatives.

The time evolution of the dynamical gyrofrequency can then be written as:

$$\frac{d\Omega_0}{dt} = \frac{-\Omega_0\,\Omega_1\,p_\Phi}{m^2\gamma^2c^2}\Big(p'_x\cos(k\,z') - p'_y\sin(k\,z') + \eta\left(p_\phi + p'_z\right)\sin(k\,z' + \Psi)\Big). \tag{4}$$

We can now proceed by studying the dynamical system properties in terms of the fixed points and their stability as well as the dependence on wave parameters such as wave obliquity θ and normalized wave amplitude $\delta_1 = \delta B/B_0$. More importantly, we would like to know whether the surfatron acceleration mechanism, the processes by which a particle is trapped along the wave vector and accelerated uniformly along the background magnetic field by the parallel electric field (Osmane and Hamza, 2012b), is also available when a longitudinal component is added. The surfatron is a trapping effect, and is therefore only achievable if a parallel electric field of sufficiently large amplitude is present. If the propagation of the wave is parallel, the parallel electric field is zero, and neither trapping along the wave vector nor acceleration along the background magnetic field is possible. We therefore assume that the longitudinal component can for various parameters either enhance or destroy the surfatron.

A quick look at the dynamical system (Eq. 3) shows that the fixed point for the electromagnetic case ($p'_x = p'_z = 0$, $p'_y = -p_\phi\tan(\theta)$, $k\,z' = 0$) exists for $\Psi = 0$. For $\Psi \neq 0$ no fixed points exist. Hence, one would expect the acceleration mechanisms associated with Hopf–Hopf bifurcation to be available for sufficiently large wave amplitude capable of trapping when $\Psi = 0$. Whenever $\Psi \neq 0$, one can think of the longitudinal component as a perturbation to uniformly accelerated particles (orbits). Hence, particles (orbits) could still be energized, but the longitudinal component could break the locking eventually. Additionally, if the longitudinal component of the wave, for given parameters η and Ψ, cancels the parallel component of the electric field with respect to the background magnetic field due to the electromagnetic component, no uniform acceleration should take place. This condition can be written as follows:

$$\delta\mathbf{E}\cdot\mathbf{b}_0 = -\delta E_y\sin(\theta) + \delta E_k\cos(\theta) = 0. \tag{5}$$

Setting $\Psi = 0$, one finds that this condition translates to $\eta = -\tan(\theta)$, for which neither trapping nor uniform acceleration should be possible. We therefore expect the lack of a parallel electric field component to translate into unstable orbits for linear perturbation around the fixed points. In the next section we apply the stability analysis to the dynamical system (Eq. 3) for the fixed point ($p'_x = p'_z = 0$, $p'_y = -p_\phi\tan(\theta)$, $k\,z' = 0$).

2.1 Stability analysis

We now proceed similarly as for the electromagnetic case (Osmane and Hamza, 2012b) to quantify the linear stability of the fixed point[1]. In order to do so we linearize the dynamical system and assume a perturbation of the form $\sum_{i=1}^{4}\xi_i\,e^{\lambda_i t}$, for which an imaginary eigenvalue translates into marginally stable orbits, negative eigenvalues into linearly stable orbits, and positive eigenvalues linearly into unstable orbits. Computing the resulting Jacobian evaluated at the fixed point $\mathbf{J} = \frac{\partial F}{\partial x_i}|_{\mathbf{x}_0}$ for the longitudinal case, we find the following matrix:

$$\mathbf{J} = \begin{pmatrix} 0 & a & b & 0 \\ c & 0 & 0 & 0 \\ d & 0 & 0 & e \\ 0 & 0 & f & 0 \end{pmatrix}$$

for the parameters δ_1, $\delta_2 = \frac{m\,\omega\,c}{e\,B_0}$, θ and η:

$$a = \frac{\cos(\theta)}{\delta_2\,\gamma_0}\left(1 - \frac{\tan(\theta)^2}{n^2 - 1}\right) \tag{6}$$

$$b = \frac{\mp\delta_1 + \sin(\theta)\frac{n^2}{n^2-1}}{\delta_2\,\gamma_0} \tag{7}$$

$$c = -\frac{\cos(\theta)}{\delta_2\,\gamma_0} \tag{8}$$

$$d = \left(-\sin(\theta) \pm \delta\frac{n^2 - 1}{n^2}\right)\frac{1}{\delta_2\,\gamma_0} \tag{9}$$

$$e = \mp\frac{\delta_1}{\delta_2}\frac{n^2 - 1}{n^2}(\tan(\theta) + \eta) - \lambda \tag{10}$$

$$f = \frac{1}{\gamma_0} = \sqrt{1 - \frac{v_\Phi^2}{c^2}\left(1 + \tan^2(\theta)\right)}. \tag{11}$$

Once again the dark $\pm$ and $\mp$ correspond to the fixed points components for $Z = (0,\pi)$, that is the upper sign for $Z = 0$ and the lower one for $Z = \pi$. Hence, all four fixed points are represented in this Jacobian matrix and their stability can be analyzed by choosing the right $\pm$ symbols. In order to find the eigenvalues, we need to solve the characteristic polynomial given by the following expression:

$$\chi(\lambda) = \begin{vmatrix} -\lambda & \frac{\cos(\theta)}{\delta_2\gamma_0}\left(1 - \frac{\tan(\theta)^2}{n^2-1}\right) & \frac{\mp\delta_1 + \sin(\theta)\frac{n^2}{n^2-1}}{\delta_2\gamma_0} & 0 \\ -\frac{\cos(\theta)}{\delta_2\gamma_0} & -\lambda & 0 & 0 \\ \left(-\sin(\theta) \pm \delta_1\frac{n^2-1}{n^2}\right)\frac{1}{\delta_2\gamma_0} & 0 & \mp\frac{\delta_1}{\delta_2}\frac{n^2-1}{n^2}(\tan(\theta) + \eta) & \\ 0 & 0 & \frac{1}{\gamma_0} & -\lambda \end{vmatrix}.$$

A little algebra results in the following bi-quadratic expression:

$$\chi(\lambda) = \lambda^4 + \zeta_1\lambda^2 + \zeta_2 = 0 \tag{12}$$

with the values ζ_1 and ζ_2 given by the following expressions:

[1]The fixed points are located by setting the four equations of motions as zero and resolving the remaining algebraic equations.

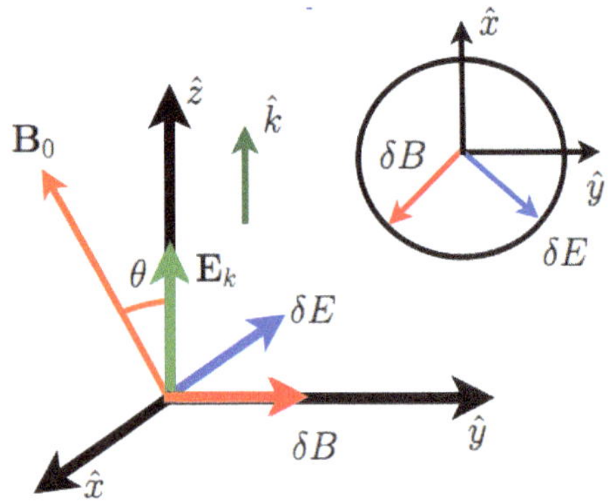

Fig. 2. Eigenvalues' dependence on the propagation angle θ for fixed parameters $\delta_1 = 0.07$, $\delta_2 = 0.21$, $n^2 = 2$, $\eta = 0.9$ and the fixed point of component $Z_0 = 0$. A Hopf–Hopf bifurcation takes place for $\tan(\theta_c)^2 = n^2 - 1$ and a second bifurcation takes place for $\eta = -\tan(\theta)$, corresponding to a null parallel electric field. The fixed point is stable for $\theta < \theta_c$ and $\eta > -\tan(\theta)$ and unstable for $\theta > \theta_c$ and $\eta > -\tan(\theta)$.

$$\zeta_1 = \frac{\delta_1}{\delta_2 \gamma_0} \frac{n^2 - 1}{n^2} (\tan(\theta) + \eta) + \frac{1}{\delta_2^2 \gamma_0^2} \left(\delta_1 \frac{n^2 - 1}{n^2} + 1 \mp 2\delta_1 \sin(\theta) \right)$$

$$\zeta_2 = \frac{\delta_1}{\delta_2^3 \gamma_0^5} \frac{n^2 - 1}{n^2} \left(\sin(\theta)\cos(\theta) + \eta \cos(\theta)^2 \right).$$

One can compare the Jacobian matrix as well as the characteristic equation for the purely electromagnetic case with the expressions above for an additional longitudinal component with $\Psi = 0$. It is clear that minimal differences arise, as denoted in the appearance of a factor of η in the Jacobian and the characteristic equation. The characteristic equation once again has four roots given by the following equation:

$$\lambda_{1,2,3,4} = \pm \sqrt{\frac{-\zeta_1 \pm \sqrt{\zeta_1^2 - 4\zeta_2}}{2}}. \quad (13)$$

Figure 2 shows the dependence of all four eigenvalue solutions for typical parameters relevant to space plasmas on the propagation angle. It is clear that the Hopf–Hopf bifurcation takes place once again for parameters resulting in $\gamma_0 = 0$. That is, whenever parameters are such that $n^2 - 1 = \tan^2(\theta)$, the fixed point evolves from marginally stable to linearly unstable when we add a longitudinal component with $\psi = 0$. Additionally, a second bifurcation takes place when $\eta = -\tan(\theta)$. Setting $\eta = -\tan(\theta)$ in the coefficients ζ_1 and ζ_2 results in the following characteristic equation: $\lambda^4 + \zeta_1 \lambda^2 = 0$. With $\zeta_1 > 0$ for $\delta_1 \sim O(1)$, it is clear that two eigenvalues are null and two eigenvalues are imaginary. As noted in the previous section, this expression denotes a null parallel electric field resulting in the destructive interference of the parallel longitudinal component and the parallel electromagnetic component. We now investigate the nonlinear effects of the longitudinal components on the surfatron process for various parameters η and ψ.

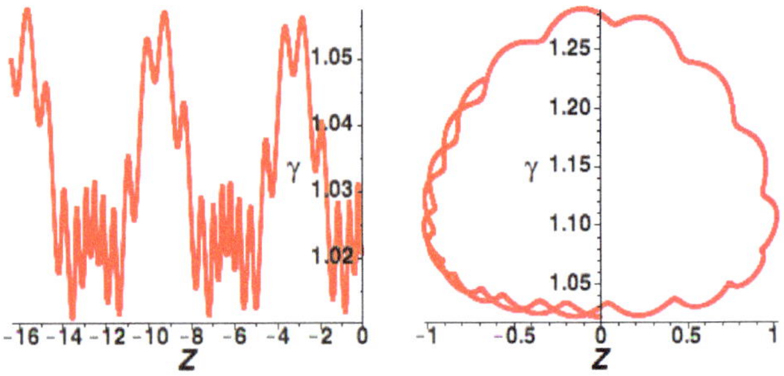

Fig. 3. Orbit for parameters $\delta_1 = 0.1$, $v_\phi = 0.33\,c$, $\theta = -31°$, $\delta_2 = 0.1$ for $\eta > -\tan(\theta)$ (right panel) and $\eta < -\tan(\theta)$ (left panel). A transition from untrapped to trapped orbit is observed as we evolve the parameter η.

2.2 Landau resonant orbits

In this section we determine whether the addition of the longitudinal component enhances or prevents the uniform acceleration for orbits caught in the basin of attraction centered at Landau resonance. As noted in the stability analysis, a Hopf–Hopf bifurcation does indeed take place when a longitudinal component is added. The main difference in the linear stability around the fixed points resides in the addition of a parallel electric field capable of canceling the electromagnetic component parallel to the background field. Hence, whenever $\eta = -\tan(\theta)$, the parallel component of the electric field is zero and the surfatron process cannot take place. Indeed, choosing the parameter η to coincide with $\sqrt{n^2 - 1}$ results in reducing the Hopf–Hopf bifurcation to a single Hopf bifurcation (one pair of imaginary eigenvalues crossing the real plane instead of two pairs). For such a parameter the surfatron process is not expected to be applicable because a parallel electric field causing the uniform acceleration is now set to a null value.

Figure 3 shows two orbits for $\eta > -\tan(\theta)$ (in the right panel) and $\eta < -\tan(\theta)$ (in the left panel). A transition from untrapped to trapped orbit is observed as we evolve the parameter η. For $\eta = \tan(\theta)$ the particle located sufficiently close to the fixed point (Landau resonant velocity) is trapped, but small perturbation results in untrapped orbits.

In Fig. 4, the four panels represent a seeded particle with energy of the order of 100 keV but for $\theta = 40°$ (up and left), $\theta = 55°$ (up and right), $\theta = 70°$ (down and left) and $\theta = 85°$ (down and right). The particle gains a maximum amount of energy for $\theta = 70°$, which corresponds to a propagation angle close to θ_c and to the surfatron process. The longitudinal component for parameters $\delta_1 = 0.045$, $v_\phi = 0.33\,c$, $\eta = -1$, and $\delta_2 = 0.1$ enhances the parallel electric field component and results in smaller range of values in $\delta_1 = \delta B / B_0$

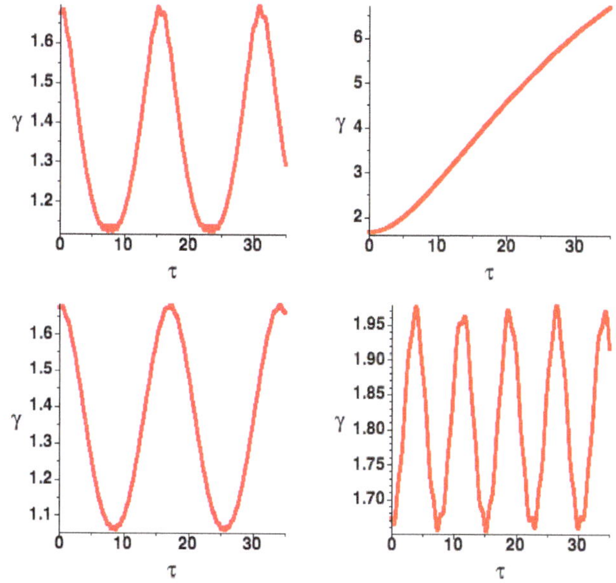

Fig. 4. Orbit for parameters $\delta_1 = 0.045$, $v_\phi = 0.33\,c$, $\eta = -1$, $\delta_2 = 0.1$ and $\theta = 40°$ (up and left), $\theta = 55°$ (up and right), $\theta = 70°$ (down and left) and $\theta = 85°$ (down and right). Each case is seeded with a particle of energy of the order of 100 keV. The particle gains a maximum amount of energy for $\theta = 70°$, which corresponds to a propagation angle close to θ_c. The longitudinal component can enhance the parallel electric field component and result in a smaller range of values $\delta_1 = \delta B/B_0$ for which the surfatron process is accessible. In this case the ratio $\delta_1 \sim 4\,\%$.

for which the surfatron process is accessible. In this case the ratio δ_1 is of the order of 4 %. This order of magnitude for the electromagnetic wave component is comparable to large-amplitude bursty waves observed in the radiation belts (Catell et al., 2008; Kellog et al., 2010; Kersten et al., 2011; Wilson III et al., 2011).

However, before concluding that the longitudinal component preserves the surfatron mechanism for $\theta = \theta_c$, we need to evaluate the effects of the phase difference Ψ. In Fig. 4 the parameter Ψ has been set to zero. Yet, when $\Psi \neq 0$, the dynamical system possesses no fixed points and the acceleration observed for the surfatron process should not arise uniformly. Panels in Fig. 5 show particles for $\eta = 1$, $\delta_1 = 0.06$, $\delta_2 = 0.1$, $n^2 = 9$, and $\theta = -71°$ and three different values for $\Psi = -\pi/4$, 0 and $\pi/4$. It is seen that depending on the phase difference, the surfatron process can still take take place for sufficiently long time to energize the particle. As for the purely electromagnetic case, a charged particle in this field would gain a significant amount of energy (from keV levels to MeV) on small kinetic time scales $\omega t \sim 0.1$ $\Omega_0 t \sim 10$. Hence a particle can be energized in such a field on time scales of the order of the 1/100 of a second for a wave frequency $\omega \sim 3$ kHz. Since the acceleration takes place on very small time scales, inhomogeneous effects should not prevent the mechanism entirely. We therefore conclude this section by suggesting that the large-amplitude electromagnetic waves observed in the radiation belts can energize particles efficiently on kinetic time scales for propagation angles close to the critical Hopf–Hopf bifurcation value θ_c. If the propagation angle is not sufficiently close to θ_c, then the particle will just oscillate back and forth in the potential of the electric field and no significant gain in energy should be observed. We now proceed in the next section by quantifying the inhomogeneous effects on the surfatron herein described.

3 Inhomogeneous magnetic field effects on the surfatron

In this section we want to include the effects of a non-homogeneous magnetic field on the acceleration process described in the previous section. We first discuss the motion of a particle in an inhomogeneous magnetic field with no wave-particle interaction. We then provide an approximation for the time scales for which inhomogeneous effects can result in surfatron breaking and a discussion on numerical integration of a particle trajectory interacting with a large-amplitude electromagnetic wave in an inhomogeneous magnetic field.

3.1 Particle orbits in an inhomogeneous magnetic field

Relativistic electrons trapped in the radiation belts bounce back and forth along the (approximately) dipolar magnetic field of the Earth. Before addressing the more complicated motion of relativistic electrons bouncing back and forth in the Earth's magnetic field and at the same time interacting with an obliquely propagating wave, we would like to quantify the impact of the magnetic field inhomogeneities on the relativistic motion. Assuming that the magnetic moment

$$\mu = \frac{p_\perp^2}{2B} = \frac{mc^2(\gamma^2 - 1)\sin(\alpha)^2}{2B} \tag{14}$$

is an adiabatic invariant, we can derive the forces due to the magnetic field inhomogeneities as follows.

The magnetic force perpendicular to the magnetic field can be deduced from the conservation of the magnetic moment:

$$\dot{p}_\perp = m\gamma\,\mu\,\dot{B}/p_\perp. \tag{15}$$

Replacing μ in terms of B and $p_\perp$, and $\dot{B} = v_\parallel \nabla_\parallel B$, we find the following expression for the force perpendicular to the magnetic field:

$$\dot{p}_\perp = \frac{p_\parallel\, p_\perp\, \nabla_\parallel B}{2m\,B\,\gamma}. \tag{16}$$

Using the above equation and assuming that the energy of the particle is conserved to first order in μ for a particle moving in an inhomogeneous magnetic field, we can write:

$$\gamma\dot{\gamma} = p_\parallel \dot{p}_\parallel + p_\perp \dot{p}_\perp = 0. \tag{17}$$

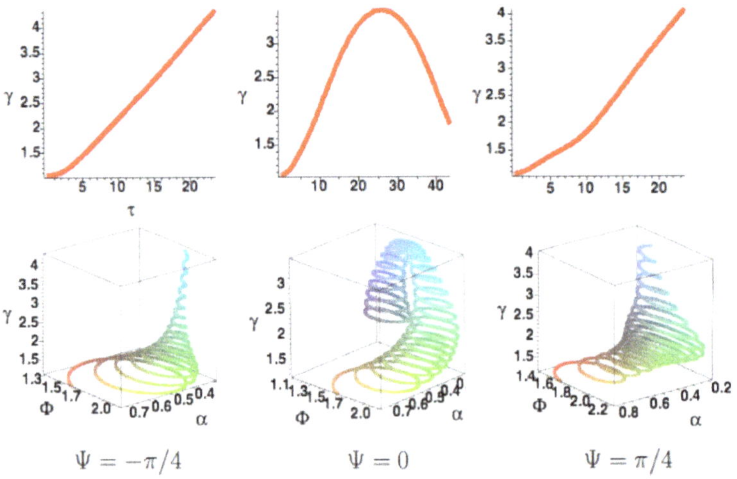

Fig. 5. Effect of the phase angle ψ for the longitudinal component on the surfatron process for parameters ($\delta_1 = 0.06$, $\delta_2 = 0.1$, $n^2 = 9$, $\theta = -71°$). The phase angle for a given parameter η determines whether the longitudinal parallel component is enabling or breaking the locking of particles into the surfatron.

Therefore, the equation of motion in the parallel direction can be written as:

$$\dot{p}_\parallel = -p_\perp \dot{p}_\perp / p_\parallel = -\frac{p_\perp^2 \nabla_\parallel B}{2 m B \gamma} = -\mu \nabla_\parallel B. \tag{18}$$

Setting $\gamma = 1$ in Eqs. (18) and (16) recovers the expressions for non-relativistic particles (Bell and Inan, 1981). We can show that similarly to the non-relativistic case, the conservation of magnetic moment results in magnetic trapping. Assuming a slab geometry for the magnetic field as shown in Eq. (6) for $z \sim s_\parallel$, i.e., the component along the parallel coordinate, and the magnetic field of the form $B(s_\parallel) = B_0(1 + \frac{9}{2} s_\parallel^2 / R^2)$ to mimic the dipolar field, for which R is equal to the Earth's radius, we can solve both Eqs. (18) and (16). Hence, replacing the expression for $B(s_\parallel)$ in Eq. (18) results in the following equation:

$$\frac{d^2 s_\parallel}{dt^2} + \frac{9 \mu B_0}{m \gamma R^2} s_\parallel = 0. \tag{19}$$

The solution of the above equation is therefore of the form $s_\parallel \sim \cos(\sqrt{\frac{9 \mu B_0}{m \gamma R^2}} t)$. Replacing the expression for the magnetic field in Eq. (16) for the perpendicular momentum, we find the following differential:

$$\frac{dp_\perp}{p_\perp} = \frac{9}{2 R^2} d\left(s_\parallel^2\right) \tag{20}$$

with a perpendicular momentum solution

$$p_\perp = p_{\perp 0} e^{\left(\frac{9}{2R^2} s_\parallel^2\right)}. \tag{21}$$

Replacing the solution for $s_\parallel$ in the above equation provides for a complete solution for the particle motion in the inhomogeneous field $B(s_\parallel) = B_0(1 + \frac{9}{2} s_\parallel^2 / R^2)$. We see that the particle oscillates back and forth along the parallel direction, while the perpendicular momentum increases as the particle reaches regions of larger magnetic field strength corresponding to $s_\parallel \sim R$.

3.2 Surfatron breaking due to inhomogeneous B field effect

Numerous effects can cause the breaking of the surfatron process: dispersive wave effects, dissipation of the wave amplitude, inhomogeneous magnetic fields damping the acceleration along the field line, or simply the result of precipitation into the atmosphere. Because of the slow time scales upon which the surfatron process becomes interesting to sustain particle precipitation, and since other effects would take place on longer time scales, we now focus solely on the inhomogeneous effect. That is, we want to obtain time scales for which the surfatron would not be prevented by field inhomogeneities.

The surfatron results in the parallel acceleration of a particle caused by the parallel component of the electric field. As demonstrated in the previous section, magnetic field inhomogeneities result in a $-\mu \nabla B$ force that can reduce the surfatron process, in the same way that the parallel electric field from a longitudinal component can prevent parallel acceleration. Hence, when $e \delta E_\parallel \sim \mu \nabla B$, parallel acceleration becomes marginal. This condition translates as follows:

$$\frac{s_\parallel}{R} \sim \frac{\delta_1}{n} \frac{\Omega_0 R}{\left(\gamma_0^2 - 1\right) \sin(\alpha_0)^2 c} \tag{22}$$

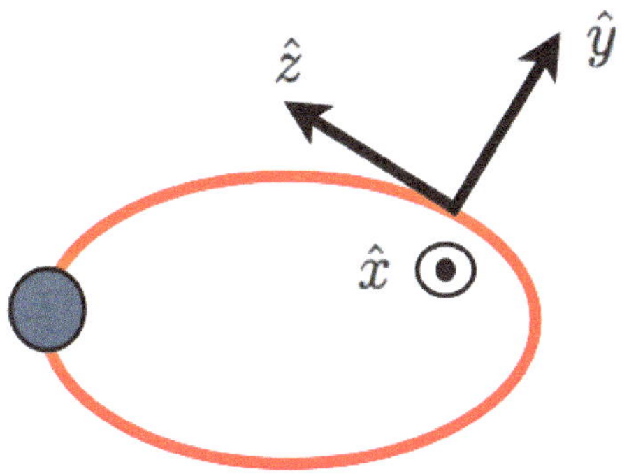

Fig. 6. Slab geometry for the dipolar magnetic field.

for the parameters $\delta_1 = \delta B/B_0$ and $n = c/v_\phi$, the gyrofrequency $\Omega_0 = e B_0/m c$ as well as the pitch angle α_0 and Lorentz factor γ_0 of the particle at the equatorial region. Setting $n = 3$, $\delta_1 \sim 0.01$, $R \sim 6000$ km, $\Omega_0 \sim 3 \times 10^4$ for a particle with an initial energy of the order of 100 keV and pitch angle of 45°, we find $s_\parallel \sim 10^4$ km. Therefore, we can conclude that a particle would gain energy of the order of $W \sim e\delta E_\parallel s_\parallel$, which corresponds to a gain in energy of the order of 100 keV for an electron interacting with an electric field of 100 mV m^{-1}. This approximation is comparable to energization of electrons reported by Artemyev et al. (2013) who found gains of the order of 80–100 keV for particles going through several Landau resonance in an inhomogeneous field. The difference with our result hereafter, however, is that the particle gains energy during the time of one Landau resonance, making the process much more efficient, even though less probable.

We now write a dynamical system for a relativistic charged particle interacting with an obliquely propagating wave in an inhomogeneous field. In order to make the set of equations more transparent to the reader, we write them in a coordinate axis for which $\hat{z} \parallel \hat{b}$, that is the background magnetic field is parallel to the z axis. We denote $\hat{y} = \hat{\perp}_1$ and $\hat{x} = \hat{\perp}_2$. Rewriting the magnetic field in terms of this coordinate system, we obtain:

$$\begin{cases} \delta B_x = \delta B \sin(\Phi) \\ \delta B_y = \delta B \cos(\Phi)\cos(\theta) \\ \delta B_\parallel = -\delta B \cos(\Phi)\sin(\theta) \end{cases} \tag{23}$$

and similarly, using Faraday's laws, we obtain the following components for the electric field:

$$\begin{cases} \delta E_x = -v_\Phi \delta B \cos(\Phi)/c \\ \delta E_y = v_\Phi \delta B \sin(\Phi)\cos(\theta)/c \\ \delta E_\parallel = -v_\Phi \delta B \sin(\Phi)\sin(\theta)/c \end{cases} \tag{24}$$

for the phase $\Phi = k_\parallel z + k_{\perp 1} y - \omega t = k z - k y - \omega t$. We choose the background magnetic field to be written as $\mathbf{B}_0 = -y B_0 g'(z)\hat{y} + B_0 g(z)\hat{z}$, for the function $g(z) = 1 + z^2/R^2$ and its partial derivative with respect to z, $g'(z)$, denoting the background magnetic field variation

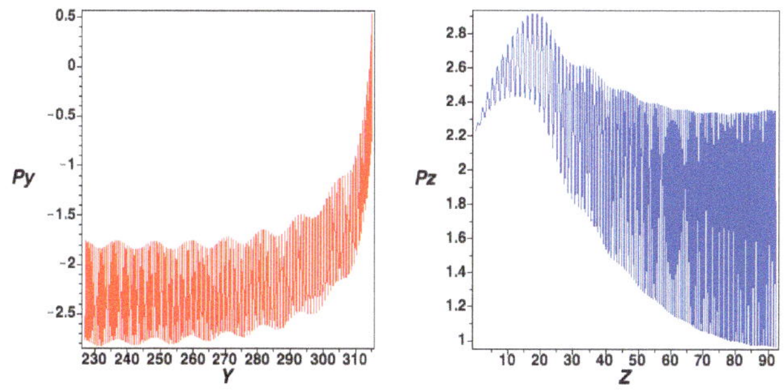

Fig. 7. Orbits for parameters $\delta_1 - 0.01$, $v_\phi = 0.33\,c$, $\delta_2 = 0.1$ and $\theta = 71.9°$. The left panel shows the perpendicular component of momentum P_y against the perpendicular position Y. The right panel shows the parallel component of momentum P_z against the parallel position Z. Initially the parallel component of momentum increases because of surfatron, until the $-\mu\nabla B$ force becomes sufficiently strong to break the trapping due to the parallel electric field. The particle gains energy of the order of 76 keV during the process.

as the particle propagates along the field line and away from the equatorial region $z \sim 0$. We can then write the dynamical system in terms of the variables (p_x, p_y, p_z, z, y, $\Omega_0 = e B_0/m\gamma c$, $\Omega_1 = e\delta B/m\gamma c$) and the functions $g(z)$ and $q(y, z) = y g'(z)$:

$$\begin{cases} \dot{p}_x = -p_\phi \Omega_1 \cos(\Phi) + p_y \Omega_0 g(z) - p_z \Omega_0 q(y,z) - p_y \Omega_1 \cos(\Phi)\sin(\theta) \\ -p_z \Omega_1 \cos(\Phi)\cos(\theta) \\ \dot{p}_y = p_\phi \Omega_1 \sin(\Phi)\cos(\theta) - p_x \Omega_0 g(z) + p_z \Omega_1 \sin(\Phi) + p_x \Omega_1 \cos(\Phi)\sin(\theta) \\ \dot{p}_z = p_x \Omega_0 q(y,z) - p_\phi \Omega_1 \sin(\Phi)\sin(\theta) + p_x \Omega_1 \cos(\Phi)\cos(\theta) - p_y \Omega_1 \sin(\Phi) \\ \dot{y} = p_y v_\Phi / p_\Phi \\ \dot{z} = p_z v_\Phi / p_\Phi \end{cases} \tag{25}$$

and

$$\begin{aligned} \dot{\Omega}_0 &= \frac{\mathrm{d}}{\mathrm{d}t}\left(\frac{e B_0}{m c \gamma}\right) \\ &= -\left(\frac{e B_0}{m c \gamma}\right)\frac{1}{\gamma}\frac{\mathrm{d}\gamma}{\mathrm{d}t} \\ &= -\Omega_0 \frac{p \dot{p}}{m^2 \gamma^2 c^2} \\ &= -\Omega_0 \frac{p c^2}{m^2 c^4 + p^2 c^2}\dot{p} \\ &= \frac{-\Omega_0 \Omega_1 p_\Phi}{m^2 \gamma^2 c^2}\left(\sin(\Phi)\left(p_y \cos(\theta) - p_z \sin(\theta)\right) - p_x \cos(\Phi)\right). \end{aligned} \tag{26}$$

We then proceed by normalizing the variables as follows: $p_i/m v_\phi = P_i$, $kz = Z$, $ky = Y$, $\Omega_0/\omega = \delta_3$ and $\omega t = \tau$, and write the dynamical system in terms of the normalized variables P_x, P_y, P_z, Y, Z, and δ_3 and the parameters δ_1, δ_2 and n as previously defined.

$$\begin{cases} \dot{P}_x = -\delta_1 \cos(\Phi)/\delta_2 + P_y \delta_3 g(Z) - P_z \delta_3 q(Y,Z) - P_y \delta_1 \delta_3 \cos(\Phi)\sin(\theta) \\ -P_z \delta_1 \delta_3 \cos(\Phi)\cos(\theta) \\ \dot{P}_y = \delta_1 \sin(\Phi)\cos(\theta)/\delta_2 - P_x \delta_3 g(Z) + P_z \delta_1 \delta_3 \sin(\Phi) + P_x \delta_1 \delta_3 \cos(\Phi)\sin(\theta) \\ \dot{P}_z = P_x \delta_3 q(Y,Z) - \delta_1 \sin(\Phi)\sin(\theta)/\delta_2 + P_x \delta_1 \delta_3 \cos(\Phi)\cos(\theta) - P_y \delta_1 \delta_3 \sin(\Phi) \\ \dot{Y} = \delta_2 \delta_3 P_y \\ \dot{Z} = \delta_2 \delta_3 P_z \\ \dot{\delta}_3 = -\frac{\delta_1 \delta_2 \delta_3^3}{n^2}\left(\sin(\Phi)\left(P_y \cos(\theta) - P_z \sin(\theta)\right) - P_x \cos(\Phi)\right) \end{cases} \tag{27}$$

We can now integrate this dynamical system for parameters relevant to radiation belt electrons with large-amplitude obliquely propagating waves to study the surfatron process in an inhomogeneous magnetic field mimicking the Earth's magnetic field. Figure 8 show orbits for parameters $\delta_1 = 0.01$, $v_\phi = 0.33\,c$, $\delta_2 = 0.1$ and $\theta = 71.9°$. The left panel shows the perpendicular component of momentum P_y against the perpendicular position Y. The right panel shows the parallel component of momentum P_z against the parallel position Z. Initially the parallel component of momentum increases because of surfatron, until the $-\mu\,\nabla B$ force becomes sufficiently strong to break the trapping due to the parallel electric field. The particle gains energy of the order of 76 keV during the process. This gain in energy is of similar order (~ 100 keV) as the one computed above for a simple balance of forces. Even though this gain appears modest, it should be kept in mind that it takes place on a time scale of the order of $\tau \sim 12$, hence $t \sim 10$ ms for a wave frequency $\omega \sim 3$ kHz. As the surfatron acceleration is lost, the conservation of the adiabatic invariant leads to a transfer of energy from the parallel direction to the perpendicular as noted by the continuous increase of the perpendicular momentum for $\tau > 12$. Since the gain in energy is irreversible, the particle uplifted by tens of keV can only oscillate back and forth in the potential of the wave and transfer energy along the perpendicular or parallel direction to preserve adiabatic invariance.

This point is clearly demonstrated in the right panel of Fig. 8. As the particle is uniformly accelerated, γ increases. Once the surfatron is broken, the particle resides in a state of higher energy. The left panel of the figure shows the three-dimensional orbit in velocity space Vx, Vy, Vz. For $\tau < 12$, the particle is uniformly accelerated through the surfatron process along the parallel direction. Once the surfatron is made inoperable, the magnetic field gradient dictates the particle orbits and conservation of μ leads to transfer of energy to the perpendicular direction. As denoted by the left panel of Fig. 8, the gyroradius of the particle increases ($V_y \gg V_z$) as the particle comes out of the surfatron.

Even though the surfatron accelerates particles parallel to the magnetic field, the inhomogeneous field results in redistributing the energy perpendicularly to the magnetic field. Hence, such a process, if statistically common in the radiation belts, could provide for an explanation to anisotropic distribution resulting in whistler wave turbulence without the need to resort to cyclotron resonance. Indeed, both resonant and nonresonant electron whistler instabilities require an initial anisotropy with $K_\perp$, the perpendicular kinetic energy density, to exceed $K_\parallel$, the parallel kinetic energy density by a certain amount. As the waves are being triggered by the instability and $K_\parallel / K_\perp$ reaching marginal stability levels, particles in the tail can be accelerated through the surfatron, travel toward a region of larger magnetic field and gain greater gyroradius, bounce back to equatorial region and contribute to the breaking of the marginal stability state of the whistler turbulence. This back and forth mechanism could then be stopped by precipitating the particles in the atmosphere, instead of having them bounce back toward the equatorial region.

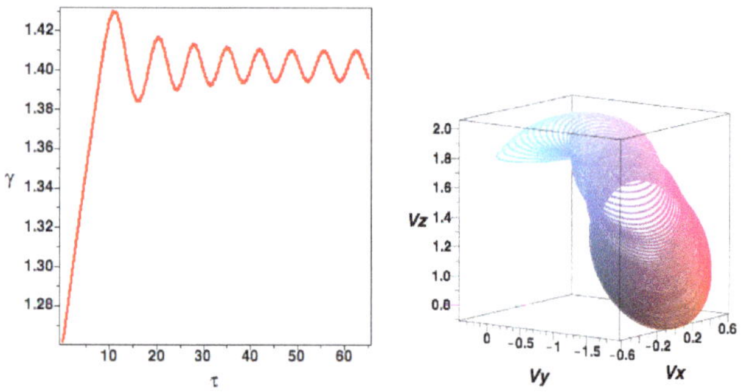

Fig. 8. Orbits for parameters $\delta_1 = 0.01$, $v_\phi = 0.33\,c$, $\delta_2 = 0.1$ and $\theta = 71.9°$. The left panel shows the relativistic Lorentz factor γ against normalized time $\tau = \omega t$. The right panel shows the three-dimensional orbit in velocity space Vx, Vy, Vz. For $\tau < 12$, the particle is uniformly accelerated through the surfatron process along the parallel direction. Once the surfatron is made inoperable, the magnetic field gradient dictates the particle orbits and conservation of μ leads to transfer of energy to the perpendicular direction. The gyroradius of the particle increases ($V_y \gg V_z$) as the particle comes out of the surfatron.

4 Conclusions

We extended a previous theoretical study (Osmane and Hamza, 2012b) of nonlinear wave-particle interactions for the study of electrons in radiation belts by including longitudinal wave effects and inhomogeneous magnetic fields. We found that, similarly than for the electromagnetic case in a uniform background field, the acceleration of particles along the background magnetic field, for propagation angles in close proximity to a critical propagation θ_c and associated with a Hopf–Hopf bifurcation condition, can arise on sufficiently small timescales to be of relevance to radiation belt dynamics. Even though longitudinal wave components and inhomogeneous magnetic fields can limit the surfatron acceleration of electrons in radiation belts, gains in energy of the order of 100 keV, taking place on one tenth of a millisecond, are sufficiently strong for the mechanism to sustain particle energization on timescales of the order of 0.1 ms. Future studies will investigate the effect of wave obliquity and field inhomogeneities on electron distribution functions for parameters consistent with radiation belt dynamics.

Appendix A

Dynamical system derivation for a transverse electromagnetic wave

Our starting point is the relativistic Lorentz equation for the motion of a particle in an electromagnetic field. The force is therefore written as

$$\frac{\mathrm{d}\boldsymbol{p}}{\mathrm{d}t} = e\left[\boldsymbol{E}(\boldsymbol{x}, t) + \frac{\boldsymbol{v}}{c} \times \boldsymbol{B}(\boldsymbol{x}, t)\right] \tag{A1}$$

for a particle of momentum $\boldsymbol{p} = m\,\gamma\,\boldsymbol{v}$, rest mass m and charge e. The Lorentz contraction factor γ is defined as follows:

$$\gamma = \frac{1}{\sqrt{1 - \frac{v^2}{c^2}}}. \tag{A2}$$

In order to avoid dealing with both the velocity and the momentum in the dynamical system, we simply write the equations in terms of the more physically relevant quantity of the two, that is the relativistic momentum $\boldsymbol{p}$:

$$\frac{\mathrm{d}\boldsymbol{p}}{\mathrm{d}t} = e\left[\boldsymbol{E}(\boldsymbol{x}, t) + \frac{\boldsymbol{p}}{m\,\gamma\,c} \times \boldsymbol{B}(\boldsymbol{x}, t)\right]. \tag{A3}$$

Similarly, the Lorentz factor can be written as follows:

$$\gamma = \frac{\sqrt{m^2\,c^4 + p^2\,c^2}}{m\,c^2}. \tag{A4}$$

The electromagnetic field is superposed onto a background magnetic field B_0.

$$\boldsymbol{E}(\boldsymbol{x}, t) = \delta\,\boldsymbol{E}(\boldsymbol{x}, t) \tag{A5}$$

$$\boldsymbol{B}(\boldsymbol{x}, t) = \boldsymbol{B}_0 + \delta\,\boldsymbol{B}(\boldsymbol{x}, t) \tag{A6}$$

The electromagnetic wave vector points in the $\hat{z}$ direction and the background magnetic field lies in the $y - z$ plane.

$$\boldsymbol{k} \cdot \boldsymbol{B}_0 = \boldsymbol{k}\,B_0\cos(\theta) \tag{A7}$$

$$\begin{cases} \delta\,\boldsymbol{E} = \delta\,E_x\,\hat{\boldsymbol{x}} + \delta\,E_y\,\hat{\boldsymbol{y}} \\ \delta\,\boldsymbol{B} = \delta\,B_x\,\hat{\boldsymbol{x}} + \delta\,B_y\,\hat{\boldsymbol{y}} \end{cases} \tag{A8}$$

where

$$\begin{cases} \delta\,B_x = \delta\,B\sin(k\,z - \omega\,t) \\ \delta\,B_y = \delta\,B\cos(k\,z - \omega\,t) \end{cases}. \tag{A9}$$

Faraday's law, expressed in terms of the Fourier components gives the relation between the components of the electromagnetic fields.

$$c\boldsymbol{k} \times \delta\,\boldsymbol{E}(\boldsymbol{k}, \omega) = \omega\,\delta\,\boldsymbol{B}(\boldsymbol{k}, \omega) \tag{A10}$$

The electric force is therefore written as

$$\begin{cases} F_{Ex} = e\,v_\phi\,\delta\,B\cos(k\,z - \omega\,t)/c \\ F_{Ey} = -e\,v_\phi\,\delta\,B\sin(k\,z - \omega\,t)/c \\ F_{Ez} = 0 \end{cases} \tag{A11}$$

for which $v_\phi = \omega/k$ is the phase velocity. Taking the cross product of the momentum and the magnetic field, the magnetic force is written as

$$\begin{cases} F_{Bx} = \frac{1}{m\,\gamma\,c}\left(p_y\,e\,B_0\cos(\theta) + p_z\,e\,B_0\sin(\theta) - p_z\,e\,\delta\,B_y\right) \\ F_{By} = \frac{1}{m\,\gamma\,c}\left(-p_x\,e\,B_0\cos(\theta) + p_z\,e\,\delta\,B_x\right) \\ F_{Bz} = \frac{1}{m\,\gamma\,c}\left(-p_x\,e\,B_0\sin(\theta) + p_x\,e\,\delta\,B_y - p_y\,e\,\delta\,B_x\right) \end{cases}. \tag{A12}$$

We can write the dynamical system equations in terms of the following variables: $p_\Phi = m\,\gamma\,v_\Phi$; $\Omega_1 = e\,\delta\,B/m\,c\,\gamma$; $\Omega_0 = e\,B_0/m\,c\,\gamma$, which results in the following equations:

$$\begin{cases} \dot{p}_x = p_y\,\Omega_0\cos(\theta) + (p_\Phi - p_z)\,\Omega_1\cos(k\,z - \omega\,t) + p_z\,\Omega_0\sin(\theta) \\ \dot{p}_y = -p_x\,\Omega_0\cos(\theta) + (p_z - p_\Phi)\,\Omega_1\sin(k\,z - \omega\,t) \\ \dot{p}_z = -p_x\,\Omega_0\sin(\theta) + p_x\,\Omega_1\cos(k\,z - \omega\,t) - p_y\,\Omega_1\sin(k\,z - \omega\,t) \\ \dot{z} = p_z\,v_\Phi/p_\Phi \end{cases}. \tag{A13}$$

In the classical case we have 4 equations to integrate, the three components of the velocity plus the position coordinate along $\boldsymbol{k}$. In the relativistic case, for a non-zero propagation angle, the energy of the particle is not a constant of the motion, that is, $\dot{\gamma} \neq 0$. Hence, we can think of the relativistic dynamical system as composed of 4 equations, the three components of the momentum plus the position coordinate along $\boldsymbol{k}$, and one constraint relating γ and the momentum components. Without any loss of generality we take the constraint into consideration by writing an equation for the dynamical gyrofrequency:

$$\begin{aligned} \dot{\Omega}_0 &= \frac{\mathrm{d}}{\mathrm{d}t}\left(\frac{e\,B_0}{m\,c\,\gamma}\right) \\ &= -\left(\frac{e\,B_0}{m\,c\gamma}\right)\frac{1}{\gamma}\frac{\mathrm{d}\gamma}{\mathrm{d}t} \\ &= -\Omega_0\frac{p\,\dot{p}}{m^2\,\gamma^2\,c^2} \\ &= -\Omega_0\frac{p\,c^2}{m^2\,c^4 + p^2\,c^2}\,\dot{p} \\ &= -\frac{\Omega_0\,\Omega_1\,p_\Phi}{m^2\gamma^2c^2}\left(p_x\cos(k\,z - \omega\,t) - p_y\sin(k\,z - \omega\,t)\right). \end{aligned} \tag{A14}$$

If we define the constant $\delta = \Omega_1/\Omega_0$, it is straightforward to see that

$$\dot{\Omega}_1 = \delta\,\dot{\Omega}_0. \tag{A15}$$

Since $p_\Phi = p_\Phi(\gamma)$, the time evolution of this quantity is written as:

$$\begin{aligned} \dot{p}_\Phi &= m\,v_\Phi\frac{\mathrm{d}\gamma}{\mathrm{d}t} \\ &= m\,v_\Phi\frac{p}{\gamma\,m^2\,c^2}\,\dot{p} \\ &= -m\,v_\Phi\,\gamma\,\frac{\dot{\Omega}_0}{\Omega_0}. \end{aligned} \tag{A16}$$

We can now eliminate the explicit time dependence of the equations by making a transformation of variables. Even though this transformation corresponds to a translation in the wave frame for low phase speed of the wave ($v_\phi \ll c$), it does not correspond to a physical frame of reference for phase speeds similar to the speed of light $v_\phi \sim c^2$. The explicit time dependence can therefore be eliminated by the following change of variables:

[2]The transformation in the position coordinates along z and v_z are Lorentz transformations in the wave frame, but because we do not also transform the time component into the wave frame ($t' = \gamma\,(t + v_\phi\,z/c^2)$), the dynamical system does not correspond to a particle orbit in the wave frame for relativistic regimes.

$$p'_x = p_x,\ p'_y = p_y,\ p'_z = \gamma_w \left(p_z - p_\phi\right),\ z' = \gamma_w (z - v_\phi t) \quad \text{(A17)}$$

for the Lorentz factor:

$$\gamma_w = \frac{1}{\sqrt{1 - \frac{v_\Phi{}^2}{c^2}}}. \quad \text{(A18)}$$

We can then write the equations of motion in terms of the new variables as follows:

$$\begin{cases} \dot{p}'_x = \Omega_0 p'_y \cos(\theta) - \Omega_1 p'_z \cos\left(k z'/\gamma_w\right)/\gamma_w + \Omega_0 \left(p'_z/\gamma_w + p_\phi\right) \sin(\theta) \\ \dot{p}'_y = -\Omega 0 p'_x \cos(\theta) + \Omega_1 p'_z \sin\left(k z'/\gamma_w\right)/\gamma_w \\ \dot{p}'_z/\gamma_w = -\Omega_0 p'_x \sin(\theta) + \Omega_1 p'_x \cos\left(k z'/\gamma_w\right) - \Omega_1 p'_y \sin\left(k z'/\gamma_w\right) - \dot{p}_\Phi \\ \dot{z}' = p'_z v_\Phi / p_\Phi \end{cases} . \quad \text{(A19)}$$

If we absorb the Lorentz factor γ_w into p'_z and k, that is, we write $p'_z \rightarrow p'_z/\gamma_w$ and $k \rightarrow k/\gamma_w$, and write $\dot{p}_\Phi$ in terms of $(p'_x, p'_y, p'_z, z', \Omega_0)$, we can write the dynamical system as follows:

$$\begin{cases} \dot{p}'_x = \Omega_0 p'_y \cos(\theta) - \Omega_1 p'_z \cos(k z') + \Omega_0 \left(p'_z + p_\phi\right) \sin(\theta) \\ \dot{p}'_y = -\Omega_0 p'_x \cos(\theta) + \Omega_1 p'_z \sin(k z') \\ \dot{p}'_z = -\Omega_0 p'_x \sin(\theta) + \Omega_1 \left(\frac{n^2-1}{n^2}\right) \left(p'_x \cos(k z') - p'_y \sin(k z')\right) \\ \dot{z}' = p'_z v_\Phi / p_\Phi \end{cases} \quad \text{(A20)}$$

with the refractive index $n^2 = c^2/v_\Phi^2$. The magnitude of the momentum is now written as $p' = \sqrt{p'^2_x + p'^2_y + (p'_z/\gamma_w)^2}$, hence the Lorentz contraction factor also transforms from $\gamma(p) \rightarrow \gamma(p')$.

Acknowledgements. This work was supported by the Natural Sciences and Engineering Research Council of Canada (NSERC). One of the authors, A. M. Hamza, wishes to acknowledge CSA (Canadian Space Agency) support. Computational facilities are provided by ACEnet, the regional high-performance computing consortium for universities in Atlantic Canada.

Edited by: T. Hada

References

Albert, J. M.: Nonlinear interaction of outer zone electrons with VLF waves, Geophys. Res. Lett., 29, 116-1–116-3, doi:10.1029/2001GL013941, 2002.

Artemyev, A. V., Krasnoselskikh, V. V., Agapitov, O. V., Mourenas, D. and Rolland, G.: Non-diffusive resonant acceleration of electrons in the radiation belts, Phys. Plasmas, 19, 122901, doi:10.1063/1.4769726, 2013.

Bell, T. F. and Inan, U.: Transient nonlinear pitch angle scattering of energetic electrons by coherent vlf wave packets in the magnetosphere, J. Geophys. Res., 86, 9047–9063, 1981.

Cattell, C., Wygant, J. R., Goetz, K., Kersten, K., Kellogg, P. J., von Rosenvinge, T., Bale, S. D., Roth, I., Temerin, M., Hudson, M. K., Mewaldt, R. A., Wiedenbeck, M., Maksimovic, M., Ergun, R., Acuna, M., and Russell, C. T.: Discovery of very large amplitude whistler-mode waves in Earth's radiation belts, Geophys. Res. Lett. 35, L01105, doi:10.1029/2007GL032009, 2008.

Chen, Y., Reeves, G. D., and Friedel, R. H. W.: The energization of relativistic electrons in the outer van allen radiation belt, Nature, 3, 614–617, 2007.

Friedel, R. H. W., Reeves, G. D., and Obara, T.: Relativistic electron dynamics in the inner magnetosphere – a review, J. Atmos. Sol.-Terr. Phy., 64, 265–282, 2002.

Green, J. C. and Kivelson, M. G.: Relativistic electrons in the outer radiation belt: Differentiating between acceleration mechanisms, J. Geophys. Res., 109, A03213, doi:10.1029/2003JA010153, 2004.

Horne, R. B., Thorne, R. M., Shprits, Y. R., Merredith, N. P., Glauert, S. A., Smith, A. J., Kanekal, S. G., Baker, D. N., Engebretson, M. J., Posch, J. L., Spasojevic, M., Inan, U. S., Pickett, J. S., and Decreau, P. M.: Wave acceleration of electrons in the van allen radiation belts, Nature, 437, 227–230, 2005.

Kellogg, P. J., Cattell, C. A., Goetz, K., Monson, S. J., and Wilson III, L. B.: Electron trapping and charge transport by large amplitude whistlers, Geophys. Res. Lett., 37, L20106, doi:10.1029/2010GL044845, 2010.

Kersten, K., Cattell, C. A., Breneman, A., Goetz, K., Kellogg, P. J., Wygant, J. R., Wilson III, L. B., Blake, J. B., Looper, M. D., and Roth, I.: Observation of relativistic electron microbursts in conjunction with intense radiation belt whistler-mode waves, Geophys. Res. Lett., 38, 8107, doi:10.1029/2011GL046810, 2011.

Kulsrud, R. M.: Plasma physics for astrophysics, Princeton University Press, New Jersey, 2005.

Lorentzen, K. R., Blake, J. B., Inan, U. S., and Bortnik, J.: Observations of relativistic electron microburts in association with vlf chorus, J. Geophys. Res., 106, 6017–6027, 2001.

Miyoshi, Y., Morioka, A., and Obara, T.: Large enhancementof the outer belt electrons during magnetic storms, Earth Planets Space, 53, 1163–1170, 2002.

O'brien, T. P., Lorentzen, K. R., Mann, I. R., Meredith, N. P., Blake, J. B., Fennell, J. F., Looper, M. D., Milling, M. K., and Anderson, R. R.: Energization of relativistic electrons in the presence of ulf power and mev microbursts: Evidence for dual vlf and vlf acceleration, J.Geophys. Res., 108, 1329, doi:10:1029/202JA009784, 2003.

Omura Hikishima, M. and Summers, D.: Microburst precipitation of energetic electrons associated with chorus wave generation, Geophys. Res. Lett., 37, L07103, doi:10.1029/2010GL042678, 2010.

Osmane, A. and Hamza, A. M.: Relativistic acceleration of Landau resonant particles as a consequence of Hopf bifurcations, Phys. Plasmas, 19, 030702, doi:10.1063/1.3692234, 2012a.

Osmane, A. and Hamza, A. M.: Dynamical-systems approach to relativistic nonlinear wave-particle interaction in collisionless plasmas, Phys. Rev. E, 85, 056410, doi:10.1103/PhysRevE.85.056410, 2012b.

Shprits, Y. Y., Thorne, R. M., Horne, R. B., Glauert, S. A., Cartwright, M., Russell, C. T., Baker, D. N., and Kanekal, S. G.: Acceleration mechanism responsible for the formation of the new radiation belt during the 2003 halloween solar storm, J. Geophys. Res., 33, L05104, doi:10.1029/2005GL024256, 2006.

Summers, D. and Omura, Y.: Ultra-relativistic acceleration of electrons in planetary magnetospheres, Geophys. Res. Lett., 34, 24205, doi:10.1029/2007GL032226, 2007.

Thorne, R. M., Horne, B., Glauert, S., Meredith, N., Shprits, Y. Y., Ss, D., and Anderson, R.: The influence of wave-particle interactions on relativistic electron dynamics during storms, in: vol. Inner Magnetosphere Interactions: New Perspectives From Imaging, Geophysical Monograph 159, edited by: Burch, J., Schulz, M., and Spence, H., American Geophysical Union, Washington, D.C., 2005.

Wilson III, L. B., Cattell, C. A., Kellogg, P. J., Wygant, J. R., Goetz, K., Breneman, A., and Kersten, K.: The properties of large amplitude whistler mode waves in the magnetosphere: Propagation and relationship with geomagnetic activity, Geophys. Res. Lett., 38, L17107, doi:10.1029/2011GL048671, 2011.

Yoon, P. H.: Large-amplitude whistler waves and electron acceleration, Geophys. Res. Lett., 38, L12105, doi:10.1029/2011GL047893, 2011.

17

A non-Gaussian analysis scheme using rank histograms for ensemble data assimilation

S. Metref[1], E. Cosme[1], C. Snyder[2], and P. Brasseur[1]

[1]CNRS – Univ. Grenoble Alpes, LGGE (UMR5183), 38041 Grenoble, France
[2]National Center for Atmospheric Research, Boulder, Colorado, USA

Correspondence to: S. Metref (sammy.metref@legi.grenoble-inp.fr)

Abstract. One challenge of geophysical data assimilation is to address the issue of non-Gaussianities in the distributions of the physical variables ensuing, in many cases, from nonlinear dynamical models. Non-Gaussian ensemble analysis methods fall into two categories, those remapping the ensemble particles by approximating the best linear unbiased estimate, for example, the ensemble Kalman filter (EnKF), and those resampling the particles by directly applying Bayes' rule, like particle filters. In this article, it is suggested that the most common remapping methods can only handle weakly non-Gaussian distributions, while the others suffer from sampling issues. In between those two categories, a new remapping method directly applying Bayes' rule, the multivariate rank histogram filter (MRHF), is introduced as an extension of the rank histogram filter (RHF) first introduced by Anderson (2010). Its performance is evaluated and compared with several data assimilation methods, on different levels of non-Gaussianity with the Lorenz 63 model. The method's behavior is then illustrated on a simple density estimation problem using ensemble simulations from a coupled physical–biogeochemical model of the North Atlantic ocean. The MRHF performs well with low-dimensional systems in strongly non-Gaussian regimes.

1 Introduction

The principal goal of data assimilation is to estimate the state of a dynamical system, based on prior information and a time series of observations, while calculating probabilistic measures corresponding to the accuracy of this estimation. Kalman filter theory (Kalman, 1960) became a reference in data assimilation as it provides the optimal solution to the linear and Gaussian filtering problem (Cohn, 1997). The ensemble Kalman filter (EnKF, Evensen, 1994) and closely related methods (Lermusiaux, 1999; Pham, 2001; Whitaker and Hamill, 2002, to cite only a few) are different implementations of the Kalman filter relying on ensembles. For the analysis step, they transform the prior ensemble into a posterior ensemble, using a function that is optimal (an optimal map, Cotter and Reich, 2013) under the assumption of Gaussianity of the prior ensemble and the observation errors. Those methods are applicable to high-dimensional systems in meteorology (Whitaker et al., 2008; Buehner et al., 2010) and oceanography (Lermusiaux, 2006; Sakov et al., 2012). This success is – in part – due to the fact that the dynamics of these systems are weakly nonlinear, that is, do not strongly deviate from a linear evolution within the space and timescales characterizing the density of available observations. Briefly, a weak nonlinearity transforms a Gaussian distribution into a weakly non-Gaussian distribution, with which the EnKF still performs well. Many recipes have been developed to enforce the good behavior of the EnKF with such systems, including localization techniques (Sakov and Bertino, 2010; Greybush et al., 2011), sampling strategies (Pham, 2001; Anderson, 2012), and observational targeting (Bishop et al., 2001); see Bocquet et al. (2010) for more details and other examples. Nevertheless, EnKF-based methods remain sensitive to the violation of the Gaussian assumption (Lawson and Hansen, 2004; Lei et al., 2010) and may lead to unwanted phenomena such as inaccurate estimations, failure to respect nonlinear physical balances, or more dramatically to instability of the filter.

Along with the developments of the EnKF, there is a growing need for non-Gaussian ensemble data assimilation methods. Data assimilation is no longer a tool solely for meteorology and oceanography. Other disciplinary fields with stronger nonlinearities and much sparser data networks (i.e., geomagnetism; Fournier et al., 2010) increasingly depend upon data assimilation. Even in the traditional fields of application, models' nonlinearity tends to increase along with their complexity. Even with linear models, non-Gaussian observation error densities make the assimilation problems non-Gaussian. Intrinsically, non-Gaussian variables are common in the atmosphere and the ocean, such as humidity (Dee and Da Silva, 2003) and concentrations of sea ice or phytoplankton (Brankart et al., 2012).

The ensemble data assimilation methods can be sorted in two categories: those that transform the prior ensemble particles using a deterministic map (transform methods), and those that sample the posterior probability density (sampling methods). They can also be classified as parametric, non-parametric, or semi-parametric, depending on the assumptions on the shape of the probability densities they use; the EnKF falls in the parametric (with the Gaussian assumption) transform methods. The EnKF with Gaussian anamorphosis, further described in Sect. 2 of the present paper, transforms variables to make their densities Gaussian before applying the EnKF analysis. It can be considered as a semi-parametric transform method, since it is not a fully non-parametric method able to deal with any kind of probability density, as illustrated in Sect. 2. The truncated-Gaussian EnKF as described by Lauvernet et al. (2009) is of the parametric, sampling type. Reich (2013) introduces a sequential method of the non-parametric, transform category. The particle filter (Gordon et al., 1993; van Leeuwen, 2009) is the most popular method of the non-parametric, sampling type, but is well known to be particularly subject to the curse of dimensionality, which makes it difficult to use with high-dimensional systems (Snyder et al., 2008). Finding solutions to make the particle filter applicable to high-dimensional systems is a very active topic of research (Nakano et al., 2007; van Leeuwen, 2010; Morzfeld et al., 2012; Snyder, 2012). A few of them actually rely on some hybridization with the EnKF (Bocquet et al., 2010; Lei et al., 2010; Hoteit et al., 2012).

Ensemble methods of the non-parametric, transform category have rarely been explored, although they could be less sensitive to the curse of dimensionality than the sampling methods, due to the transformation step that helps enforce a better fit of particles to the observations. In this respect, the approach proposed by Reich (2013) would deserve further examination, in particular with high-dimensional systems. Another method that could be somewhat classified as a partly non-parametric transform method is the rank histogram filter (RHF, Anderson, 2010). The RHF is a hybrid between the EnKF and a fully non-Gaussian approach, named that way because it is based on a statistical processing similar to the rank histograms (Anderson, 1996; Hamill, 2001) used for ensemble forecast evaluation. The RHF corrects observed variables by representing their prior densities and the observation likelihoods as piecewise continuous functions in order to directly apply Bayes' rule. This theoretically solves the generalized problem for a single observed variable. However, the other variables are still corrected using a linear regression onto the corrections of observed variables, as in the EnKF. We believe these advantages justify a more detailed exploration of the RHF philosophy. The main objective of this paper is to present an extension of the rank histogram approach of Anderson (2010) to unobserved variables yielding a fully non-parametric transform scheme for ensemble data assimilation, in the spirit of the method of Reich (2013). Throughout the paper, this multivariate RHF is referred to as MRHF.

The article is outlined as follows. In Sect. 2, we present some considerations on the joint non-Gaussianity of two variables, and how ensemble analysis schemes perform with such densities. Emphasis is given to the EnKF and to the RHF of Anderson (2010). Section 3 develops the extension of the RHF to unobserved variables called MRHF along with an approximation of the latter. Numerical experiments are presented in Sect. 4, where the MRHF and its approximation are evaluated with the highly nonlinear and non-Gaussian Lorenz 63 model, and compared in different setups (corresponding to different levels of nonlinearities) to the EnKF, to the RHF and to a particle filter. In Sect. 5, the new schemes are finally illustrated with a density estimation problem based on a realistic ensemble from a coupled marine biogeochemical model. Even though this last experiment is not a data assimilation problem, the results give an insight into the behavior of the method. A discussion and a conclusion are given in the last section.

2 Gaussian and non-Gaussian analysis in ensemble filtering

The increasing popularity of ensemble filters is largely due to the relative simplicity of their implementation. They basically alternate propagation steps and analysis steps. During a propagation step, each particle of the ensemble is advanced in time using the dynamical system model, possibly including some parameterization of the model error. An analysis step occurs after a propagation step, when an observation $\boldsymbol{Y}^{\mathrm{m}}$ is available. The observation $\boldsymbol{Y}^{\mathrm{m}}$ is a realization of the (random) measurement vector $\boldsymbol{Y}^{\mathrm{o}} = h(\boldsymbol{X}) + \epsilon$, where h is a forward observation operator, $\boldsymbol{X}$ the state vector to be estimated, and ϵ the observation error. The analysis step conflates the prior ensemble $\{\boldsymbol{X}_i^f\}_{i=1,\ldots,N_\mathrm{e}}$, composed of N_e particles resulting from the previous forecast, and the available observation $\boldsymbol{Y}^{\mathrm{m}}$, to provide a posterior (analysis) ensemble $\{\boldsymbol{X}_i^{\mathrm{a}}\}_{i=1,\ldots,N_\mathrm{e}}$. Observation errors are often assumed temporally and spatially uncorrelated so that each one can be independently assimilated (Houtekamer and Mitchell, 2001; Evensen, 2003). If the spatial correlations cannot be neglected, a linear transformation of the observation vector is theoretically possible (and exact in the linear and Gaussian

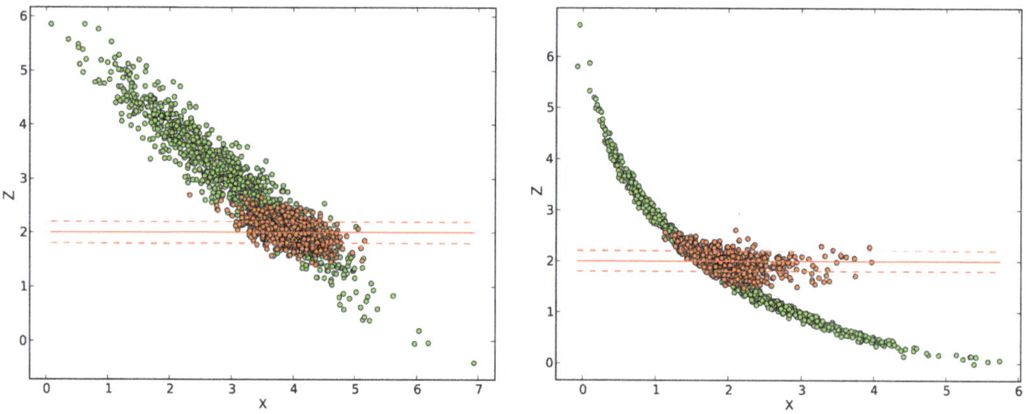

Figure 1. Illustration of EnKF analyses with joint Gaussian (left panel) and weakly non-Gaussian (right panel) prior distributions. The prior ensemble is represented by the green dots and the posterior ensemble is represented by the red dots. The variable corresponding to the z axis is observed with a value shown by the red solid line. Its uncertainty (identical in all the illustrations of Sect. 2) is assumed Gaussian with standard deviations symbolized with the red dashed lines.

context), in which the observation error covariance matrix is diagonal (Anderson, 2003).

A popular implementation of the ensemble analysis is the EnKF with serial processing of observations (Houtekamer and Mitchell, 2001). Following the description given by Anderson (2003), the ensemble update by each observation is performed in two steps, namely, the update of the observed variable followed by the update of unobserved variables. If the observation is not a direct observation of a state variable, then the state vector $\boldsymbol{X}$ can be augmented with the observed function of state variables, $\boldsymbol{Y}^{\mathrm{o}}$, to introduce a directly observed variable. In what follows, only the case of the direct observation of a variable is considered. With two scalar variables, $\boldsymbol{X}=(x, z)^T$, z being subject to a direct measurement z^{o} with realization z^{m}, decomposing the EnKF analysis equations shows that the correction of the observed variable z for ensemble particle i is

$$\delta z_i = \frac{\mathrm{Var}(z)}{\mathrm{Var}(z)+\mathrm{Var}(\epsilon)}\left(z^{\mathrm{m}} - z_i - \epsilon_i\right), \tag{1}$$

where ϵ_i is a perturbation that takes the observation error ϵ into account. The correction for the unobserved variable x, for particle i, is

$$\delta x_i = \frac{\mathrm{Cov}(x, z)}{\mathrm{Var}(z)}\delta z_i. \tag{2}$$

It is clear from the latter equation that the correction of any unobserved variable is a function of the linear correlation between the variable and the observed variable. Such formulation must be questioned when the linear correlation is not a relevant measure of the statistical relationship between these variables, as may occur when the statistics are non-Gaussian. This issue is illustrated in Fig. 1: in the left panel, a prior ensemble of a bivariate Gaussian state (x, z) is depicted (green dots); the second variable z is observed. The corrections for the unobserved variable, based on a relevant linear correlation with the observed variable, leads to an analysis ensemble (red dots) fitting the bivariate Gaussian probability density function (pdf) that would be produced by implementing Bayes' theorem. This analysis ensemble is consistent with both the physics, introduced through the prior information, and the observation. In the right panel, the two (non-jointly Gaussian) variables exhibit a nonlinear statistical relationship, that cannot be fully captured by a linear regression. Consequently, even if the corrections of the observed variable are somewhat correct, those of the unobserved variable can be erroneous. This results in a rather poor analysis ensemble, where particles appear in unexpected parts of the phase space, in violation of the inter-variable relationship, as described by the prior ensemble.

This problem is not new; it falls under the general designation of non-Gaussian data assimilation. Solutions exist, among the "resampling" methods in particular, but their effective application in high dimensions is either impossible or requires further development; see Bocquet et al. (2010) for a review. Some non-Gaussian schemes derive from refinements to the EnKF. A promising one, mostly studied in oceanography, is the Gaussian anamorphosis (Bertino et al., 2003; Simon and Bertino, 2009; Béal et al., 2010; Brankart et al., 2012). Anamorphosis consists in transforming the initial physical variables to make them fit Gaussian distributions. The standard EnKF analysis can be applied to these transformed variables. Then, the physical analysis variables are recovered by the inverse transformation. The transformation can be either analytical or numerical (Bocquet et al., 2010). The following illustration is performed with the numerical transformation described by Brankart et al. (2012). The left panel of Fig. 2 shows the same non-Gaussian prior ensemble as the right panel of Fig. 1 (green dots), along with the analysis ensemble obtained using Gaussian

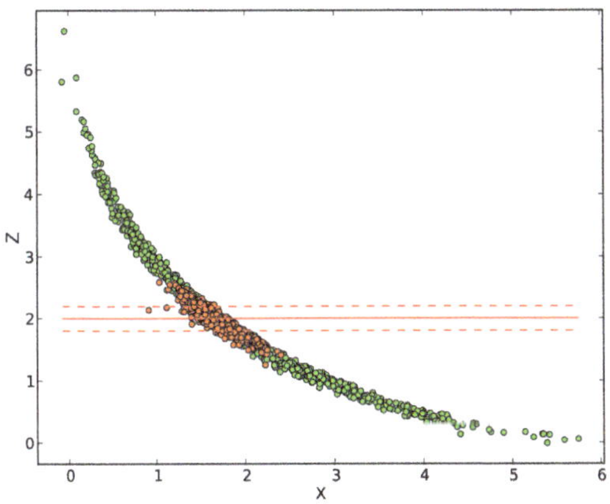

Figure 2. Same as Fig. 1 but for the EnKF with Gaussian anamorphosis and for joint weakly non-Gaussian (left panel) and strongly non-Gaussian (right panel) prior distributions. The weakly non-Gaussian prior is the same as in Fig. 1 (right panel).

anamorphosis (red dots). Anamorphosis clearly improves the EnKF. But anamorphosis presents several limitations, one of which is that it is based on a one-to-one correspondence between the prior and the target (Gaussian) distributions involved in the transformation. If this is not verified, Gaussian anamorphosis fails. Such failure is depicted on the right panel of Fig. 2, where the prior ensemble follows a strongly non-Gaussian density law which exhibits bimodality under conditioning by z. The EnKF with Gaussian anamorphosis (or without; not shown) provides a very poor analysis ensemble.

Fully non-Gaussian ensemble analysis schemes, that is, schemes derived without any assumption on the shape of the prior ensemble density, implement Bayes' rule to solve the analysis step:

$$p\left(\boldsymbol{X}|\boldsymbol{Y}^{\mathrm{o}}\right) \propto p(\boldsymbol{X})\,p\left(\boldsymbol{Y}^{\mathrm{o}}|\boldsymbol{X}\right), \qquad (3)$$

where $p(\boldsymbol{X})$ is the prior probability density for the state vector $\boldsymbol{X}$ to estimate, $p(\boldsymbol{Y}^{\mathrm{o}}|\boldsymbol{X})$ the observation likelihood (identical to the observation density for Gaussian observation errors), and $p(\boldsymbol{X}|\boldsymbol{Y}^{\mathrm{o}})$ the posterior density, that is, the density of the state given the observations. A detailed Bayesian description of data assimilation is provided by Wikle and Berliner (2007) for instance. The fully non-Gaussian ensemble data assimilation problem is usually solved by resampling methods of the particle filter type. The particle filter (Gordon et al., 1993; Doucet et al., 2001) is subject to very active developments for its future application with high-dimensional geophysical problems (Nakano et al., 2007; van Leeuwen, 2009, 2010; Morzfeld et al., 2012). The key point in implementing a particle filter is to beat the curse of dimensionality, and that will probably not be solved shortly for large applications (Snyder et al., 2008). One major reason, we believe, is that particle filters have yet to implement localization, in which any given observation affects the update only in a spatially local region near the observation location and as is common in the EnKF (Houtekamer and Mitchell, 1998; Hamill et al., 2001).

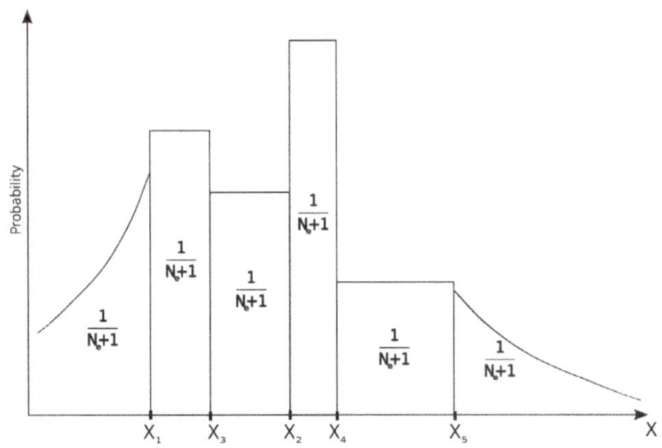

Figure 3. Reconstruction of a density from an ensemble using the rank histogram approach.

Here, we explore a new non-Gaussian ensemble analysis scheme of the "transform" type, in which localization can be implemented. We start from the the rank histogram filter (RHF), a partially non-Gaussian transform scheme, that has been proposed by Anderson (2010). The RHF processes observations serially. For the direct observation z^{o} of variable z, the continuous prior density for z is represented as a rank histogram (by analogy with the rank histogram diagnostic used to evaluate ensemble predictions, as introduced in geophysics by Anderson (1996), and later discussed in Hamill (2001), and Candille and Talagrand (2005)). The histogram is composed of $N_{\mathrm{e}}-1$ bounded regions partitioned by the sorted ensemble particles (the order statistics of the problem) and two unbounded regions on the tails. In each inner region, a density value is assigned so that the region contains a probability mass of $\frac{1}{N_{\mathrm{e}}+1}$ (Fig. 3). The two outer regions are covered by tails of probability mass $\frac{1}{N_{\mathrm{e}}+1}$ as well; their shape may be chosen freely, and this may actually be a key element for the success of the RHF (Anderson, 2010). In particular,

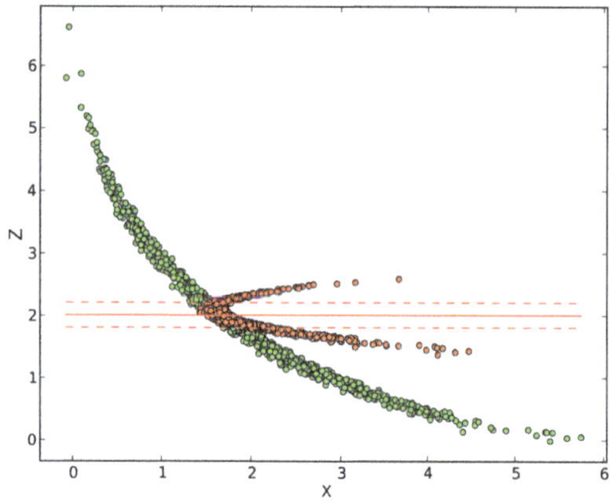

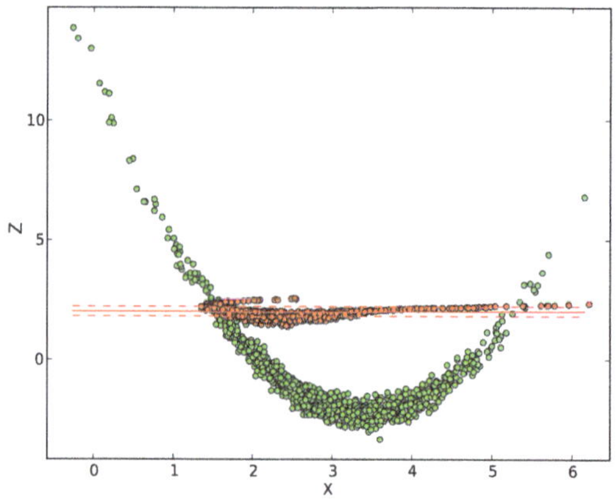

Figure 4. Same as Fig. 2 but for the RHF.

very long tails can help to correct biases and to make the filter more resilient to divergence. More precisely, the prior density of z is written:

$$p(z) = \frac{1}{N_e + 1} \sum_{i=1}^{N_e - 1} \frac{1_{[z_i, z_{i+1}[}(z)}{(z_{i+1} - z_i)} + T(z), \tag{4}$$

with $1_{[z_i, z_{i+1}[}(z)$ the indicator function on the interval $[z_i, z_{i+1}[$ (yielding 1 if z belongs to this interval, 0 otherwise) and $T(z)$ also a combination of indicator functions representing the tails term applied to the two outer regions. The likelihood $p(z^o|z)$ is known analytically from the observation error density. It is discretized on the same grid as $p(z)$, and the two functions are multiplied pointwise to provide a constant piecewise expression (after normalization) of the posterior density $p(z|z^o)$. The analysis ensemble is finally obtained using a (deterministic) procedure of inversion of the cumulative distribution function. Corrections on z particles are calculated by the difference between the posterior and the prior values of z. These corrections are used to compute the corrections for the unobserved variables with Eq. (2), that is, applying a linear regression. This latter step, inherited from the EnKF, is perhaps the main weakness of the RHF, as illustrated in Fig. 4: the RHF analysis performs rather poorly in both weakly and strongly non-Gaussian cases addressed in the previous illustrations. However, considering the many positive aspects of this scheme (non Gaussian, robust, deterministic, possible to localize), it seems worth trying to correct this weakness and extend the rank histogram approach to unobserved variables.

3 Multivariate rank histogram filter

3.1 Principle

We wish to generalize the RHF to the general Bayesian framework. The RHF first addresses the analysis of the observed variable, then deals with the others. It is thus an implementation of the *sequential realization method* to sample a multivariate probability density, for example, as presented by Tarantola (2005). This method leans on the Knothe–Rosenblalt rearrangement, a decomposition of the joint probability density into a product of marginal and conditional univariate densities. With three scalar variables $\boldsymbol{X} = (x, y, z)$, this decomposition is

$$p(x, y, z) = p(z)\, p(x|z)\, p(y|x, z). \tag{5}$$

A sample from the joint density is obtained by deterministically sampling $p(z)$ first, then $p(x|z)$ (using the result for $p(z)$), and $p(y|x, z)$ (using the previous two results). The purpose of data assimilation here is to condition the joint density to an observation z^o of z. Following the usual Markovian memoryless assumption for the observation process, which implies $p(z^o|x, y, z) = p(z^o|z)$, and using the decomposition (Eq. 5) it is straightforward to find that

$$p\left(x, y, z|z^o\right) = p(z|z^o)\, p(x|z)\, p(y|x, z). \tag{6}$$

Here again, the deterministic sampling of both densities conditioned on z are based on the previously sampled densities; hence, the sampling of variables x and y depends on the observation z^o. To obtain the first factor on the right-hand side, the EnKF uses Eq. (1); the RHF implements Bayes' rule for z: $p(z|z^o) \propto p(z)\, p(z^o|z)$. But for both methods, the second and third terms on the right-hand side are computed using Eq. (2), which comes from a Gaussian, Kalman filtering perspective. We propose below a new non-Gaussian approach to sample scalar particles from these conditional densities to implement Eq. (6) with non-parametric densities. This scheme is deterministic, in the sense that no random number need be generated during the analysis process. The analyses are then reproducible and the method is of the "transform" type.

3.2 Implementation of the MRHF analysis

Let $\{z_i^a\}_{i=1,\dots,N_e}$ be the posterior ensemble of the observed variable z, that is, a sample of $p(z|z^o)$. Consider the first

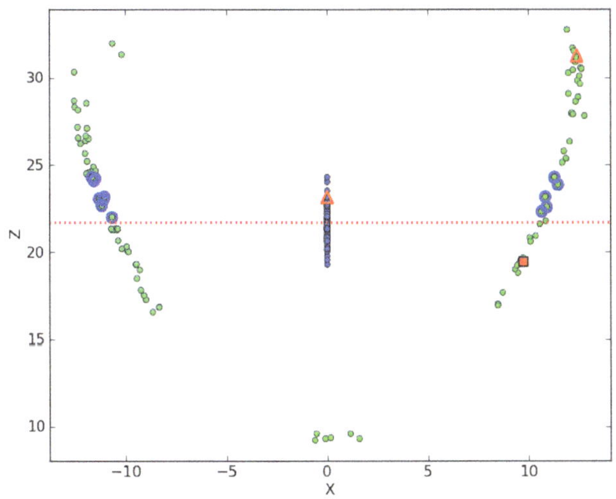

Figure 5. Illustration of the MRHF analysis step for the first unobserved variable: green dots represent the prior ensemble; blue dots vertically aligned at $X = 0$ represent the posterior z ensemble. The red dotted line is the observation of z, the red square is the true state (not used for the analysis). The red empty triangles show the particle being processed and its corresponding z analysis value. Blue circles show the selected particles to form the posterior density $p_i^{\mathrm{a}}(x)$; see text for details.

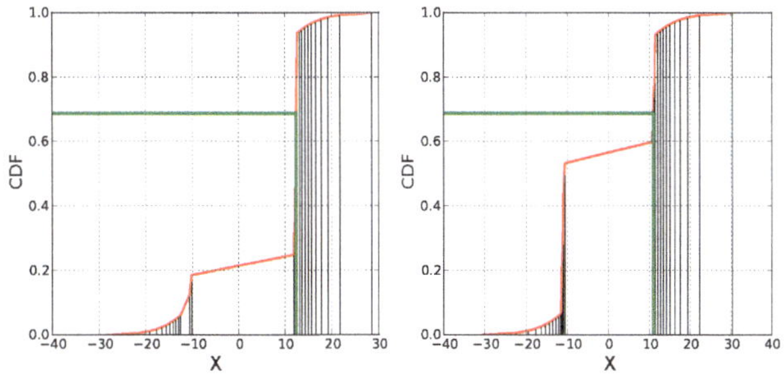

Figure 6. Illustration of the MRHF analysis step for the first unobserved variable (one particle): red lines represent the cumulative distribution functions (cdf) of the prior density $p_i^{\mathrm{b}}(x)$ (left panel) and the posterior density $p_i^{\mathrm{a}}(x)$ (right panel). The vertical black lines show the selected particles used to build these densities. To compute the x analysis value for the prior x particle near 11 on the left panel, the green line must be followed: the cdf for the prior x is computed with the prior cdf (Result is near 0.68); from this cdf value, the x analysis value is obtained on the right panel; see text for details.

unobserved variable, x in Eq. (6). The MRHF analysis determines x_i^{a}, the x analysis value for particle i, by deterministically sampling the conditional density $p_i^{\mathrm{a}}(x) \equiv p(x|z = z_i^{\mathrm{a}})$. This density must first be formed. Some steps of the procedure are illustrated in Fig. 5. The green dots represent the prior ensemble in the X–Z plane; the blue dots at $X = 0$ represent the z analysis ensemble $\{z_i^{\mathrm{a}}\}_{i=1,\dots,N_{\mathrm{e}}}$. The red line is the observation realization. The following process is repeated for $i = 1, \dots, N_{\mathrm{e}}$. For a given i, a subset of particles is selected in the prior ensemble (green dots with blue circles), whose z values lie in the neighborhood of z_i^{a} (blue dot with red triangle) along the z direction. The selection process is discussed later. Applying the rank histogram approach to the x values of the selected particles, a one-dimensional density is then formed to represent $p_i^{\mathrm{a}}(x)$.

To follow Eq. (6), one approach would be to draw a random realization from this density to provide x_i^{a}. This, however, is far from optimal from the physical viewpoint, because it can generate large corrections resulting in physical instabilities and imbalances, as previously observed by Anderson (2003) in the EnKF context. In Fig. 5 for instance, the prior particle (green dot with red triangle) is in the right-hand side mode of the distribution. Since the observation does not enable one to know in which mode the truth (red dot) actually is, it makes sense to try to keep this particle in its mode of origin, thus minimizing its modification.

Instead of a random draw in $p_i^{\mathrm{a}}(x)$ which could arbitrarily move the particle to the left-hand side mode, the following steps are proposed:

- with a similar selection and a rank histogram process, form the density of x conditioned to the *background* value of z: $p_i^{\mathrm{b}}(x) \equiv p(x|z = z_i^{\mathrm{b}})$;
- compute the cumulative distribution functions $C_i^{\mathrm{b}}(x)$ and $C_i^{\mathrm{a}}(x)$ from $p_i^{\mathrm{b}}(x)$ and $p_i^{\mathrm{a}}(x)$, respectively;
- compute the position of the prior particle in the prior density: $c_i = C_i^{\mathrm{b}}(x_i^{\mathrm{b}})$;
- preserve the rank of the particle in the posterior density by taking $x_i^{\mathrm{a}} = C_i^{\mathrm{a}(-1)}(c_i)$ as analysis value for x and particle i. This is illustrated in Fig. 6.

Although this method does not always prevent a particle shifting from one mode to another, two neighboring particles (i.e., close to each other) in the prior ensemble are likely to remain neighbors in the posterior ensemble. In a multimodal case for instance, two particles in the same mode are more likely to remain in the same mode after analysis. Figure 7 illustrates the successful behavior of the MRHF analysis in the strongly non-Gaussian case introduced in Sect. 2 (Figs. 2 and 4, right panels).

Once the z and x analysis values are computed for each particle, the analysis values for the third variable y can be computed. The process is strictly similar to the one described above, but for the variable y, and with an additional $x = x_i^{\mathrm{a}}$ term in the conditional statement. In practice, this reduces to selecting particles from the prior distribution in the neighborhood of $(x_i^{\mathrm{a}}, z_i^{\mathrm{a}})$ in the two-dimensional plane (x, z), to form the density $p(y|x, z, z^{\mathrm{o}})$. The other steps remain unchanged.

As a remark, one may notice that this analysis method brings some similarities with the heuristic method presented in Anderson (2003) in the EnKF context. His idea is to compute the covariance term in Eq. (2) using a subset of neighboring particles. The MRHF goes one step further by considering the nonlinear relationship between those particles.

3.3 Selection of particles and mean-field approximation

We now come back to the selection of particles, and start with the first unobserved variable x. To represent the target density $p_i^{\mathrm{a}}(x)$ accurately, the selected particles must have a z value close to z_i^{a} used in the conditional statement. At the same time, there may be few or no particles whose z values differ from z_i^{a} by less than a specified, small amount, since N_{e} is finite. Thus, there is a trade-off between selecting particles that are very close to z_i^{a} and selecting a sufficient number of particles to represent $p_i^{\mathrm{a}}(x)$.

In the numerical experiments with the Lorenz 63 system, presented in Sect. 4, a maximal distance ($d_{\max}$) is prescribed, along with a minimal and a maximal number of particles. Specifications of these parameters are detailed in Sect. 3.4. No attempt has been made in this work to make the scheme independent of the variables, for example by normalizing all variable by their prior variance.

For the second unobserved variable y, the difference is computed as a distance in the two-dimensional plane (x, z). The maximal distance to consider as a threshold must be prescribed accordingly. As the algorithm proceeds to additional unobserved variables, the curse of dimensionality becomes apparent: each time a new unobserved variable is analyzed, a dimension is added and the volume of states within the maximal distance decreases so fast that each region defined by one particle and a finite radius around it has negligible probability to hold another particle.

As a schematic illustration, assume that $z_i^{\mathrm{a}} = 30$ in Fig. 5, and $d_{\max} = 1$. There exist prior particles with $29 < z < 31$. Thus, the analysis for x can be conducted accurately. Assume now that the x analysis provides $x_i^{\mathrm{a}} = -10$. Prior particles in the two-dimensional neighborhood of $(x_i^{\mathrm{a}}, z_i^{\mathrm{a}})$ (for example, in a circle centered on this point with radius $\sqrt{2}$) are sparse or nonexistent. More distant particles must therefore be included and the accuracy of the analysis for y may be poor.

This obstacle leads us to introduce an approximation, termed the mean-field approximation by Cotter and Reich (2013), which consists in dropping the unobserved variables in the conditional statements in Eq. (6), and thus computing the posterior density as

$$p\left(x, y, z|z^{\mathrm{o}}\right) \simeq p\left(z|z^{\mathrm{o}}\right) p(x|z)\, p(y|z). \qquad (7)$$

This amounts to processing each unobserved variable in the same way as the first one of the series. The approximation (Eq. 7) limits the scheme's ability to handle complex, jointly multimodal densities, as will be illustrated with the Lorenz (1963) model in Sect. 4. An important advantage of the approximation is that it makes the analyses of the different unobserved variables independent. Thus, they can be parallelized on a computer.

3.4 MRHF parameters and possible tuning

Several of the MRHF parameters are related to the computation of rank histograms that is used to update both the observed and unobserved variables. Building a pdf with the rank histogram approach implies a division by the distance between two consecutive particles (Anderson, 2010). To avoid possible computational overflow, it is important to set a minimum spacing ϵ_{RHF} between two consecutive particles. This is done by moving each particle at a distance of ϵ_{RHF} from its closest neighbor when necessary. The particles are processed sequentially from the mean toward the tails of the distribution. For the following experiments with the Lorenz 63 system, a wide range of ϵ_{RHF} values are tested, from 10^{-6} to 10^{-2}, and the values that provide the smallest errors are retained. The main and expected conclusion is that experiments with the RHF and MRHF with large ensemble sizes are sensitive to ϵ_{RHF} within this range, because large ϵ_{RHF} tend to excessively diffuse peaked probability densities when those are built by a large number of particles close to each other.

As stressed by Anderson (2010), several choices are possible for the shape of the tails. With the Lorenz 63 system, sensitivity tests have shown that the MRHF with small ensembles and frequent observations ($\Delta t = 10$) is sensitive to the shape of the tails, and performs better with Gaussian tails. For larger ensembles, results are similar with different shapes of tails. Constant tails are specified because it is a bit cheaper computationally. Constant tails extend to prescribed values slightly beyond the model phase space boundaries. Tails are introduced only for the observed variables, because their probability densities are multiplied by the observation densities before resampling. No tail is introduced for unobserved variables. For the Lorenz 63 model, the minimal and maximal values are set to [−20, −30, 0] and [20, 30, 50], respectively.

An additional parameter controls the scheme's behavior when the prior distribution has multiple, well-separated modes. Figure 6 (left panel) shows the cumulative distribution function of a probability density made of two disjoint modes. Between the two modes, this function increases, although it should remain constant because the modes are disjoint. This is due to the rank histogram approach to build the probability density, and emphasized by the limited number of particles in the ensemble. In the analysis step, unrealistic particles may then appear in the region between the two modes. In the Lorenz 63 experiments, the probability density has been set to zero when below a threshold of $1/6 \times 1/(N_{\mathrm{e}} + 1)$.

As discussed in Sect. 3.3, the particle selection depends on three additional parameters. The first is the maximal distance $d_{\max}$. The specification of $d_{\max}$ should account for magnitude and the variability of the variables, the dimension of the space in which the difference is defined, and the ensemble size. With the Lorenz 63 system, we take $d_{\max} = d\sqrt{n}$, where n is the dimension of the space ($n = 1$ for the first unobserved variable; $n=2$ for the second unobserved variable) and d is prescribed according to the ensemble size: 1 for small ensembles ($N_e = 8, 16, 32$), 0.1 for medium ensembles ($N_e = 64$, 128), and 0.01 for the larger ensembles ($N_e = 256, 512$). To ensure a sufficient but not too large number of selected particles, it is wise to fix a minimum and a maximum number of particles. In the following experiments, those are set to 5 and 15, respectively.

Finally, like many ensemble methods, the MRHF suffers from sampling errors in the description of the densities. With the EnKF, this is usually corrected with covariance inflation. This does not make real sense with the MRHF, since the analysis does not rely on covariances. After the analysis, the particles are slightly perturbed with a white Gaussian noise, as it is often done with particle filters to avoid collapse toward one single particle. For the Lorenz 63 experiments that follow, a few values of variance have been tested in the range 0.01–0.05 for this noise, and the experiments yielding the smallest errors have been retained.

3.5 Localization

Localization consists in reducing the corrections to some variables, as computed during the analysis step, according to their distance from the observation. Beyond a certain distance, all corrections are set to 0. With the MRHF, localization is straightforward because it is a transform method. Similar to the EnKF with serial processing of observations, the analysis corrections are multiplied by coefficients that are functions of the distance to the observation. Therefore the correction may be restricted to elements of the state vector that are spatially close to the location of the observed variable, and changes to the state vector can then be gracefully tapered to zero as distance from the observation location increases. In the present work, no localization has been applied in the experiments since it is not needed on those assimilation problems. Localization will be studied in future works.

3.6 Connections with other methods

The MRHF is a non-parametric transform ensemble data assimilation method. We have presented it as a generalization of the RHF, but it also has connections with other non-parametric transform methods discussed in the Introduction, such as the method introduced by Reich (2013). This method derives from the theory of optimal transportation (or optimal mapping), whose application to sequential data assimilation has been suggested by El Moselhy and Marzouk (2012), Cotter and Reich (2013), and Reich (2013). Instead of trying to compute an approximation of the posterior pdf, the crucial idea of this theory is to find a "transfer map" f such that $f(\boldsymbol{X})$ is distributed according to the posterior pdf when $\boldsymbol{X}$ is distributed according to the prior pdf. Once such a transfer map is identified, a posterior sample can be generated from the prior sample. The transfer map is a mathematical expression of the consistency of the posterior sample depending on both the prior sample and the observation. However a transfer map is not unique (Villani, 2009). To find one, one may require the map to satisfy some additional optimality condition. Cotter and Reich (2013) and Reich (2013) propose to find the map that minimizes the expected squared distance between $\boldsymbol{X}$ and $f(\boldsymbol{X})$, so as to make the smallest possible changes to go from the prior to the posterior sample. As described in Sect. 3.1, the MRHF uses a transfer map. This map is particularly simple, since only one-dimensional probability densities are involved: we choose the map that preserves the position of the particle during its transfer from the prior to the posterior density. This also makes the smallest possible changes to go from the prior to the posterior sample.

4 Numerical experiments with the Lorenz 63 model

The Lorenz 63 model (L63) is a well-known system of three ordinary differential equations based on a simplification of atmospheric cellular convection (Lorenz, 1963). It is also a widely used test case for developments in data assimilation. The state variables are denoted X, Y, Z. The usual configuration of the model is adopted here: the parameters are set to $\sigma = 10$, $\rho = 28$ and $\beta = 8/3$; the integration time step is $\delta t = 0.01$. In the following experiments a simulation of reference is performed and considered as the true trajectory to recover.

4.1 Fully observed state vector

4.1.1 Experimental setup and diagnostics

In this subsection, the full state (X, Y, Z) is observed. The observations are created by adding to the true trajectory independent perturbations drawn from a white Gaussian noise with standard deviation $\sigma_o = 2$ as in Harlim and Hunt (2007) and Bocquet (2011). Three different experiments are conducted for different observation time intervals: $\Delta t = 0.10$, $\Delta t = 0.25$, and $\Delta t = 0.50$. These observation time intervals are expected to provide mild, medium, and strong nonlinear test cases (Bocquet, 2011).

Each experiment is run over 10^5 assimilation cycles. To avoid any spin-up issues, a burn-in period of 1000 analysis cycles is used. Five filters are compared: the stochastic EnKF, the RHF, a particle filter, the MRHF and the MRHF with mean-field approximation (see Sect. 3). The EnKF and the RHF are tested with a large set of inflation factors and the best in terms of root mean square error are retained for comparisons. The particle filter is implemented in its sequential importance resampling (SIR particle filter) version (Gordon et al., 1993). Resampling is performed using the universal resampling method described by Whitley (1994). After resampling, the particles are perturbed with a white Gaussian noise with variance selected in the $[0.01, 0.05]$ interval to provide the smallest errors.

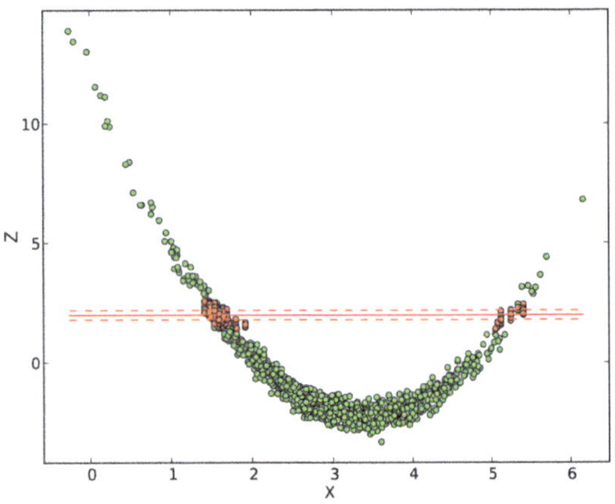

Figure 7. Same as Fig. 4, right panel, but for the MRHF.

The filters are tested for different ensemble sizes: $N_e = [8, 16, 32, 64, 128, 256, 512]$. The filters are first evaluated by the time-averaged value of the root mean square error (RMSE) between the analysis and the simulation of reference. We also evaluate the filters' approximation of the full posterior distribution using the Kullback–Leibler (KL) divergence, or relative entropy (Kullback, 1959), which measures a distance between two probability densities P and Q according to the formula:

$$\mathrm{KL}(P, Q) = \int \log \frac{P}{Q} \mathrm{d}P. \quad (8)$$

Density P refers to the density described by the ensemble after an analysis step. Ideally, the reference density Q should be the analytical solution to the problem. Since this is not available to us, we take the SIR particle filter solution with 2048 particles as a reference. It proves to be the best of all in terms of RMSE, as it will be shown in the results section. Also, by construction, the SIR particle filter provides a physically balanced solution, assuming that the noise added to each particle during the resampling step is prescribed small enough not to affect this balance significantly. For a given assimilation method, a small KL divergence guarantees that the solution is physically balanced. For a perfect computation of the KL divergence, the joint probability densities should be used. To limit the problematic effects of subsampling, especially when the ensemble sizes are small, we stick to the marginal densities of the first Lorenz variable, X. Using Y or Z provides very similar results. The marginal densities of X are recovered from the updated ensembles of each filter by using rank histograms.

4.1.2 Results

When $\Delta t = 0.10$, Fig. 8, upper panel, shows that the EnKF outperforms all the other methods for $N_e \leq 128$. This is because the system is fully and accurately observed, and frequently enough to make the analysis problem close to Gaussian. However, with $N_e \geq 256$, $\Delta t = 0.10$, the fully nonlinear methods perform slightly better. In a rather similar setup, but with an ensemble transform Kalman filter instead of a stochastic EnKF, Bocquet (2011) also concludes that the Kalman filter, perfectly designed for such problems, is extremely hard to beat. Nonetheless, the RHF and MRHFs behave rather well, even if not as well as the EnKF. In a medium nonlinear case ($\Delta t = 0.25$, central panel of Fig. 8) and for $N_e \geq 32$, both the MRHF in its full formulation and the mean-field approximated MRHF produce a smaller RMSE than the EnKF and the RHF. The SIR particle filter needs 128 particles to perform as well as the MRHFs. Finally, in a case of strong nonlinearities (Fig. 8 bottom-panel), the MRHFs outperforms the EnKF and the RHF for any ensemble size. The SIR particle filter needs in this case more than 256 particles in order to achieve similar performance. In all cases, the MRHF and the MRHF with mean-field approximation behave very similarly, the latter being even slightly better most of the time. This suggests that the mean-field approximation has a small negative impact, and that the dimensionality issue that may affect the MRHF, as described in Sect. 3.3, is already present in a three-variable system.

Figure 9 shows the counterpart of Fig. 8 for the KL divergence for the X-variable. The KL divergences are computed with respect to a SIR particle filter solution with 2048 particles. It is remarkable that other, independent 2048-SIR particle filter solutions do not provide null KL divergences. This is because the random perturbations introduced after resampling are different in the test and reference experiments. In all observation scenarios, this KL divergence approaches 1; this can then be considered as the target score for the other methods. The MRHFs perform very well, even in the mildly nonlinear case, with large ensembles. As nonlinearities grow stronger, they perform increasingly well in comparison with the others. In particular, the ensemble size required by the SIR particle filter to reach the performance of the MRHFs increases dramatically. In the strongly nonlinear case (bottom panel), the MRHFs perform better than the EnKF and RHF for any ensemble size and are only outperformed by the SIR particle filter for very large ensemble sizes ($N_e \geq 256$). In any case, the SIR particle filter performs better than the others for large ensembles.

4.2 Bimodal case – Z observed

4.2.1 Experimental setup

The L63 attractor is characterized by two lobes centered on points of attraction and connected to each other at their bottom (where the minimal values of Z are encountered). Figure 10 displays horizontal slices through the L63 attractor represented in its phase space. The two lobes are easily identified in the region $Z > 24$ (bottom row), exhibiting two or four distinct modes. In a data assimilation framework without any prior information other than the whole attractor itself,

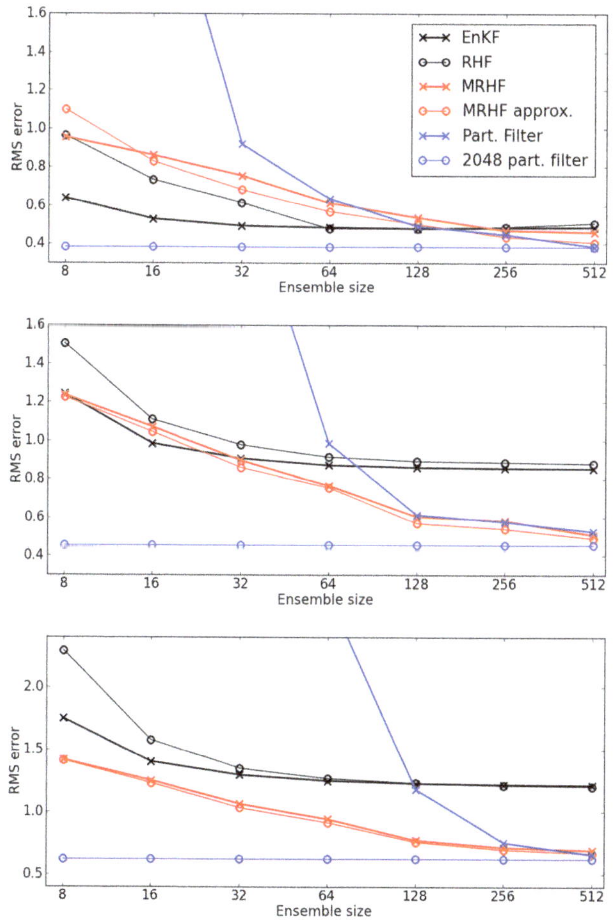

Figure 8. Time-averaged analysis root mean square error (RMSE) for the EnKF (thick black line with crosses), the RHF (thin black line with open circles), the SIR particle filter (thick blue line with crosses), the full MRHF (thick red line with crosses), and the mean-field approximated MRHF (thin red line with open circles); for experiments on the fully observed Lorenz 63 with observation time intervals $\Delta t = 0.10$ (upper panel), $\Delta t = 0.25$ (center panel), and $\Delta t = 0.50$ (bottom panel). The thin blue line with open circles represents the time-averaged analysis RMSE for the SIR particle filter with 2048 particles, which can be considered as a target score.

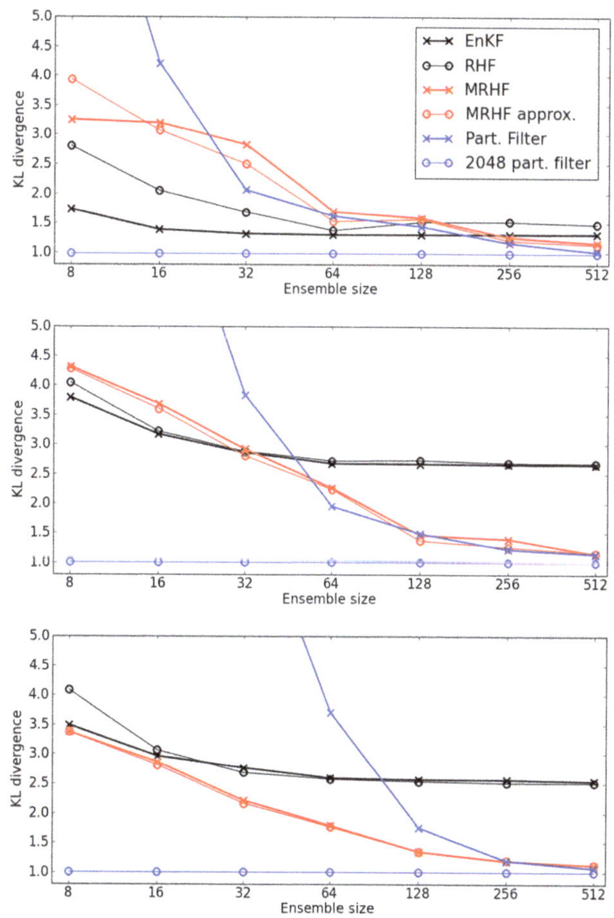

Figure 9. Same as Fig. 8 but for the mean Kullback–Leibler divergence on the X variable (results are similar on the Y and Z variables).

a single observation of Z does not enable us to determine the mode where the truth actually is. The dynamics often help to determine whether it is in an ascending branch or a descending branch of the attractor so we expect bimodal posteriors. It is thus a strongly non-Gaussian data assimilation problem.

In the following experiment, observations of Z are extracted from a "true" trajectory, perturbed with a white Gaussian noise of variance 1, and assimilated every 40 time steps ($\Delta t = 0.40$). The assimilation is conducted over 10^5 analysis cycles after a burn-in period of 1000 time steps.

The evaluation of the MRHF performance is strictly similar to the previous experiments, except that the reference solution to compute the KL divergence comes from the SIR particle filter with 4096 particles, instead of 2048. As it is argued in Sect. 4.2.2 below, the RMSE is not a meaningful diagnostic in this case, making it difficult to verify that the 4096-SIR particle filter provides an accurate solution. However, the objective of that test relies on the fact that an appropriate data assimilation method should be able to maintain the representation of the bimodality in this particular case. The SIR particle filter, given a substantial number of particles, might generate overly dispersive ensembles but does maintain the bimodality (Fig. 12). It has been checked that it is true for the whole integration period. Hence, a small KL divergence between the methods and the 4096-SIR particle filter will confirm that the bimodality is respected.

4.2.2 Results

The time-averaged RMSE is a classical performance diagnostic in data assimilation. However, when the solution is bimodal, the RMSE does not tell much. Here, the RMSE vary between 3 and 7, whatever the method and the ensemble size are, and they do not decrease with increasing ensemble sizes.

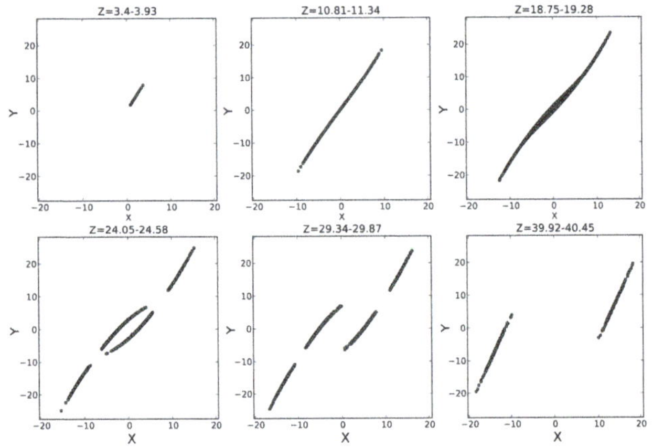

Figure 10. Horizontal slices (in X–Y planes, Z intervals indicated on tops) through L63 attractor. In the two bottom left graphs, the two modes in the center are parts of the descending branches of the lobes (Z decreases with time), while the modes in the corners are the ascending branches.

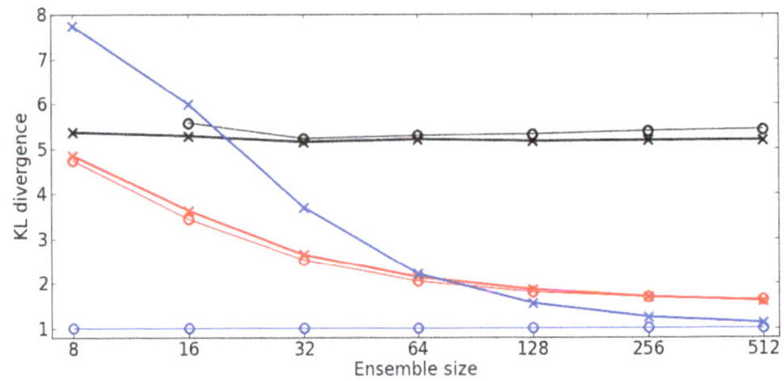

Figure 11. Time-averaged Kullback–Leibler divergence on the X variable for the EnKF (thick black line with crosses), the RHF (thin black line with open circles), the SIR particle filter (thick blue line with crosses), the full MRHF (thick red line with crosses), and the mean-field approximated MRHF (thin red line with open circles); for the experiment on Lorenz 63 when only the third variable (Z) is observed every 40 time steps during 10^5 analysis cycles. The thin blue line with open circles represents the time-averaged Kullback–Leibler divergence for the SIR particle filter with 2048 particles, which can be considered as a target score.

Since they provide no relevant information, the RMSE are not discussed further.

The KL divergence is a much more appropriate diagnostic in this case, especially using the marginal densities of X, since bimodality is expected in the X direction. Figure 11 displays the KL divergences as a function of ensemble size, for the 5 methods considered in the previous setup. The RHF with 8 particles proved to be unstable with any inflation factor. The fully non-parametric methods (SIR particle filter, MRHFs) yield much better scores than the others (EnKF and RHF). The KL divergence of the EnKF and RHF does not depend on the ensemble size, confirming that these methods are not designed to deal with such multimodal problems. The SIR particle filter needs at least $N_e = 128$ particles to obtain a smaller KL divergence than the MRHFs.

To illustrate the differences in the behaviors of the various methods, Fig. 12 depicts scatter plots of the analysis ensembles at the 389th cycle, given by the reference SIR particle filter (with 4096 particles), and the other five methods with 256 particles. The scatter plots are drawn in the X–Y plane, and the true state is shown by the red squares and red dashed lines. The 389th cycle has been chosen arbitrarily for this illustration. Similar behavior of the filters is observed throughout the experiment.

While the correct posterior distribution is bimodal, the EnKF and the RHF tend to create particles between the two modes. This explains their large KL divergences with respect to the reference solution. The SIR particle filter and the MRHF, on the other hand, provide visually correct, bimodal solutions in this case (but they can be subject to mode leakage too, at other cycles). The MRHF with mean-field approximation is again broadly similar to the full MRHF. Because the correction to Y is independent of the correction to X, however, a few particles can switch from one mode to the other along the X (resp. Y) direction without switching along the Y (resp. X) direction. These particles appear in unrealistic regions of the model phase space (regions that cannot be visited by the attractor, near $X = -10, Y = 10$ or $X = 10, Y = -10$). This illustrates the limitation of the MRHF with mean-field approximation to deal with bimodal distribution. Mode leakage of this kind is significantly reduced by the deterministic resampling method used in the MRHF, in comparison with a stochastic method (result not shown). This is because the MRHF method, described in Sect. 3.2, preserves the relative rank of particles at the analysis step. Also, the Lorenz 63 system does not seem affected by a few outliers, and the KL divergence scores of the MRHF with mean-field approximation are good.

5 An illustration of density estimation

We next consider the analysis step for each scheme in a more complex model, with no forecast loop. The interest of this illustration is to observe their behavior in a first realistic context.

5.1 The marine biogeochemical context

The interactions between ocean dynamics and biogeochemistry are complex. The variations of the mixed layer depth (MLD) strongly influences the nutrient supply, hence the phytoplankton production in the euphotic layer (Dutkiewicz et al., 2001). MLD variations are themselves controlled, at least for a large part, by variations in the wind forcing. With the growing interest in understanding ocean biogeochemical cycles and thanks to the increasing amount of dedicated satellite missions (SeaWiFS, MERIS, MODIS), the

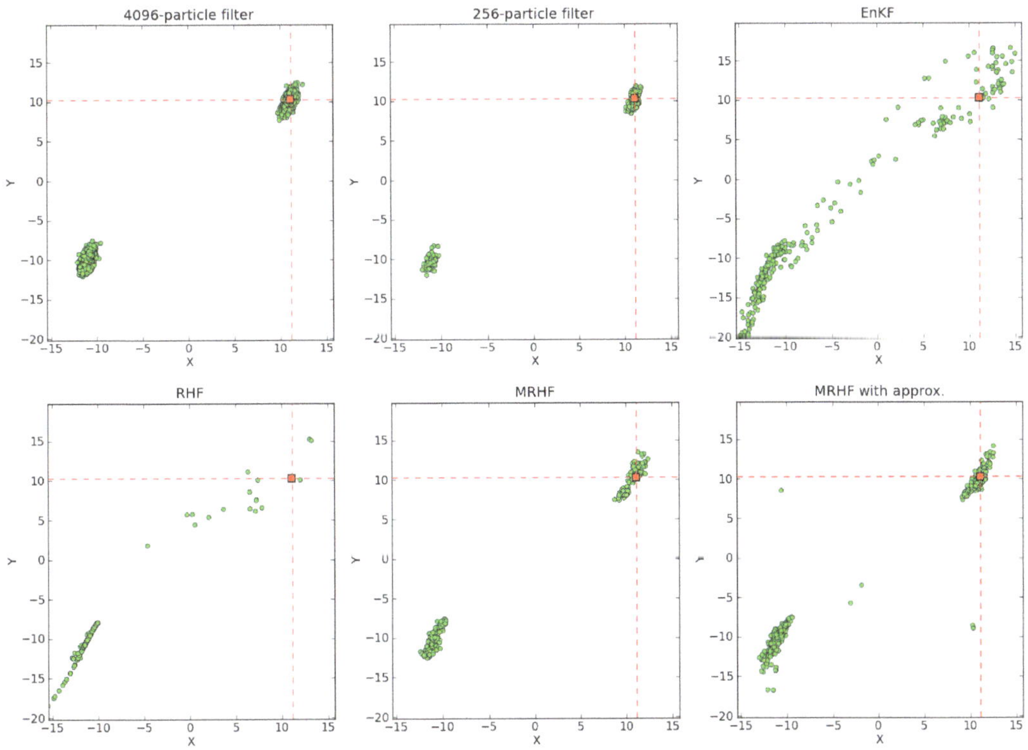

Figure 12. Scatter plots of the analysis ensembles with 256 particles (except upper-left) at the 389th analysis cycle, in the X–Y plane of Lorenz 63 model phase space: 4096-SIR particle filter (considered as the reference solution) (upper left panel); Particle filter (upper center panel); EnKF (upper right panel); RHF (bottom left panel); MRHF (bottom center panel); Mean-field approximation MRHF (bottom right panel). The red dashed lines indicate the truth from which the observation of Z is produced.

assimilation of ocean color data, a proxy of chlorophyll concentration, has taken off in the last few years (Gregg et al., 2009). A better estimation of the dynamical variables, including wind forcing, appears as an interesting by-product. Because of the nonlinear relationships of biogeochemical variables between each other and with dynamical variables, ocean biogeochemical data assimilation is a fundamentally non-Gaussian problem. This is well demonstrated by Béal et al. (2010). They use a three-dimensional coupled physical–biogeochemical model of the North Atlantic with a 1/4° horizontal resolution. A 200-particle ensemble simulation is run, perturbing the wind forcing, during the 1998 North Atlantic spring bloom. The nonlinear model response, expressed in the non-Gaussianity of the bivariate distributions of the ensemble variables, is clearly demonstrated.

5.2 The density estimation experiment

The data used for the illustration below comes from the ensemble simulation of Béal et al. (2010). We focus on the 1 day forecast at the Gulf Stream station (47° W, 40° N).

The upper-left graphs of Figs. 13, 14, and 15 respectively show the forecast ensemble in the chlorophyll (CHL)–mixed layer depth (MLD) plane, the chlorophyll (CHL)–biogeochemical detritus (DET) plane and the biogeochemical detritus (DET)–mixed layer depth (MLD) plane. Each green dot represents an ensemble particle and the blue dot shows the values from a reference model run. A chlorophyll observation is created by perturbing the value from this reference with a Gaussian white noise with variance $\sigma_o = 0.001$.

Each of the assimilation schemes from the previous section is applied, at a fixed time step and a fixed grid point, to the seven-variable control vector. The control vector is composed of seven prognostic variables of the model: one dynamical variable (MLD) and six biological variables (chlorophyll, detritus, dissolved organic matter, NH_4, NO_3 and phytoplankton). The 200-particle forecast ensemble is used as a prior distribution to test various analysis schemes using the chlorophyll observation. The other panels of Figs. 13, 14, and 15 display the estimated ensembles, with red triangles, on, respectively, the chlorophyll (CHL)–mixed layer depth (MLD) plane, the chlorophyll (CHL)–biogeochemical detritus (DET) plane and the biogeochemical detritus (DET)–mixed layer depth (MLD) plane, obtained using the EnKF (upper right graphs), the RHF of Anderson (center left graphs), the full MRHF (center right graphs), the mean-field approximation MRHF (bottom-left graphs), and the SIR particle filter (bottom right graphs) analysis steps. The chlorophyll observation is represented by the blue full line;

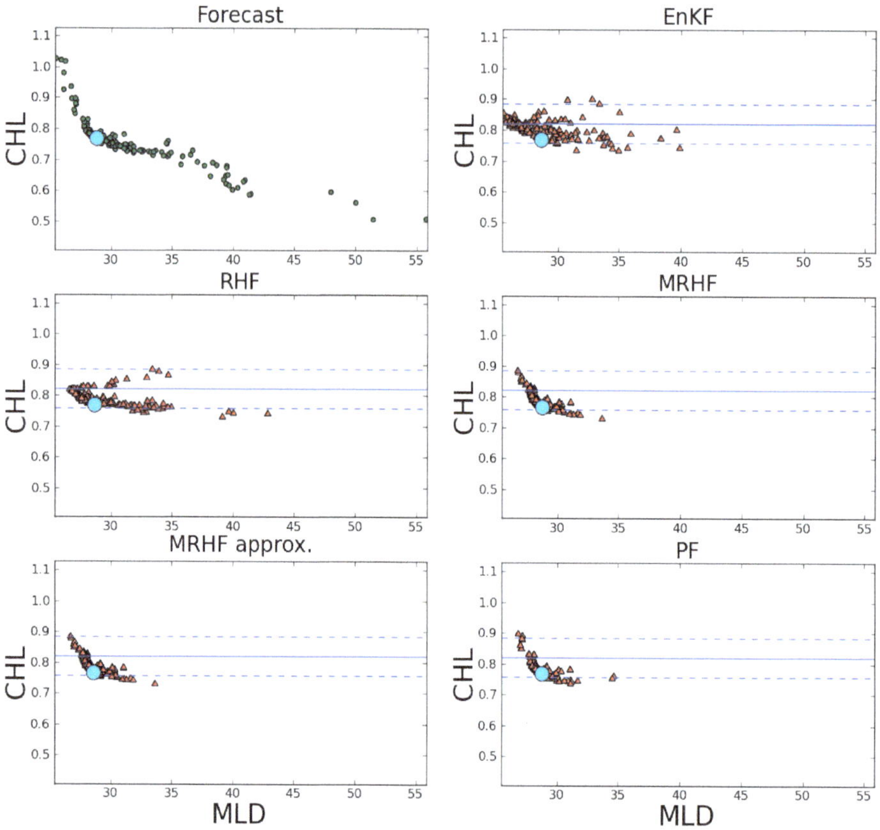

Figure 13. Prior forecast ensemble (upper left panel) and posterior analysis ensembles (red triangles), on the chlorophyll (CHL)–mixed layer depth (MLD) plane, produced by the EnKF (upper right panel), the RHF (center left panel), the full MRHF (center right panel), the mean-field approximation MRHF (bottom-left), and the SIR particle filter (bottom right panel). Only chlorophyll is observed with the value CHL_o indicated by the blue line on each panel. The two dashed blue lines represent the interval $[CHL_o - 2\,\sigma_o, CHL_o + 2\,\sigma_o]$, where $\sigma_o^2 = 0.001$ is the observation error standard deviation. The blue dot represents the true state, from which the chlorophyll value has been extracted and perturbed to form the observation CHL_o.

the blue dashed lines are at a distance of $2 \times \sigma_o$ from the blue full line.

A data assimilation method is obviously expected to move the prior particles close to the observations. However, a major requirement for those methods, especially in the non-Gaussian context, is also to maintain the information on the model attractor contained in the shape of the prior ensemble dispersion. In Figs. 13 and 14, the prior ensemble clearly says that the observed variable (CHL) and the unobserved variables (i.e., MLD and DET) have an obvious statistical connection, but this connection is not linear. Thus, the methods using a linear regression in the physical space to correct the unobserved variable, that is, EnKF and RHF, fail to maintain the non-Gaussian shape of the prior density. The SIR particle filter analysis step is a very good approximated implementation of Bayes' theorem (modulo sampling errors). Hence, with a sufficiently large ensemble, it provides a posterior ensemble consistent with the observation and the prior ensemble. In comparison the full MRHF analysis step also manages to produce a posterior distribution consistent with both the observation and the prior ensemble. Nevertheless, it clearly appears in Fig. 15 that the curse of dimensionality striking during the particle selection process (as discussed in Sect. 3.3) degrades the correction on the second unobserved variable. The mean-field approximation discussed in Sect. 3.3, appears to overcome this issue and produces a posterior distribution very similar to the SIR particle filter.

Figure 15 displays the posterior ensembles in the mixed layer depth (MLD)–biogeochemical detritus (DET) plane. Those graphs allow us to observe the bivariate densities between two unobserved state variables. It is known that the EnKF and the RHF appropriately maintain the covariances between all pairs of variables in a linear and Gaussian context. However, in a non-Gaussian case such as this one, Fig. 15 shows that both filters do not provide an appropriate ensemble update (in the sense of Bayes' rule). Meanwhile, the MRHF and the SIR particle filter, provide an estimated ensemble taking into accout the relationship between all pairs of variables by respecting the information contained in the prior distribution.

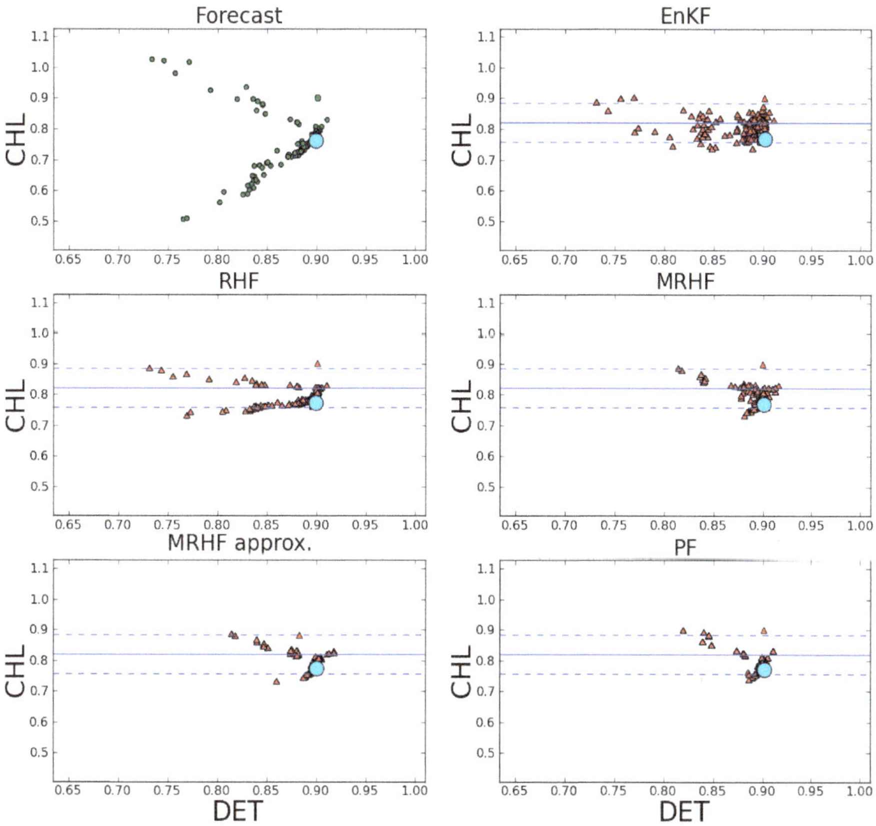

Figure 14. Same as Fig. 13 but on the chlorophyll (CHL)–biogeochemical detritus (DET) plane.

6 Discussion and conclusions

This paper has introduced the multivariate rank histogram filter (MRHF), a fully non-Gaussian analysis scheme for ensemble data assimilation. This filter is an extension of the rank histogram filter introduced by Anderson (2010). In order to set the MRHF in the wider context of non-Gaussian analysis methods, the behavior of some of these methods has been examined in idealized, bivariate frameworks where the variables were jointly Gaussian, weakly non-Gaussian, or strongly non-Gaussian (Sect. 2). The MRHF clearly falls into the category of methods able to deal with strong non-Gaussianity, along with the particle filters. An approximated version of the MRHF, based on the mean-field approximation (Cotter and Reich, 2013), is also proposed.

Numerical experiments with the Lorenz 63 model, in a data assimilation problem with fully observed state vector but for different observation time intervals corresponding to different levels of nonlinearity (Sect. 4.1), confirm that the MRHF does not perform better than the EnKF in quasi-linear context with small ensemble sizes. Nevertheless, in this context, the MRHF performs slightly better with larger ensembles ($N_e \geq 256$). When nonlinearity is stronger, the MRHF considerably reduces the root mean square error and the Kullback–Leibler divergence in comparison with other methods when given a sufficiently large number of particles ($N_e \geq 64$). The MRHF with mean-field approximation exhibits very similar performance. Experiments in the most nonlinear regime, characterized by the bimodality of the state density, confirm the ability of the MRHF to handle strong non-Gaussianities (Sect. 4.2). Finally, an experiment with prior data from a coupled physical–biogeochemical model (Sect. 5) illustrates the behavior of the MRHF analysis (in its full and approximated forms) when facing strongly non-Gaussian densities in a more explicitly geophysical context. This illustration is not a full data assimilation problem but the posterior densities produced by the MRHF analysis show that Bayes' theorem is correctly approximated in this formulation.

Aside from documenting the performance of the MRHF, the experiments in this paper show the importance of matching the assimilation method to the level of non-Gaussianity in the problem at hand. Even though a general method such as the MRHF should perform well in any situation, the fact is that the EnKF (or other linear methods) are perfectly adequate in many applications and often much less expensive computationally. An advantage of the MRHF in this respect is that it is easily hybridized with other serial transform methods, such as the EnKF, because such schemes process observations serially and update observed and unobserved

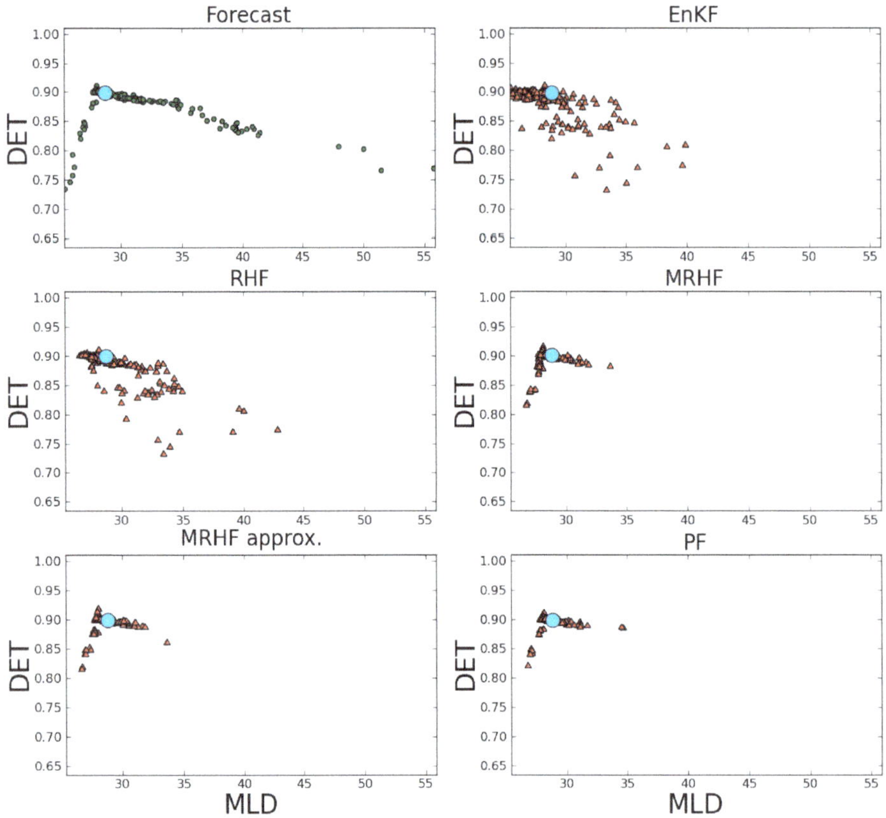

Figure 15. Same as Fig. 13 but on the biogeochemical detritus (DET)–mixed layer depth (MLD) plane.

variables serially. One can then consider, for example, updating wind components with the EnKF but the more non-Gaussian humidity with the MRHF.

Relative to Kalman filters, the MRHF's deterministic scheme has assets to preserve physical balances during the analysis. Relative to particle filters, the MRHF also has advantages in addressing the curse of dimensionality. As explained in Sect. 3.5, spatial localization of the update step comes naturally with the MRHF, although it has not been tested here. Also, effects of the curse of dimensionality on the MRHF can be greatly reduced using the mean-field approximation, as illustrated in Sect. 3.3. In fact, use of this approximation is likely necessary for successful implementation of the MRHF in many realistic problems. Finally, it is relatively easy to reduce any sensitivity of the MRHF to biases in observed variables, and to decrease the associated tendency for filter divergence, because of the freedom to choose the tails of the prior density for these variables (Sect. 2). Then, contrary to the particle filters (at least in their bootstrap formulation, Gordon et al., 1993), a large number of observations helps in avoiding divergence.

According to our experiments, the MRHF is much more expensive than the EnKF in terms of computation. The MRHF analysis proves to be approximately 50, 100, and 200 times more expensive than the EnKF for typical ensemble sizes of 64, 128, and 256, respectively. The MRHF might then be best suited for rather specific problems with strong non-Gaussianity and few observations or as part of hybrid schemes where the MRHF is used only for certain variables. Nevertheless, spatial localization and a more efficient implementation of the mean-field approximation have the potential to greatly reduce computation cost and expand the range of problems for which the MRHF is feasible. Both aspects need to be investigated before the application of the MRHF to realistic problems and these are the next steps of this work.

Acknowledgements. This work was supported by the Région Rhônes-Alpes and NCAR. It is also a contribution to the SANGOMA project supported by European Comissions Seventh Framework Programme FP7/2007–2013, grant agreement no. 283580, and to the CNRS/INSU/LEFE program. The authors are grateful to Sebastian Reich and two other anonymous reviewers for their relevant and constructive comments.

Edited by: T. Gneiting

References

Anderson, J. L.: A Method for Producing and Evaluating Probabilistic Forecasts from Ensemble Model Integrations, J. Climate, 9, 1518–1530, 1996.

Anderson, J. L.: A local least squares framework for ensemble filtering, Mon. Weather Rev., 131, 634–642, 2003.

Anderson, J. L.: A non-Gaussian ensemble filter update for data assimilation, Mon. Weather Rev., 138, 4186–4198, 2010.

Anderson, J. L.: Localization and Sampling Error Correction in Ensemble Kalman Filter Data Assimilation, Mon. Weather Rev., 140, 2359–2371, 2012.

Béal, D., Brasseur, P., Brankart, J.-M., Ourmières, Y., and Verron, J.: Characterization of mixing errors in a coupled physical biogeochemical model of the North Atlantic: implications for nonlinear estimation using Gaussian anamorphosis, Ocean Sci., 6, 247–262, doi:10.5194/os-6-247-2010, 2010.

Bertino, L., Evensen, G., and Wackernagel, H.: Sequential Data Assimilation Techniques in Oceanography, Internat. Stat. Rev., 71, 223–241, 2003.

Bishop, C. H., Etherton, B. J., and Majumdar, S. J.: Adaptive sampling with the ensemble transform Kalman filter, Part I: Theoretical aspects, Mon. Weather Rev., 129, 420–436, 2001.

Bocquet, M.: Ensemble Kalman filtering without the intrinsic need for inflation, Nonlin. Processes Geophys., 18, 735–750, doi:10.5194/npg-18-735-2011, 2011.

Bocquet, M., Pires, C. A., and Wu, L.: Beyond Gaussian statistical modeling in geophysical data assimilation, Mon. Weather Rev., 138, 2997–3023, 2010.

Brankart, J.-M., Testut, C.-E., Béal, D., Doron, M., Fontana, C., Meinvielle, M., Brasseur, P., and Verron, J.: Towards an improved description of ocean uncertainties: effect of local anamorphic transformations on spatial correlations, Ocean Sci., 8, 121–142, doi:10.5194/os-8-121-2012, 2012.

Buehner, M., Houtekamer, P. L., Charette, C., Mitchell, H. L., and He, B.: Intercomparison of Variational Data Assimilation and the Ensemble Kalman Filter for Global Deterministic NWP, Part I: Description and Single-Observation Experiments, Mon. Weather Rev., 138, 1550–1566, 2010.

Candille, G. and Talagrand, O.: Evaluation of probabilistic prediction systems for a scalar variable, Q. J. Roy. Meteorol. Soc., 131, 2131–2150, 2005.

Cohn, S. E.: An introduction to estimation theory, J. Meteorol. Soc. Jpn., 75, 257–288, 1997.

Cotter, C. J. and Reich, S.: Ensemble filter techniques for intermittent data assimilation, in Large Scale Inverse Problems, Radon Ser. Comput. Appl. Math., 13, 91–134, 2013.

Dee, D. and Da Silva, A. M.: The choice of variable for atmospheric moisture analysis, Mon. Weather Rev., 131, 155–171, 2003.

Doucet, D., de Freitas, N., and Gordon, N.: An introduction to sequential Monte Carlo methods, in: Sequential Monte Carlo Methods in Practice, edited by: Doucet, D., de Freitas, N., and Gordon, N., Statistics for Engineering and Information Science, Springer-Verlag, New York, 2001.

Dutkiewicz, S., Follows, M., Marshall, J., and Gregg, W. W.: Interannual variability of phytoplankton abundances in the North Atlantic, Deep-Sea Res., Pt. II, 48, 2323–2344, 2001.

El Moselhy, T. A. and Marzouk, Y. M.: Bayesian inference with optimal maps, J. Comput. Phys., 231, 7815–7850, 2012.

Evensen, G.: Sequential data assimilation with a nonlinear quasi-geostrophic model using Monte Carlo methods to forecast error statistics, J. Geophys. Res., 99, 10143–10162, 1994.

Evensen, G.: The Ensemble Kalman Filter: theoretical formulation and practical implementation, Ocean Dynam., 53, 343–367, 2003.

Fournier, A., Hulot, G., Jault, D., Kuang, W., Tangborn, A., Gillet, N., Canet, E., Aubert, J., and Lhuillier, F.: An Introduction to Data Assimilation and Predictability in Geomagnetism, Space Sci. Rev., 155, 247–291, 2010.

Gordon, N. J., Salmond, D. J., and Smith, A. F. M.: Novel approach to nonlinear/non-Gaussian Bayesian state estimation, IEEE Proc. F., 140, 107–113, 1993.

Gregg, W. W., Friedrichs, M. A., Robinson, A. R., Rose, K. A., Schlitzer, R., Thompson, K. R., and Doney, S. C.: Skill assessment in ocean biological data assimilation, J. Mar. Syst., 76, 16–33, 2009.

Greybush, S. J., Kalnay, E., Miyoshi, T., Ide, K., and Hunt, B. R.: Balance and Ensemble Kalman Filter Localization Techniques, Mon. Weather Rev., 139, 511–522, 2011.

Hamill, T. M.: Interpretation of Rank Histograms for Verifying Ensemble Forecasts, Mon. Weather Rev., 129, 550–560, 2001.

Hamill, T. M., Whitaker, J. S., and Snyder, C.: Distance-Dependent Filtering of Background Error Covariance Estimates in an Ensemble Kalman Filter, Mon. Weather Rev., 129, 2776–2790, 2001.

Harlim, J. and Hunt, B.: A non-Gaussian ensemble filter for assimilating infrequent noisy observations, Tellus A, 59, 225–237, 2007.

Hoteit, I., Luo, X., and Pham, D.-T.: Particle Kalman Filtering: A Nonlinear Bayesian Framework for Ensemble Kalman Filters, Mon. Weather Rev., 140, 528–542, 2012.

Houtekamer, P. L. and Mitchell, H. L.: Data Assimilation Using an Ensemble Kalman Filter Techniqu, Mon. Weather Rev., 126, 796–811, 1998.

Houtekamer, P. L. and Mitchell, H. L.: A Sequential Ensemble Kalman Filter for Atmospheric Data Assimilation, Mon. Weather Rev., 129, 123–137, 2001.

Kalman, R.: A new approach to linear filtering and prediction problems, Trans. Am. Soc. Mech. Mech. Eng. J. Basic Eng., 82, 35–45, 1960.

Kullback, S.: Information Theory and Statistics, Wiley, New York, 1959.

Lauvernet, C., Brankart, J.-M., Castruccio, F., Broquet, G., Brasseur, P., and Verron, J.: A truncated Gaussian filter for data assimilation with inequality constraints: Application to the hydrostatic stability condition in ocean models, Oceanogr. Meteorol., 27, 1–17, 2009.

Lawson, W. and Hansen, J.: Implications of stochastic and deterministic filters as ensemble-based data assimilation methods in varying regimes of error growth, Mon. Weather Rev., 132, 1966–1981, 2004.

Lei, J., Bickel, P., and Snyder, C.: Comparison of Ensemble Kalman Filters under Non-Gaussianity, Mon. Weather Rev., 138, 1293–1306, 2010.

Lermusiaux, P. F. J.: Data assimilation via error subspace statistical estimation, Part II: Middle Atlantic Bight shelfbreak front simulations and ESSE validation, Mon. Weather Rev., 127, 1408–1432, 1999.

Lermusiaux, P. F. J.: Uncertainty estimation and prediction for interdisciplinary ocean dynamics, J. Comp. Phys., 217, 176–199, 2006.

Lorenz, E.: Deterministic Nonperiodic Flow, J. Atmos. Sci., 20, 130–141, 1963.

Morzfeld, M., Tub, X., Atkinsa, E., and Chorin, A.: A random map implementation of implicit filters, J. Comp. Phys., 231, 2049–2066, 2012.

Nakano, S., Ueno, G., and Higuchi, T.: Merging particle filter for sequential data assimilation, Nonlin. Processes Geophys., 14, 395–408, doi:10.5194/npg-14-395-2007, 2007.

Pham, D. T.: Stochastic Methods for Sequential Data assimilation in Strongly nonlinear systems, Mon. Weather Rev., 129, 1194–1207, 2001.

Reich, S.: A nonparametric ensemble transform method for Bayesian inference, SIAM J. Sci. Comput., 35, A2013–A2024, 2013.

Sakov, P. and Bertino, L.: Relation between two common localisation methods for the EnKF, Comput. Geosci., 15, 225–237, 2010.

Sakov, P., Counillon, F., Bertino, L., Lisæter, K. A., Oke, P. R., and Korablev, A.: TOPAZ4: an ocean-sea ice data assimilation system for the North Atlantic and Arctic, Ocean Sci., 8, 633–656, doi:10.5194/os-8-633-2012, 2012.

Simon, E. and Bertino, L.: Application of the Gaussian anamorphosis to assimilation in a 3-D coupled physical-ecosystem model of the North Atlantic with the EnKF: a twin experiment, Ocean Sci., 5, 495–510, doi:10.5194/os-5-495-2009, 2009.

Snyder, C.: Particle filters, the optimal proposal and high-dimensional systems, Proc. Seminar on Data Assimilation for Atmosphere and Ocean, ECMWF, Reading, Berkshire, 161–170, 2012.

Snyder, C., Bengtsson, T., Bickel, P., and Anderson, J.: Obstacles to high-dimensional particle filtering, Mon. Weather Rev., 136, 4629–4640, 2008.

Tarantola, A.: Inverse problem theory and methods for model parameter estimation, SIAM, Philadelphia, USA, 2005.

van Leeuwen, P.: Particle filtering in geophysical systems, Mon. Weather Rev., 137, 4089–4114, 2009.

van Leeuwen, P.: Nonlinear data assimilation in geosciences: an extremely efficient particle filter, Q. J. Roy. Meteorol. Soc., 136, 1991–1999, 2010.

Villani, C.: Optimal transportation: Old and new, Springer-Verlag, Berlin, Heidelberg, 2009.

Whitaker, J. S. and Hamill, T. M.: Ensemble data assimilation without perturbed observations, Mon. Weather Rev., 1913–1924, 2002.

Whitaker, J. S., Hamill, T. M., Wei, X., Song, Y., and Toth, Z.: Ensemble data assimilation with the NCEP Global Forecast System, Mon. Weather Rev., 136, 463–482, 2008.

Whitley, D.: A genetic algorithm tutorial, Statist. Comput., 4, 65–85, 1994.

Wikle, C. and Berliner, L.: A Bayesian tutorial for data assimilation, Physica D, 230, 1–16, 2007.

18

Propagation regimes of interfacial solitary waves in a three-layer fluid

O. E. Kurkina[1], A. A. Kurkin[1], E. A. Rouvinskaya[1], and T. Soomere[2,3]

[1]Nizhny Novgorod State Technical University n.a. R. Alekseev, Nizhny Novgorod, Russia
[2]Institute of Cybernetics, Tallinn University of Technology, Akadeemia tee 21, 12618 Tallinn, Estonia
[3]Estonian Academy of Sciences, Kohtu 6, 10130 Tallinn, Estonia

Correspondence to: A. A. Kurkin (aakurkin@gmail.com)

Abstract. Long weakly nonlinear finite-amplitude internal waves in a fluid consisting of three inviscid layers of arbitrary thickness and constant densities (stable configuration, Boussinesq approximation) bounded by a horizontal rigid bottom from below and by a rigid lid at the surface are described up to the second order of perturbation theory in small parameters of nonlinearity and dispersion. First, a pair of alternatives of appropriate KdV-type equations with the coefficients depending on the parameters of the fluid (layer positions and thickness, density jumps) are derived for the displacements of both modes of internal waves and for each interface between the layers. These equations are integrable for a very limited set of coefficients and do not allow for proper description of several near-critical cases when certain coefficients vanish. A more specific equation allowing for a variety of solitonic solutions and capable of resolving most near-critical situations is derived by means of the introduction of another small parameter that describes the properties of the medium and rescaling of the ratio of small parameters. This procedure leads to a pair of implicitly interrelated alternatives of Gardner equations (KdV-type equations with combined nonlinearity) for the two interfaces. We present a detailed analysis of the relationships for the solutions for the disturbances at both interfaces and various regimes of the appearance and propagation properties of soliton solutions to these equations depending on the combinations of the parameters of the fluid. It is shown that both the quadratic and the cubic nonlinear terms vanish for several realistic configurations of such a fluid.

1 Introduction

There are many important topics related to internal waves (IWs) in the ocean. Large IWs are highly significant for sediment resuspension and transport (Bogucki and Redekopp, 1999; Stastna and Lamb, 2008; Reeder et al., 2011) and for the biology on the continental shelf (Sandstrom and Elliott, 1984). The currents forced by large or breaking IWs cause powerful forces on marine platforms and submersibles. The associated strong distortion of the density field has a severe impact on acoustic signaling (Apel et al., 2007; Chin-Bing et al., 2009; Warn-Varnas et al., 2009; Sridevi et al., 2011). Their capacity to break and impact the local microstructure has major consequences for the understanding of interior ocean mixing (Muller and Briscoe, 2000). Internal waves are believed to be responsible for substantial damage (Osborne, 2010). High water velocities in intense IWs can create enormous local loads and bending moments and represent a potential danger to off-shore structures, such as oil platforms, drill rigs, etc. The danger from IWs is considered so critical that, similar to the systems of tsunami warning, the potential for automated detection systems for large-amplitude IWs (internal soliton early warning systems) is being discussed now. Such systems were even tested to support drilling campaigns and guarantee the safety of drilling platforms (Stober and Moum, 2011).

Most of the studies of IWs focus on large-amplitude localized IWs, which propagate in relatively shallow areas of oceans, shelf regions or semi-sheltered seas. A fascinating feature of many waves of this kind is that they propagate for a long time without any significant change in their energy or shape as solitons do (Jeans, 1995). They can be interpreted and adequately described as internal solitary waves (ISWs) and are often referred to as internal solitons (Ostrovsky and Stepanyants, 1989; Grimshaw, 2001). Observations of large-amplitude ISWs are presented in many original research papers and reviews (Holloway et al., 2001; Jackson, 2004; Ostrovsky and Stepanyants, 2005; Sabinin and Serebryany, 2005; Vlasenko et al., 2005; Apel et al., 2007; Grimshaw et al., 2007).

The concept of layered fluid, widely used for both theoretical research and applications, is convenient for the description of IWs because wave motion in such environments is described by equations containing a small number of parameters and often allowing for analytical studies of the properties of solutions. These simple models are able to mirror the basic properties of the actual internal wave systems. The efficiency of the two-layer model – the simplest system basically representing the key properties of IWs – has been established in numerous analytical and numerical studies as well as by means of in situ observations and laboratory experiments (Funakoshi, 1985; Funakoshi and Oikawa, 1986; Mirie and Pennell, 1989; Choi and Camassa, 1996, 1999; Pullin and Grimshaw, 1998; Craig et al., 2004; Guyenne, 2006; Zahibo et al., 2007; Camassa et al., 2010).

The two-layer model only conditionally represents the vertical structure of seas and oceans. Its direct extension – three-layerstratification – has proved to be a proper approximation of the sea water density profile in some basins in the world ocean with specific hydrological conditions (Knauss, 1996; Leppäranta and Myrberg, 2009; Kurkina et al., 2011a). Three-layer models are more complex than two-layer systems, because of the increased number of set-up parameters, but they represent new dynamical effects and allow for much more analytical progress compared to the arbitrary stratified medium. Another advantage of three-layer models in contrast to two-layer ones is the ability to describe two modes of IWs, their properties and possible interactions.

Nonlinear theory of IWs in a three-layer fluid with constant density in each layer is presented in the literature in the framework of different approaches. Simplified equations for three-layer conjugate flows are derived in Lamb (2000) and Rusås and Grue (2002), and some analytic results are obtained to understand conjugate flows and, in consequence, large-amplitude, flat-centered (table-like) solitary internal waves of both modes. Comparison with solutions obtained for continuous stratifications is also given in Lamb (2000). A fully nonlinear numerical method for the calculation of solitary waves in a three-layer fluid (Rusås and Grue, 2002) allows for the investigation of both mode-1 and mode-2 waves, including broad flat-centered waves and extreme (overhanging) waves. The similarity of mode-1 waves to the interfacial waves in a two-layer fluid is the probable reason why the analysis is mainly performed for situations when one of the layers is relatively thin. These models have greatly contributed to the understanding of nonlinear internal wave properties in three-layer fluids and flows. They are, however, often very complex, and the results are not easy to visualize. The drawback is that the analytical progress is limited and numerical computations should be involved to obtain final localized solutions and to analyze their properties.

Although contemporary numerical methods and fully nonlinear approaches such as the above-discussed method of conjugate flows allow for extensive studies into properties of highly nonlinear IWs, many specific features can still be recognized, analyzed and understood using classical methods for analytical studies into IWs in the weakly nonlinear framework. Such fully analytical methods make it possible to establish exactly the appearance of disturbances of different shapes and amplitudes, and, more importantly, to understand the specific features of the behavior of waves corresponding to the situation where a substantial change in the overall regime of wave propagation is possible (Kurkina et al., 2011a).

Weakly nonlinear analytical models of different levels for IWs in a three-layer fluid have been developed for some specific situations. The properties of mode-1 IWs in a symmetric three-layer fluid (when the undisturbed state is symmetric about the mid-depth) were analyzed up to the second order in nonlinearity by Grimshaw et al. (1997). An extension involving up to the fourth-order nonlinear terms was derived in Kurkina et al. (2011b). The model of Yang et al. (2010) involved both modes of IWs in a general three-layer ocean, the first order in nonlinearity and dispersion (Korteweg–de Vries (KdV) approximation; only mode-2 waves were analyzed). The weakly nonlinear theory for a continuously stratified fluid can be used to derive the parameters of evolution equations for IWs in a three-layer fluid (Grimshaw et al., 1997; Yang et al., 2010). These parameters for continuous stratification are given in the form of integral expressions including the vertical mode (whose shape is determined by the density stratification), its higher-order corrections and their derivatives (Lamb and Yan, 1996; Pelinovskii et al., 2000; Grimshaw et al., 2002). When this technique is adapted to a layered model (with a piecewise continuous density profile), the relevant integrands contain very complex expressions (especially for higher-order corrections and coefficients) involving generalized functions, and analytical progress in their analysis is not always possible. Therefore, it is preferable initially to use equations for layered models and to apply asymptotic expansions to them directly. This approach has been used, for instance, in Kurkina et al. (2011b) and Koop and Butler (1981).

A specific feature of weakly nonlinear equations for a three-layer fluid is that some coefficients at the nonlinear terms may vanish for certain modes and certain specific density stratifications (Kakutani and Yamasaki, 1978; Grimshaw et al., 1997; Yang et al., 2010). This feature is common for several wave classes in stratified environments (Soomere, 2003). Some coefficients at the nonlinear terms may vanish even in a two-layer fluid with the surface tension between the

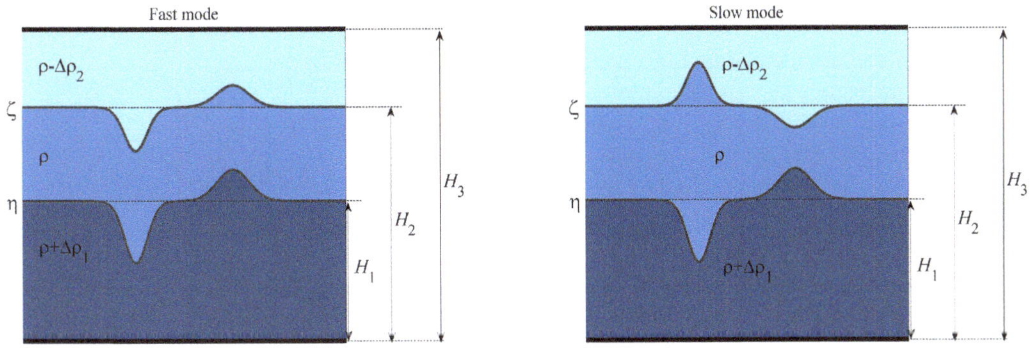

Figure 1. Definition sketch of the three-layer fluid and internal waves of mode-1 (left) and mode-2 (right).

layers (Giniyatullin et al., 2014; Kurkina et al., 2015). In such cases, it is often necessary to account for higher-order nonlinear terms to describe the motion adequately. In the case of IWs in a three-layer fluid, it is necessary to produce a second-order weakly nonlinear theory to resolve some of such situations. The attempts in this direction have so far been limited to the specific case of symmetric stratification (Kurkina et al., 2011b).

We develop here an analytic model for long IWs of finite amplitude for the three-layer fluid with an arbitrary combination of the layers' thicknesses. Such a vertical structure much better matches the properties of real seas and oceans. The relevant nonlinear evolutionary equations are obtained first with the use of an asymptotic procedure from the governing equations for three homogeneous fluid layers. The derivation of these equations relies on the second order of the perturbation theory for both interfaces and for the waves of each mode. The developed framework makes it possible to investigate the properties of both modes of long internal gravity waves in a three-layer fluid. The key development is that, for each equation, the coefficients at the nonlinear terms and terms expressing linear and nonlinear dispersion of waves are expressed explicitly via parameters of the fluid configuration. This makes it possible to analyze comprehensively the behavior and signs of all coefficients. The necessary order of the equations is discussed and determined for each case. Special attention is paid to the situations when the nonlinear terms of the lowest order of the perturbation theory can vanish. For such situations, a particular rescaling is performed in order to balance the nonlinear and dispersion terms in the equations.

2 Governing equations

As usual in the analysis of long internal waves, let us consider a model situation of irrotational motions in a three-layer inviscid fluid with undisturbed interface positions $z = H_{1,2}$ ($H_2 \geq H_1$) and total thickness H_3 ($H_3 \geq H_2$) overlying a flat horizontal bottom located at $z = 0$ in the approximation of a rigid lid on the surface of the fluid (Fig. 1). The reference density of the middle layer is $\rho_2 = \rho$ and the densities in the lowermost and uppermost layers are $\rho_1 = \rho + \Delta\rho_1$ and $\rho_3 = \rho - \Delta\rho_2$, respectively. We employ the Boussinesq approximation and assume that densities in the layers differ insignificantly ($\Delta\rho_{1,2}/\rho \ll 1$). In this case, the equations of the motion are Laplace equations for the velocity potential Φ_i in each layer:

$$\nabla^2\Phi_i = 0, \quad i = 1, 2, 3. \tag{1}$$

The kinematic boundary conditions at the bottom and at the fixed upper boundary reduce to the condition that vertical velocities at these boundaries vanish:

$$\Phi_{1_z} = 0 \text{ at } z = 0, \; \Phi_{3_z} = 0 \text{ at } z = H_3. \tag{2}$$

Here and further on, the subscripts x, y, z or t denote a partial derivative along the respective coordinate (x, y, z) or with respect to time t. The classical kinematic and dynamic boundary conditions at the interfaces between the layers are as follows (Kurkina et al., 2011b):

$$\left.\begin{array}{l} \eta_t + \Phi_{1_x}\eta_x - \Phi_{1_z} = 0 \\ \eta_t + \Phi_{2_x}\eta_x - \Phi_{2_z} = 0 \\ \rho_1\left(\Phi_{1_t} + \frac{1}{2}(\nabla\Phi_1)^2 + g\eta\right) \\ = \rho_2\left(\Phi_{2_t} + \frac{1}{2}(\nabla\Phi_2)^2 + g\eta\right) \end{array}\right\} z = H_1 + \eta(x,t), \tag{3}$$

$$\left.\begin{array}{l} \zeta_t + \Phi_{2_x}\zeta_x - \Phi_{2_z} = 0 \\ \zeta_t + \Phi_{3_x}\zeta_x - \Phi_{3_z} = 0 \\ \rho_2\left(\Phi_{2_t} + \frac{1}{2}(\nabla\Phi_2)^2 + g\zeta\right) \\ = \rho_3\left(\Phi_{3_t} + \frac{1}{2}(\nabla\Phi_3)^2 + g\zeta\right) \end{array}\right\} z = H_2 + \zeta(x,t). \tag{4}$$

Here, the unknown functions $\eta(x,t)$ and $\zeta(x,t)$ denote the instantaneous position of the interface between the bottom and middle layers and between the upper and middle layers, respectively.

3 Asymptotic procedure

The appearance of the proper evolution equation for a phenomenon in a realistic environment may substantially depend on the characteristic scales of the processes to be highlighted or analyzed. Similarly to Kurkina et al. (2011b), our goal is to derive an equation for small (but finite) amplitude long-wave motions in a layered fluid. This goal naturally implies two small parameters. Firstly, the typical horizontal scale of wave motion L (equivalently, the typical wave length for linear or nonlinear wave trains) considerably exceeds the fluid depth H. This assertion $L \gg H$ not only ensures that a small parameter H/L is present in the system, but also matches the usual properties of environments supporting the IW motion in natural water bodies. It also fits the basic assumptions of common derivations of the KdV equation and its generalizations. Another small parameter naturally emerges from the restriction that the amplitude a of the disturbances to the interfaces' positions from their undisturbed location is small compared to the fluid depth. This assertion gives rise to the parameter of nonlinearity $\varepsilon = a/H \ll 1$.

Our specific interest is to describe long-living solitary waves, for which the dispersion in some sense balances the impact of nonlinearity. As will be shown below, a convenient parameter that characterizes the role of dispersion in wave propagation is $\mu = H^2/L^2$. As the existence of long-living nonlinear wave motions and solitary waves usually presumes a specific balance between the terms representing nonlinear and dispersive effects, we assume that $\varepsilon \sim \mu^0$.

These assumptions and introduced parameters make it possible to perform the analysis of the problem Eqs. (1)–(4) in nondimensional form. The standard procedure of introducing nondimensional coordinates into Eqs. (1)–(4) under these assumptions leads to the following boundary problem:

$$\Phi_{1_{zz}} + \varepsilon\Phi_{1_{xx}} = 0, \quad 0 \le z \le H_1, \tag{5}$$

$$\Phi_{2_{zz}} + \varepsilon\Phi_{2_{xx}} = 0, \quad H_1 \le z \le H_2, \tag{6}$$

$$\Phi_{3_{zz}} + \varepsilon\Phi_{3_{xx}} = 0, \quad H_2 \le z \le H_3, \tag{7}$$

$$\Phi_{1_z} = 0 \text{ at } z = 0, \quad \Phi_{3_z} = 0 \text{ at } z = H_3 \tag{8}$$

$$\left.\begin{aligned} &\eta_t + \varepsilon\Phi_{1_x}\eta_x - \Phi_{1_z} = 0,\\ &\eta_t + \varepsilon\Phi_{2_x}\eta_x - \Phi_{2_z} = 0,\\ &\rho_1\left(\Phi_{1_t} + \tfrac{1}{2}\varepsilon(\Phi_{1_x})^2 + \tfrac{1}{2}(\Phi_{1_z})^2\right.\\ &\left.+g\eta\right) = \rho_2\left(\Phi_{2_t} + \tfrac{1}{2}\varepsilon(\Phi_{2_x})^2\right.\\ &\left.+\tfrac{1}{2}(\Phi_{2_z})^2 + g\eta\right) \end{aligned}\right\} z = H_1 + \eta(x,t), \tag{9}$$

$$\left.\begin{aligned} &\zeta_t + \varepsilon\Phi_{2_x}\zeta_x - \Phi_{2_z} = 0,\\ &\zeta_t + \varepsilon\Phi_{3_x}\zeta_x - \Phi_{3_z} = 0,\\ &\rho_2\left(\Phi_{2_t} + \tfrac{1}{2}\varepsilon(\Phi_{2_x})^2 + \tfrac{1}{2}(\Phi_{2_z})^2\right.\\ &\left.+g\zeta\right) = \rho_3\left(\Phi_{3_t} + \tfrac{1}{2}\varepsilon(\Phi_{3_x})^2\right.\\ &\left.+\tfrac{1}{2}(\Phi_{3_z})^2 + g\zeta\right) \end{aligned}\right\} z = H_2 + \zeta(x,t). \tag{10}$$

The asymptotic analysis of the resulting equations is straightforward. As the procedure is fairly cumbersome, largely follows a well-known approach (Koop and Butler, 1981) and provides almost no instructive aspects, we omit its details and present only its basic steps. First, all unknown functions are expanded into Taylor series in the vicinity of one of the interfaces:

$$f(x, z = H_1 + \eta(x,t), t) = \sum_{j=0}^{\infty} \frac{\eta^j}{j!} \frac{\partial^j f}{\partial z^j}\bigg|_{z=H_1},$$

$$f(x, z = H_2 + \zeta(x,t), t) = \sum_{j=0}^{\infty} \frac{\zeta^j}{j!} \frac{\partial^j f}{\partial z^j}\bigg|_{z=H_2}.$$

All constituents of the resulting series are then expanded into power series with respect to powers of the parameter $\varepsilon \ll 1$:

$$\begin{aligned} \eta &= \varepsilon\left(\eta_0 + \varepsilon\eta_1 + \varepsilon^2\eta_2 + \ldots\right),\\ \zeta &= \varepsilon\left(\zeta_0 + \varepsilon\zeta_1 + \varepsilon^2\zeta_2 + \ldots\right). \end{aligned} \tag{11}$$

It is easy to show that all the potentials $\Phi_i \sim \sqrt{\varepsilon}$; consequently, the power series for the potentials are as follows:

$$\Phi_i = \sqrt{\varepsilon}\left(\varphi_{i,0} + \varepsilon\varphi_{i,1} + \varepsilon^2\varphi_{i,2} + \ldots\right), \quad i = 1,2,3, \tag{12}$$

Finally, we employ the technique of multiple temporal and spatial scales (Engelbrecht et al., 1988; Nayfeh, 2000). The "slow" time and "stretched" reference frame coordinate are introduced as follows:

$$\xi = \varepsilon^{1/2}(x - ct), \quad \tau = \varepsilon^{3/2}t. \tag{13}$$

Here, c denotes the (as yet unknown) phase speed of free linear internal waves. This substitution is equivalent to the following replacement of the operators of partial derivatives:

$$\frac{\partial}{\partial x} = \varepsilon^{1/2}\frac{\partial}{\partial \xi}, \quad \frac{\partial}{\partial t} = -\varepsilon^{1/2}c\frac{\partial}{\partial \xi} + \varepsilon^{3/2}\frac{\partial}{\partial \tau}. \tag{14}$$

Substitution of expansions (Eqs. 11–12) and definitions of modified coordinates (Eqs. 13–14) into Eqs. (5)–(10) leads to an infinite system of equations with respect to the elements of expansions (Eqs. 11–12). Equations of the resulting system are scaled by one of the multiples $\varepsilon^{1/2}$ or $\varepsilon^{3/2}$, where they arise. The obtained system can be solved recursively until any desired order. The system of equations obtained in the leading (lowest) order ($\sim \varepsilon^0$) for the elements of expansions (Eqs. 11–13) has the form (Ruvinskaya et al., 2010)

$$\varphi_{i,0_{zz}} = 0, \quad i = 1,2,3, \tag{15}$$

$$\varphi_{1,0_z} = 0 \text{ at } z = 0, \quad \varphi_{3,0_z} = 0 \text{ at } z = H_3, \tag{16}$$

$$\left.\begin{aligned} &\varphi_{1,0_z} = 0, \quad \varphi_{2,0_z} = 0 \\ &-(\rho + \Delta\rho)c\varphi_{1,0_\xi} + \Delta\rho g \eta_0 = -\rho c \varphi_{2,0_\xi} \end{aligned}\right\} z = H_1, \tag{17}$$

$$\left.\begin{aligned} &\varphi_{2,0_z} = 0, \quad \varphi_{3,0_z} = 0 \\ &-\rho c\varphi_{2,0_\xi} + r\Delta\rho g \zeta_0 = -(\rho - r\Delta\rho) c \varphi_{3,0_\xi} \end{aligned}\right\} z = H_2, \tag{18}$$

where $\Delta\rho_1 = \Delta\rho$, $\Delta\rho_2 = r\Delta\rho$. It follows from Eqs. (15)–(18) that the functions $\varphi_{i,0}, i = 1,2,3$ do not depend on the coordinate z.

The system of equations obtained in the first order ($\sim \varepsilon^1$) is

$$\varphi_{i,1_{zz}} + \varphi_{i,0_{\xi\xi}} = 0, \quad i = 1,2,3, \tag{19}$$

$$\varphi_{1,1_z}(z=0) = 0, \quad \varphi_{3,1_z}(z = H_3) = 0,$$

$$\left.\begin{aligned} &c\eta_{0_\xi} + \varphi_{1,1_z} = 0, \quad c\eta_{0_\xi} + \varphi_{2,1_z} = 0 \\ &(\rho + \Delta\rho)\left(\varphi_{1,0_\tau} - c\varphi_{1,1_\xi} + \tfrac{1}{2}\varphi_{1,0_\xi}\right) + \Delta\rho g \eta_1 \\ &= \rho\left(\varphi_{2,0_\tau} - c\varphi_{2,1_\xi} + \tfrac{1}{2}\varphi_{2,0_\xi}^2\right) \end{aligned}\right\} z = H_1, \tag{20}$$

$$\left.\begin{aligned} &c\zeta_{0_\xi} + \varphi_{2,1_z} = 0, \quad c\zeta_{0_\xi} + \varphi_{3,1_z} = 0 \\ &\rho\left(\varphi_{2,0_\tau} - c\varphi_{2,1_\xi} + \tfrac{1}{2}\varphi_{2,0_\xi}^2\right) + \Delta\rho g \zeta_1 \\ &= (\rho - r\Delta\rho)\left(\varphi_{3,0_\tau} - c\varphi_{3,1_\xi} + \tfrac{1}{2}\varphi_{3,0_\xi}^2\right) \end{aligned}\right\} z = H_2. \tag{21}$$

From Eqs. (19)–(21), we obtain the following relationships:

$$\varphi_{1,1_z} = -\varphi_{1,0_{\xi\xi}} z, \quad \varphi_{3,1_z} = \varphi_{3,0_{\xi\xi}}(H_3 - z), \tag{22}$$

$$\varphi_{1,0_{\xi\xi}} H_1 = c\eta_{0_\xi}, \quad \varphi_{2,0_{\xi\xi}}(H_1 - H_2) = c(\eta_0 - \zeta_0)_\xi, \tag{23}$$

$$\varphi_{3,0_{\xi\xi}}(H_2 - H_3) = c\zeta_{0_\xi}.$$

Equations (22) and (23) together with Eqs. (17) and (18) allow one to define the phase speed c of the linear IWs. The relevant equation is bi-quadratic and, not unexpectedly, reveals that two wave modes exist in this system. One of them results in synchronous in-phase movements of both the interfaces. The other mode is antisymmetric: the motions of the interfaces are synchronous but have opposite directions (Fig. 1). The corresponding expressions for the phase speeds are

$$\begin{aligned} c^{\pm 2} &= \frac{1}{2}\frac{\Delta\rho}{\rho} g H_3 \left(l_1(1-l_1) + r l_2(1-l_2) \pm d\right) \\ &= \frac{1}{2}\frac{\Delta\rho}{\rho} g H_3 \frac{1}{q^{\pm}}, \end{aligned} \tag{24}$$

where

$$d = \sqrt{r^2 l_2^2 (1-l_2)^2 - 2r l_1 (1-l_2)(l_2 - l_1(2-l_2)) + l_1^2 (1-l_1)^2}$$

and $l_{1,2} = H_{1,2}/H_3$. The symmetric mode of internal waves has the larger phase speed c^+, corresponding to the "+" sign before d in Eq. (24). For this reason, we call it fast mode, or mode-1. The antisymmetric mode, whose phase speed is c^-, is called slow mode, or mode-2. Below we shall consider the nonlinear motions of both modes.

The first-order and second-order corrections to the linear solution satisfy the following equations:

$$\begin{aligned} \varepsilon^2 : \; &\eta_{1_\tau}^\pm + \alpha^\pm \eta_1^\pm \eta_{1_\xi}^\pm + \beta^\pm \eta_{1_{\xi\xi\xi}}^\pm = 0, \\ &\zeta_{1_\tau}^\pm + \tilde{\alpha}^\pm \zeta_1^\pm \zeta_{1_\xi}^\pm + \tilde{\beta}^\pm \zeta_{1_{\xi\xi\xi}}^\pm = 0, \end{aligned} \tag{25}$$

$$\begin{aligned} \varepsilon^3 : \; &\eta_{2_\tau}^\pm + \alpha^\pm (\eta_1^\pm \eta_2^\pm)_\xi + \beta^\pm \eta_{2_{\xi\xi\xi}}^\pm + \alpha_1^\pm \eta_1^{\pm 2} \eta_{1_\xi}^\pm \\ &+ \beta_1^\pm \eta_{1,5_\xi}^\pm + \gamma_1^\pm \eta_1^\pm \eta_{1_{\xi\xi\xi}}^\pm + \gamma_2^\pm \eta_{1_\xi}^\pm \eta_{1_{\xi\xi}}^\pm = 0, \\ &\zeta_{2_\tau}^\pm + \tilde{\alpha}^\pm (\zeta_1^\pm \zeta_2^\pm)_\xi + \tilde{\beta}^\pm \zeta_{2_{\xi\xi\xi}}^\pm + \tilde{\alpha}_1^\pm \zeta_1^{\pm 2} \zeta_{1_\xi}^\pm + \tilde{\beta}_1^\pm \zeta_{1,5_\xi}^\pm \\ &+ \tilde{\gamma}_1^\pm \zeta_1^\pm \zeta_{1_{\xi\xi\xi}}^\pm + \tilde{\gamma}_2^\pm \zeta_{1_\xi}^\pm \zeta_{1_{\xi\xi}}^\pm = 0. \end{aligned} \tag{26}$$

The coefficients of Eqs. (25) and (26) are presented in the Appendix. The first-order equations (Eq. 25) are, as expected, the well-known KdV equations that describe the motions of both the interfaces. The second-order equations (Eq. 26) are linear equations with respect to $\eta_2^\pm$ and $\zeta_2^\pm$. Note that equations for $\eta_n^\pm$ and $\zeta_n^\pm$ have a similar structure for each order n and differ from each other only by values of their coefficients (note also that the coefficients in the equations for $\zeta_n^\pm$ have tildes). Furthermore, we omit indices "±" where possible, having in mind different expressions for different modes.

4 Equations for interfaces

In each order, we also obtain relationships between the corresponding terms in series (Eqs. 11–12). These relationships for the first-order and second-order terms are

$$\zeta_1 = s\eta_1, \quad \zeta_2 = s\eta_2 + s_{\text{quad}}\eta_1^2 + s_{\text{disp}}\eta_{1_{\xi\xi}} x, \tag{27}$$

where

$$s = (l_2 - 2ql_1(l_2 - l_1))/l_1,$$

$$s_{\text{quad}} = \frac{2q(l_2 - l_1)}{l_1 H_3}\left((\alpha^* + 3q)l_1 - 3\right),$$

$$\alpha^* = \alpha H_3/c,$$

$$s_{\text{disp}} = \frac{(l_2 - l_1)H_3^2}{6l_1}\left(2ql_1\left((l_2 - l_1)^2 + 12\beta^*\right) - l_2(l_1 + l_2)\right),$$

$$\beta^* = \beta/\left(cH_3^2\right).$$

Note that $s^{\pm} = \pm 1$ for a symmetric stratification ($H_1 = H_3 - H_2$, $\Delta\rho_1 = \Delta\rho_2$, or $r = 1$) analyzed in Kurkina et al. (2011b) for mode-1 waves.

Substituting ζ_1, ζ_2 into (Eq. 11), we obtain

$$\begin{aligned}\zeta &= \varepsilon(\zeta_1 + \varepsilon\zeta_2) + O\left(\varepsilon^3\right)\\ &= \varepsilon s(\eta_1 + \varepsilon\eta_2) + \varepsilon^2\left(s_{\text{quad}}\eta_1^2 + s_{\text{disp}}\eta_{1\xi\xi}\right) + O\left(\varepsilon^3\right) \quad (28)\\ &= s\eta + s_{\text{quad}}\eta^2 + s_{\text{disp}}\eta_{xx} + O\left(\varepsilon^3\right).\end{aligned}$$

Combining Eqs. (25) and (26) leads to the following generalizations of the KdV equation for the interfaces (presented here in the original coordinates):

$$\eta_t + c\eta_x + \alpha\eta\eta_x + \beta\eta_{xxx} + \alpha_1\eta^2\eta_x + \beta_1\eta_{5_x} + \gamma_1\eta\eta_{xxx} + \gamma_2\eta_x\eta_{xx} = 0, \quad (29)$$

$$\zeta_t + c\zeta_x + \widetilde{\alpha}\zeta\zeta_x + \widetilde{\beta}\zeta_{xxx} + \widetilde{\alpha}_1\zeta^2\zeta_x + \widetilde{\beta}_1\zeta_{5_x} + \widetilde{\gamma}_1\zeta\zeta_{xxx} + \widetilde{\gamma}_2\zeta_x\zeta_{xx} = 0. \quad (30)$$

Although Eqs. (29) and (30) can be derived separately, they are not independent. Every such derivation results in a link to the solutions of the other equation. This link is presented explicitly in Eq. (28) for the equation for the lower interface $\eta(x,t)$. It allows for a direct calculation of the disturbances for the upper interface $\zeta(x,t)$ with an accuracy of $O\left(\varepsilon^3\right)$. A similar link (not shown) exists for the determination of the disturbances of the lower interface using the solution for the upper interface. These couplings do not transform an equation for one interface to an equation for the other interface: e.g., substituting Eq. (28) into Eq. (30) does not yield Eq. (29), but the difference includes terms which are higher order than those retained in Eq. (29). The equation for one interface and the equation obtained by substituting the connection between the two displacements into the relationship for the other interface are only equivalent asymptotically as the small parameters go to zero. Similar interrelations become evident for a similar pair of Gardner equations below.

The coefficients of Eqs. (29) and (30) account for the main properties of the environment supporting the internal wave motion such as the location of the undisturbed interfaces, the magnitude of the density jumps between the layers and the total fluid depth. The main development here is the derivation of explicit expressions for the coefficients of these equations. The expressions for $\alpha^{\pm}, \widetilde{\alpha}^{\pm}, \beta^{\pm}, \widetilde{\beta}^{\pm}, \alpha_1^{\pm}, \widetilde{\alpha}_1^{\pm}$ are quite complex and given in the Appendix. As the coefficients β_1, $\widetilde{\beta}_1$, γ_1, $\widetilde{\gamma}_1$, γ_2, $\widetilde{\gamma}_2$ are not analyzed in this paper, we do not provide their explicit expressions here. Note that $\beta^{\pm} \equiv \widetilde{\beta}^{\pm}$, $\beta_1^{\pm} \equiv \widetilde{\beta}_1^{\pm}$ as they are just coefficients of k^3 and k^5 in the Taylor expansion at small k (long-wave limit) of the dispersion relation $\omega(k)$ for linear IWs in a three-layer fluid. This dispersion relation is the same for both interfaces for a fixed mode.

The coefficients of the equation for the lower interface mode-1 displacement η^+ in the limits $H_1 \to H_2$ or $H_2 \to H_3$ are reduced into known expressions for internal waves in a two-layer fluid with density jumps $\Delta\rho_1 + \Delta\rho_2$ and $\Delta\rho_1$, respectively. For the mode-1 displacement of the upper interface ζ^+, the vanishing of the lowermost or middle layer (in the limiting process $H_1 \to 0$ or $H_1 \to H_2$) leads to the corresponding formulations for parameters of IWs in two-fluid systems with density jumps $\Delta\rho_2$ or $\Delta\rho_1 + \Delta\rho_2$, respectively.

The similarity of the structures of the equations for disturbances of mode-1 and mode-2 considerably simplifies further analysis; namely, it is sufficient for solving only one of those equations (say, η) and then for using the above-derived relationships to obtain an approximate solution to the other equation. The relationship (Eq. 28) connecting interfacial displacements should be used when the initial conditions for Eq. (29) are being set up. A similar expression exists for $\eta(\zeta)$. This interdependence is, however, effective only during limited time intervals. The relevant nonlinear and solitary-wave solutions have different wave speeds (see Eq. 24) and therefore will split over longer time intervals (considerably larger than $1/\varepsilon$). This process of separation of initially basically equivalent solutions for different modes implicitly expresses the limitations of the applicability of the used asymptotic procedure.

5 Gardner equations for interfacial displacements

Equations (29) and (30) are integrable using the inverse scattering method only for two specific sets of nontrivial values of their coefficients (Newell, 1985; Zwillinger, 1997; Weisstein, 2002a, b). These sets, however, are not applicable to the problem of internal wave motion in stratified environments, for which this equation apparently remains nonintegrable. In this case, stationary (in a properly chosen moving coordinate system) solitary waves normally do not interact elastically. They may be generated from a suitably chosen wave system and they often radiate their energy during their motion and interact with other components of the wave field in a complicated manner (Marchant and Smyth, 1996; Osborne et al., 1998; Osborne, 2010).

It was recognized already in the first studies of internal waves using the KdV equation several decades ago that the coefficient at the (first order in terms of the asymptotic procedure, or quadratic in the appearance) nonlinear term in this equation may vanish or change its sign (Kakutani and Yamasaki, 1978; Miles, 1979; Gear and Grimshaw, 1983). The sign of the dispersive term, however, is always the same. Importantly, the sign of the nonlinear term governs the polarity of solitary waves and soliton solutions to the relevant equation. The event of vanishing of the relevant coefficient or a change in its sign therefore leads to substantial rearrangement of the wave propagation in the vicinity of the vanishing location. A natural consequence is that an approaching soliton of, say, elevation will be transformed into a soliton of depression, and vice versa (Knickerbocker and Newell, 1980; Talipova et al., 1997; Grimshaw et al., 1999).

In order to understand how this process functions, it is necessary to build a higher-order equation that is able to describe the phenomena in the area where the coefficient at the quadratic nonlinear term vanishes. Moreover, such equations are crucial for the description of the wave motion in domains where this term is much smaller than other terms and where higher-order terms govern the wave evolution. The derivation of such equations is possible through a detailed analysis of the role of higher-order contributions (represented by coefficients $\alpha_1, \widetilde{\alpha}_1$ in Eqs. 29 and 30) of the presented asymptotic procedure into Eqs. (29) and (30) in cases when the quadratic terms vanish or are small enough to be ignored.

The relevant generalization of the KdV equation is called the Gardner equation. This equation contains a cubic nonlinear term, the coefficient at which may also be sign-variable, depending on the variations in the stratification (Talipova et al., 1999; Grimshaw et al., 2002, 2010; Kurkina, 2011a). If this term is negative (which is the case in a two-layer environment), the solitons' heights are limited and, in the process of reaching the limiting height, the soliton widens infinitely into a table-like disturbance. If the cubic term has a positive coefficient, the Gardner equation possesses several types of solitons with different polarities.

If the quadratic and cubic terms of Eqs. (29) and (30) have the same order of magnitude, the above-described asymptotic procedure is not applicable. A possibility is to make explicit use of the smallness of the coefficient at the quadratic nonlinear term and to assume that $\alpha \sim \delta$, where $\delta \ll 1$. A suitable procedure may be constructed using asymptotic expansions similar to Eqs. (10) and (11) but built for two independent small parameters μ and ε (see, e.g., Lamb and Yan, 1996). It is now necessary to account for second-order terms with respect to both parameters. This can be done by re-expanding the above series in terms of $\alpha \sim \delta$. A meaningful evolution equation can then be obtained using the condition that three terms (quadratic and cubic nonlinearity, and the leading term responsible for linear dispersion) are of the same order of magnitude. This is possible if $\delta \sim \varepsilon$ and $\mu \sim \varepsilon^2$.

The procedure itself is similar to the above-described technique, and sorting out the terms with similar powers of ε leads to the desired evolution equations. These equations, at the lowest order, for the disturbances at the interfaces of the three-layer environment are equivalent to a pair of interrelated Gardner equations:

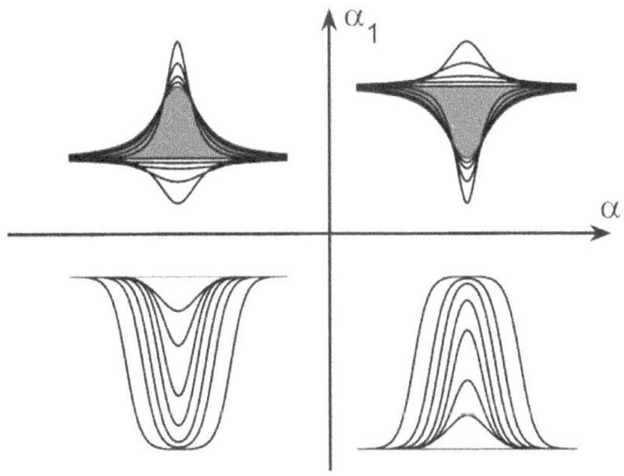

Figure 2. Shapes of soliton solutions to the Gardner equation for different combinations of the signs of coefficients at its nonlinear terms (idea of representation by R. Grimshaw, E. Pelinovsky and T. Talipova).

$$\eta_t + c\eta_x + \alpha\eta\eta_x + \alpha_1\eta^2\eta_x + \beta\eta_{xxx} = 0, \qquad (31)$$
$$\zeta_t + c\zeta_x + \widetilde{\alpha}\zeta\zeta_x + \widetilde{\alpha}_1\zeta^2\zeta_x + \beta\zeta_{xxx} = 0.$$

Formally, they only differ from Eqs. (29) and (30) by the absence of terms $\beta_1\eta_{5x}$, $\gamma_1\eta\eta_{xxx}$ and $\gamma_2\eta_x\eta_{xx}$. The actual difference is, however, much deeper, as Eq. (31) expresses the dynamics on another scale, governed by two independent small parameters. This difference becomes to some extent evident via another relationship between the soliton amplitude and its width. These equations govern the interfacial motion of weakly nonlinear finite-amplitude unidirectional waves neglecting mode interactions. Their coefficients are called environmental parameters, because they account for the background conditions (configuration of the medium). These coefficients are functions of the layers' positions and density jumps between the layers (or their ratio r).

Similarly to Eqs. (29) and (30), Eq. (31) is interrelated within the accuracy of $O(\varepsilon^3)$:

$$\zeta = s\eta + s_{\text{quad}}\eta^2. \qquad (32)$$

An inverse of Eq. (32), $\eta = \eta(\zeta)$, can be derived similarly in a straightforward manner.

The analytical one-soliton solution of the Gardner equation is well known (Helfrich and Melville, 2006; Pelinovsky et al., 2007):

$$\eta(x,t) = \frac{A}{1 + B\cosh(\gamma(x - Vt))}. \qquad (33)$$

The soliton velocity $V = c + \beta\gamma^2$ is expressed through the inverse width of the soliton γ. The parameters A and B depend on the coefficients of Eq. (31) and determine the soliton amplitude a as the extreme value of the function $\eta(x,t)$ in Eq. (33):

$$a = \frac{A}{1+B}, \quad A = \frac{6\beta\gamma^2}{\alpha}, \quad B^2 = 1 + \frac{6\alpha_1\beta\gamma^2}{\alpha^2}. \tag{34}$$

The parameters of the family of solutions can also be expressed through its amplitude a:

$$\gamma^2 = a\frac{2\alpha + a\alpha_1}{6\beta}, \quad A = a\left(2 + a\frac{\alpha_1}{\alpha}\right), \quad B = 1 + a\frac{\alpha_1}{\alpha}. \tag{35}$$

There are different branches of the soliton solutions depending on the signs of coefficients at the nonlinear terms; see Fig. 2. The asymptotics of the solution (Eq. 33) is described for instance in Pelinovsky et al. (2007).

6 Environmental parameters of IWs

Whilst the mathematical properties of the Gardner equation and its solutions are quite well known, these properties may vary radically along the propagation of realistic IWs in natural environments (Nakoulima et al., 2004). Therefore, it is important to analyze the behavior of the environmental parameters for possible ranges of parameters of the medium, which govern the kinematic and nonlinear characteristics of the IW field. The above-described explicit expressions for the coefficients of Eqs. (29) and (30) (see Appendix) remain valid also for Eq. (31) and allow the production of the estimates of nonlinear IW shapes, limiting amplitudes and projections of possible transformations of solitary IWs in an inhomogeneous medium (Kurkina et al., 2011a).

To describe the behavior of the environmental parameters, we use their graphs on the "phase" plane of scaled undisturbed interface positions ($l_1 = H_1/H_3$, $l_2 = H_2/H_3$) for the fixed value of the ratio r of density jumps. Along the diagonal $l_1 = l_2$, the two interfaces coincide, and this line represents a two-layer fluid. The other diagonal $l_1 = 1 - l_2$ corresponds to the symmetric geometric configuration of layers that has been addressed in detail in Kurkina et al. (2011b). As we always require $l_2 \geq l_1$, the triangle to the left of the diagonal $l_2 = l_1$ represents all possible ranges of l_1 and l_2. Therefore, we use this triangle to analyze the environmental parameters for the upper interface (coefficients in the equations for $\zeta^{\pm}$). For a Boussinesq fluid, there is symmetry: switching the depths of the upper and lower layers and interchanging the density jumps will result in exactly the same behavior with the waves flipped over. The coefficients for the equation for η given in terms of l_1 and l_2 should be the same as the equations for ζ given in terms of $1 - l_2$ and $1 - l_1$ and replacing r with $1/r$. When the Boussinesq approximation is involved, no preferred vertical direction exists; therefore, if the solution of a fixed polarity exists in this framework, its mirror disturbance is also a solution.

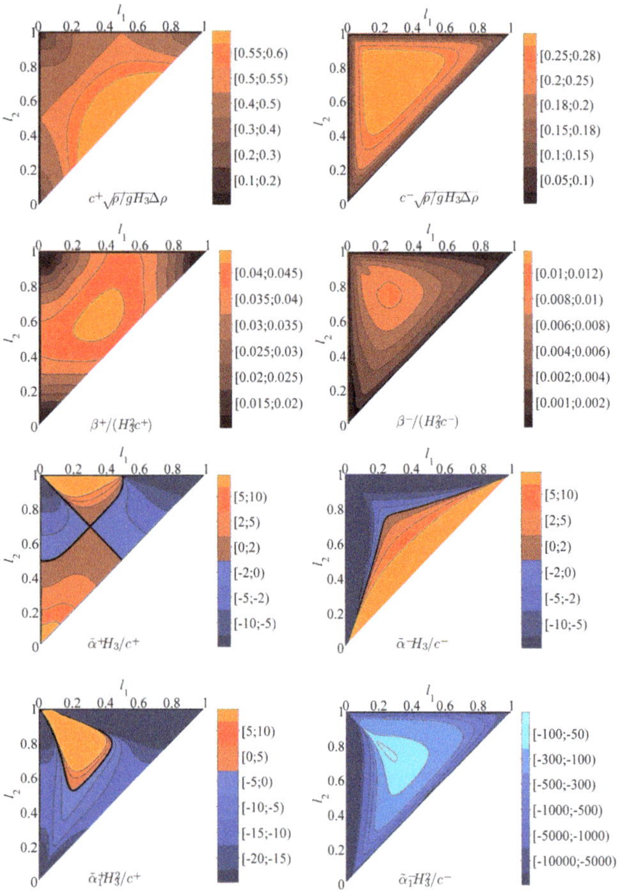

Figure 3. Variation in environmental parameters for IWs of fast (left column) and slow (right column) modes in the particular case $r = 1$ ($\Delta\rho_1 = \Delta\rho_2$). Zero contours are given by bold black lines.

Figure 3 shows the maps of values of parameters $c^{\pm}$, $\beta^{\pm}$, $\tilde{\alpha}^{\pm}$, $\tilde{\alpha}_1^{\pm}$ as functions of l_1 and l_2 when $r = \Delta\rho_2/\Delta\rho_1 = 1$. The parameters $c^{\pm}$ and $\beta^{\pm}$ are always positive for the three-layer fluid. They tend to zero only when one of the layers vanishes for mode-2 waves (c^-, β^-) or when the thicknesses of two out of the three layers simultaneously tend to zero for mode-1 waves (c^+, β^+). The parameters c^- and β^- both have one maximum for the symmetric configuration $l_2 = 1 - l_1 = 3/4$. Their counterparts c^+ and β^+ have the maxima at the point $l_1 = l_2 = 1/2$, i.e., when the three-layer fluid degenerates into a two-layer fluid with equal layers and with a density jump of $2\Delta\rho$.

A specific feature of the coefficients at the quadratic nonlinear term $\alpha^{\pm}$, $\tilde{\alpha}^{\pm}$ is that they may vanish for some nondegenerate three-layer configurations. These parameters for mode-1 IWs vanish along the diagonal $l_2 = 1 - l_1$ due to the symmetry of the wave motion. They also have other coinciding zero contours passing through the points (0; 0.5), (0.3; 0.7), (0.5; 1) on the plane (l_1, l_2). The signs of α^+ and $\tilde{\alpha}^+$ always coincide. Each of the quadratic nonlinear coefficients α^- and $\tilde{\alpha}^-$ for mode-2 IWs has a smooth zero contour

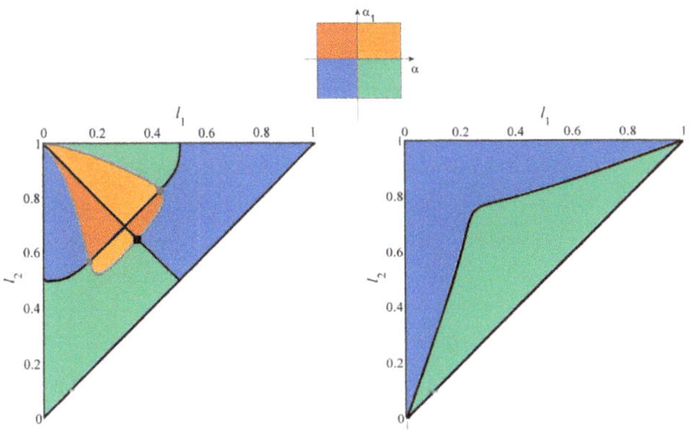

Figure 4. Map of the appearance of possible soliton solutions to the Gardner equation for a three-layer fluid for fast (left panel) and slow (right panel) internal wave modes in the particular case $r = 1(\Delta\rho_1 = \Delta\rho_2)$. The legend of patterns is given at the top. Bold black curves correspond to $\widetilde{\alpha} = 0$. Bold grey curves correspond to $\widetilde{\alpha}_1 = 0$. The markers show points at which $\widetilde{\alpha} = \widetilde{\alpha}_1 = 0$. Black squares indicate the combinations $(l_1, l_2) = (9/26;\ 17/26) \approx (0.3462;\ 0.6538)$, and grey circles the combinations $(l_1, l_2) \approx (0.1703;\ 0.5717)$ and $(l_1, l_2) \approx (0.4283;\ 0.8297)$.

passing through the points (0; 0), (0.25; 0.75), (1; 1) on the plane (l_1, l_2). The signs of these parameters, determining the polarity of solitary IWs in the KdV approximation, are always opposite, reflecting the structure of mode-2 waves that always create oppositely directed interface displacements.

The coefficients at the cubic nonlinear terms $\alpha_1^{\pm}$, $\widetilde{\alpha}_1^{\pm}$ behave differently for IWs of different modes (Fig. 4). The coefficients α_1^{+} and $\widetilde{\alpha}_1^{+}$ in the Gardner equation for mode-1 IWs have two distinct closed zero contours passing through the points (0; 1) and (9/26; 17/26) on the plane (l_1, l_2). The latter point corresponds to the symmetric stratification. The possibility of the change in the sign of the cubic nonlinear parameter at this point for IWs in a three-layer fluid was first mentioned in Grimshaw et al. (1997). For such a stratification, both coefficients at the leading nonlinear terms in the weakly nonlinear theory for mode-1 IWs vanish simultaneously. The wave dynamics near such a regime is described by an extended (2+4) KdV equation (Kurkina et al., 2011b). There exist two other such points (Fig. 4) in each half-plane. The parameters α_1^{-} and $\widetilde{\alpha}_1^{-}$ of mode-2 waves are always negative. Their minimal absolute values occur for a symmetric stratification, when all layers are relatively thick. When one of the layers vanishes, α_1^{-} and $\widetilde{\alpha}_1^{-}$ tend to negative infinity, reflecting the fact that the three-layer mode-2 wave regime does not have a valid asymptotic process towards a two-layer dynamics.

For both modes, the absolute values of the coefficients at the quadratic and cubic terms of the equations for the upper and lower interfaces are not equal except if they are zero or if the stratification is symmetric. This feature exemplifies the asymmetry of the interfacial displacements for asymmetric three-layer stratifications.

Figure 4 illustrates possible soliton branches of IWs of both modes for $r = \Delta\rho_2/\Delta\rho_1 = 1$. It is clear that all possible regimes shown in Fig. 2 can be implemented for mode-1 waves. Waves of both polarities can exist (waves of elevation on both interfaces or waves of depression on both interfaces), but properties of such waves can be different for different combinations of the parameters of the medium. The left panel of Fig. 4 also presents information about all points at which both the coefficients at the cubic and quadratic terms are zero. Higher-order extensions of the weakly nonlinear models have to be produced to provide a balance between the nonlinear and dispersive terms and to describe IWs in the vicinity of such points properly. These models for the points corresponding to asymmetric stratifications (and not located at the diagonal of the (l_1, l_2) plane) are supposed to be similar to those described in Kurkina et al. (2011b) for a symmetric situation.

The pool of solitonic mode-2 waves (Fig. 4, left panel) consists of solitons with a limited amplitude. As described above, these solitons broaden while approaching this limit and form table-like structures. Their polarity on each interface is determined by the sign of the coefficient at quadratic nonlinear terms α_1^{-} and $\widetilde{\alpha}_1^{-}$. Thus, two types of mode-2 waves can exist in the three-layer fluid: convex waves displacing the upper interface upward and the lower interface downward, and concave waves doing the opposite. Both types of mode-2 IWs were observed in the South China Sea (Yang et al., 2010), but concave waves occur fairly rarely. Convex waves were found much more often. They were experimentally observed in Mehta et al. (2002) and their nature can be explained as an intrusion of water into the middle layer.

The fluid at the mid-depth of a symmetric three-layer fluid (with equal density jumps) is not displaced in the case of mode-2 waves (Lamb, 2006), so a rigid boundary placed at the mid-depth will not be affected by the flow. The solution in the upper or lower half of the fluid is simply the solution for a two-layer fluid involving half of the middle layer; these "partial" solutions of course only exist together, and must be synchronized. The middle layer plays the role of a fluid that both underlies the upper layer and overlies the lower layer. If half of the middle layer is shallower than the lower and upper layers, a hump of elevation may exist at the upper interface, and a hump of depression at the lower one. This feature was used to interpret mode-2 solitary IWs experimentally observed by Mehta et al. (2002) with the help of the Gardner model for a two-layer fluid.

A summary of the results of some calculations of fully nonlinear and higher-order weakly nonlinear solitary IWs is presented in Fig. 5. These data are in qualitative agreement with the predictions of the model based on the Gardner equation. Some differences can be explained by the use of smoothed three-layer stratifications in numerical calculations in Lamb (2006).

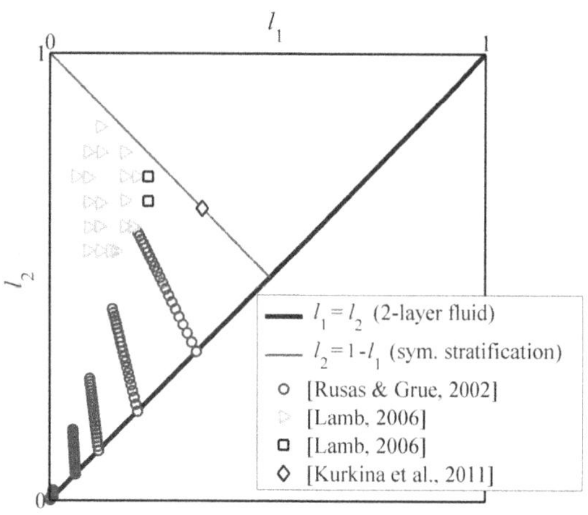

Figure 5. Points corresponding to stratifications for which mode-1 ISWs were considered in Boussinesq approximation with equal density jumps in different approaches. Circles: positive polarity waves (fully nonlinear numerical method for ISWs in a three-layer fluid) (Rusås and Grue, 2002). Triangles: waves of elevation with a minimum amplitude greater than zero; squares: waves of elevation with no minimum amplitude (fully nonlinear numerical model, continuous almost three-layerstratification) (Lamb, 2006). Left vicinity of the diamond on the line $l_2 = 1 - l_1$: waves of both polarities broaden and flatten while approaching an amplitude limit, (2+4) KdV model for IWs in a symmetric three-layer fluid (Kurkina et al., 2011a, 2012).

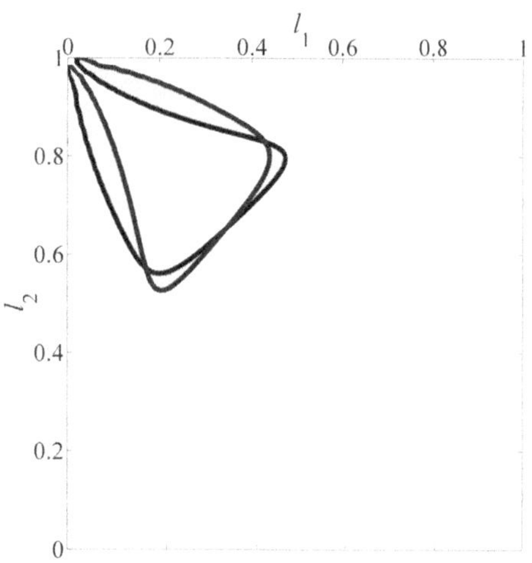

Figure 6. Comparison of curves $\alpha_1^+ = 0$ and $\tilde{\alpha}_1^+ = 0$ when $\Delta\rho_1 = \Delta\rho_2$. The intersection points are the same as marked points in Fig. 5, left panel, where $\alpha^+ = 0$ and $\tilde{\alpha}^+ = 0$. The shaded regions represent the configurations where the displacements of the interfaces belong to different branches of the solution of Eq. (33) shown in Fig. 2.

The contours of $\alpha_1^+ = 0$ and $\tilde{\alpha}_1^+ = 0$ in Gardner equations for the lower and the upper interface, respectively, are shown in Fig. 6 for the case $\Delta\rho_1 = \Delta\rho_2$. The difference between zero contours as shown in Fig. 6 is related to the fact that the second-order coefficients are not uniquely determined until some choice is made as to what the dependent variable in the weakly nonlinear equation represents: it could be the displacement of the lower interface, the upper interface, the average of the two, the velocity at some level between the two interfaces, etc. This was discussed, for example, in Lamb and Yan (1996). While the first-order coefficients are not affected by this choice, the second-order coefficients are. As the signs of α^+ and $\tilde{\alpha}^+$ are always the same for any fixed point of the plane (l_1, l_2), both of these solutions are taken from the left half-plane (α, α_1) or both from the right half-plane. So, different choices give different predictions when we are close to critical values where the cubic nonlinear coefficient is zero. For the fully nonlinear dispersive equations, we will only be in one regime.

The waves from mode-1 produce disturbances of the interfaces of the same polarity. Therefore, if $\alpha_1 > 0$, the polarity of the solution matches the sign of α. The shaded areas in Fig. 6 (the configurations where the displacements of the interfaces belong to different branches of the solution; Eq. 33) are located fairly close to the contours $\alpha_1^+ = 0, \tilde{\alpha}_1^+ = 0$. Consequently, the absolute values of the coefficients α_1^+ and $\tilde{\alpha}_1^+$ are small in these areas, and the corresponding solution apparently resembles the KdV soliton. This regime is the limiting one for the solutions corresponding to both "upper" and "lower" families of small-amplitude solitons depicted in Fig. 2. This feature confirms the correctness of the presented constructions and the applicability of the solutions (η^+, ζ^+) to internal waves of mode-1, even in these areas.

In this case, it should be decided to derive a Gardner equation for the lower interface or for the upper interface. The fact that the relationship (Eq. 32) does not convert the second of Eq. (31) to the first of Eq. (31) is connected to the fact that deriving the Gardner equation for the upper interface amounts to a different choice and hence results in different coefficients. As for stratifications with the same density jump across each interface, the interface furthest to the mid-depth is, probably, the best choice for the dependent variable in the Gardner equation. In this case, one would use the equation for ζ and the relationship connecting interfacial displacements if $l_1 < 1 - l_2$ and the equation for η with connection Eq. (32) otherwise. For different density jumps, the interface with the largest density jump should be used.

In the seas and oceans, the assertion $r = \Delta\rho_2/\Delta\rho_1 = 1$ is normally not valid, and the magnitudes of density jumps at different interfaces can be considerably different. To depict the changes associated with the above "phase diagrams" of various solutions, we chose the values $r = 2$ and $r = 1/2$ to characterize the variations in the coefficients in question and the wave propagation regimes. Different values of r significantly influence the patterns for mode-1 environmental parameters. The modifications are the largest for the values of the coefficients at the nonlinear terms. These changes obviously affect the possible regimes for soliton appearance and propagation (Figs. 7 and 8). Interestingly, the parameters

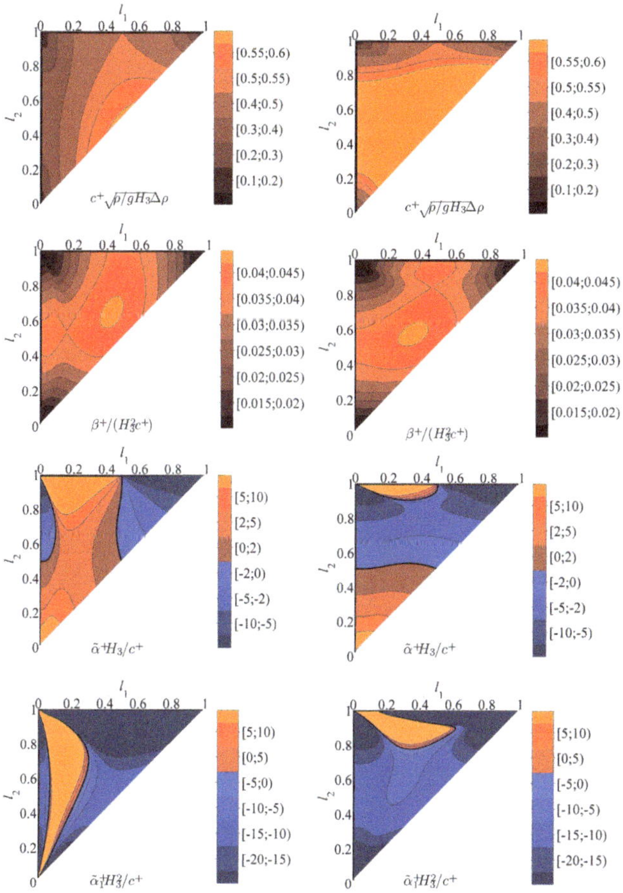

Figure 7. Variation in environmental parameters for mode-1 solitary IWs in a three-layer fluid with unequal density jumps at the interfaces. Left column $\Delta\rho_1 = \Delta\rho_2/2$. Right column: $\Delta\rho_1 = 2\Delta\rho_2$.

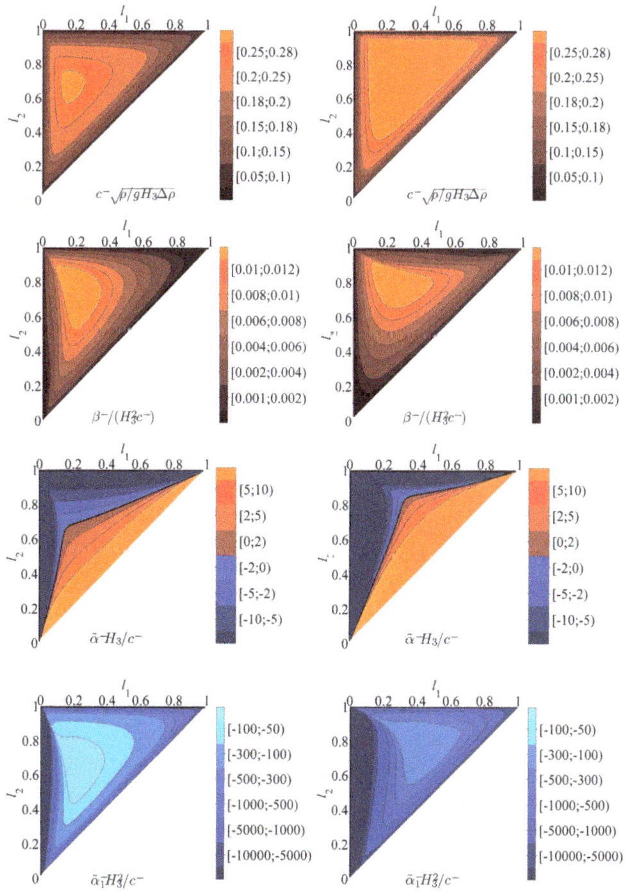

Figure 9. Variation in the environmental parameters for mode-2 IWs in a three-layer fluid with unequal density jumps at the interfaces. Left column: $\Delta\rho_1 = 2\Delta\rho_2$. Right column: $\Delta\rho_1 = \Delta\rho_2/2$.

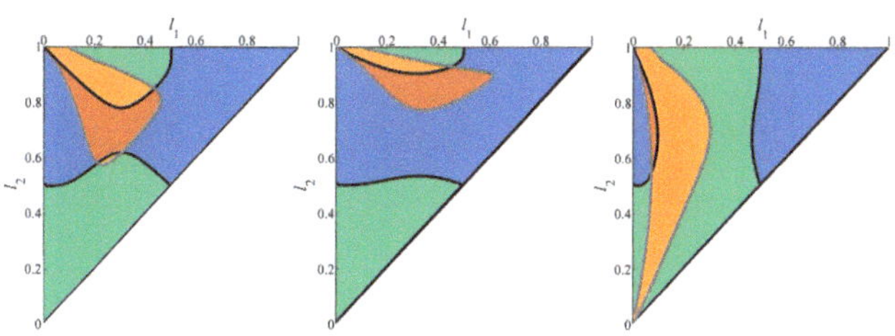

Figure 8. Map of the appearance of possible soliton solutions for mode-1 IWs in a three-layer fluid with unequal density jumps at the interfaces. Left column: $\Delta\rho_1 = 1.1\Delta\rho_2$. Middle column: $\Delta\rho_1 = 2\Delta\rho_2$. Right column: $\Delta\rho_1 = \Delta\rho_2/2$.

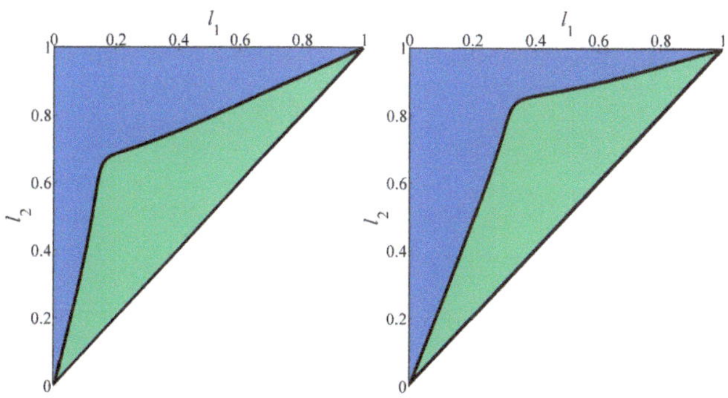

Figure 10. Map of the appearance of possible soliton solutions for mode-2 IWs in a three-layer fluid with unequal density jumps at the interfaces. Left column: $\Delta\rho_1 = 2\Delta\rho_2$. Right column: $\Delta\rho_1 = \Delta\rho_2/2$.

characterizing solutions of waves of mode-2 and the relevant propagation regimes are only slightly affected (Figs. 9 and 10).

It is worth noting that thermocline thicknesses can be considerably different under natural conditions. This is another way to destroy symmetry in a continuous stratification and change the wave propagation regime. While the fluid with the stratification $\rho = \rho_0 - \frac{1}{2}\Delta\rho_1\tanh((z-H_1)/d_1) - \frac{1}{2}\Delta\rho_2\tanh((z-H_2)/d_2)$ for small d_1, d_2 could be modeled numerically with a smoothed three-layer model, the considered three-layer model completely misses this.

7 Conclusions

The performed analytical investigation of equations governing the propagation of weakly nonlinear internal waves in a relatively simple but frequently occurring in nature and rich in content three-layer environment has highlighted several interesting features that distinguish this environment from the similar but symmetric case (Kurkina et al., 2011b). We have derived a system of nonlinear evolution equations for long, finite-amplitude internal waves, valid for both possible modes of wave motion at the interfaces of a three-layer fluid of arbitrary ratios of the depths between the layers and arbitrary (but small) values of the density jumps at the interfaces in the commonly used framework of Boussinesq approximation. Although each of the equations of this system is known, the analysis of interrelations of the equations for disturbances at different interfaces and the description of the possible wave propagation regimes provides new insight into the dynamics of internal waves in natural environments.

As expected, the classical Korteweg–de Vries equation is only conditionally a proper tool for the description of the internal wave motion, because the coefficient at its nonlinear (quadratic) term may vanish for certain combinations of the environmental parameters (layer depths and densities) of the three-layer fluid. This shortage affects the description of both mode-1 and mode-2 waves (with the same and opposite polarities of the disturbances to the interfaces, respectively). To resolve such situations, the derivation of the evolution equations is extended by systematically accounting for the next (second) order in magnitude terms in the relevant asymptotic procedure. This procedure leads to two implicitly interrelated Gardner equations (one for each interface) with possibly spatially varying coefficients.

The derived explicit analytical (algebraic) expressions for their coefficients as functions of the properties of the fluid made it possible to perform full analysis of the possible appearances, propagation regimes and transformation domains of solitary solutions of both modes to these equations in horizontally inhomogeneous environments. In particular, we perform a comprehensive analysis of the situations in which the coefficients of the quadratic and cubic nonlinearity simultaneously vanish. This may happen for three different combinations of the properties of the fluid. On such occasions, Gardner equations fail to describe the internal wave dynamics, and it is necessary to extend the derivation of the proper evolution equation to even higher-order terms. While the relevant analysis has recently been performed for the special case of a symmetric three-layer fluid (Kurkina et al., 2011b), the multitude of options for the failure of the Gardner equation to describe the dynamics of three-layer flows calls for further in-depth analysis of this problem. More generally, the situations corresponding to any line representing vanishing coefficients in Fig. 5 require deeper analysis of the associated dynamics.

Appendix A

Coefficients of Eqs. (29), (30) as well as the Gardner equations, namely, Eq. (31) for interfacial waves of both modes propagating on both interfaces.

$$\frac{\beta^{\pm}}{c^{\pm}H_3^2}=\frac{\widetilde{\beta}^{\pm}}{c^{\pm}H_3^2}=$$
$$12\frac{2rq^{\pm}(1-l_2)(l_2-l_1)\Big(2l_1(l_2-l_1)^2q^{\pm}-l_2(l_2+l_1)\Big)+2q^{\pm}l_1\Big(l_1^3-3l_1^2+\big(3-(1-l_2)^2\big)l_1-2l_2(2-l_2)\Big)-l_1^2+l_2(2-l_2)}{1-4rq^{\pm 2}l_1(1-l_2)(l_2-l_1)}$$

$$\frac{\alpha^{\pm}H_3}{c^{\pm}}=\frac{3}{2(1-l_2)l_1}\frac{8rq^{\pm 2}l_1(1-l_2)^2(l_2-l_1)\big(l_1q^{\pm}-1\big)-4q^{\pm 2}l_1^2\big(l_1^2-(1+l_2)l_1+l_2\big)+4q^{\pm}l_1(l_2-l_1)-2l_2+1}{1-4rq^{\pm 2}l_1(1-l_2)(l_2-l_1)}$$

$$\alpha_1^{\pm}=\frac{1}{2(1-l_2)^2H_3^2c^{\pm}l_1^2\Big(1-4rq^{\pm 2}(1-l_2)l_1(l_2-l_1)\Big)}\times\Big(\alpha^{\pm 2}H_3^2l_1^2(1-l_2)^2\Big(-12rq^{\pm 2}l_1(1-l_2)(l_2-l_1)-1\Big)\Big)$$
$$+\alpha^{\pm}H_3c^{\pm}(1-l_2)l_1\Big(4rq^{\pm 2}(1-l_2)^2(l_2-l_1)l_1\big(-20l_1q^{\pm}+11\big)+16q^{\pm 2}l_1^2(1-l_1)(l_2-l_1)+2q^{\pm}l_1(l_2-l_1)+5(1-2l_2)\Big)$$
$$+12rq^{\pm 2}c^{\pm 2}(1-l_2)^3(l_2-l_1)l_1\Big(-10q^{\pm 2}l_1^2+15q^{\pm}l_1-3\Big)$$
$$+12q^{\pm 3}c^{\pm 2}l_1^3\Big(-4l_1^3+3l_1^2(3+l_2)+l_1\Big((1-l_2)^2+6(1-l_2)-12\Big)+(5-l_2)l_2\Big)$$
$$-18q^{\pm 2}c^{\pm 2}l_1^2\Big(2l_1^2(3-l_2)+l_1\Big(2(1-l_2)^2+(1-l_2)-8\Big)+l_2(5-l_2)\Big)+18q^{\pm}c^{\pm 2}l_1(1+l_2)(l_2-l_1)$$
$$-6c^{\pm 2}\Big(3(1-l_2)^2-3(1-l_2)+1\Big)\Big)$$

$$\gamma_1^{\pm}=\frac{1}{1}2c^{\pm}H_3l_1(1-l_2)\Big(1-4rq^{\pm 2}l_1(1-l_2)(l_2-l_1)\Big)\times\Big(\Big(144rq^{\pm}H_3l_1^2(1-l_2)^2(l_1-l_2)-12H_3l_1(1-l_2)\Big)\beta^{\pm}$$
$$+\Big(\Big(8H_3^3l_1^2c^{\pm}(1-l_2)^2\Big(l_1^3-l_2^3\Big)+24H_3^3l_1^3c^{\pm}(1-l_2)^2(l_1-l_2)\Big)\Big)rq^{\pm}\ -2H_3^3l_1c^{\pm}(1-l_2)\Big(l_1^2-l_2^2\Big)\Big)\alpha^{\pm}$$
$$+\Big(384l_1^2c^{\pm}q^{\pm 3}r(1-l_2)^2(l_1-l_2)-240l_1c^{\pm}q^{\pm 2}r(1-l_2)^2(l_1-l_2)+48l_1^2c^{\pm}q^{\pm 2}\Big(l_1^2-l_1(1+l_2)+l_2\Big)$$
$$24c^{\pm}\big(q^{\pm}l_1(l_2-l_1)+1+2l_2\big)\big)\beta^{\pm}+\Big(40H_3^2\,c^{\pm 2}(1-l_2)^2\Big(q^{\pm 3}l_1^2\Big(l_1^3-l_2^3\Big)-q^{\pm 2}l_1\Big(l_1^3-l_2^3\Big)\Big)+120H_3^2c^{\pm 2}(1-l_2)^2$$
$$\times q^{\pm 3}l_1^3\Big(l_2^2-l_2l_1\Big)-96H_3^2c^{\pm 2}q^{\pm 2}(1-l_2)^2l_1^2\Big(l_2^2-l_2l_1\Big)+4H_3^2c^{\pm 2}q^{\pm}(1-l_2)^2\Big(l_2l_1^2-l_1^3\Big)\Big)r+(12\,l_1^4$$
$$-4l_1^3(5+7l_2)+4l_1^2\Big(11(1-l_2)^2-3-21l_2\Big)+4l_1\Big(7(1-l_2)^3-22(1-l_2)^2+21(1-l_2)-12\Big)$$
$$-4l_2\Big(7(1-l_2)^2-4(1-l_2)+3\Big)\Big)H_3^2c^{\pm 2}q^{\pm 2}l_1^2+(20\,l_1^3(1-l_2)-12l_1^2(3-2l_2)-4l_1(5(1-l_2)^3$$
$$-11(1-l_2)^2+6(1-l_2)-6\Big)H_3^2c^{\pm 2}q^{\pm}l_1+\Big(l_1^2(1-4l_2)-l_2^2(1-4l_2)\Big)H_3^2c^{\pm 2}\Big)$$

$$\gamma_2^{\pm} = \frac{1}{12c^{\pm} H_3 l_1 (1-l_2)\left(1-4rq^{\pm 2} l_1 (1-l_2)(l_2-l_1)\right)} \times \Big(\Big(216rq^{\pm} H_3 l_1^2 (1-l_2)^2 (l_1-l_2) - 18H_3 l_1 (1-l_2)\Big)\beta^{\pm}$$
$$+\Big(\Big(24H_3^3 l_1^2 c^{\pm}(1-l_2)^2\left(l_1^3-l_2^3\right)+72H_3^3 l_1^3 c^{\pm}(1-l_2)^2(l_1-l_2)\Big) rq^{\pm} - 6H_3^3 l_1 c^{\pm}(1-l_2)\left(l_1^2-l_2^2\right)\Big)\alpha^{\pm}$$
$$+\Big(192 l_1^2 c^{\pm} q^{\pm 3} r(1-l_2)^2(l_1-l_2) - 48 l_1 c^{\pm} q^{\pm 2} r(1-l_2)^2(l_1-l_2) + 96 l_1^2 c^{\pm} q^{\pm 2}\left(l_1^2 - l_1(1+l_2)+l_2\right)$$
$$+12c^{\pm}\left(2q^{\pm} l_1 (l_2-l_1)+1+2l_2\right)\Big)\beta^{\pm} + \Big(56H_3^2 c^{\pm 2}(1-l_2)^2\Big(q^{\pm 3} l_1^2\left(l_1^3-l_2^3\right) - q^{\pm 2} l_1\left(\frac{13}{7} l_1^3 - l_2^3\right)\Big) + 168H_3^2 c^{\pm 2}(1-l_2)^2$$
$$\times q^{\pm 3} l_1^3\left(l_2^2 - l_2 l_1\right) - 96H_3^2 c^{\pm 2} q^{\pm 2}(1-l_2)^2 l_1^2\left(\frac{5}{2} l_2^2 - l_2 l_1\right) - 4H_3^2 c^{\pm 2} q^{\pm}(1-l_2)^2\left(l_2 l_1^2 - l_1^3\right)\Big) r + (12\, l_1^4$$
$$-4l_1^3(7+5l_2)+4l_1^2\left(19(1-l_2)^2+3+15l_2\right)+4l_1\left(17(1-l_2)^3-38(1-l_2)^2+15(1-l_2)-12\right)$$
$$-4l_2\left(17(1-l_2)^2-2(1-l_2)+3\right)\Big) H_3^2 c^{\pm 2} q^{\pm 2} l_1^2 + (52\, l_1^3(1-l_2) - 12 l_1^2 (9-8l_2) - 4l_1 (13\,(1-l_2)^3$$
$$-19(1-l_2)^2 - 6(1-l_2) - 6\Big) H_3^2 c^{\pm 2} q^{\pm} l_1 + \left(l_1^2(5-8l_2) - l_2^2(5-8l_2)\right) H_3^2 c^{\pm 2}\Big)$$

$$\beta_1^{\pm} = \frac{1}{7} 20c^{\pm}\left(1-4rq^{\pm 2} l_1 (1-l_2)(l_2-l_1)\right) \times \Big(\Big(-360+4320 l_1 (1-l_2)(l_1-l_2) rq^{\pm 2}\Big)\beta^{\pm 2}$$
$$+\Big(480H_3^2 c^{\pm} rq^{\pm 2} l_1 (1-l_2)\left(l_1^3-l_2^3\right) + 1440H_3^2 c^{\pm} rq^{\pm 2} l_1^2 (1-l_2)\left(l_2^2 - l_2 l_1\right) - 120H_3^2 c^{\pm}\left(l_1^2 - 2l_2 + l_2^2\right)\Big)\beta^{\pm}$$
$$+\Big(\Big(12 l_1\left(l_1^5-l_2^5\right)+60 l_1^2\left(l_2^4 - l_2 l_1^3\right)+120 l_1^3\left(l_1 l_2^2 - l_2^3\right)\Big) q^{\pm 2} + \left(14 l_2 l_1^4 - 20 l_2^3 l_1^2 + 6 l_2^5\right) q^{\pm}\Big) H_3^4 c^{\pm 2}(1-l_2) r$$
$$+\Big(-6l_1^6+30l_1^5+20l_1^4\left(-3+(1-l_2)^2\right)+60 l_1^3 l_2 (2-l_2) - 2l_1^2\left(7(1-l_2)^4-30(1-l_2)^2+15\right)$$
$$-2l_1 l_2 (2-l_2)\left(7(1-l_2)^2-3\right)\Big) H_3^4 c^{\pm 2} q^{\pm} l_1 + \Big(-7l_1^4 + 10 l_1^2 l_2 (2-l_2) + l_2 (2-l_2)\left(7(1-l_2)^2-3\right)\Big) H_3^4 c^{\pm 2}\Big)$$

$$\widetilde{\alpha}^{\pm}\left(1-l_2, 1-l_1, \frac{1}{r}\right) = \alpha^{\pm}(l_1, l_2, r),$$
$$\widetilde{\alpha}_1^{\pm}\left(1-l_2, 1-l_1, \frac{1}{r}\right) = \alpha_1^{\pm}(l_1, l_2, r),$$
$$\widetilde{\gamma}_1^{\pm}\left(1-l_2, 1-l_1, \frac{1}{r}\right) = \gamma_1^{\pm}(l_1, l_2, r),$$
$$\widetilde{\gamma}_2^{\pm}\left(1-l_2, 1-l_1, \frac{1}{r}\right) = \gamma_2^{\pm}(l_1, l_2, r)$$

Note that the coefficients of the equations for both modes have the same appearance. All the differences are in the choice of the corresponding expressions of $c^{\pm}, q^{\pm}, \alpha^{\pm}$. The coefficients at terms describing linear dispersion $\beta^{\pm}$, $\widetilde{\beta}^{\pm}$ as well as the coefficients $\beta_1^{\pm}$ and $\widetilde{\beta}_1^{\pm}$ are equal for the equations for different interfaces within the same mode.

Acknowledgements. The research was partially supported by task 5.30.2014/K of the national targets of scientific research in Russia, Estonian Science Foundation (grant no. 7413), the targeted financing by the Estonian Ministry of Education and Science (grant no. SF0140007s11) and through support of the ERDF to the Centre of Excellence in Non-linear Studies (CENS) and the KESTA program.

Edited by: R. Grimshaw

References

Apel, J. R., Ostrovsky, L. A., Stepanyants, Y. A., and Lynch, J. F.: Internal solitons in the ocean and their effect on underwater sound, J. Acoust. Soc. Am., 121, 695–722, 2007.

Bogucki, D. J. and Redekopp, L. G.: A Mechanism for sediment resuspension by internal solitary waves, Geophys. Res. Lett., 26, 1317–1320, 1999.

Camassa, R., Rusas, P.-O., Saxen, A., and Tiron, R.: Fully nonlinear periodic internal waves in a two-fluid system of finite depth, J. Fluid Mech., 652, 259–298, 2010.

Chin-Bing, S. A., Warn-Varnas, A., King, D. B., Hawkins, J., and Lamb, K. G.: Effects on acoustics caused by ocean solitons – Part B: Acoustics, Nonlinear. Anal.-Theor., 71, 2194–2204, 2009.

Choi, W. and Camassa, R.: Weakly nonlinear internal waves in a two-fluid system, J. Fluid Mech., 313, 83–103, 1996.

Choi, W. and Camassa, R.: Fully nonlinear internal waves in a two-fluid system, J. Fluid Mech., 396, 1–36, 1999.

Craig, W., Guyenne, P., and Kalisch, H.: A new model for large amplitude long internal waves, C. R. Mechanique, 332, 525–530, 2004.

Engelbrecht, J. K., Fridman, V. E., and Pelinovsky, E. N.: Nonlinear Evolution Equations: Pitman Research Notes in Mathematics Series, 180, Longman, London, UK, 122 pp., 1988.

Funakoshi, M.: Long internal waves in a two-layer fluid, J. Phys. Soc. Jpn., 54, 2470–2476, 1985.

Funakoshi, M. and Oikawa, M.: Long internal waves of large amplitude in a two-layer fluid, J. Phys. Soc. Jpn., 55, 128–144, 1986.

Gear, J. A. and Grimshaw, R. H. J.: A second-order theory for solitary waves in shallow fluids, Phys. Fluids, 26, 14–29, 1983.

Giniyatullin A. R., Kurkin A. A., Kurkina O. E., and Stepanyants Y. A.: Generalised Korteweg–de Vries equation for internal waves in two-layer fluid, Fund. Appl. Hydrophys., 7, 16–28, 2014 (in Russian).

Grimshaw, R. H. J. (Ed.): Internal solitary waves, in: Environmental Stratified Flows, Kluwer, Boston, 1–28, 2001.

Grimshaw, R., Pelinovsky, E., and Talipova, T.: The modified Korteweg – de Vries equation in the theory of large – amplitude internal waves, Nonlin. Processes Geophys., 4, 237–250, doi:10.5194/npg-4-237-1997, 1997.

Grimshaw, R., Pelinovsky, E., and Talipova, T.: Solitary wave transformation in a medium with sign-variable quadratic nonlinearity and cubic nonlinearity, Physica D, 132, 40–62, 1999.

Grimshaw, R., Pelinovsky, E., and Poloukhina, O.: Higher-order Korteweg-de Vries models for internal solitary waves in a stratified shear flow with a free surface, Nonlin. Processes Geophys., 9, 221–235, doi:10.5194/npg-9-221-2002, 2002.

Grimshaw, R., Pelinovsky, E., and Talipova, T.: Modeling internal solitary waves in the coastal ocean, Surv. Geophys., 28, 273–298, 2007.

Grimshaw, R., Slunyaev, A., and Pelinovsky, E.: Generation of solitons and breathers in the extended Korteweg–de Vries equation with positive cubic nonlinearity, Chaos, 20, 013102, doi::10.1063/1.3279480, 2010.

Guyenne, P.: Large-amplitude internal solitary waves in a two-fluid model, CR Mecanique, 334, 341–346, 2006.

Helfrich, K. R. and Melville, W. K.: Long nonlinear internal waves, Annu. Rev. Fluid Mech., 38, 395–425, 2006.

Holloway, P., Pelinovsky, E., and Talipova, T.: Internal tide transformation and oceanic internal solitary waves, in: Environmental Stratified Flows, edited by: Grimshaw, R. H. J., Kluwer, Boston, 29–60, 2001.

Jackson, C. R.: An atlas of internal solitary-like waves and their properties. Prepared under contract with the Office of Naval Research Code 322PO, Contract N00014-03-C-0176, Global Ocean Associates, VA, Alexandria, VA, USA, 2004.

Jeans, D. R. G.: Solitary internal waves in the ocean. A literature review completed as part of the internal wave contribution to Morena, Rep. U-95, UCES, Marine Science Labs, University of North Wales, 1995.

Kakutani, T. and Yamasaki, N.: Solitary waves in a two-layer fluid, J. Phys. Soc. Jpn., 45, 674–679, 1978.

Knauss, J. A.: Introduction to Physical Oceanography, Prentice Hall, New Jersey, 309 pp., 1996.

Knickerbocker, C. and Newell, A.: Internal solitary waves near a turning point, Phys. Lett., 75, 326–330, 1980.

Koop, C. G. and Butler, G.: An investigation of internal solitary waves in a two-fluid system, J. Fluid Mech., 112, 225–251, 1981.

Kurkina, O. E., Pelinovsky, E., Talipova, T., and Soomere, T.: Mapping the internal wave field in the Baltic Sea in the context of sediment transport in shallow water, J. Coastal Res., Special Issue 64, 2042–2047, 2011a.

Kurkina, O. E., Kurkin, A. A., Soomere, T., Pelinovsky, E. N., and Rouvinskaya, E. A.: Higher-order (2+4) Korteweg–de Vries-like equation for interfacial waves in a symmetric three-layer fluid, Phys. Fluids, 23, 116602, doi:10.1063/1.3657816, 2011b.

Kurkina O., Singh N., and Stepanyants Y.: Structure of internal solitary waves in two-layer fluid at near-critical situation, Comm. Nonlin. Sci. Num. Simulation, 22, 1235–1242, 2015.

Lamb, K. G.: Conjugate flows for a three-layer fluid, Phys. Fluids, 12, 2169–2185, 2000.

Lamb, K. G.: Extreme Internal Solitary Waves in the Ocean: Theoretical Considerations, Preprint, University of Waterloo, 109–117, 2006.

Lamb, K. G. and Yan, L.: The evolution of internal wave undular bores: comparisons of a fully nonlinear numerical model with weakly nonlinear theory, J. Phys. Oceanogr., 26, 2712–2734, 1996.

Leppäranta, M. and Myrberg, K.: Physical Oceanography of the Baltic Sea, Springer Praxis, Berlin, Heidelberg, New York, 378 pp., 2009.

Marchant, T. R. and Smyth, N. F.: Soliton interaction for the extended Korteweg–de Vries equation, IMA J. Appl. Math., 56, 157–176, 1996.

Mehta, A. P., Sutherland, B. R., and Kyba, P. J.: Interfacial gravity currents. II. Wave excitation, Phys. Fluids, 14, 3558–3569, 2002.

Miles, J. W.: On internal solitary waves, Tellus, 31, 456–462, 1979.
Mirie, R. M. and Pennell, S. A.: Internal solitary waves in a two-fluid system, Phys. Fluids A-Fluid, 1, 986–991, 1989.
Muller, P. and Briscoe, M.: Diapycnal mixing and internal waves, Oceanography, 13, 98–103, 2000.
Nakoulima, O., Zahibo, N., Pelinovsky, E., Talipova, T., Slunyaev, A., and Kurkin, A.: Analytical and numerical studies of the variable-coefficient Gardner equation, Appl. Math. Comput., 152, 449–471, 2004.
Nayfeh, A. H.: Perturbation Methods, Wiley, New York, 437 pp., 2000.
Newell, A. C.: Solitons in Mathematics and Physics, SIAM, Philadelphia, 244 pp., 1985.
Osborne, A. R.: Nonlinear Ocean Waves and the Inverse Scattering Transform, Elsevier, San Diego, 917 pp., 2010.
Osborne, A. R., Onorato, M., Serio, M., and Bergamasco, L.: Soliton creation and destruction, resonant interactions, and inelastic collisions in shallow water waves, Phys. Rev. Lett., 81, 3559–3562, 1998.
Ostrovsky, L. A. and Stepanyants, Y. A.: Do internal solitions exist in the ocean?, Rev. Geophys., 27, 293–310, 1989.
Ostrovsky, L. A. and Stepanyants, Y. A.: Internal solitons in laboratory experiments: comparison with theoretical models, Chaos, 15, 037111, doi:10.1063/1.2107087, 2005.
Pelinovskii, E. N., Polukhina, O. E., and Lamb, K.: Nonlinear internal waves in the ocean stratified in density and current, Oceanology, 40, 757–766, 2000.
Pelinovsky, E. N., Polukhina, O., Slunyaev, A., and Talipova, T.: Internal solitary waves, in: Solitary Waves in Fluids, edited by: Grimshaw, R. H. J., WIT Press, Southampton, Boston, 85–110, 2007.
Pullin, D. I. and Grimshaw, R. H. J.: Finite-amplitude solitary waves at the interface between two homogeneous fluids, Phys. Fluids, 31, 3550–3559, 1998.
Reeder, D. B., Ma, B. B., and Yang, Y. J.: Very large subaqueous sand dunes on the upper continental slope in the South China Sea generated by episodic, shoaling deep-water internal solitary waves, Mar. Geol., 279, 12–18, 2011.
Rusås, P.-O. and Grue, J.: Solitary waves and conjugate flows in a three-layer fluid, Eur. J. Mech. B-Fluid., 21, 185–206, 2002.
Ruvinskaya, E. A., Kurkina, O. E., and Kurkin, A. A.: A modified nonlinear evolution equation for internal gravity waves in a three-layer symmetric fluid, Transactions of the Nizhny Novgorod R. E. Alekseev State Technical University, 83, 30–39, 2010 (in Russian).
Sabinin, K. and Serebryany, A.: Intense short-period internal waves in the ocean, J. Mar. Res., 63, 227–261, 2005.
Sandstrom, H. and Elliott, J. A.: Internal tide and solitons on the Scotian Shelf: A nutrient pump at work, J. Geophys. Res., 89, C6415, doi:10.1029/JC089iC04p06415, 1984.
Soomere, T.: Coupling coefficients and kinetic equation for Rossby waves in multi-layer ocean, Nonlin. Processes Geophys., 10, 385–396, doi:10.5194/npg-10-385-2003, 2003.
Sridevi, B., Murty, V. R., Sadhuram, Y., and Murty, V. S. N.: Impact of internal waves on the acoustic field at a coastal station off Paradeep, east coast of India, Nat. Hazards, 57, 563–576, doi:10.1007/s11069-010-9567-9, 2011.
Stastna, M. and Lamb, K. G.: Sediment resuspension mechanisms associated with internal waves in coastal waters, J. Geophys. Res., 113, C10016, doi:10.1029/2007JC004711, 2008.
Stober, U. and Moum, J. N.: On the potential for automated real-time detection of nonlinear internal waves from seafloor pressure measurements, Appl. Ocean Res., 33, 275–285, 2011.
Talipova, T., Pelinovskii, E., and Grimshaw, R. H. J.: Transformation of a soliton at a point of zero nonlinearity, JETP Lett., 65, 120–125, 1997.
Talipova, T., Pelinovsky, E., Lamb, K., Grimshaw, R. H. J., and Holloway, P.: Cubic nonlinearity effects in the propagation of intense internal waves, Dokl. Earth Sci., 365, 241–244, 1999.
Vlasenko, V., Stashchuk, N., and Hutter, K.: Baroclinic Tides: Theoretical Modeling and Observational Evidence, Cambridge University Press, Cambridge, 351 pp., 2005.
Warn-Varnas, A., Chin-Bing, S. A., King, D. B., Hawkins, J., and Lamb, K. G.: Effects on acoustics caused by ocean solitons, Part A: Oceanography, Nonlinear Anal.-Theor., 71, 1807–1817, 2009.
Weisstein, E. W.: "Sawada-Kotera Equation", available at: http://mathworld.wolfram.com/Sawada-KoteraEquation.html (last access: 19 February 2014), 2002a.
Weisstein, E. W.: "Kupershmidt Equation", available at: http://mathworld.wolfram.com/KupershmidtEquation.html (last access: 19 February 2014), 2002b.
Yang, Y. J., Fang, Y. C., Tang, T. Y., and Ramp, S. R.: Convex and concave types of second baroclinic mode internal solitary waves, Nonlin. Processes Geophys., 17, 605–614, doi:10.5194/npg-17-605-2010, 2010.
Zahibo, N., Slunyaev, A., Talipova, T., Pelinovsky, E., Kurkin, A., and Polukhina, O.: Strongly nonlinear steepening of long interfacial waves, Nonlin. Processes Geophys., 14, 247–256, doi:10.5194/npg-14-247-2007, 2007.
Zwillinger, D.: Handbook of Differential Equations, 3rd editon, Academic Press, Boston, USA, 834 pp., 1997.

19

FLIP-MHD-based model sensitivity analysis

C. Skandrani[1], M. E. Innocenti[2], L. Bettarini[3], F. Crespon[1], J. Lamouroux[1], and G. Lapenta[2]

[1]NOVELTIS, Space and Remote Sensing Department, Space Weather Unit, rue du Lac 153, 31670 Labège, France
[2]Center for Mathematical Plasma Astrophysics, Department of Mathematics, KULeuven – University of Leuven, Celestijnenlaan 200B, bus2400, 3001 Leuven, Belgium
[3]Fluid and Plasma Dynamics group, Department of Physics, Université Libre de Bruxelles, Campus Plaine – CP231 Boulevard du Triomphe, 1050 Brussels, Belgium

Correspondence to: C. Skandrani (chafih.skandrani@noveltis.fr)

Abstract. The state of the art in the forecast of the background solar wind speed and of the interplanetary magnetic field at Earth is based on the use as boundary conditions for heliospheric models of the input data provided by solar observations. Magnetogram synoptic maps are used to obtain information on the magnetic field configuration at the solar source surface. Magnetic field inputs at the solar source surface thus constitute one of the main external sources of errors in solar wind models. The assimilation of data into forecasting models used in the terrestrial domain showed the ability to control model state errors. A sensitivity study performed through the analysis of the ensemble variances and the representers technique is used here to assess how process and model state errors propagate in a nonlinear two-dimensional MagnetoHydro Dynamic (MHD) system. The aim is to understand the impact of the source surface input parameters on the evolution of MHD heliospheric models and the potentialities of data assimilation techniques in solar wind forecasting. The representer technique in fact allows one to understand how far from the observation point the improvement granted from the assimilation of a measure propagates.

1 Introduction

The interplanetary medium, although once considered a perfect vacuum, is actually a turbulent area dominated by the solar wind, which flows at the average velocities of 400 km s^{-1} (slow wind) and 750 km s^{-1} (fast wind) (McComas et al., 1998). Among the events which may enhance and provoke a jump in the solar wind speed, coronal mass ejections (CMEs) are the most disrupting. CMEs are one of the most energetic and interesting solar phenomena. They propel magnetic clouds with a mass up to 10^{17} g to speeds up to and above 2600 km s^{-1} into the heliosphere (Dwivedi and Mohan, 2005), influencing near-Earth plasma conditions, causing fluctuations in the magnetic field near Earth and in the ionospheric density and driving geomagnetic storms and auroras. However, CMEs are not the only source of fluctuations in the solar wind parameters. Coronal holes produce high-speed streams (Levine et al., 1977; Sheeley and Harvey, 1981) and the alternation of slow and fast flows generates perturbations named co-rotating interaction regions (CIRs) (Gosling and Pizzo, 1999). CIRs can themselves impact the near-Earth space. The understanding of these slow and fast flows is of prime importance for the forecast of interplanetary CME (ICME) propagation and associated threats (Vršnak and Zic, 2007). Such threats are relevant for a variety of human assets, both in space, air borne and ground based. Space weather impacts society in diverse ways (Baker, 1998; Siscoe, 2000; Bothmer and Daglis, 2007): apart from posing direct threats to astronauts' lives, space weather events may damage satellite equipment and alter the atmospheric drag on satellites. This causes satellite displacements and losses and disrupts communication and positioning signals. Airline passengers may be exposed to ionising radiations and induced currents at ground level may damage ground-based operations such as pipelines, the power grid and telecommunication cables. Thus there will be a continued and intensified need for space environment services to address health, safety and commercial issues. In recent years, the increased availability of solar wind measurements has allowed the scientific community to develop models and tools aimed both at better understanding the dynamics of solar wind propagation and at enabling solar wind now-casting and forecasting at 1 astronomical unit (AU) and beyond.

Two main kinds of heliospheric models exist: MagnetoHydro Dynamics (MHD) based models, such as MHD-Around-A-Sphere (MAS, Riley et al., 2001), ENLIL (Odstrcil, 2003) and SWMF/SC/IH (Space Weather Modelling Framework/ Solar Corona/ Inner Heliosphere, Toth et al., 2005), and empirically based models, such as the Wang–Sheeley–Arge (WSA) (Wang and Sheeley, 1990) and the model described in Vršnak et al. (2007).

Physics-based kinetic models are still too computationally expensive to be used for large-scale heliospheric simulations. However, recently, methods such as the implicit adaptive Multi Level Multi Domain method (Innocenti et al., 2013; Beck et al., 2013) have been developed with the aim of increasing the portion of space which can be simulated with kinetic codes at a reasonable computational cost.

In the first two kinds of models observations are used as inputs or boundary conditions. See, for example, Wu et at. (2006), where a data-driven MHD model is continuously fed with SOHO/MDI magnetogram data to study the evolution of active regions starting from the observed magnetic field configuration.

A third, less commonly employed direction may be explored in heliospheric modeling: data assimilation (DA) techniques (Bouttier and Courtier, 1999; Evensen, 2009) may be integrated into existing models to enhance their forecasting abilities, as recently explored in Innocenti et al. (2011). There, an empirical model is enhanced with the application of a Kalman filtering DA technique (Kalman, 1960; Welch and Bishop, 2001). It is noticed, among other results, that the assimilation of solar wind temperature observations into the model dramatically improves the forecasts of the solar wind speed and also grants the baseline model some resilience against CME activity.

In Schrijver and Derosa (2003), instead, SOHO/MDI magnetograms are assimilated into a flux dispersal model for the evolution of the photospheric magnetic field, which is then mapped to the solar source surface through a Potential Field Source Surface model. In Barrero Mendoza et al. (2006), an ensemble Kalman filtering (EnKF) technique is applied on top of an MHD model. The variation of the root mean square error for the state estimate is studied as a function of the position and number of the assimilation points. The case study examined is a magnetic storm in the terrestrial magnetosphere. A similar approach is used here for the region of space extending from the source surface to 1 AU. No planetary magnetospheres are considered at this point of the investigation.

The study presented aims at further investigating the topic of DA in the area of space weather forecast. DA combines observational data with the outputs of numerical simulations to produce an optimal estimate of the evolving state of the system. Thus, by assimilating space environment data into global numerical models, a better understanding of the physics of the interplanetary space and increased forecasting capabilities can be expected. DA has been successfully used by meteorologists and oceanographers in the last twenty years. On the other hand, the space physics community has been more reluctant to implement such techniques.

Applications of DA techniques to the near-Earth system can be found for electron dynamics in the radiation belts (Kondrashov et al., 2007; Rigler et al., 2004) and for global assimilation of ionospheric measurements (GAIM) (Schunk et al., 2003, 2004). Such a rather limited use may be traced to the fact that the space system is much larger and that there is a shortage of measurements when compared to meteorological and oceanographic systems. A massive amount of information for terrestrial weather exists, while the number of satellites collecting space weather-relevant data in the magnetosphere and in the region of space between the Sun and the Earth is rather low. The most prominent of them are SOHO (Solar and Heliospheric Observatory – Fleck et al., 1995), ACE (Advanced Composition Explorer – Stone et al., 1998), the Cluster mission (Laakso et al., 2010) and the PROBA-2 (Project for On-Board Autonomy-2, Gantois et al., 2010) ESA mission. Both SOHO and ACE are located at L1, while the Cluster mission consists of four satellites in the Earth's magnetosphere. PROBA2, instead, is located in a Sun-synchronous dusk–dawn orbit. The two STEREO spacecrafts, located in the Earth's orbit but in orbital positions ahead and behind the Earth, can be used for space weather data collection as well, as shown in Turner and Li (2011).

However, more and more observational data will be available soon. The NASA Living With a Star programme promises to provide an incredible amount of data regarding all of the different aspects of the Sun–Earth connection. The first mission, Solar Dynamics Observatory (Pesnell et al., 2012), was launched in 2010 and the community is just starting to use the huge amount of data it delivers.

The application of data assimilation to space sciences will benefit not only from this soon-to-be-available amount of observational data, but also from the recent development of heliospheric models.

One of the main goals of the present paper is to assess the kind of benefits that DA could bring to MHD heliospheric models. These preliminary feasibility studies are essential for understanding the potentialities of DA in the specific field and thus for deciding if more effort should be put into the activity.

As a starting point, the first setting considered is a simplified two-dimensional MHD model of the solar wind between the Sun and the Earth including the Lagrange point L1. The physics-based model used here is the FLIP MHD model (Brackbill, 1991) in a simplified two-dimensional configuration. A description of the model is provided in Sect. 2.

It is well known (see, for example, Wu et al., 2006) that a major source of forecast errors for heliospheric models is the use of non-optimal boundary conditions at the photosphere or at the source surface.

Another purpose of the present paper is to investigate if and how the use of statistical techniques for data assimilation could lead to obtaining a more accurate estimate of magnetogram input parameters and an improved initialisation state, in order to predict the future model variables in the whole area better.

The methodology adopted is reviewed in Sect. 3 and is based on ensemble Monte Carlo techniques (Evensen, 2009). Multiple runs of the model are initialised with statistically guided modifications of the input. Then, the sensitivity study is conducted by diagnosing process and model state errors for the nonlinear two-dimensional MHD system simulated with FLIP-MHD.

The results of the ensemble statistics are presented in Sect. 4.

First, in Sect. 4.1, it is checked if the model errors are distributed according to a Gaussian distribution function. Then, in Sect. 4.2, the model ensemble variances are used to characterise the model sensitivity. The ensemble variance analysis provides an objective tool for evaluating how initial condition errors affect a forecast and whether it would be desirable to gather additional observations to reduce model forecast errors. The variances in time and space of the errors for the solar wind magnetic field and velocity are in fact calculated for model runs which differ only for their initial conditions and can therefore be considered to be the model response to the uncertainties in the source surface inputs.

Section 4.3 studies the domain of influence of observations made at a given location on the whole forecasting domain and especially at the source surface. This so-called "representers analysis" allows one to understand at a glance how the system augmented with assimilation propagates the information spatially. Then, it becomes easier to see the potential impact of model improvement using DA.

In the Conclusion a summary of the work presented is provided and further considerations are drawn on the role of data assimilation in space weather forecasting models.

2 The FLIP-MHD model

The numerical model used in this study is the FLIP-MHD model, a 2-D resistive MHD code described in Brackbill (1991).

Two directions, z and x, are simulated, as depicted in Fig. 1. The direction z is the longitudinal or streamwise direction, oriented away from the Sun. Open boundary conditions are set at $z = 0$ and $z = L_z$, the domain length in the z direction. In the transverse or cross-stream direction x reflecting boundary conditions are set. All three velocity components are retained. The following viscous-resistive compressible MHD equations in dimensionless form, written in the Lagrangian frame, are solved:

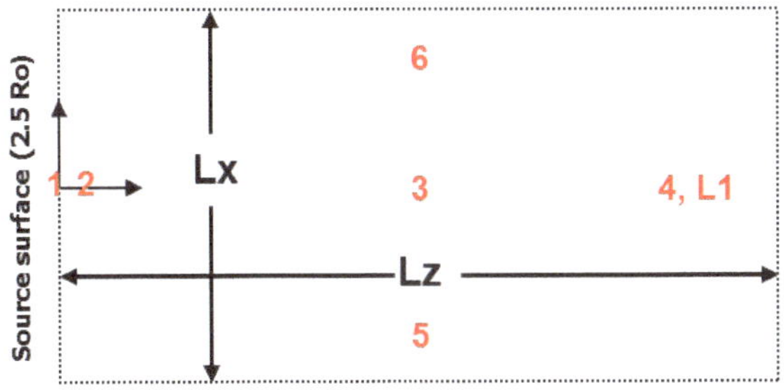

Fig. 1. Sketch of the domain simulated with FLIP-MHD. The left boundary corresponds to the source surface, conventionally located at 2.5 solar radii. The right boundary is placed at the terrestrial atmosphere. L_x and L_z are the box size in the x and z directions. The red numbers mark the diagnoses locations, as explained in Sect. 3.

Mass continuity:

$$\frac{\partial \rho}{\partial t} = -\nabla \cdot \rho \boldsymbol{v} \tag{1}$$

Momentum equation:

$$\rho \frac{\partial \boldsymbol{v}}{\partial t} = -\rho\,(\boldsymbol{v} \cdot \nabla)\,\boldsymbol{v} - \nabla p - \left[\nabla \frac{|\boldsymbol{B}|^2}{2} - (\boldsymbol{B} \cdot \nabla)\,\boldsymbol{B}\right] + \frac{1}{\Re_\upsilon} \Delta \boldsymbol{v} \tag{2}$$

Magnetic flux equation (Faraday's law):

$$\frac{\partial \boldsymbol{B}}{\partial t} = \nabla \times (\boldsymbol{v} \times \boldsymbol{B}) + \frac{1}{S} \Delta \boldsymbol{B} \tag{3}$$

Energy equation

$$\rho \frac{\partial I}{\partial t} = -p \nabla \cdot \boldsymbol{v} + \frac{(\nabla \cdot \boldsymbol{v})^2}{\Re_\lambda} + \frac{\prod \cdot \prod}{\Re_\upsilon} + \frac{\boldsymbol{J} \cdot \boldsymbol{J}}{S}, \tag{4}$$

where ρ is the mass density, $\boldsymbol{B}$ is the magnetic field which obeys the Gauss law constraint $\nabla \cdot \boldsymbol{B} = 0$, $\boldsymbol{v}$ is the fluid velocity, I is the specific internal energy, $\boldsymbol{J}$ is the current density, p is the fluid pressure, and $\prod$ the symmetric rate-of-strain tensor defined as

$$\prod = \frac{1}{2}\left[\nabla \boldsymbol{v} + \nabla \boldsymbol{v}^T\right]. \tag{5}$$

In Eqs. (2)–(4) $\Re_\lambda$ and $\Re_\upsilon$ are the Reynolds numbers measuring respectively the global kinematic and dynamic viscosity in the numerical box; in the simulations shown, $\Re_\lambda = \Re_\upsilon = 10^3$ is always assumed. The quantity $S = 10^3$ is the Lundquist number measuring the global explicit resistivity set within the domain.

2.1 Definition of the source function

Currently, operational prediction models are driven by input data at the source surface, which is conventionally located

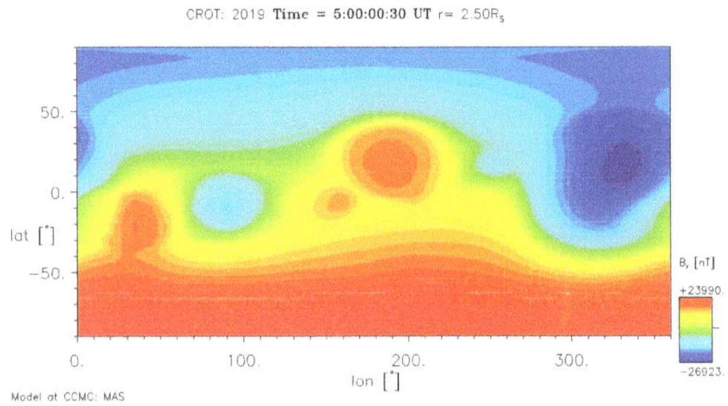

Fig. 2. Example of an extrapolated magnetic field at the source surface. From CCMC, http://ccmc.gsfc.nasa.gov.

at a distance of 2.5 solar radii from the solar surface. The boundary conditions at the source surface are in turn obtained from observations at the photosphere, such as GONG magnetogram synoptic maps. Figure 2 shows an example of the extrapolated magnetic field at the source surface obtained using the MAS tool (Riley et al., 2001) at the Community Coordinated Modeling Center, CCMC (http://ccmc.gsfc.nasa.gov).

In this paper, the FLIP-MHD model is driven by a synthetic analytical function providing the input conditions for the magnetic and velocity fields at the source surface. For simplicity, the velocity at the source surface is taken as constant. Instead, the boundary condition for the z component of the magnetic field at the source surface is imposed as follows:

$$b_z = b_0 \tanh(x - x_{\mathrm{centre}}) + \frac{b_0}{f(t)} \sin\left(\frac{2\pi(x - x_{\mathrm{centre}})}{L_B}\right) \quad (6)$$

x_{centre} is the middle point of the domain in the x direction.

The equations solved in the code and reported in the results below use normalised units. The magnetic field is normalised to b_0, the unit length is defined by the thickness L_B, and the unit time is computed as the time required by an Alfvén wave to travel one unit distance L_B, the so-called Alfvén time: $\tau_A = L_B/v_A$. Note that in this choice of units the reference Alfvén speed is $v_A = 1$.

The magnetic field at the source surface varies in time as determined by the function $f(t)$ defined as

$$f(t) = \begin{cases} 0.5(1 - \cos(\pi(1 - t/t_\alpha))t < t_\alpha \\ 1 \ \text{Otherwise} \end{cases} \quad (7)$$

$t_\alpha = t_{\mathrm{final}} \cdot \alpha$. The maximum value of t_α is reached before the final time, $t_{\mathrm{final}}/\tau_A = 400$, when alpha is less then 1. In particular, we select a value $\alpha = 0.4$, to introduce an initial phase of dynamic perturbation and subsequently observe the relaxation of the system to a constant drive.

The y and x components of the magnetic field and the velocity in the longitudinal direction are selected to ensure an initial force-free equilibrium (that is, no force is acting on the plasma):

$$b_y = -\frac{b_0}{\cosh(x\_x_{\mathrm{centre}})}, \quad (8)$$

$$b_x = 0, \quad (9)$$

and

$$u = 1. \quad (10)$$

With these boundary conditions, the evolution of the solar wind magnetic field and velocity is studied between the source surface and 1 AU.

The real heliospheric case is of course spherical. However, a full 3-D description is beyond the scope of the present paper that aims instead at ascertaining the sensitivity of MHD models to boundary drives. For this reason and at the cost of losing the effects of the geometric expansion, a Cartesian grid is used instead. Figure 1 shows the domain used. L_x and L_z are the system dimensions in the cross stream wise and stream wise direction respectively. The out of plane y dimension is assumed ignorable. In code units, $L_x/L_B = 3.6$ and $L_z/L_B = 98.4$. The red numbers mark the diagnoses locations defined later in Sect. 3.

The solar wind speed is assumed to be Alfvénic, $V_{\mathrm{SW}} = V_A$, at the injection boundary, and the total simulation time allows roughly 4 transit times in the z direction. When compared with the real heliospheric case, the distance L_z represents the Sun to Earth distance and therefore the total simulation time corresponds to approximately 4 times the total time required for the solar wind to reach the Earth. For typical solar wind speeds this corresponds to roughly 12 days. Of course, given the choice of a Cartesian grid this is only an indicative number.

3 Methodology

The model sensitivity analysis consists in describing the uncertainties of a reference simulation from a set of simulations over a chosen time period. Information about both the reliability of the model and about which input parameters contribute most to the variability of the model response is provided through such analysis. From the ensemble method (Evensen, 2009), ensemble variances and ensemble covariances can help improve the knowledge of how boundary conditions affect the evolution of the system and which areas are particularly sensitive to their variations. This could be helpful for the identification of the areas where dedicated observations may be useful for the data assimilation process.

Model errors can be simulated in a number of ways. The different options include using a different model, varying the resolution of the discretisation grid or altering some parameters in the model (Evensen, 2009). One of the available alternatives is to characterise the model sensibilities through ensemble variances (Evensen, 2009). This leads to the depiction of the model errors. In the framework of DA, this ensemble technique is widely used to characterise the inaccuracies in the initialisation of the state of atmospheric and oceanographic forecasting models.

Several initial states are obtained by imposing small variations on top of the initial conditions. Then, starting from each of these initial states, calculations for future states are made. At a given time, all of the calculated states can be represented by a dispersion or probability distribution.

One of the preliminary steps before applying DA to a model is to derive this probability distribution of the state vector. It can be obtained through a Monte Carlo sampling method which describes the uncertainties with respect to a reference simulation of a set of similar simulations over a chosen period of time. The error covariances can thus be specified. The sample characterises the sensitivity of the model forecast to parameter errors around the current state of the system. Based on this ensemble technique, sensitivity analysis and representers study constitute the cornerstone of the approach presented here, leading to the preliminary understanding of how much DA will be effective.

The analysis includes three main stages:

1. an ensemble of 100 members is generated by perturbing the source function as described in Eq. (6);

2. the integration of the FLIP-MHD model over this set of perturbed initial states provides access to a set of model solutions, representing the probability density of the model state (model errors and uncertainties);

3. diagnoses on the simulated sets are conducted around some specific locations such as the Lagrange point L1 (location 4), near the source surface (location 2), in the middle of the domain (location 3) and to both sides of the longitudinal axis near the borders of the domain study (locations 5 and 6) (see Fig. 1).

In Sect. 3.1 more information regarding the generation of the perturbations is provided, while in Sect. 3.2, the ensemble diagnostic statistical tools are introduced in greater detail.

3.1 Generation of perturbations

Within the framework of this study, evolved forecast error covariances are constructed using an ensemble of model runs that start from perturbed initial states. Perturbations are built by adding uncertainty to the synthetic source surface magnetic field inputs in the streamwise z direction. The velocity, instead, is kept constant.

If $F(x,t)$ is the analytical function at the source surface, a particular perturbed field $F^{\text{pert}}(x,t)$ is obtained as in

$$F^{\text{pert}}(x,t) = F(x,t) + \gamma(t)F(x,t), \tag{11}$$

where $\gamma(t)$ is a time-dependent coefficient generated from a Gaussian distribution with mean zero and standard deviation of 0.05. An error of about 5 % with respect to the reference field is imposed.

For each member of the ensemble, the simulation is carried out for 100 time steps up to a final time of $t_{\text{final}}/\tau_{\text{A}} = 400$.

3.2 Ensemble theory and statistical tools

In this section the ensemble statistical tools used in this study are presented in detail. The concepts of representers technique and domain of influence are introduced, based on the ensemble Kalman filter method.

3.2.1 Ensemble Kalman filter and representers

The representers technique has been used widely for oceanic or tidal circulation problems as reported in the literature (Bennet, 1992; Egbert and Bennet, 1996; Lyard, 1999; Echevin et al., 2000; Evensen, 2009). At this stage of the preliminary analysis, it is a simple and robust tool for investigating how the system with DA will spatially propagate the information brought by dedicated observations and how strong the correlation between the observation taken at a particular spatial location and the other grid points is. A brief summary of the ensemble Kalman filter technique is provided here with the aim of facilitating the reading of the rest of the paper.

In the framework of the ensemble Kalman filter, a series of forecasting and analysis steps are executed in a sequence.

First, the forecasting step provides model predictions from the last analysis state and calculates the forecast errors of the model by propagating the analysis errors appropriately.

Then the analysis step at time t_i corrects the forecasted state $\boldsymbol{x}^{\text{f}}(t_i)$ with the information obtained from the observation vector $\boldsymbol{y}^{\text{o}}(t_i)$. The corrected state $\boldsymbol{x}^{\text{a}}(t_i)$ is calculated according to the following formula:

$$\boldsymbol{x}^{\text{a}}(t_i) = \boldsymbol{x}^{\text{f}}(t_i) + \mathbf{K}(t_i)\left[\boldsymbol{y}^{\text{o}}(t_i) - \mathbf{H}(t_i)\boldsymbol{x}^{\text{f}}(t_i)\right], \tag{12}$$

where $\boldsymbol{x}^{\text{f}}(t_i)$ is the forecast of the state at time t_i, $\boldsymbol{y}^{\text{o}}(t_i)$ is the vector of observations at the same time and the matrix $\mathbf{H}(t_i)$ is the so-called "observation operator". It ensures the projection of the model space onto the observation space. In this particular case, the state vector $\boldsymbol{x}$ includes the solar wind magnetic and velocity fields at all grid points of the simulated domain; $\boldsymbol{y}^{\text{o}}(t_i) - \mathbf{H}(t_i)\boldsymbol{x}^{\text{f}}(t_i)$ is the "innovation" matrix, which carries the information relative to the difference between the actual observations and the corresponding values for the forecasted fields; $\mathbf{K}(t_i)$ is the Kalman gain matrix at time t_i, given by:

$$\mathbf{K}(t_i) = \mathbf{P}^{\text{f}}(t_i)\mathbf{H}^T(t_i)\left[\mathbf{H}(t_i)\mathbf{P}^{\text{f}}(t_i)\mathbf{H}^T(t_i) + \mathbf{r}(t_i)\right]^{-1} \tag{13}$$

where $\mathbf{P}^{\text{f}}(t_i)$ and $\mathbf{r}(t_i)$ are the forecast error covariance and the observation error covariance matrices, respectively, at time t_i. The gain matrix $\mathbf{K}(t_i)$ is the optimal least square gain at time t_i, i.e. the solution of the minimisation problem of the model observations error cost function (Ghil and

Malanotte-Rizzoli, 1991; Bouttier and Courtier, 1999). The matrix $\mathbf{K}(t_i)$ weights the forecast and the observations according to their respective error statistics, $\mathbf{P}^{\mathrm{f}}(t_i)$ and $\mathbf{r}(t_i)$. Notice that Eqs. (12) and (13) are recalled to provide a general background to the Kalman filtering procedure used for data assimilation: indeed, the actual assimilation of heliospheric observations has not been carried out yet. The present work focuses in fact on the concept of representers, which is described below. The rest of this section aims at explaining the role of the representer matrices in this sensitivity study without actually performing the assimilation procedure and without having observations at our disposal.

Representers or influence functions are obtained by reformulating the analysis step of the Kalman filter described in Eq. (12), as in

$$\boldsymbol{x}^{\mathrm{a}}(t_i) = \boldsymbol{x}^{\mathrm{f}}(t_i) + \sum_{j=1}^{j=p_i} b_j(t_i)\mathbf{r}_j(t_i), \tag{14}$$

where $\boldsymbol{x}^{\mathrm{f}}(t_i)$ is the forecasted state vector at time t_i, $\mathbf{r}_j(t_i)$ is the representer or influence function relative to the jth observation at time t_i and p_i is the number of observations available at time t_i. The representer $\mathbf{r}_j(t_i)$ is a vector which has the dimensions of the state vector and $b_j(t_i)$ is the scalar coefficient associated with the observation j at time t_i.

Equation (14) translates in matrix form into

$$\boldsymbol{x}^{\mathrm{a}}(t_i) = \boldsymbol{x}^{\mathrm{f}}(t_i) + \mathbf{r}(t_i)\boldsymbol{b}(t_i), \tag{15}$$

with the matrix $\mathbf{r}(t_i)$ defined at time t_i as

$$\mathbf{r}(t_i) = \mathbf{P}^{\mathrm{f}}(t_i)\mathbf{H}^T(t_i), \tag{16}$$

and the vector $\boldsymbol{b}(t_i)$ as

$$\boldsymbol{b}(t_i) = \left[\mathbf{H}(t_i)\mathbf{P}^{\mathrm{f}}(t_i)\mathbf{H}^T(t_i) + \mathbf{r}(t_i)\right]^{-1} \left[\boldsymbol{y}^{\mathrm{o}}(t_i) - \mathbf{H}(t_i)\boldsymbol{x}^{\mathrm{f}}(t_i)\right]. \tag{17}$$

The matrix $\mathbf{r}(t_i)$ contains as many columns as there are observations available at time t_i and each column constitutes the representer $\mathbf{r}_j(t_i)$ relative to a given observation. The representer formulation of the analysis step of Eq. (14) can be interpreted as the correction of the forecasted state $\boldsymbol{x}^{\mathrm{f}}(t_i)$ done by using a combination of representers, depending on the forecast error covariance matrix $\mathbf{P}^{\mathrm{f}}(t_i)$. Note that in the case of a multivariate state vector (e.g. solar wind magnetic and velocity fields), the information from the jth observation can be spread to the other model state variables, thanks to the error cross-covariance of these variables.

Notice that from this point onwards, the time dependence is dropped for ease of reading. It is also assumed that the model grid is composed of n nodes and that the state vector contains two variables (namely magnetic field and solar wind velocity). The dimension of the state vector is then $2n$.

Now, focusing on a single observation, Eq. (16) becomes, at a given time:

$$\mathbf{r}_j = \mathbf{P}^{\mathrm{f}}\mathbf{H}_j^T \tag{18}$$

$\mathbf{H}_j$ is the raw of the matrix $\mathbf{H}$ corresponding to the jth observation. For the purpose of the demonstration, let us assume that the jth observation is *(1)* exactly located at a grid point k and *(2)* associated with one of the state vector variables. Thus, this observation is identified in the state vector by its index k_j. Then, $\mathbf{H}_j\ (k_j)$ is the only non-zero element of $\mathbf{H}_j$, that is

$$\mathbf{H}_j = (0, \ldots, 0, 1, 0, \ldots, 0), \tag{19}$$

with only one "1" at index k_j. Thus, Eq. (18) gives, from the definition of $\mathbf{P}^{\mathrm{f}}$,

$$\mathbf{r}_{k_j} = \mathbf{P}^{\mathrm{f}}\mathbf{H}_{k_j}^T = \begin{pmatrix} \mathrm{cov}(\varepsilon_{k_j}^{\mathrm{f}}, \varepsilon_1^{\mathrm{f}}) \\ \vdots \\ \mathrm{cov}(\varepsilon_{k_j}^{\mathrm{f}}, \varepsilon_n^{\mathrm{f}}) \\ \mathrm{cov}(\varepsilon_{k_j}^{\mathrm{f}}, \varepsilon_{n+1}^{\mathrm{f}}) \\ \vdots \\ \mathrm{cov}(\varepsilon_{k_j}^{\mathrm{f}}, \varepsilon_{2n}^{\mathrm{f}}) \end{pmatrix} \tag{20}$$

ε^{f} is the model forecast error, defined as the difference between the forecasted and true states $(\boldsymbol{x}^{\mathrm{f}} - \boldsymbol{x}^{t})$, which is not known. The abbreviation "cov" stands for "covariance".

Consequently, the representer of observation j contains the covariances between the model forecast errors at the observation point and all the grid nodes, for all the state variables (e.g. magnetic field, solar wind velocity).

As explained in Sect. 3, the model forecast errors can be estimated through the stochastic modelling approach: if an ensemble of simulations generated by perturbing one or several sources of errors of the model is available, these model forecast errors are approximated by statistics computed from this ensemble. Thus, the covariances of the model forecast errors are estimated by the ensemble covariances of the model forecast errors. Therefore

$$\begin{aligned} \mathrm{cov}(\varepsilon_{k_j}^{\mathrm{f}}, \varepsilon_i^{\mathrm{f}}) &\approx \mathrm{cov}^{\mathrm{ens}}(\varepsilon_{k_j}^{\mathrm{f}}, \varepsilon_i^{\mathrm{f}}) \\ &= \frac{1}{m}\sum_{s=1}^{m}(\varepsilon_{k_j,s}^{\mathrm{f}} - \frac{1}{m}\sum_{s=1}^{m}\varepsilon_{k_j,s}^{\mathrm{f}})(\varepsilon_{i,s}^{\mathrm{f}} - \frac{1}{m}\sum_{s=1}^{m}\varepsilon_{i,s}^{\mathrm{f}}) \\ &= \overline{(\varepsilon_{k_j}^{\mathrm{f}} - \overline{\varepsilon_{k_j}^{\mathrm{f}}})(\varepsilon_i^{\mathrm{f}} - \overline{\varepsilon_i^{\mathrm{f}}})}, \end{aligned} \tag{21}$$

where "ens" stands for "ensemble", m is the number of members in the ensemble and the horizontal bar represents the ensemble average. Under the assumption of the BLUE (best linear unbiased estimator) hypotheses (Bouttier and Courtier, 1999; Evensen 2009), the model forecast errors are assumed unbiased. This implies that

$$\begin{aligned}\operatorname{cov}(\varepsilon_{k_j}^{\mathrm{f}},\varepsilon_i^{\mathrm{f}}) &\approx \overline{(\varepsilon_{k_j}^{\mathrm{f}})(\varepsilon_i^{\mathrm{f}})}\\ &\approx \overline{(\boldsymbol{x}_{k_j}^{\mathrm{f}}-\overline{\boldsymbol{x}_{k_j}^{\mathrm{f}}})(\boldsymbol{x}_i^{\mathrm{f}}-\overline{\boldsymbol{x}_i^{\mathrm{f}}})}\\ &\approx \operatorname{cov}(\boldsymbol{x}_{k_j}^{\mathrm{f}},\boldsymbol{x}_i^{\mathrm{f}}). \end{aligned} \tag{22}$$

Thus, more generally, the variances, covariances and correlations of the model forecast errors are identifiable as the ensemble variances, covariances and correlations of the state variables.

3.2.2 Domain of influence of a single observation

Here an explicit expression for the domain of influence of a single observation is presented. Consider an isolated observation $\boldsymbol{y}_{k_j}$ relative to the forecasted state variable $\boldsymbol{x}^{\mathrm{f}}$ (magnetic field components or solar wind velocity components) and located on a node k of the mesh. The correction applied on the forecasted state $\boldsymbol{x}^{\mathrm{f}}$ calculated by assimilation of the observation $\boldsymbol{y}_{k_j}$ is expressed by

$$\delta\boldsymbol{x} = \boldsymbol{x}^{\mathrm{a}} - \boldsymbol{x}^{\mathrm{f}} = b_{k_j}\mathbf{r}_{k_j} \tag{23}$$

with

$$\mathbf{r}_{k_j} = \mathbf{P}^{\mathrm{f}}\mathbf{H}_{k_j}^{T} = (\operatorname{cov}(\varepsilon_{k_j}^{\mathrm{f}},\varepsilon_i^{\mathrm{f}}))_{i=1,2n} \tag{24}$$

$$\begin{aligned} b_{k_j} &= \frac{\boldsymbol{y}_{k_j}-\boldsymbol{x}_{k_j}}{\operatorname{cov}(\varepsilon_{k_j}^{\mathrm{f}},\varepsilon_{k_j}^{\mathrm{f}})+\operatorname{cov}(\varepsilon_{k_j}^{\mathrm{o}},\varepsilon_{k_j}^{\mathrm{o}})}\\ &= \frac{\delta\boldsymbol{y}_{k_j}}{\operatorname{var}(\varepsilon_{k_j}^{\mathrm{f}})+\operatorname{var}(\varepsilon_{k_j}^{\mathrm{o}})} \end{aligned} \tag{25}$$

where $\delta\boldsymbol{y}_{k_j}$ and $\varepsilon_{k_j}^{\mathrm{o}}$ are respectively the innovation and the observation error associated with the observation $\boldsymbol{y}_{k_j}$; the abbreviation "var" stands for the ensemble variance.

Combining Eqs. (23), (24) and (25), the correction at node i brought by the observation $\boldsymbol{y}_{k_j}$ is then expressed as follows:

$$\delta\boldsymbol{x}_i = \frac{\operatorname{cov}(\varepsilon_{k_j}^{\mathrm{f}},\varepsilon_i^{\mathrm{f}})}{\operatorname{var}(\varepsilon_{k_j}^{\mathrm{f}})+\operatorname{var}(\varepsilon_{k_j}^{\mathrm{o}})}\delta\boldsymbol{y}_{k_j} \tag{26}$$

Introducing the formal definition of the correlation, the previous equation becomes

$$\begin{aligned}\delta\boldsymbol{x}_i &= \frac{\frac{\operatorname{cov}(\varepsilon_{k_j}^{\mathrm{f}},\varepsilon_i^{\mathrm{f}})}{\sqrt{\operatorname{var}(\varepsilon_{k_j}^{\mathrm{f}})\operatorname{var}(\varepsilon_i^{\mathrm{f}})}}}{\frac{\operatorname{var}(\varepsilon_{k_j}^{\mathrm{f}})+\operatorname{var}(\varepsilon_{k_j}^{\mathrm{o}})}{\sqrt{\operatorname{var}(\varepsilon_{k_j}^{\mathrm{f}})\operatorname{var}(\varepsilon_i^{\mathrm{f}})}}}\delta\boldsymbol{y}_{k_j} = \frac{\operatorname{correl}(\varepsilon_{k_j}^{\mathrm{f}},\varepsilon_i^{\mathrm{f}})}{1+\frac{\operatorname{var}(\varepsilon_{k_j}^{\mathrm{o}})}{\operatorname{var}(\varepsilon_{k_j}^{\mathrm{f}})}}\sqrt{\frac{\operatorname{var}(\varepsilon_i^{\mathrm{f}})}{\operatorname{var}(\varepsilon_{k_j}^{\mathrm{f}})}}\delta\boldsymbol{y}_{k_j}\\ &\approx \operatorname{correl}(\boldsymbol{x}_{k_j}^{\mathrm{f}},\boldsymbol{x}_i^{\mathrm{f}})\frac{\sqrt{\frac{\operatorname{var}(\varepsilon_i^{\mathrm{f}})}{\operatorname{var}(\varepsilon_{k_j}^{\mathrm{f}})}}}{1+\frac{\operatorname{var}(\varepsilon_{k_j}^{\mathrm{o}})}{\operatorname{var}(\varepsilon_{k_j}^{\mathrm{f}})}}\delta\boldsymbol{y}_{k_j}\\ &\approx \operatorname{correl}(\boldsymbol{x}_{k_j}^{\mathrm{f}},\boldsymbol{x}_i^{\mathrm{f}})\times \text{modulation factor}(\varepsilon_{k_j}^{\mathrm{o}},\varepsilon_{k_j}^{\mathrm{f}},\varepsilon_i^{\mathrm{f}})\delta\boldsymbol{y}_{k_j}, \end{aligned} \tag{27}$$

where the abbreviation "correl" refers to correlation.

Therefore, the correction at point i, obtained by the assimilation of the isolated observation $\boldsymbol{y}_{k_j}$, depends on the correlation between the model forecast errors associated with the observed variable $\boldsymbol{x}_{k_j}$ and the variable to be corrected $\boldsymbol{x}_i$. According to Eq. (22), this correlation is approximated by the ensemble correlation of the observed variable $\boldsymbol{x}_{k_j}$ and the variable to be corrected $\boldsymbol{x}_i$. This correlation term reflects the spatial extent of the correction and thus represents the *domain of influence* (hereafter "doi") of the observation $\boldsymbol{y}_{k_j}$. Note that the magnitude of the correction $\delta\boldsymbol{x}_i$ is represented by the *modulation factor* term that depends on the observation error and on the ratio of the model error variances at the point i and at the observation point k: if the observation error is strong, the correction has to remain low; if the model error variance is higher/lower at the point i than at the observed point, the correction must be significant/lower.

As a first step in the implementation of a dedicated DA system, this study will focus on the doi of isolated observations (magnetic or solar wind velocity fields) rather than on the *modulation factor*.

4 Results on ensemble statistics

In this section the statistical ensemble variance and domain of influence analysis outlined above are presented. With these experiments, we inspect the potential impact of the DA technique on the improvement of the state at the source surface and, more in general, on the entire simulated domain. All parameters (spatial, time, magnetic and velocity field) in the paper should be considered normalised to the normalisation factors described earlier in Sect. 2.1.

4.1 Gaussian character of error statistics

For the model error statistics exploration process of this study, a set of 100 ensemble members is generated. This number is usually chosen for stochastic modelling to ensure a statistical significance at a reasonable computational cost.

When implementing real DA experiment it should be necessary to conduct a convergence analysis to determine the optimal number of members leading to a sufficient stability of the statistics computed from the ensemble members.

The Gaussian nature of the perturbed states is very important for the data assimilation schemes since it ensures that the optimal state calculated in the analysis step is the most probable in the sense of maximum likelihood. In the context of ensemble methods, the forecast error of the model can be estimated from the dispersion of the perturbed model trajectories around the overall average at a given time step.

Figure 3 shows the distribution of the error for the z components of the solar wind velocity and magnetic field for the different members of the ensemble at the diagnosis locations of Fig. 1. Each time plot at each location depicts

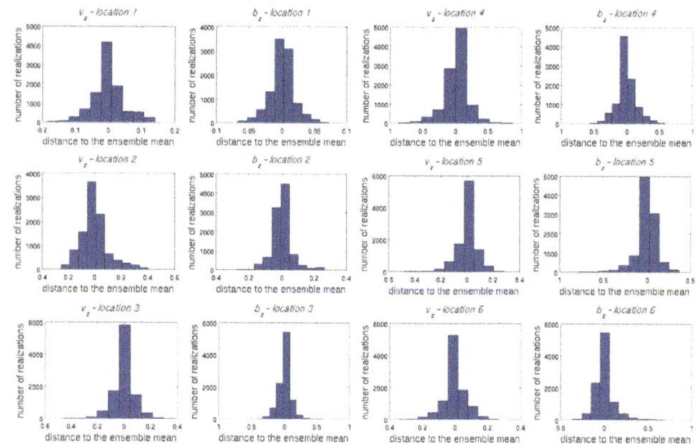

Fig. 3. Distribution of the forecast errors for the z component of the solar wind velocity and magnetic field at the different test locations illustrated in Fig. 1: location 1 (source surface), location 2, location 3, location 4 (Lagrange point, L1), location 5 and location 6. The errors are calculated from the dispersion of the perturbed ensemble member trajectories around the ensemble average.

the information relative to all the ensemble members at all the time steps available. These histograms are thus computed on a sample of 10 000 (100 members times 100 outputs) achievements and show the empirical distributions of the model error as a function of the perturbed source surface input.

It can be noted that the shape of the error distribution is slightly broader for locations 4, 5 and 6 compared to locations 1, 2 and 3. This is due to a higher level of non-linear interactions and unsteady state conditions further away from the source surface.

Overall, the sampling distributions can be fitted, in a first approximation, with a quasi-Gaussian. This ensures that the BLUE (cf. Sect. 3.2) hypothesis can be made and that the classical formulation of data assimilation, on which the present study relies, is pertinent. However, it should be mentioned that some reconnection events are triggered in some ensemble members. This could explain, at some locations, the deviation from the Gaussian shape of the distribution of the errors.

4.2 Ensemble variances

As part of the sensitivity study, ensemble variance diagnoses for the magnetic field and velocity at the previously designated locations of the domain are discussed. The objective is to estimate the error structures of the model, in order to point out the areas where gathering dedicated observations may be useful. In this framework, the model response to perturbations of the boundary condition inputs at the source surface has been calculated. This will highlight the areas where physical processes are poorly modelled and will also allow estimating of the reliability of the model. In the first subsection, the temporal evolution of the ensemble variances is analysed

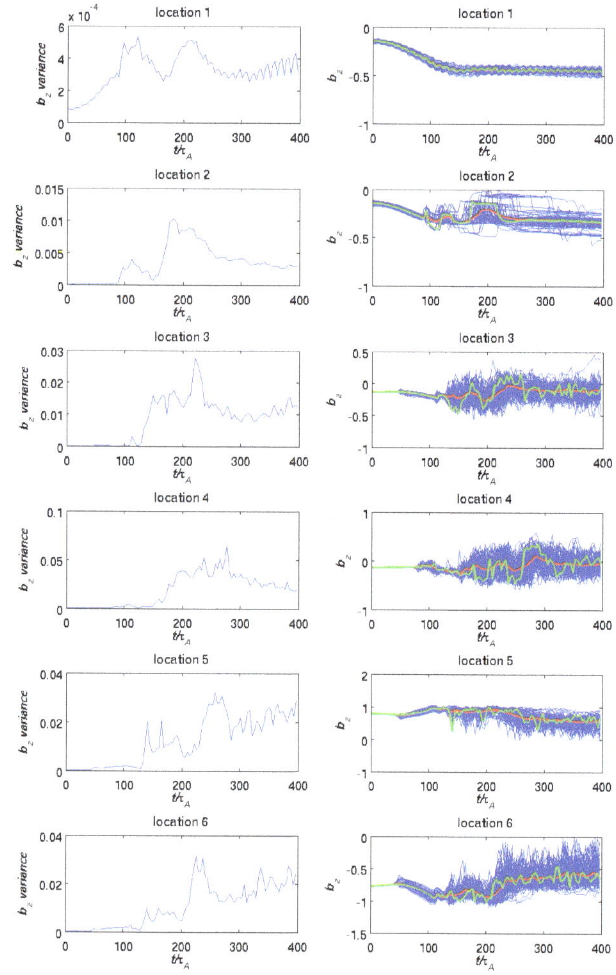

Fig. 4. Time evolution of the ensemble variance for the normalised b_z (left panels) and time evolution of the normalised b_z for all the ensemble members (right panels) at the selected diagnosis locations. The red curve represents the average of the ensemble and the green curve refers to the reference simulation, where no perturbation is applied to the source surface inputs. b_z is normalised to b_0 and the time to the Alfvén time τ_A. Note that the left and right b_z scales are not common to the six diagnosis points.

at fixed points in space. In the second subsection, the spatial structures of the ensemble variances are studied in the whole domain simulated.

4.2.1 Local temporal evolution of ensemble variances

The temporal evolution of the various perturbed simulations composing the ensemble provides empirical information about the time evolution of the error variances. It has to be recalled here that, while generating the ensemble, only b_z is perturbed, while v_z is left unperturbed. Figures 4 and 5 refer to the z component of the solar wind magnetic field and velocity respectively. They present a comparison of the temporal evolution of the ensemble variances (left panels)

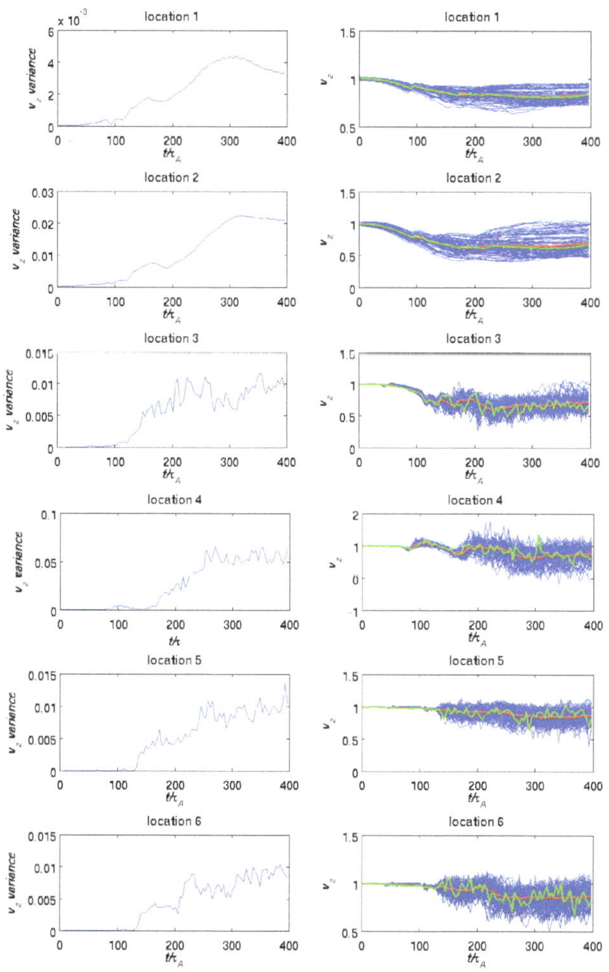

Fig. 5. Time evolution of the ensemble variance for the normalised v_z (left panels) and time evolution of v_z for all the ensemble members (right panels) at the selected diagnosis locations. The red curve represents the average of the ensemble and the green curve refers to the reference simulation, where no perturbation is applied to the source surface inputs. v_z is normalised to the Alfvén speed v_A and the time to the Alfvén time τ_A. Note that the left and right v_z scales are not common to the six diagnosis points.

and their associated maps of evolution for all of the ensemble members (right panels), for six representative locations chosen in the whole domain as depicted in Fig. 1. In the right panels, the red curve represents the temporal evolution of the ensemble average, while the green curve refers to the reference simulation whose source surface input is left unperturbed.

Considering the rate of perturbation that has been applied in this stochastic modelling, one can see that the evolution of the variability in the ensemble remains quite admissible and maintains fairly reasonable error bounds for both b_z and v_z.

Except for location 1 near the source surface, where the variance remains very low, it can be noted here that the ensemble variance evolution is quite similar for the six locations: the variance of the ensemble is extremely low at the beginning of the time series. Reasonably, though, the locations far from the source surface show slightly higher differences in the evolution than the closer ones, as depicted in the right panels of Figs. 4 and 5. At later time steps, $t/\tau_A \approx 88$ for location 2, $t/\tau_A \approx 120$ for location 3, $t/\tau_A \approx 140$ for locations 5 and 6 and $t/\tau_A \approx 160$ for location 4, the variance starts increasing for all of the locations: the ensemble experiences rather different evolutions. These differences in the time evolution of the ensemble variances are the signature of a dynamical process which crosses the domain from the source surface to L1.

This notwithstanding, it can be noticed that ensemble averages and reference simulations behave quite similarly; this confirms that the stochastic modelling protocol is mostly linear, and thus corroborates the quasi-Gaussian feature of the distribution of the errors stated in Sect. 4.1.

4.2.2 Spatial structure of the ensemble variance

The spatial distribution of the ensemble variance structures at a given instant reflects the level of model variability over the ensemble and hence gives an efficient estimation of the errors of the model in response to the perturbations of the z component of the magnetic field. The time evolution of the velocity variances is quite stable; however, a rather interesting time evolution is noticed for the magnetic field. In Fig. 6 several snapshots taken at different times show the spatial extent of the ensemble variance of the model for the z component of the magnetic field. Figure 7 shows the z component of the magnetic field with superimposed magnetic field lines, for the reference simulation, at the same time steps as in Fig. 6. These two figures give an efficient insight into the time evolution of the spatial distribution of the model error structures, in response to the perturbation of the z component of the magnetic field. In Fig. 6, some "bursts" can be observed in the ensemble dispersion, from time step $t/\tau_A \approx 140$ to the final time step $t/\tau_A \approx 400$. These spatial structures correspond to the ones observed in the temporal evolution of the ensemble variances in Fig. 4. Figure 7, which depicts the magnetic field lines for the reference simulation, shows that the major increases in the ensemble variance seem to be due to some reconnection events in the simulated solar wind. For example, the reference simulation shows signs of reconnection in correspondence with the first model error's bubble at $t/\tau_A \approx 140$ and also with the model error's bubble visible in the ensemble variance at $t/\tau_A \approx 160$ on the right side of the domain, but not in correspondence with the model error's bubble visible on the left-hand side. The disturbance, when present, is then advected in time to L1, thus provoking the time-shifted variance signals in the observation points closer to L1. The high variances in correspondence with reconnection events show that not all ensemble instances experience reconnection at the same time and in the same locations. The evolution over time shows that more of these events occur.

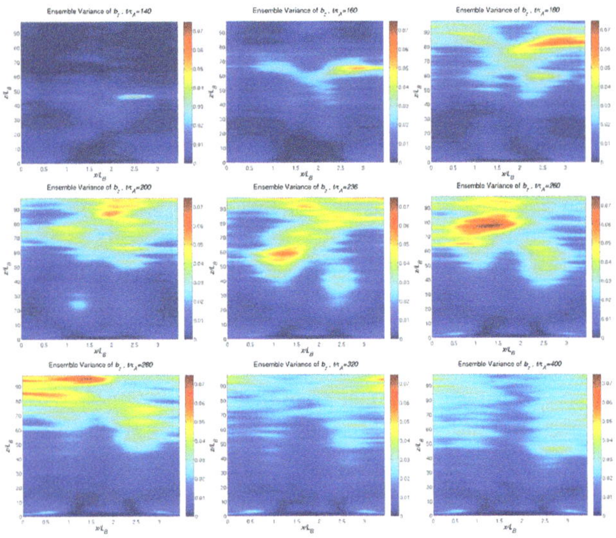

Fig. 6. Ensemble variance of the normalised z component of the magnetic field at different time steps and on the entire simulated domain. The spatial dimensions are normalised to L_B, the length scale of variation of the magnetic field. Time is normalised to the Alfvén time τ_A.

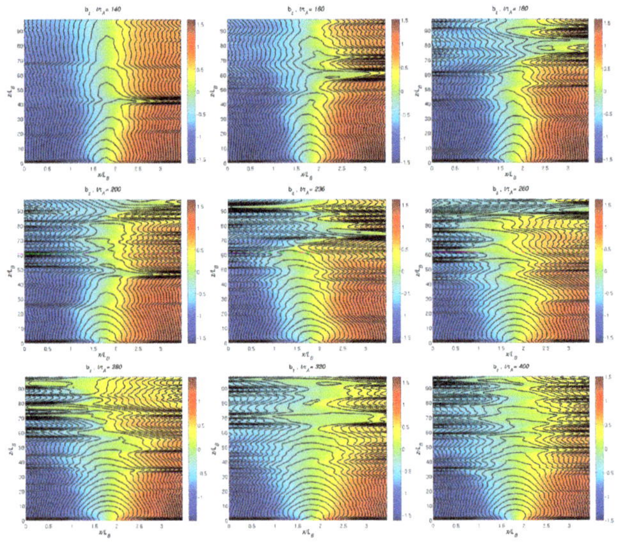

Fig. 7. z component of the magnetic field and, superimposed, the field lines for the reference simulation at different time steps and on the entire simulated domain. The spatial dimensions are normalised to L_B, the length scale of variation of the magnetic field. Time is normalised to the Alfvén time τ_A, b_z is normalised to b_0.

The simulation instances reconnecting "together" (i.e. at similar time and space) are presumably the ones excited with a similar magnetic field at the boundary.

4.3 Domains of influence

The study of the ensemble variance of the model allowed one to describe how the physical fields respond differently to a perturbation of the model inputs and to characterise the model forecast error. The ensemble modelling protocol that has been set up in the present study provides an efficient way of investigating the model forecast error covariances. These covariances indeed play a leading role in the problem of data assimilation since they ensure the propagation of the information provided by the observations to all model variables. Therefore they determine the spatial extent and magnitude of the state corrections provided by the assimilation of observations. Here the objective is indeed to illustrate how observations taken at a measurement point located in one particular position of the domain would affect the overall model evolution if those observations were assimilated into the model.

For real heliospheric application, the availability of observation points is currently very limited. The main observation point is the ACE spacecraft located at L1 position corresponding to location 4 in this study. As mentioned before, additional important information can be obtained by the STEREO mission.

In Fig. 8, the panels illustrate the domain of influence "doi" as defined in Eq. (27), for b_z, for observations taken at different locations. The effects of reconnection at the different locations of the simulation domain (see Fig. 1) and at different times are shown.

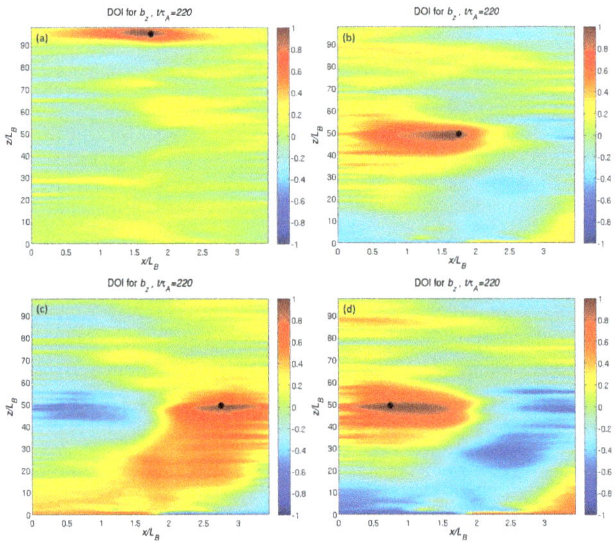

Fig. 8. Domain of influence for the z component of the magnetic field b_z for the assimilation of a magnetic field observation made at different locations at the same time step, $t/\tau_A = 220$. The black circle indicates the data observation location. **(a)** refers to location 4 (L1, 1 AU), **(b)** to location 3, **(c)** to location 5 and **(d)** to location 6. The spatial dimensions are normalised to L_B, the length scale of variation of the magnetic field; time is normalised to the Alfvén time τ_A.

For example, for panel a, the doi of one virtual observation located at L1 is depicted at $t/\tau_A \approx 220$. Both the observation

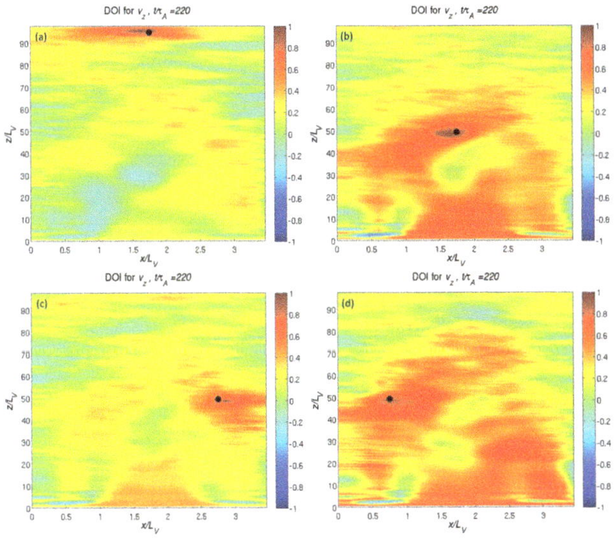

Fig. 9. Domain of influence for the z component of the solar wind velocity v_z for the assimilation of a velocity observation taken at different locations at the same time step $t/\tau_A = 220$. The black circle indicates the data observation location. **(a)** refers to location 4 (L1, 1 AU), **(b)** to location 3, **(c)** to location 5 and **(d)** to location 6. The spatial dimensions are normalised to $L_V = L_B$, the length scale of variation of the velocity; time is normalised to the Alfvén time τ_A.

and the domain of influence are referred to the z component of the magnetic field.

Notice that the amplitude of the correlation at the observation point location is equal to "1", which is the maximum positive correlation. Then, the correlation progressively decreases when moving away from the observation point. When the ensemble correlation approaches the maximum negative correlation, value "−1", degradation may occur. This can actually happen when moving too far away from the observation area. In this case, when implementing DA technique, the analysis step should be restricted to the area with positive correlation. As explained in Sect. 3.2.2, the correlations here reflect the potential spatial extent of the correction around a given location with an isolated observation. Then, the correlation amplitude (Eq. 27) will be modulated, taking into account the observation error (with strong observation errors, the model correction remains low) and the ratio of error variances at the point to be corrected (correction could be significant or lower).

Notice that the domain of influence of the observation is rather localised and decreases to 0 when moving towards the Sun. This result makes sense, since the driver of the simulation is located at the source surface, i.e. far away from the L1 location, and means that observations of the magnetic field taken at L1 can only slightly improve the model evolution closer to the source surface.

It is also important to observe that the occurrence of magnetic reconnection results in the fact that some regions of the domain remain magnetically isolated. The beneficial feedback from assimilation is thus prevented from spreading into those regions. This effect is particularly severe for the observations taken at L1.

The doi plots from Fig. 8 for location 3 (panel b), location 5 (panel c) and location 6 (panel d) respectively, which again refer to the space and time evolutions of the domain of influence of various magnetic field observations, show more reassuring results: observations taken at those locations would significantly improve the model structure at greater distances, thus making the use of data assimilation a worthy tool for making the model evolution closer to reality. Notice again that the influence of an observation point on the rest of the domain is strictly bound to the occurrence of reconnection, which temporarily magnetically isolates the reconnection area from the rest of the domain. Note that these findings are independent of time and remain similar throughout the whole period.

The same doi analyses are performed for the solar wind velocity v_z. Figure 9 shows the domain of influence for the solar wind velocity in the case of the assimilation of a velocity observation made at location 4 (panel a), location 3 (panel b), location 5 (panel c) and location 6 (panel d) respectively at time $t/\tau_A \approx 220$. Notice that the correlation impact is wider than for magnetic field measures and extends over time to cover the simulation domain entirely, as visible in Fig. 10, which refers to the same spatial locations but to a later time, $t/\tau_A \approx 344$. This ensures that data assimilation would significantly minimise the model errors in the velocity field when velocity observations are assimilated. Panel (a) is of particular relevance, since it shows that observations taken at 1 AU, where most of the space weather relevant satellites are located, also have a very strong effect in improving the solar wind velocity representation over almost the entire domain.

The results thus show that data assimilation should improve the forecast of velocity and magnetic fields not only near and around the observation location, but also for grid points rather far from the measurement location.

It is reminded that, as from Eq. (14), the correction of the forecasted state is a linear combination of representers associated with the observations assimilated. The representers include both contributions from covariances for the same fields (magnetic field or velocity field) and from cross-covariances between different fields (magnetic field with velocity field): each representer propagates information from the observation to the different variables of the state vector. This means, in this case, that assimilation of magnetic field observations might improve the state for the solar wind velocity as well, and vice versa.

Here, the ensemble cross correlations between the z components of the magnetic and velocity fields are analysed in the whole domain of study. A virtual b_z observation is supposed to be assimilated at the same previous location points and the doi on v_z is computed. Figure 11 shows snapshots of cross correlations for locations 4, 3, 5 and 6, selected

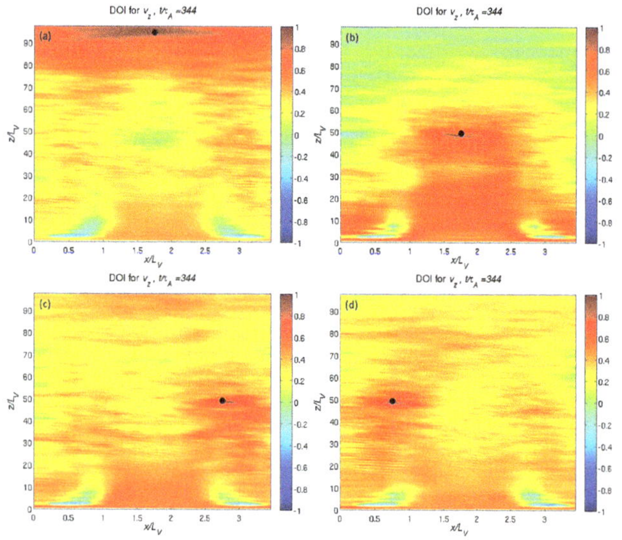

Fig. 10. Domain of influence for the z component of the solar wind velocity v_z for the assimilation of a velocity observation taken at different locations at the same time step $t/\tau_A = 334$. The black circle indicates the data observation location. **(a)** refers to location 4 (L1, 1 AU), **(b)** to location 3, **(c)** to location 5 and **(d)** to location 6. The spatial dimensions are normalised to $L_V = L_B$, the length scale of variation of the velocity; time is normalised to the Alfvén time τ_A.

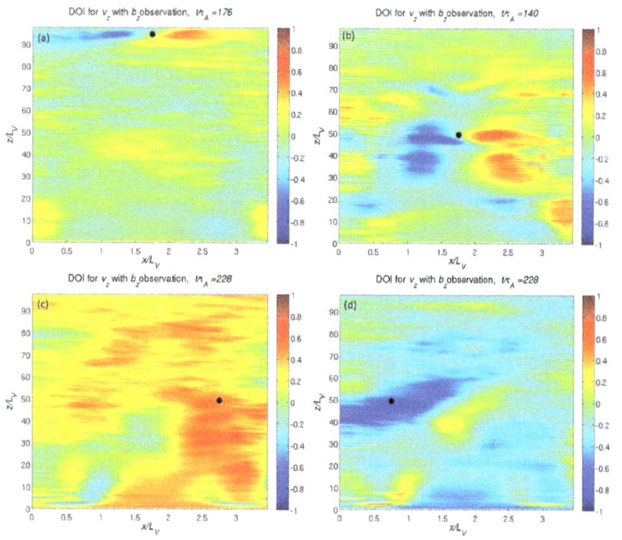

Fig. 11. Domain of influence for the z component of the solar wind velocity v_z for the assimilation of a magnetic field observation taken at different locations and different times. The black circle indicates the data observation location. **(a)** refers to location 4 (L1) and time $t/\tau_A = 176$, **(b)** to location 3 and time $t/\tau_A = 140$, panel **(c)** to location 5 and **(d)** to location 6, both at time $t/\tau_A = 228$. The spatial dimensions are normalised to $L_V = L_B$, the length scale of variation of the velocity; time is normalised to the Alfvén time τ_A.

throughout the simulation at different time steps. It should be noted that only specific cases are shown. The less interesting

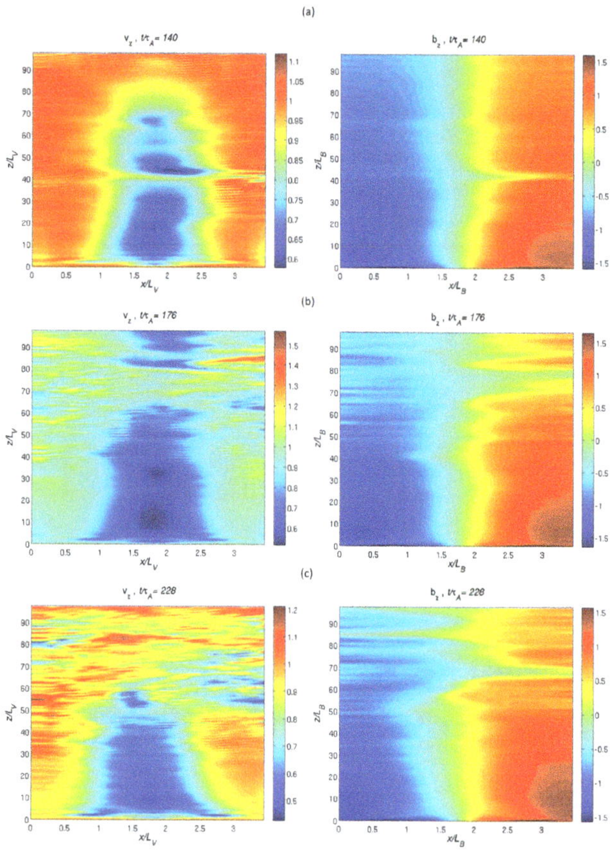

Fig. 12. z components of the solar wind velocity v_z (left panel) and of the magnetic field b_z (right panel) of the reference simulation in correspondence with doi plots of Fig. 11, for times **(a)** $t/\tau_A = 176$, **(b)** $t/\tau_A = 140$ and **(c)** $t\tau_A = 228$. The spatial dimensions are normalised to $L_V = L_B$, the length scales of variation of the magnetic field and of the velocity; time is normalised to the Alfvén time τ_A. b_z is normalised to b_0 and v_z is normalised to the Alfvén speed v_A.

cases with no correlation are dropped. For both locations 4 and 3 the domain of influence is characterised by the appearance of a dipole around the observation location with a separation of correlated and anti-correlated areas. For location 6 the doi shows an anti-correlated area, while for location 5 the doi is characterised by a positive ensemble correlation. As a reference, Fig. 12 shows the z components of the solar wind velocity v_z (left panel) and of the magnetic field v_z (right panel) of the reference simulation for the same times of the doi plots of Fig. 11. Preliminarily, this figure indicates that the cross correlations are related to the sign of the magnetic and velocity fields, respectively.

It is known in ensemble methods (Evensen, 2009; Bouttier and Courtier, 1999) that ensemble correlations are related to the ensemble member structures of model errors. The correlation/anti-correlation space extent seems to be related to the sign of the z components of the magnetic and velocity fields. While the z component of velocity is positive for all ensemble members at all locations, as can be noticed form

Fig. 5 (right panels), the behaviour of b_z at the different locations is more complicated. From the right panels of Fig. 4 it is possible to notice that at locations 4 and 3, where dipoles are present in Fig. 10, some of the ensemble members exhibit positive values of b_z, while the rest of the ensemble members experience negative values of the magnetic field. At location 6, where most of the domain shows negative correlation, all the ensemble members have negative magnetic fields. At location 5, which is mostly positively correlated, b_z is positive for all the ensemble members.

Further studies need to be conducted to better understand this behaviour.

5 Conclusions

Investigations were made to assess if and how the use of data assimilation techniques could lead to more accurate estimates of state variables (i.e. solar wind velocity and magnetic field) for solar wind models in different regions of the heliosphere. A location of particular interest is the source surface, since boundary conditions for solar wind models are usually taken there. More in general, the potential domains of influence due to assimilation of virtual measurements taken at different locations of space on the evolution of the FLIP-MHD model for the solar wind were investigated. The general aim of the study was to understand the processes and dynamics in terms of spatial and temporal features of an MHD heliospheric model. A sensitivity study has been conducted through ensemble variances and representer "influence functions" analysis, both performed with multiple runs of the model on statistically guided modifications of the input.

First, the sensitivity study permits us to characterise the sensibility of the model to the variation of the input parameters. This allowed us to better identify the reaction of the model to small variations in the boundary conditions. It is now understood how minimal differences in the boundaries may give rise to a different pattern of reconnection events and that, as a consequence, a faithful representation of the source surface conditions is paramount for the evolution of such models. Hence, the necessity of reliable boundary conditions arises. Another result is the estimate of the error structures of the model, which helps to emphasise the domain areas where dedicated observations may be useful to be collected.

Second, the representer analysis has been used to anticipate and estimate the potential contribution of DA to the model evolution. Indeed, the ensemble modelling led us to investigate the model error covariances with the aim of identifying the potential domains of influence through the assimilation of one virtual observation. Such an analysis was performed by computing the ensemble correlations and cross correlations, both for magnetic field and velocity measurements at different spatial locations. It was shown that, while for magnetic field measurements the model improvement is rather bound to the location of the observation and also to the occurrence of magnetic reconnection, the assimilation of velocity measures grants model benefits which span further away from the observation location in space and which are less affected by the occurrence of reconnection. Velocity measurements taken at L1 proved to be strongly correlated with the model evolution, also at large distances.

It should be remarked that calculating ensemble correlations also has the advantage of considering the observations one by one, thus isolating the influence of each observation in incrementing the correction. In the intermediate goal of optimising a monitoring network for space weather events, this approach is very interesting and helpful. Indeed, the domain of influence analysis can already provide useful information about the spatial extent of the expected correction obtained by assimilating one isolated observation. This allows one to identify the spatial coverage required for the design of a space weather observation network.

The approach and the results presented here do not assess the success or the failure of the assimilation system. However, the positive results obtained by computing the domains of influence of virtual observation are a further hint of the potential benefits of DA on space weather MHD models. The next step would be to implement and run the complete DA, by applying the correction brought by the observations to the forecasted state $\boldsymbol{x}^{\mathrm{f}}$ (Eq. 14).

From this study, two main strategies for applying DA to heliospheric models emerge. One consists in updating directly the model fields in the simulation space. If magnetic field values are updated, special attention must be dedicated to the solenoidal condition for the magnetic field. The other one consists in correcting the boundary values at the source surface in order to feed the model continuously with optimal boundary conditions. The study focused on the second strategy, since poor magnetograms are known as a major source of errors in ambient solar wind modelling.

Acknowledgements. The authors thank their colleagues for continuing support and discussion. The research leading to these results has received funding from the European Commission's Seventh Framework Programme (FP7/2007-2013) under grant agreement eHeroes (project no. 284461, www.eheroes.eu). Figure 2 has been provided by the Community Coordinated Modeling Center at Goddard Space Flight Center through their public Runs on Request system (http://ccmc.gsfc.nasa.gov). The CCMC is a multi-agency partnership between NASA, AFMC, AFOSR, AFRL, AFWA, NOAA, NSF and ONR. The MAS model was developed by J. Linker, Z. Mikic, R. Lionello, P. Riley, N. Arge and D. Odstrcil at PSI, AFRL, U. Colorado.

Edited by: S. Sharma

References

Baker, D. N.: What is space weather?, Adv. Space Res., 22, 7–16, 1998.

Barrero Mendoza, O., De Moor, B., and Bernstein, D. S.: Data assimilation for magnetohydrodynamics systems, J. Comput. Appl. Mathe., 189, 242–259, 2006.

Beck, A., Innocenti, M. E., Lapenta, G., and Markidis, S.: Multi-level multi-domain algorithm implementation for two-dimensional multiscale particle in cell simulations, doi:10.1016/j.jcp.2013.12.016, 2013.

Bennet, A. F.: Inverse Methods in Physical Oceanography, Monographs on Mechanics and Applied Mathematics, Cambridge University Press, Cambridge, 1992.

Bothmer, V. and Daglis, I. A.: Space Weather – Physics and effects, Springer, Praxis Publishing, Chichester, 2007.

Bouttier, F. and Courtier, P.: Data assimilation concepts and methods, Meteorological Training Course Lecture Series, ECMWF, 1999.

Brackbill, J. U.: FLIP MHD – A particle-in-cell method for magnetohydrodynamics, J. Comput. Phys., 96, 163–192, 1991.

Dwivedi, B. N. and Mohan, A.: First tomographic view of coronal mass ejections, Current Sci., 88, 688–690, 2005.

Echevin, V., De Mey, P., and Evensen, G.: Horizontal and vertical structure of the representer functions for sea surface measurements in a coastal circulation model, J. Phys. Oceanogr., 30, 2627–2635, 2000.

Egbert, G. D. and Bennett, A. F.: Data assimilation methods for ocean tides, Modern Approaches to Data Assimilation in Ocean Modeling, edited by: Malonotte-Rizzoli, P., Elsevier Science, 147–179, 1996.

Evensen, G.: Data Assimilation: The Ensemble Kalman Filter, 2nd Edn., Springer, Berlin, 2009.

Fleck, B., Domingo, B., and Poland, A. I.: The SOHO Mission, Solar Phys., 162, 1–531, 1995.

Gantois, K., Santandrea, S., Teston, F., Strauch K., Zender, J., Tilmans, E., and Gerrits, D.: Big year for small satellite – ESA's second in-orbit technology demonstrator mission: PROBA-2, ESA Bulletin, 144, 22–33, 2010.

Ghil, M. and Malanotte-Rizzoli, P.: Data assimilation in meteorology and oceanography, Adv. Geophys., 33, 141–266, 1991.

Gosling, J. and Pizzo, V.: Formation and evolution if corotating interaction regions and their three dimensional structure, Space Sci. Rev., 89, 21–52, 1999.

Innocenti, M. E., Lapenta, G., Vrsnak, B., Crespon, F., Skandrani, C., Temmer, M., Veronig, A., Bettarini, L., Markidis, S., and Skender, M.: Improved forecasts of solar wind parameters using the Kalman filter, Space Weather, 9, 10005, doi:10.1029/2011SW000659, 2011.

Innocenti, M. E., Lapenta, G., Markidis, S., Beck, A., and Vapirev, A.: A multi-level multi domain method for particle in cell plasma simulations, J. Comput. Phys., 238, 115–140, doi:10.1016/j.jcp.2012.12.028, 2013.

Kalman, R.: A new approach to linear filtering and prediction problems, J. Basic Eng., 82, 35–45, 1960.

Kondrashov, D., Shprits, Y., Ghil, M., and Thorne, R.: A Kalman filter technique to estimate relativistic electron lifetimes in the outer radiation belt, J. Geophys. Res., 112, A10227, doi:10.1029/2007JA012583, 2007.

Laakso, H., Taylor, M. G. G. T., and Escoubet, C. P.: The Cluster active archive – Studying the Earth's space plasma environment, Springer, Astrophysics and Space Science Proceedings, 2010.

Levine, R. H., Altschuler, M. D., and Harvey, J. W.: Solar Sources of the Interplanetary Magnetic Field and Solar Wind, J. Geophys. Res., 82, 1061–1065, 1977.

Lyard, F.: Data assimilation in a wave equation: a variational representer approach for the Grenoble tidal model, J. Comput. Phys., 149, 1–31, 1999.

McComas, D. J., Bame, S. J., Barraclough, B. L., Feldman, W. C., Funsten, H. O., Gosling, J. T., Riley, P.; Skoug, R., Balogh, A., Forsyth, R., Goldstein, B. E., and Neugebauer, M.: Ulysses' return to the slow solar wind, Geophys. Res. Lett., 25, 1-4, 1998.

Odstrcil, D.: Modeling 3-D solar wind structure, Adv. Space Res., 32, 497–506, 2003.

Pesnell, W. D., Thompson, B. T., and Chamberlin, P. C.: The Solar Dynamics Observatory (SDO), Solar Phys., 275, 3–15, doi:10.1007/978-1-4614-3673-7_2, 2012.

Rigler, E., Baker, D., and Weigel, R.: Adaptive linear prediction of radiation belt electrons using the Kalman filter, Space Weather, 2, S03003, doi:10.1029/2003SW000036, 2004.

Riley, P., Linker, J. A., and Mikic, Z.: An empirically-driven global MHD model of the solar corona and inner heliosphere, J. Geophys. Res., 106, 15889–15901, 2001.

Schunk, R. W., Scherliess, L., and Sojka, J. J.: Recent approaches to modeling ionospheric weather, Adv. Space Res., 31, 819–828, 2003.

Schunk, R. W., Scherliess, L., Sojka, J. J., Thompson, D. C., Anderson, D. N., Codrescu, M., Minter, C., Fuller-Rowell, T. J., Heelis, R. A., Hairston, M., and Howe, B. M.: Global Assimilation of Ionospheric Measurements (GAIM), Radio Sci., 39, doi:10.1029/2002RS002794, 2004.

Schrijver, C. J. and Derosa, M. L.: Photospheric and heliospheric magnetic fields, Solar Phys., 212, 165–200, 2003.

Sheeley, N. R. and Harvey, J. W.: Coronal holes, solar wind streams, and geomagnetic disturbances during 1978 and 1979, Solar Phys., 70, 237–249, 1981.

Siscoe, G.: The space-weather enterprise: past, present, and future, J. Atmos. Solar-Terrest. Phys., 62, 1223–1232, 2000.

Stone, E. C., Frandsen, A. M., Mewaldt, R. A., Christian, E. R., Margolies, D., Ormes, J. F., and Snow, F.: The Advanced Composition Explorer, Space Sci. Rev., 86, 1–22, doi:10.1007/978-94-011-4762-0_1, 1998.

Toth, G., Sokolov, I. V., Gombosi, T. I., Chesney, D. R., Clauer, C. R., De Zeeuw, D. L., Hansen, K. C., Kane, K. J., Manchester, W. B., Oehmke, R. C., Powell, K. G., Ridley, A. J., Roussev, I. I., Stout, Q. F., Volberg, O., Wolf, R. A., Sazykin, S., Chan, A., Yu, B., and Kta, J.: Space weather modeling framework: A new tool for the space science community, J. Geophys. Res., 110, A12226, doi:10.1029/2005JA011126, 2005.

Turner, D. L. and Li, X.: Using spacecraft measurements ahead of Earth in the Parker spiral to improve terrestrial space weather forecasts, Space Weather, 9, S01002, doi:10.1029/2010SW000627, 2011.

Vršnak, B. and Žic, T.: Transit times of interplanetary coronal mass ejections and the solar wind speed, Astro. Astrophys., 472, 937–943, 2007.

Vršnak, B., Temmer, M., and Veronig, A.: Coronal Holes and Solar Wind High-Speed Streams: I. Forecasting the Solar Wind Parameters, Solar Phys., 240, 315–330, 2007.
Wang, Y. and Sheeley Jr., N. R.: Solar wind speed and coronal flux-tube expansion, Astrophys. J., 355, 726–732, 1990.
Welch, G. and Bishop, G.: An introduction to the Kalman filter, TR 95-041, University of North Carolina, Department of Computer Science, 2001.
Wu, S. T., Wang, A. H., Liu, Y., and Hoeksema, J. T.: Data Driven Magnetohydrodynamic model for active region evolution, Astrophys. J., 652, 800–811, 2006.

20

On the entrainment coefficient in a forced plume: quantitative effects of source parameters

A. Matulka[1], P. López[2], J. M. Redondo[1], and A. Tarquis[3]

[1]Department of Applied Physics, UPC, Barcelona, Spain
[2]Department of Applied Mathematics, UCM, Madrid, Spain
[3]Department of Applied Mathematics, UPM, Madrid, Spain

Correspondence to: P. López (maplopez@bio.ucm.es)

Abstract. The behavior of a forced plume is mainly controlled by the source buoyancy and momentum fluxes and the efficiency of turbulent mixing between the plume and the ambient fluid (stratified or not). The interaction between the plume and the ambient fluid controls the plume dynamics and is usually represented by the entrainment coefficient α_E. Commonly used one-dimensional models incorporating a constant entrainment coefficient are fundamental and very useful for predictions in geophysical flows and industrial situations. Nevertheless, if the basic geometry of the flow changes, or the type of source or the environmental fluid conditions (e.g., level of turbulence, shear, ambient stratification, presence of internal waves), new models allowing for variable entrainment are necessary. The presented paper is an experimental study based on a set of turbulent plume experiments in a calm unstratified ambient fluid under different source conditions (represented by different buoyancy and momentum fluxes). The main result is that the entrainment coefficient is not a constant and clearly varies in time within the same plume independently of the buoyancy and the source position. This paper also analyzes the influence of the source conditions on the mentioned time evolution. The measured entrainment coefficient α_E has considerable variability. It ranges between 0.26 and 0.9 for variable Atwood number experiments and between 0.16 and 0.55 for variable source position experiments. As is observed, values are greater than the traditional standard value of Morton et al. (1956) for plumes and jets, which is about 0.13.

1 Introduction

Forced plumes play a fundamental role in a large variety of natural phenomena and industrial processes. Understanding the dynamics of plumes issuing from industrial chimneys or those generated by forest fires or volcanoes is a major goal for environmental sciences because they are able to transport toxic gas and fine particles into the high atmosphere. River plumes are another natural plume phenomenon. These are turbid freshwaters flowing from land and generally in the distal part of a river outside the bounds of an estuary or river channel. Submarine plumes are another example of natural flows. The hot fluid rises into the cold ocean (Carazzo et al., 2008). Plumes, jets, and thermals are important concepts (Squires and Turner, 1962) for describing particular cases of atmospheric convection. Looking at a large thunderstorm, an analogy with a steady-state buoyant turbulent plume can be suggested, having a continuous source of heat from below the cloud base. On the other hand, for small clouds, being about as deep as wide, a non-steady bubble model seems more appropriate. The growth of a cloud depends on the entrainment across the interface identifying the cloud. The experiments of Redondo et al. (1995) model this growth in a non-homogeneous turbulent environment, studying the individual importance of the buoyancy induced or internal turbulence and the environmentally induced or external turbulence. Finally, in engineering, turbulent plumes are involved in building ventilation processes to supply fresh and cool air and are essential to evaluate quality of air in rooms (Nielsen, 1993).

A forced plume is generated by the steady release of mass, momentum and buoyancy from a source situated in a region of uniform or stably stratified ambient fluid (Morton, 1959).

Initially, for a forced plume the driven mechanism is the momentum flux and later the buoyancy flux governs the dynamics (Morton, 1959). When momentum effects are more important than density differences and buoyancy effects, the forced plume is a jet. Therefore, pure jets are flows with no buoyancy as opposed to entirely buoyancy-dominated pure plumes, which have no initial momentum. Therefore, forced plumes have the characteristics of both jets and plumes and these are special cases of forced plumes.

The most popular model to describe the plume dynamics is the one-dimensional steady-state model by Morton et al. (1956), which has been extensively applied to investigate the dynamics of natural and laboratory jets and plumes under various source conditions. Morton et al. (1956) investigated the rise of turbulent buoyant plumes from point sources into a motionless but neutral or stably stratified environment. Their key hypothesis consists in a global representation of turbulence achieved by introducing an entrainment coefficient α_E, assumed to be constant. The governing equations of motion may be reduced to a set of three coupled linear differential equations (conservation of mass, momentum, and buoyancy). In all cases, the results strongly depend on the value of the entrainment coefficient that appears indeterminable by theoretical investigations. Comparison of laboratory experiments on jets and plumes with the formalism of Morton et al. (1956) showed good agreement concerning the scaling laws governing the dynamics of the flows, but this model is not sufficient to explain all the experimental (Turner, 1986; Sreenivas and Prasap, 2000; Tate, 2002) and geophysical data about plumes (Carazzo et al., 2008). Therefore, an improved description of the turbulent entrainment is necessary, as the Carazzo et al. (2008) model presents where the entrainment coefficient varies as a function of buoyancy and source distance. It is also necessary to understand the effects of source condition variations on the entrainment coefficient.

The purpose of this paper is to study the time evolution of the entrainment coefficient α_E for a given downwards plume. We also investigate the variation of this time evolution under different source conditions. Additionally, we present the vertical and entrainment velocity fields to analyze their time variation under different source parameters. In Sect. 2, we describe the entrainment coefficient definition and a brief resume of values and entrainment models. In Sect. 3, we present our experimental configuration and procedures, qualitatively describing the experiments. In Sect. 4, we present our measurements of the plume vertical velocity and entrainment velocity fields and the entrainment coefficient. In Sect. 5, we present the discussion of results, studying the variation in time and under different source conditions of vertical velocity, entrainment velocity and entrainment coefficient. Finally, we briefly present our conclusions.

2 Simple plume entrainment model

We briefly resume the governing equations in a simple uniform ambient for a forced plume. We use an axisymmetric assumption and take cylindrical polar coordinates (z, θ, r) with the z axis vertical and the source at the origin. We suppose that ρ_s, W, and U are the time-averaged source density, axial (known as vertical velocity) and radial velocities. We also consider the Atwood number A that measures the density difference (therefore, the buoyancy) across the interface between the two fluids: the plume and the ambient fluid, $\rho_a : A = \frac{\rho_s - \rho_a}{\rho_s + \rho_a}$.

We also define the mass flux Q, the buoyancy flux B and the momentum flux M as follows:

$$Q = \rho_s W b^2, \quad B = g\,(\rho_a - \rho_s)\, W b^2, \quad M = \rho_s W^2 b^2, \qquad (1)$$

where g is the gravity acceleration and b is the plume radius. We may express the buoyancy flux B in terms of the Atwood number, noting that a basic simplification may be used: the source density ρ_s has a much greater (or smaller for negative buoyancy) density than the ambient fluid ρ_a; then

$$\rho_s \gg \rho_a \Rightarrow A = \frac{\Delta\rho}{\bar{\rho}} = \frac{\Delta\rho}{\rho_s} \Rightarrow B = g \cdot A \cdot V \qquad (2)$$

where V is the volume flux, $\Delta\rho = \rho_s - \rho_a$ and $\bar{\rho} = \frac{\rho_s + \rho_a}{2}$, which represents a reference density. However, in the case that their densities are similar:

$$\rho_s \approx \rho_a \Rightarrow A = \frac{\Delta\rho}{2\bar{\rho}} \Rightarrow B = 2 \cdot g \cdot A \cdot V\,. \qquad (3)$$

The following equations are based on the same integrated equations of motion as the earlier work of Morton (1956) and Morton (1971). The same assumptions are made, which may be briefly restated as follows: plume is steady on timescale longer than eddy turnover time; the Boussinesq approximation holds throughout the flow and the profiles of mean vertical velocity and mean density deficiency remain similar at all plume cross-sections (at all heights) or self-similarity assumptions. Considering the mentioned definitions and the assumptions detailed above, using top-hat profiles to represent values of mean vertical velocity and mean buoyancy and applying conservation of mass, momentum and buoyancy, we obtain the following set of differential equations in the situation of a constant ambient density or neutral environment (then the Brunt–Väissäla frequency N is zero as in our experimental configuration):

$$\frac{\mathrm{d}Q}{\mathrm{d}z} = 2\bar{\varrho}^{1/2}\alpha_E M^{1/2}, \quad \frac{\mathrm{d}M}{\mathrm{d}z} = \frac{BQ}{M}, \quad \frac{\mathrm{d}B}{\mathrm{d}z} = 0\,. \qquad (4)$$

2.1 The entrainment coefficient

Stommel (1947) studied for the first time the entrainment concept. He studied the dilution of a cumulus cloud by entrainment of environmental air.

An important further refinement of the entrainment formulation was pointed out by Houghton and Cramer (1951). They made a distinction between dynamical entrainment due to larger scale organized inflow (advective transport across the interface) and turbulent entrainment caused by turbulent mixing at the cloud edge (diffusive nature and described with an eddy diffusivity approach).

More precise quantitative descriptions of entrainment originated from laboratory water tank experiments of thermal plumes (Morton et al., 1956; Turner, 1962) that described an increasing mass flux with height. The one-dimensional model by Morton et al. (1956) establishes the linear growth of radius with distance from the source, which implies that the mean entrainment velocity is proportional to a characteristic upward or downward velocity. This statement is called the entrainment coefficient assumption and was first used by Morton et al. (1956) in the analysis of plumes. This key hypothesis consists in a global representation of turbulence achieved by introducing an entrainment coefficient α_E, assumed constant, that defines the horizontal rate of entrainment of surrounding fluid or entrainment velocity U_E in terms of the vertical velocity W: $U_E = \alpha_E W$ (Fig. 1). Therefore, it is possible to calculate the entrainment coefficient as $\alpha_E = U_E/W$. Closure of the equations is then possible through the use of the entrainment constant α_E.

Laboratory experiments have been performed to determine a range for α_E. Morton et al. (1956) did some experiments and they found a constant value of $\alpha_E = 0.13$, and proposed this value as a universal constant. In the absence of a theory for α_E, it has become common in geophysical problems to use an arbitrary value of 0.09 without a clear physical basis. However, decades of experiments on plumes and jets do not support the assumption of a universal constant value for α_E. In fact, most authors use different entrainment coefficients for different discharge and/or environmental conditions. For example, in Tate (2002), Wright (1984) uses an entrainment coefficient of 0.30 for discharges of a jet issuing from an axisymmetric source under conditions of stratified and stagnant ambient waters, and a value of 0.14 for discharges of a plume under the same conditions. The estimates of α_E. The entrainment coefficient α_E also varies between 0.05 and 0.12 according to the direction of the buoyancy force, negatively or positively buoyant (Kaminski et al., 2005). In case of negatively buoyant jets the variation in $\alpha_E = 0.075 \pm 0.05$ is rather large as for plumes. In lazy plumes, α_E is found to be around 0.12 (Hunt and Kaye, 2001). There appears to be considerable variability in the values of the entrainment coefficient as Tate (2002) presents in his work. For more details about values for the entrainment coefficient see tables 3.4–3.7 of its work. Tate (2002) summarizes values for the entrainment coefficient for different model types that can be found in the literature and classified depending on the experimental conditions (axisymmetric or line source, jet or plume, stratified or unstratified ambient fluid, and flowing or stagnant ambient fluid). For axisymmetric jets in an unstratified and stagnant ambient fluid, the mean entrainment is 0.080 ± 0.029, and for axisymmetric plumes under the same conditions the mean entrainment is 0.110 ± 0.034.

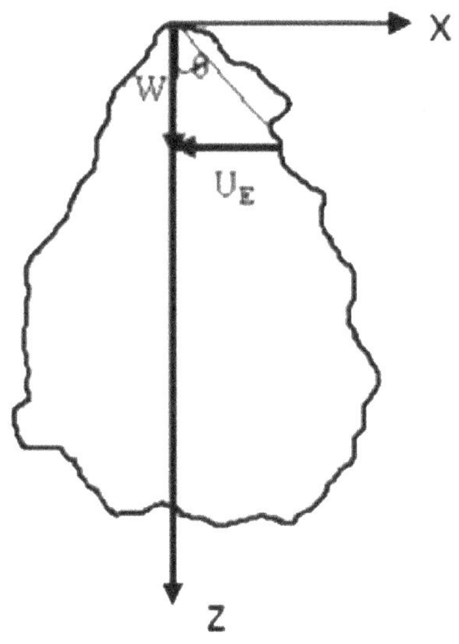

Fig. 1. Schematic view of a forced downwards plume without environmental turbulence. The mean centerline speed is the vertical velocity W and radius b. The plume entrains ambient fluid characterized by a mean entrainment velocity U_E. The figure shows the angle of the plume θ.

Therefore, the main limitation of Morton's model is its assumption of a constant value for α_E. As a consequence this model cannot explain that the values of α_E for a jet and a plume differ significantly, varying between 0.05 and 0.16, respectively (Kaminski et al., 2005), so a forced plume cannot be characterized by a single value, a constant entrainment coefficient can only give approximate predictions (at first it will exhibit jet-like characteristics and ultimately it must exhibit plume-like characteristics).

Frick (1984) recognized that a more complex entrainment function is necessary in order to reproduce the results in laboratory and field experiments. Instead of using $W = U_E/\alpha_E$, he establishes a difference between the shear entrainment, $(\alpha_E)_{shear}$, and the vortex entrainment, $(\alpha_E)_{vortex}$. Until recently, there was no theoretical explanation for the variation of α_E. The model of Kaminski et al. (2005) explains these variations and the role of both positive and negative buoyancy in the entrainment process is highlighted. The larger values of α_E in pure and lazy plumes are due to positive buoyancy which enhances the entrainment of background fluid by promoting the formation of large-scale turbulence structure. Conversely, a negative buoyancy force inhibits entrainment and reduces α_E (Kaminski et al., 2005). On the other hand, the variations within pure jets and pure plumes can be explained by the downstream evolution of each flow to a state of self-similarity (Carazzo et al., 2006).

Experimental measurements in forced plumes have shown that turbulent entrainment is limited in the near source part of the flow where inertial forces dominate compared with the buoyancy-dominated region far from the source (Morton, 1959). To account for this evolution, some authors proposed empirical or semi-empirical parametrizations allowing α_E to vary according to the ratio of the buoyancy and inertia forces:

$$\alpha_E = \alpha_j - (\alpha_j - \alpha_p)\left(\frac{Fr_p}{Fr}\right)^2, \tag{5}$$

where Fr is the Froude number and Fr_p is the constant Froude number for a pure plume in a uniform environment, and α_j and α_p are arbitrary constant values of α_E for pure jets and pure plumes, respectively (Lee and Chu, 2003). The values of the entrainment coefficient that best fit the data fall between the bounds formed by the values of α_E for pure jets and pure plumes in uniform environments (Carazzo et al., 2006) except at large distances from the source where α_E values are even smaller than those predicted for a pure jet.

The variable entrainment models may easily be adapted to a good agreement with small-scale laboratory experiments, with the Morton et al. (1956) model obtaining reasonable predictions at both small and at large distances from the source. On the other hand, they are also useful with geophysical or environmental turbulent plumes such as volcanic plumes on Earth or other planets, atmospheric sources of pollution, sewer discharge plants near the coast or submarine plumes rising from hydrothermal vents.

3 Experimental setup

The aim of the experimental procedure is to generate a turbulent axysimmetric plume, descending from a finite source in an unstratified and stagnant environment fluid and controlling its position and its physical characteristics as buoyancy and momentum fluxes (see Table 1). The experiments consisted of releasing a denser fluid downwards from a small nozzle, with a diameter D of 0.6 cm and an area S_o of 0.2827 cm^2, into a stationary body of water that was contained in a glass tank of dimensions 32 cm high by 25 cm by 25 cm. This water layer is designated as the *lighter layer* with a height $h_L = 17$ cm and a density $\rho_L = 1.0\,g\,cm^{-3}$. On top of this lighter layer, a system made of two metacrylic boxes is placed (Fig. 2). The bottoms of the boxes are pierced with one regulated orifice that is at a height H_o from the lighter layer and represents the position of the injection orifice. These boxes contain the denser layer that reaches a height $h_D = 1.134$ cm and a density ρ_D (López, 2004; López et al., 2008). The starting point of the plume is located at a height H_o, which takes the values 0, 2, 3, 6, and 8 cm. For each H_o, we use four different potassium permanganate solutions as the releasing fluids with an intense purple color and a volume of 500 cm^3. This provides the following range of Atwood numbers: 0.001, 0.0025, 0.005, and 0.01. Denser fluid with momentum and buoyancy fluxes is discharged from the orifice continuously at a flow rate of $V_o = 8.40\,cm^3\,s^{-1}$. A sketch of the experimental setup is given in Fig. 2 (López, 2004; López et al., 2008).

It is not necessary to add a dye as a passive tracer and, therefore, the flow can be directly visualized (back illuminated). The video recordings at high-velocity mode (100 fps)

Table 1. The 16 different experimental conditions of the fluid system. The density of the denser layer is ρ_D. A is the Atwood number. The location of the ejecting orifice is H_o. The buoyancy and momentum fluxes at the impinging surface are B_l and M_l.

ρ_D ($g\,cm^{-3}$)	A	H_o (cm)	M_l ($cm^4\,s^{-2}$)	B_l ($cm^4\,s^{-3}$)
1.002	0.001	2	1357.777	−19200.089
		3	1911.869	−22783.376
		6	3574.145	−31151.221
		8	4682.329	−35654.960
1.005	0.0025	2	1357.777	−19200.089
		3	1911.869	−22783.376
		6	3574.145	−31151.221
		8	4682.329	−35654.960
1.0101	0.005	2	1357.777	−19200.089
		3	1911.869	−22783.376
		6	3574.145	−31151.221
		8	4682.329	−35654.960
1.0202	0.010	2	1357.777	−19200.089
		3	1911.869	−22783.376
		6	3574.145	−31151.221
		8	4682.329	−35654.960

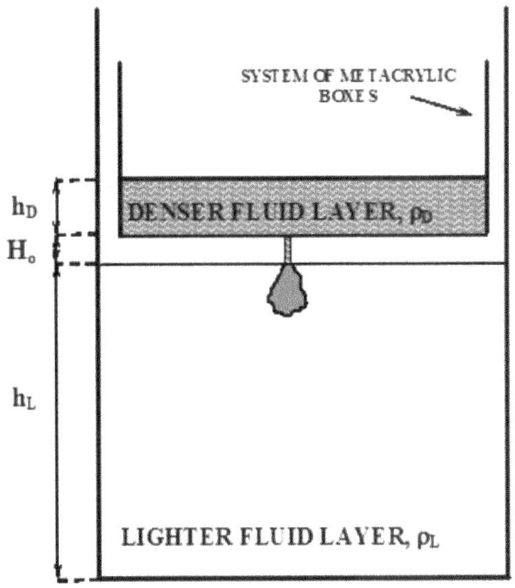

Fig. 2. Schematic diagram of the experimental configuration that shows the initial state of the fluid system with a starting forced plume. h_L and ρ_L are the height and the density of the lighter layer of fluid; h_D and ρ_D are the height and the density of the denser layer and the height H_o is the source position and varies from 0 to 8 cm.

of the experiments are sequenced into frames using frame-sequencer software (640 by 480 pixels capturing an area of 25 by 18 cm). The registered images average the concentration over the volume of the plume.

Table 1 lists the characteristics of the 16 different experimental conditions of the fluid system and the generated plume. The rate of supply of buoyancy is the buoyancy flux of the plume at source B_o, whose value is $-8232\,\mathrm{cm^4\,s^{-3}}$. The buoyancy flux is negative because the discharge fluid is heavier than the ambient fluid. The rate of supply of momentum by the source is the momentum flux at source M_o whose value is $249.593\,\mathrm{cm^4\,s^{-2}}$. The denser fluid is injected through the orifice with these fluxes and impinges on the lighter layer. B_l and M_l are the initial effective buoyancy and momentum fluxes at the free surface of the lighter layer as a consequence of the conversion from potential energy to initial kinetic energy of the plume.

3.1 Plume growth in steady environments

Plume occurs where fluid of different density from the ambient fluid is injected, as the density of the effluent is heavier than its surrounding ambient fluid and, therefore, the buoyancy force is downward and the initial flow is directed vertically downwards in the same direction as the buoyancy forces. Figure 3 shows several sequences of digitized video images from two experiments under different source conditions where t^* is the non-dimensional time defined as $t^* = \frac{t}{t_c}$. The timescale t_c is the time that a plume spends in crossing the distance $H_o + h_L$ (see Fig. 2): $t_c = \sqrt{\frac{H_o+h_L}{Ag}}$.

Upon entering the ambient fluid flow, the denser fluid becomes unstable and forms the axisymmetric plume at the center of the tank. As the plume is gravitationally unstable, it engulfs lighter fluid as it evolves, but the forced plume has different behaviors: jet-like characteristics depending on its initial mass and momentum fluxes, and plume-like characteristics depending on its initial buoyancy flux. If we use the Fischer characteristic length L_m (Fischer et al., 1979) and its related dimensionless parameter z/L_m, it is possible to analyze the degree of jet-like and plume-like behaviors of the forced plume. Our Fischer's length ranges from 10.94 for the smallest Atwood number ($A = 0.001$) to 3.46 for the greatest Atwood number ($A = 0.01$).

A forced plume can be considered as a pure jet at $\frac{z}{L_m} < 0.5$ and as a pure plume at $\frac{z}{L_m} > 5$; when $0.5 < \frac{z}{L_m} < 5$ we have the transition of the forced plume (Fischer et al., 1979). For our plume experiments, z/L_m ranges from 0.165 (for the smallest z, Atwood number A and initial Richardson number Ri_o) to 1.879 (for the greatest z, A, and Ri_o). Then, the dynamical behavior of the fluid flow modifies from a pure jet-like to a forced plume in a transition state. Moreover, for a given experiment or fixed buoyancy and source position, the non-dimensional quantity z/L_m increases with time.

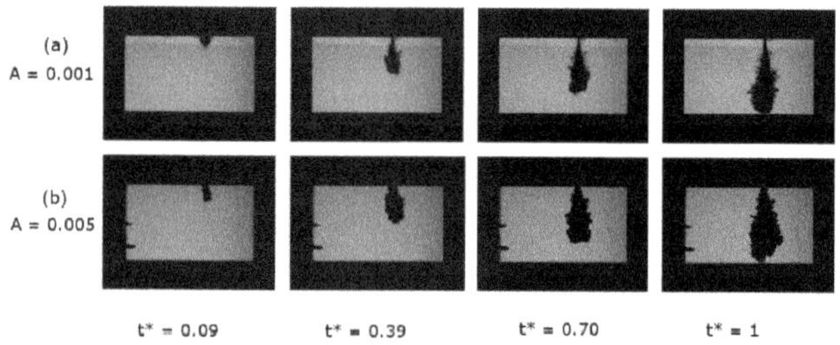

Fig. 3. Time evolution of one turbulent plume through its frame sequences corresponding to two experiments with different Atwood numbers A made with the same height for the starting point $H_o = 3$ cm. **(a)** $A = 0.001$ and **(b)** $A = 0.005$ for non-dimensional time $t^* = 0.09, 0.39, 0.70$, and 1.

Buoyancy is said to be dominant after some distance, which verifies $z > 5L_m$, which is different for each experiment. Usually, after ten to twenty source diameters the buoyancy dominates the flow. At this stage the entrainment of the ambient fluid directed through the border of the turbulent plume is more effective. The sides of the plume are zones with strong shears, which generate Kelvin–Helmholtz instabilities, and secondary Rayleigh–Taylor instabilities may also appear at the front of the convective plume.

Finally, the plume advance spreads sideways as it evolves until it reaches the physical contours of the tank which limit the development of the plume generating a gravity current and an overturn of the fluid system.

4 Quantitative results: velocity fields and entrainment coefficient variations

We use the PIV (particle image velocimetry) method to understand the behavior of turbulent plume based on our velocity field measurements. We calculate velocity PDFs with the DigiFlow program (Matulka et al., 2008) using sequences from experiments that were videotaped. We analyze the evolution of the velocity fields visualizing different graphics, which can give us an idea about vertical velocity, the entrainment velocity and the entrainment coefficient. The study of the velocity fields gives us information about the time progress of the turbulent plume, in which direction is the propagation of the plume or with what velocity is flowing. We can find its vertical velocity by looking at the length of the plume or entrainment velocity by looking at the width of the plume.

We calculate the vertical velocity as $W = z/t$ [$\mathrm{cm\,s^{-1}}$], where z is the length of the turbulent plume and t is the time of the propagation of the plume. The magnitude z_i measures the length of the plume in each time t_i. On the velocity fields we can see velocity vectors and their directions; the size of the vectors depends on the initial velocity of the plume. Figure 4 shows the elements for the vertical velocity calculation for a selected plume experiment with $A = 0.001$ and $H_o = 2$ cm.

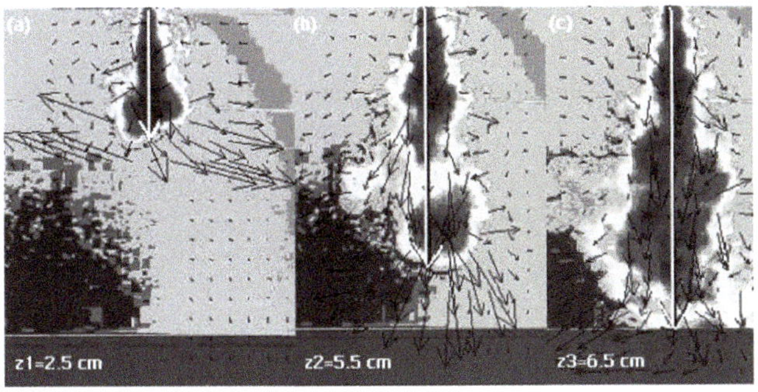

Fig. 4. Time sequence of a turbulent plume experiment with Atwood number $A = 0.001$ and $H_o = 2$ cm. The values of z_1, z_2, and z_3 are shown. (**a**) Plume at $t_1^* = 0.2$, (**b**) plume at $t_2^* = 0.5$, and (**c**) plume at $t_3^* = 1.0$.

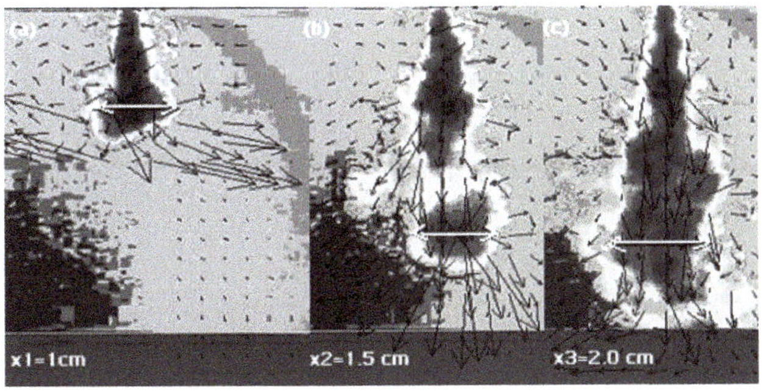

Fig. 5. Time sequence of a turbulent plume experiment with Atwood number $A = 0.001$ and $H_o = 2$ cm. The values of x_1, x_2, and x_3 are shown. (**a**) Plume at $t_1^* = 0.2$, (**b**) plume at $t_2^* = 0.5$, and (**c**) plume at $t_3^* = 1.0$.

We calculate entrainment velocity as $U_E = x/t$ [cm s^{-1}] where the x is the width of the plume at the maximum point of it, and t time is the same as for the vertical velocity. Figure 5 shows the elements for the entrainment velocity calculation for the same selected plume as in Fig. 4. Therefore, it is possible to calculate the entrainment coefficient as defined in Sect. 2.1.

As a consequence of this procedure, it is possible to analyze the time evolution of the vertical velocity W, the entrainment velocity U_E and the entrainment coefficient α_E. Figure 6 shows the time evolution of the vertical velocity W under different source conditions (varying Atwood numbers with a fixed H_o and source positions H_o with a fixed Atwood number).

The time variation of vertical velocity has a greater dispersion at early times (Fig. 6) and W ranges from 12 to 27 cm s^{-1}. Close to the end of the experiment, the vertical velocity has no dispersion and tends to a constant value close to 6.5 cm s^{-1} independently of Atwood number A and H_o. This limit vertical velocity does not depend on source conditions. Therefore, differences are only evident at early times and there are no differences at the end of the experiments for all source conditions.

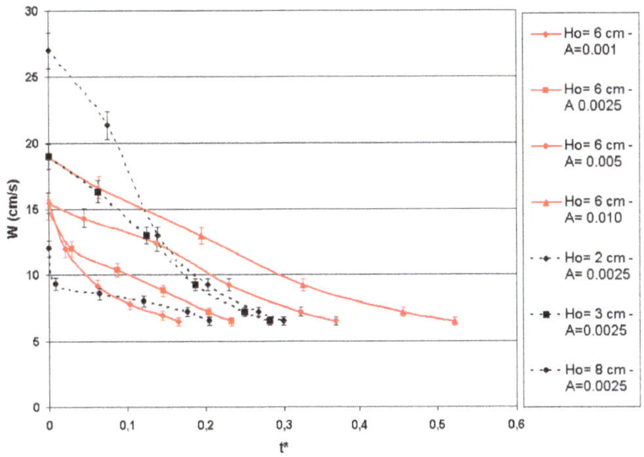

Fig. 6. Time evolution of the vertical velocity W under different source conditions.

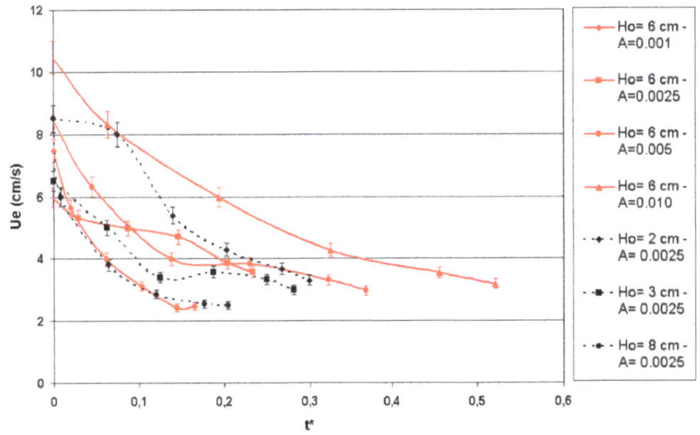

Fig. 7. Time evolution of the entrainment velocity U_E under different source conditions.

The entrainment velocity also has dispersion at initial times (Fig. 7), but the time behavior is slightly different if we compare it with the evolution of the vertical velocity, especially at the end of the experiment. The dispersion at early times ranges from 6 to 10.5 cm s^{-1} and it ranges from 2.5 to 3.3 cm s^{-1} at the end of the experiment. We observe that the data dispersion decreases in time. We observe that the entrainment velocity decreases in time in a similar way for all the source conditions (Fig. 7). The most evident difference is that there is no limit value for U_E close to the end of the experiment: different values of the source conditions (A and H_o) have different U_E.

The entrainment coefficient also has a marked dispersion at initial times (Fig. 8) and is less important close to the end of the experiment. At early times, the entrainment coefficient has an important dispersion and ranges between 0.40 and 0.55 for variable Atwood number experiments and between 0.31 and 0.71 for variable H_o experiments. Close to the end of the experiment, the entrainment coefficient ranges between (0.39, 0.55) for the first kind of experiments and

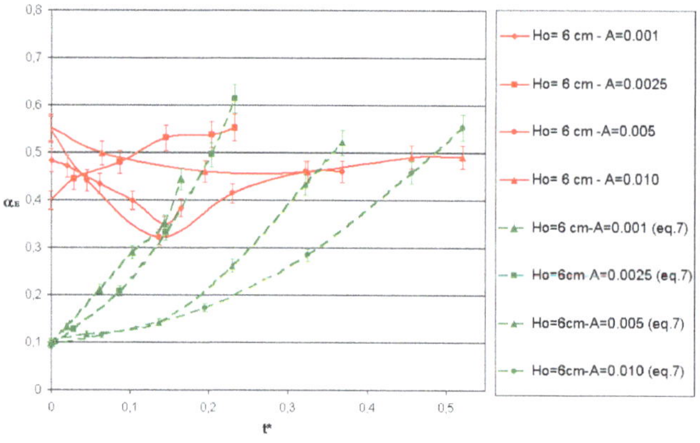

(a)

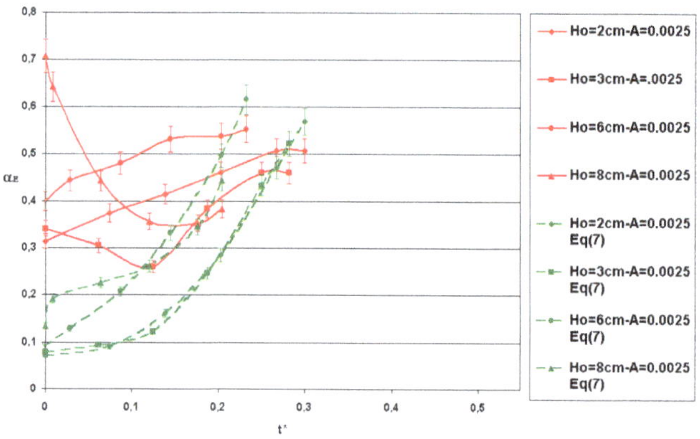

(b)

Fig. 8. Time evolution of the entrainment coefficient α_E under different experimental conditions. **(a)** With a fixed $H_o = 6$ cm; dashed lines represent the evaluation of Eq. (7) for the same source conditions. **(b)** With a fixed Atwood number $A = 0.0025$; dashed lines represent the evaluation of Eq. (7) for the same source conditions.

(0.39, 0.55) for the second ones. Therefore, the entrainment coefficient α_E does not tend to a constant value and shows a time evolution within the same plume and for all the source conditions.

A verification that the time evolution of the entrainment coefficient exits within the same plume is to measure the angle of the plume because of the following relation $\alpha_E \approx \tan(\theta)$ (Fig. 1). Figure 9 shows the time evolution of the plume angle for an experiment with $H_o = 2.5$ cm and $A = 0.03$. From these data of a measured angle, θ, the following empirical linear fit is deduced: $\theta = 8.34 + 10.27t^*$. The value of the coefficient of determination R^2 is 0.75, which indicates that the presented linear model explains 75 % of the variability in our angle data set. From the same data of the plume angle it is possible to deduce a better non-linear fit that is a log fit $\theta = 16.86 + 3.40\log(t^*)$. Figure 9 shows this log fit (dashed line). This non-linear log model explains 91 % of the variability in our angle data set. Therefore, the angle of the plume angle that is indirectly related to the entrainment coefficient is not constant in time.

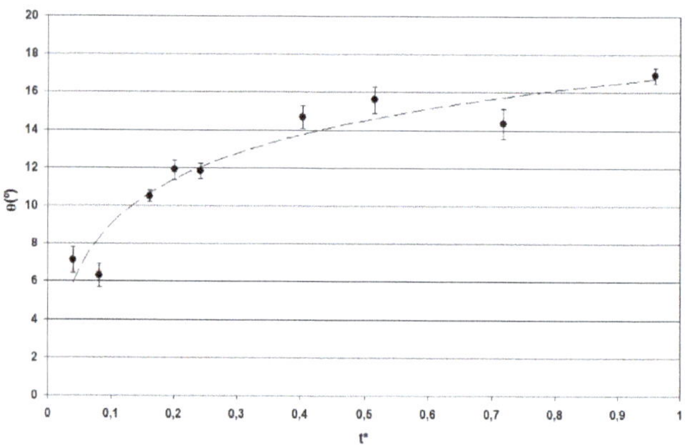

Fig. 9. Time evolution of the angle θ (°) of the plume for an experiment with a fixed $H_o = 2.5$ cm and Atwood number $A = 0.03$. The log fit (dashed line) is also shown.

5 Brief discussion and future work

The original model of Morton et al. (1956) is not sufficient to explain all the experimental and geophysical data about plumes, as was clearly stated in Sect. 2.1. The present laboratory experimental work was motivated by the idea of verifying if the entrainment coefficient varies in time within the same plume and changes with source conditions. Our results about the behavior of the entrainment coefficient verify all these assumptions.

We have considered the topological and geometrical effects affecting the initial evolution of jets/plumes that also appear in nature. The role of ambient turbulence and of environmental buoyancy as discussed by Redondo et al. (1995) and Redondo and Yagüe (1994) will be considered in future work. Here we present an experimental study where we measure the entrainment coefficient, analyzing its time behavior within the same plume and under different source conditions (Fig. 8). We also study the behavior of vertical and axisymmetric velocity at early times under different source conditions (Figs. 6 and 7).

In general, vertical velocities are greater for variable H_o experiments (Fig. 6), i.e., the source position H_o has an effect on vertical velocity. Close to the end of the experiment, there is a limit to the vertical velocity for the plume (6.5 cm s^{-1}) that is not affected by the buoyancy A and the location of the ejection orifice H_o, i.e., it is independent of the source conditions. The model of Morton et al. (1956) gives the mean vertical speed of a plume $W \propto \frac{1}{b^{1/3}}$ where b is the radius of the plume under the assumption of a constant entrainment coefficient. Therefore, if the radius grows, the plume evolves in time, goes downwards and its vertical velocity decreases. This speed decrease is not affected by the source conditions after local equilibrium at the top-heavy front is attained (Fig. 6).

The behavior of the entrainment velocity is represented by Fig. 7. The entrainment velocity decreases in time in a similar way for all source conditions. Given the definition of the entrainment coefficient and the relation for the mean vertical speed of the model of Morton et al. (1956), it is possible to suppose that the entrainment velocity decreases in a similar way ($U_E \propto \frac{1}{b^{1/3}}$) if the entrainment coefficient is supposed constant. For these experiments, there is no clear limiting value for U_E close to the end of the experiment. For variable Atwood number experiments, the entrainment velocity U_E increases with the Atwood number. As the position of the orifice H_o is fixed, only the buoyancy changes by means of the Atwood number A; therefore, the increase in buoyancy increases the entrainment velocity. For variable H_o experiments, the entrainment velocity U_E also decreases in time (Fig. 7). As buoyancy is fixed, only the source location changes, but the entrainment velocity does not clearly increase with the variation in the source location.

Finally, we conclude that the entrainment velocity increases with the Atwood number, A, and it is not so sensitive to the source location, H_o.

In this turbulent plume study we have observed that the entrainment coefficient has important variations in time within the same plume and also with the initial buoyancy and source position conditions, as Fig. 8 shows. At early times, the entrainment coefficient has an important dispersion and close to the end of the experiment, the dispersion of the entrainment coefficient is less important, as in all cases the behavior is plume-like. Although the dispersion has decreased in time, it is clear that the entrainment coefficient α_E does not tend to a constant value and has a time evolution within the same plume and for all the source conditions. This is our main result, which differs from other authors who compare different plumes or jets, or plumes and jets.

This result has an important theoretical implication because in generating the asymptotic equations for a buoyant fluid discharged into an ambient fluid (may be flowing or stagnant, stratified or unstratified), the entrainment coefficient α_E is assumed to be independent of time (Tate, 2002). Therefore, if a time variation for α_E is confirmed, the mathematical treatment to deduce the asymptotic equations should change to introduce a time dependence in the entrainment coefficient function. More experimental work is necessary to look for this time dependence and to achieve this aim, and in particular to understand the important differences between axisymmetric plumes within a 3-D space and line plumes that exhibit a 2-D geometry.

Figure 8b shows the time behavior of the entrainment coefficient for variable H_o experiments. As Atwood number A is constant, the effect of buoyancy is constant and only the effect of the source location is considered. The behavior of the entrainment coefficient is non-regular in time and varies with the source position, H_o. As the location of the source is higher, the coefficient α_E is greater at early time and also at the end of the experiment.

Moreover, the measured entrainment coefficient has considerable variability, more evident at initial steps of the experiments. We also observe that the entrainment coefficient values are greater than the typical value of Morton et al. (1956) and the values for plumes (between 0.085 and 0.16) and jets (between 0.051 and 0.07). Considering all the results, we have evaluated Eq. (5) using our data. To evaluate the equation we write the Froude number Fr in terms of the Atwood number A. The Froude number Fr is a dimensionless quantity that measures the relative importance of the inertial and gravitational effects on a fluid flow. The local densimetric Froude number of a plume based on the centerline values of the vertical velocity and reduced gravity is:

$$Fr = \frac{W}{\sqrt{\frac{\Delta\rho}{\bar{\rho}}\, g\, b}} = \frac{W}{\sqrt{2\,A\,g\,b}}, \tag{6}$$

where $\Delta\,\rho$ is the density difference between plume and ambient fluids ($\rho_D - \rho_L$) and $\bar{\rho}$ is the average fluid density. Therefore, Eq. (5) would be:

$$\alpha_E = \alpha_j - \left(\alpha_j - \alpha_p\right)\left(\frac{2Fr_p^2\,A\,g\,b}{W}\right)^2, \tag{7}$$

where $Fr_p = 4.5$, $\alpha_j = 0.057$ (for a round jet) and $\alpha_p = 0.09$ (for a round plume) (Lee and Chu, 2003). Equation (7) shows that the entrainment coefficient α_E depends on the Atwood number A, the plume vertical velocity W and the plume radius b.

Figure 8 shows the results of the evaluation of Eq. (7) for variable Atwood number A (Fig. 8a, dashed lines) and variable source position H_o (Fig. 8b, dashed lines) experiments. For both kind of experiments, a similar behavior appears: the evaluated entrainment coefficient always increases in time. Therefore, Eq. (7) also predicts a time change for α_E. Another conclusion is that our experimental entrainment coefficient is clearly greater (and has more dispersion) than that evaluated by Eq. (7) at early times. Our greater values of α_E could be related to the plume characteristics. The plume is not laminar and it presents turbulence from the starting point (with momentum and buoyancy fluxes). Another reason could be that the values of the parameters α_j and α_p in Eq. (7) are not the most adequate for our configuration. Our measurements are still greater close to the end of the experiments, but the difference in values is less important. Really, at the end of the experiment, all the values (the experimental ones and those evaluated from Eq. 7) range between 0.38 and 0.62.

We analyze in more detail the behavior of entrainment. For environmental and complex situations the initial conditions may have either 2- or 3-D basic starting topologies. In both cases there is a jet to plume transition scale and the evolution of the starting jet/plume is always slower that the traditional uniform Rayleigh–Taylor instability of $z \propto t^2$

(López et al., 2008). As dimensions of volume, momentum and buoyancy fluxes for 2-D are $[Q] = L^2T^{-1}$, $[M] = L^3T^{-2}$, $[B] = L^3T^{-3}$ and for 3-D they are $[Q] = L^3T^{-1}$, $[M] = L^4T^{-2}$, $[B] = L^4T^{-3}$, then the respective jet-plume transition length scales are (Redondo and Yagüe, 1994):

$$Ł_{JP2D} = C_{2D}MB^{2/3}, \quad Ł_{JP3D} = C_{3D}M^{3/4}B^{-1/2}. \tag{8}$$

So their mixing front time evolutions are $z \propto B^{1/3}t$ for 2-D and $z \propto B^{1/4}t^{3/4}$ for 3-D. This is verified in Landel et al. (2012) as well as in Carazzo et al. (2006) and also indicates that Rayleigh–Taylor or top-heavy initial mixing can be very efficient.

In the case of turbulent or stratified environments, local entrainment values are strongly dependent on height and it is still necessary to evaluate higher order effects such as local variability, non-homogeneity and intermittency. It is interesting to use dimensional analysis to derive a general non-dimensional parameter space for use in the environment because observations seldom correspond to equilibrium.

For 2-D or line sources we use $\frac{w}{U}$, $\frac{U^3}{NM}$ where N is the buoyancy frequency, and for 3-D axisymmetric sources we have $\frac{w}{U}$, $\frac{U^4M}{B^2}$ or $\frac{U^4}{NB}$ as a density Froude number that expresses the relative importance of kinetic to potential energies. In both cases (2- and 3-D) NM/B is a measure of the relative importance of momentum and buoyant fluxes in the presence of stratification. Finally, $U^4/(N^2M)$ represents the relative importance of the motion at buoyant fluid compared with an initial jet.

The non-dimensional parameter space is useful for generating a 4-D surface that represents the variation of the entrainment coefficient, because there is no constant value for α_E (for a single experiment under the same conditions, the entrainment coefficient changes due to intermittency, changes from jet to plume or from a 2- to a 3-D situation). Further work should analyze intermittency and extend the entrainment assumptions to include higher order moments of the velocity and density moments and structure functions (Vindel at al., 2008) following an experimental procedure described in Mahjoub et al. (1998).

6 Conclusions

In this experimental study of a turbulent plume, our main result is that the entrainment coefficient is not a constant and clearly varies in time within the same plume independently of the buoyancy and the source position. Other authors compare different plumes or different jets, or plume and jet experiments, but our result is for a given plume experiment. We also observe that the time evolution of the entrainment coefficient appears with different initial experimental forcing conditions.

Moreover, the measured entrainment coefficient has considerable variability, more evident at initial steps of the experiments. We find as in Carazzo et al. (2006) that, at early times, the entrainment coefficient has an important dispersion and ranges between 0.26 and 0.9 for variable Atwood number experiments and between 0.16 and 0.55 for variable source position experiments. It is interesting to use these alternative descriptors instead of the usual momentum fluxes of volume, momentum and buoyancy because of an easier comparison with the highest initial mixing ability of the Rayleigh–Taylor instability (López et al., 2008). It is also observed that close to the end of the experiment, the dispersion of the entrainment coefficient is less important; it ranges between (0.17, 0.46) for variable Atwood number experiments and (0.37, 0.49) for variable source location experiments. The values of the coefficient are greater than the typical value of Morton et al. (1956) (0.13) and usually measured or calculated entrainment coefficients (but it is not an exception because List (1982) presents values of 0.57 and (0.47, 0.88), or Frick (1984) a value of 0.56).

It is equivalent to consider an initial condition parameter of (Q_o, M_o, B_o) or (H_o, V_o, A_o) to determine the initial effects including the angle of spread of the jet/plume, but the variability of ambient conditions seems also to need a complex and extended parameter space as discussed above. In terms of a generalized turbulent Schmidt or Prandtl number with values between 0.5 and 2, the jets (high H_o and A) reach their equilibrium conditions faster than the plumes, in agreement with Carazzo et al. (2006) for 3-D flows. In clear 2-D situations the sideways meandering of the jets/plumes increases the effective entrainment and the angles as indicated by Landel et al. (2012). Therefore it is necessary to provide more complex models of entrainment that depend on the jet to plume lengths for 2- and 3-D as well as other factors.

Acknowledgements. The authors would like to thank the ICMAT Severo Ochoa project SEV-2011-0087 for its financial support. The authors would also like to thank the help of M. García Velarde and the facilities offered by the Pluri-Disciplinary Institute of the Complutense University of Madrid. We also acknowledge the help of the European Community under project Multi-scale complex fluid flows and interfacial phenomena (PITN-GA-2008-214919). Thanks are also due to ERCOFTAC (PELNoT, SIG 14). We acknowledge the referees for assisting in evaluating this paper.

Edited by: A. M. Mancho

References

Carazzo, G., Kaminski, E., and Tait, S.: The route to self-similarity in turbulent jets and plumes, J. Fluid. Mech., 547, 137–148, 2006.

Carazzo, G., Kaminski, E., and Tait, S.: On the rise of turbulent plumes: Quantitative effects of variable entrainment for submarine hydrothermal vents, terrestrial and extra terrestrial explosive volcanism, J. Geophys. Res., 113, 1–19, 2008.

Fischer, H. B., List, E. J., Koh, R. C., Imberger, J., and Brooks, N. H.: Mixing in inland and coastal waters, Academic Press Inc. (London) Ltd., 1979.

Frick, W. E.: Non-empirical closure of the plume equations, Atmos. Environ., 18, 653–662, 1984.

Houghton, H. G. and Cramer, H. E.: A theory of entrainment in convective currents, J. Meteorol., 8, 95–102, 1951.

Hunt, G. R. and Kaye, N. G.: Virtual origin correction for lazy turbulent plumes, J. Fluid. Mech., 435, 377–396, 2001.

Kaminski, E., Tait, S., And Carazzo, G.: Turbulent entraiment in jets with arbitrary buoyancy, J. Fluid. Mech., 526, 361–376, 2005.

Landel, J. R., Caulfield, C. P., and Woods, A.: Meandering due to large eddies and the statistically self-similar dynamics of quasi-two-dimensional jets, J. Fluid Mech., 692, 347–368, 2012.

Lee, J. H. W. and Chu, V. H.: Turbulent jets and plumes: A lagrangian approach, Kluwer Academic Publishers, the Netherlands, 2003.

List, E. J.: Turbulent jets and plumes, Annu. Rev. Fluid Mech. 14, 189–212, 1982.

López, P.: Turbulent mixing by gravitational convection: experimental modelling and application to atmospheric situations, Ph.D. thesis, Complutense University of Madrid, Spain, 2004.

López, P., Cano, J. L., and Redondo, J. M.: An experimental model of mixing processes generated by an array of top-heavy turbulent plumes, Il Nuovo Cimento, 31 C, 679–698, 2008.

Mahjoub, O. B., Redondo, J. M., and Babiano, A.: Structure functions in complex flows, Appl. Sci. Res., 59, 299–313, 1998.

Matulka, A., Redondo, J. M., and Carrillo, A.: Experiments in stratified and rotating decaying 2D flows, Il Nuovo Cimento, 31 C, 757–770, 2008.

Morton, B. R.: Forced plumes, J. Fluid Mech., 5, 151–163, 1959.

Morton, B. R.: The choice of conservation equations for plume models, J. Geophys. Res., 30, 7409–7416, 1971.

Morton, B. R., Taylor, G. I., and Turner, J. S.: Turbulent gravitational convection from maintained and instantaneous sources, P. Roy. Soc. Lond. A Mat., 234, 1–23, 1956.

Nielsen, P. V.: Displacement ventilation-theory and design, Ph.D. thesis, Aalborg University, Aalborg, Denmark, 1993.

Redondo, J. M., Fernando, J. H., And Pares, S.: Cloud entrainment by internal or external turbulence, in: Mixing in Geophysical Flows, edited by: Redondo, J. M. and Metais, O., CIMNE, Barcelona, 1995.

Redondo, J. M. and Yagüe, C.: Plume entrainment in stratified flows, in: Recent advances in the fluid mechanics of turbulent jets and plumes, edited by: Davies, P. A. and Valente-Neves, M. J., NATO ASI Series E: Applied Sciences, 255, 209–222, 1994.

Squires, P. and Turner, J. S.: An entraining jet model for cumulonimbus updraughts, Tellus, XIV, 422–434, 1962.

Sreenivas, K. R. and Prasap, A. K.: Vortex dynamics model for entrainment in jets and plumes, Phys. Fluids, 12, 2101, doi:10.1063/1.870455, 2000.

Stommel, H.: Entrainment of air into a cumulus cloud, J. Meteorol., 4, 91–94, 1947.

Tate, P. M.: The rise and dilution of buoyant jets and their behaviour in an internal wave field, Ph.D. thesis, University of New South Wales, Sydney, Australia, 2002.

Turner, J. S.: The starting plume in neutral surroundings, J. Fluid. Mech., 13, 356–368, 1962.

Turner, J. S.: Turbulent entrainment: the development of the entrainment assumption and its application to geophysical flows, J. Fluid Mech., 173, 431–471, 1986.

Vindel, J. M., Yagüe, C., and Redondo, J. M.: Relationship between intermittency and stratfication, Il Nuovo Cimento, 31 C, 669–678, 2008.

Wright, S.: Buoyant jets in density-stratified crossflow, J. Hydraul. Eng.-ASCE, 110, 643–656, 1984.

21

Four-dimensional energy spectrum for space–time structure of plasma turbulence

Y. Narita[1,2]

[1]Space Research Institute, Austrian Academy of Sciences, Schmiedlstr. 6, 8042 Graz, Austria
[2]Institut für Geophysik und extraterrestrische Physik, Technische Universität Braunschweig, 38106 Braunschweig, Germany

Correspondence to: Y. Narita (yasuhito.narita@oeaw.ac.at)

Abstract. A parametric model of the inertial-range energy spectrum is constructed for plasma turbulence in the four-dimensional wave vector and frequency domain. The model is based on that of the Eulerian wavenumber-frequency spectrum developed for describing fluid turbulence, and includes wave vector anisotropies in the three-dimensional wave vector domain by approximating the spectrum to a set of ellipses. The shape of the four-dimensional spectrum is determined by the Doppler shift, the Doppler broadening, and anisotropy coefficients. The model is applied to the magnetic energy spectrum in the near-Earth solar wind measured by four Cluster spacecraft, and the set of the spectral parameters are determined observationally. In this way, space–time structure of plasma turbulence can be condensed into a small number of parameters, which is suitable for evaluating the energy spectra in observational and numerical studies on the quantitative basis.

1 Introduction

Plasma turbulence appears in various astrophysical systems and plays an important role as an effective transport mechanism of mass, energy, and angular momentum. It can be found in the Sun and the heliosphere (Petrosyan et al., 2010; Bruno and Carbone, 2013), planetary magnetospheres (Zimbardo et al., 2010), accretion disks (Balbus, 2003; Julien and Knobloch, 2010), and interstellar medium (Elmegreen and Scalo, 2004; Scalo and Elmegreen, 2004). Above all, solar wind is particularly interesting, since it is the only fully developed plasma turbulence accessible to us, allowing detailed investigations using spacecraft in situ in space (Tu and Marsch, 1995; Bruno and Carbone, 2013). One of the recent achievements of multi-spacecraft mission Cluster (Escoubet et al., 2001) is the fact that spatial and temporal variations can be distinguished in the measurements. Furthermore, three-dimensional spatial structure can also be determined using multi-spacecraft data. Turbulence is fundamentally a spatio–temporal phenomenon, and exhibits an interplay of spatial and temporal effects (Carbone et al., 2011; Servidio et al., 2011). The large-scale flow causes Doppler shift and broadening in the measurements, and furthermore, the presence of the large-scale magnetic field imposes anisotropy in the energy spectrum. It is a challenging task both in theory and experiment to reveal the energy spectrum of plasma turbulence in the four-dimensional Fourier space spanned by frequencies and wave vectors.

There is a growing amount of evidence that the energy spectrum of plasma turbulence is intrinsically anisotropic in the wave vector domain. Earlier studies on the quasi-perpendicular wave vectors with respect to the large-scale magnetic field (Matthaeus et al., 1990; Bieber et al., 1996) have been extended to various spatial scales using numerical simulations: magnetohydrodynamic scales (Matthaeus et al., 1996), ion kinetic scales using hybrid plasma simulation (Verscharen et al., 2012; Comişel et al., 2013) or gyro-kinetic treatment (Howes et al., 2011), and electron kinetic scales using particle-in-cell simulations (Camporeale and Burgess, 2011; Gary et al., 2012; Chang et al., 2013). Vlasov–Maxwell hybrid simulation study provides evidence of electric power anisotropy (Valentini et al., 2010). Recent studies using Cluster data not only support the picture of turbulent fluctuations associated with quasi-perpendicular wave vectors, but also suggest axial asymmetry with respect to the directions around the large-scale magnetic field (Turner et al., 2011).

In this manuscript, I propose a parametric method to characterize the four-dimensional (4-D) energy spectrum of plasma turbulence using a model spectrum and observational data from Cluster in the solar wind. The model is given as analytical and contains five free parameters that need to be determined by the observations. Using the free parameters obtained by a fitting procedure, the four-dimensional energy spectrum can be reconstructed as an ideal one. Such a data-driven reconstruction has become possible by the development of the novel high-resolution spectral estimator, the MSR technique (Multi-point Signal Resonator) (Narita et al., 2011a). Complementary approach is to produce plasma turbulence in numerical simulations and to analyze the spectrum using the model. While ideal situations can easily be set in the simulations (e.g., constant mean field, choice of fluctuation amplitude relative to the mean field, and the values of plasma parameter beta), the numerical method necessarily depends on the choice of the plasma model. The energy spectrum model proposed in this manuscript assumes that the hydrodynamic treatment of frequency dependence is valid in that the effect of finite wave propagation speed or dispersion relation is negligible.

2 Model construction

The model of the 4-D energy spectrum is constructed as a generalization of the inertial-range spectrum for fluid turbulence recently derived by Wilczek and Narita (2012). In essence, the fluid turbulence model makes use of the random sweeping hypothesis developed by Kraichnan (1964), and can reproduce the elliptic model of two-time, two-point correlation function of turbulent flows (He and Zhang, 2006; Zhao and He, 2009). The model spectrum is expressed as a product of the three-dimensional (3-D) energy spectrum in the wave vector domain $E(\boldsymbol{k})$ and the Gaussian frequency distribution $F(\boldsymbol{k},\omega)$ as

$$E(\boldsymbol{k},\omega) = E(\boldsymbol{k})F(\boldsymbol{k},\omega) \tag{1}$$

$$F(\boldsymbol{k},\omega) = \frac{1}{\sqrt{2\pi k^2 V^2}} \exp\left[-\frac{(\omega - \boldsymbol{k}\cdot\boldsymbol{U})^2}{2k^2V^2}\right] \tag{2}$$

$$E(\boldsymbol{k}) \quad = E_0|\kappa(\boldsymbol{k})|^{-11/3} \tag{3}$$

$$\kappa^2(\boldsymbol{k}) \quad = a_\parallel^2 k_\parallel^2 + a_{\perp 1}^2 k_{\perp 1}^2 + a_{\perp 2}^2 k_{\perp 2}^2. \tag{4}$$

The spectrum is defined in the 4-D Fourier space spanned by the Eulerian (angular) frequency ω and the wave vector $\boldsymbol{k}$. The Gaussian frequency distribution is centered at the Doppler relation $\omega = \boldsymbol{k}\cdot\boldsymbol{U}$, and has the standard deviation kV. In this approach, the spatial structures are assumed to be swept by the mean flow $\boldsymbol{U}$ and the random sweeping flow V. The former causes the Doppler shift, and the latter causes the Doppler broadening. It is worth mentioning that the notion of the Doppler shift and broadening is valid not only in high-frequency phenomena (sound waves, electromagnetic waves), but also in low-frequency plasma turbulence. The 3-D energy spectrum $E(\boldsymbol{k})$ is modeled as a set of ellipses determined by three coefficients ($a_\parallel$, $a_{\perp 1}$, and $a_{\perp 2}$). The set of the coefficients quantifies wave vector anisotropy between the parallel and the perpendicular directions to the large-scale magnetic field by the ratios $a_{\perp 1}/a_\parallel$ or $a_{\perp 2}/a_\parallel$ and axial asymmetry with respect to the directions around the large-scale field by the ratio $a_{\perp 2}/a_{\perp 1}$. We use the large-scale magnetic field (in the direction $\boldsymbol{e}_{\text{mag}}$) and the mean flow direction ($\boldsymbol{e}_{\text{flow}}$) to establish the coordinate system in the wave vector domain. Three basis vectors are mutually orthogonal and determined by the following relations.

$$\boldsymbol{e}_{\perp 1} = \left(\boldsymbol{e}_{\text{flow}} \times \boldsymbol{e}_{\text{mag}}\right) \times \boldsymbol{e}_{\text{mag}} \tag{5}$$

$$\boldsymbol{e}_{\perp 2} = \boldsymbol{e}_{\text{flow}} \times \boldsymbol{e}_{\text{mag}} \tag{6}$$

$$\boldsymbol{e}_{\parallel} \quad = \boldsymbol{e}_{\text{mag}} \tag{7}$$

The perpendicular-1 direction is closest to the mean flow direction in the plane perpendicular to the large-scale magnetic field.

The model spectrum (Eq. 1-4) represents the simplest extension of the fluid turbulence spectrum in that the minimum set of free parameters is used. The model spectrum is two-sided, viz., it is symmetric with respect to sign reversal of frequencies and wave vectors ($\omega \to -\omega$ and $\boldsymbol{k} \to -\boldsymbol{k}$). The spectral energy is scaled to E_0, and inherits the power-law index of Kolmogorov's 3-D inertial-range spectrum (Kolmogorov, 1941), such that the isotropic limit ($a_\parallel \to 1$, $a_{\perp 1} \to 1$, and $a_{\perp 2} \to 1$) restores the inertial-range spectrum of fluid turbulence. One of the anisotropy coefficients may be set to unity by redefining the scale factor E_0 and the two other coefficients. In this work, $a_\parallel$ is set to unity.

The use of the Gaussian frequency distribution is based on the approximation that small-scale fully-developed turbulence is swept by the large-scale flow in an advective fashion. The mean flow causes Doppler shift, while the variation of the large-scale flow causes Doppler broadening around the Doppler shifted frequency (Fig. 1). In the limit of vanishing Doppler broadening ($V \to 0$) the Gaussian frequency distribution reduces to the Dirac delta function $F(\boldsymbol{k},\omega) \propto \delta(\omega - \boldsymbol{k}\cdot\boldsymbol{U})$, restoring Taylor's frozen-in flow hypothesis (Taylor, 1938). The symbol V is independently defined as a free parameter in the model. The notion of the large-scale flow is valid as far as the fluctuations is regarded as ideally convected structures. Specifically, the flow (or the plasma) should not evolve intrinsically within the measured time interval, and this can be examined, e.g., by looking for the existence of dispersion relations that deviate from the Doppler relation in the frequency-wavenumber diagram. Therefore, the model spectrum neglects the wave propagation effect such as Alfvén waves, and the model is more suited for applying to super-Alfvénic flows.

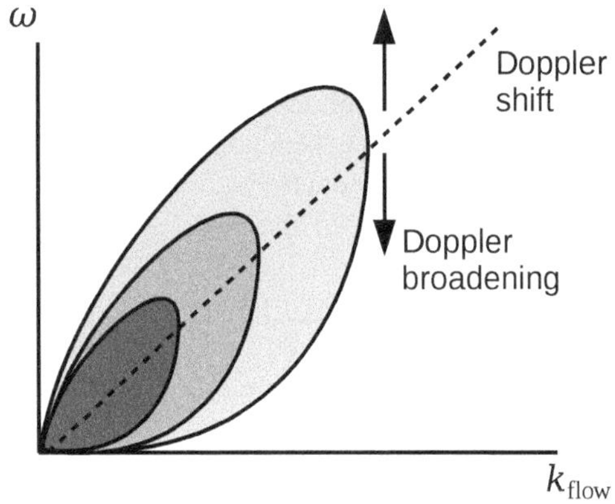

Fig. 1. Sketch of the inertial-range spectrum in the wavenumber-frequency domain. The wavenumber is aligned with the mean flow direction.

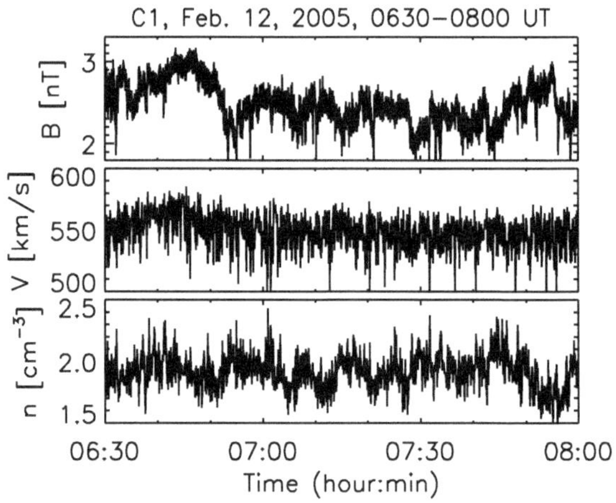

Fig. 2. Time series plot of the magnetic field magnitude, the ion bulk speed, and the ion number density measured by Cluster-1 in the solar wind.

3 Application to solar wind turbulence

The model spectrum is used as an analysis tool to characterize the space–time structure of solar wind turbulence. By doing so, it is possible to condense the turbulence spectrum into only several free parameters. Cluster spacecraft are most useful for such a task, as the 4-D energy spectrum can be estimated directly from the four-point magnetic field data (Balogh et al., 2001) by incorporating the MSR analysis technique. The data are taken on 12 February 2005, 06:30–08:00 UT, for the analysis. This interval was selected under the following criteria:

1. Cluster formed a nearly regular tetrahedron which is suitable for wave vector analysis. The tetrahedral quality parameter Q_G (von Stein et al., 1992) in the range 2.9–3.0 was set. This parameter quantifies the tetrahedral configuration as a fractal dimension.

2. Inter-spacecraft separation was about 1000 km and larger than the thermal ion gyro-radius (of the order of 100 km) in the solar wind. Turbulence may be regarded as magnetohydrodynamic on that scale, and kinetic wave dispersion effect can be neglected.

3. Fluctuations of the magnetic field and the plasma are quasi-stationary in the sense that the large-scale field may be regarded as nearly constant (cf. the relative fluctuation amplitude to the large-scale field is about 10.2 %).

4. Frequency spectra for the magnetic field fluctuations (at all four spacecraft) exhibit a power-law with the index −5/3. Fig. 2 displays the time series of the magnetic field magnitude, the ion bulk speed, and the ion number density (Rème et al., 2001) obtained by Cluster-1.

The mean values are evaluated in the temporal domain, and are 2.5 nT (magnetic field), 548.9 km s^{-1} (ion bulk speed), and 1.9 cm^{-3} (ion number density). The Alfvén Mach number is about 13.7 on this time interval, and the plasma parameter beta is estimated to be about 5.4 using ion data. This interval represents a turbulent flow in a high-beta plasma.

The 4-D energy spectrum was estimated from the Cluster magnetic field data (measured by the fluxgate magnetometer) first by Fourier transforming the four-point time series data into the frequency domain (using the Fast Fourier Transform algorithm) and then by projecting the data into the 3-D wave vector domain using the MSR technique. This method was developed particularly for studying waves and turbulence using four-point magnetic field data, and makes extensive use of the 12-by-12 covariance matrix (three components of the magnetic field measured at four spacecraft) by combining the least-square projection (known as the wave telescope or k filtering technique) and the eigenvalue analysis of the covariant matrix, known as the MUSIC algorithm (Multiple Signal Classification) (Schmidt, 1986). The MSR technique is based on the assumption that the measured fluctuations represent a set of plane waves and that the fluctuations contain small-amplitude isotropic noise. The energy spectrum is given as a function of spacecraft-frame frequencies and wave vectors. The frequency range up to 1.2 rad s^{-1} and the wavenumber range up to 0.0023 rad km^{-1} are used to determine the free parameters of the model spectrum. The upper limit of the frequency range to analyze is the spacecraft separation ($k_{\max}$), which yields, using a given relationship $\omega \propto k$, the corresponding frequency $\omega_{\max}$ that allows one to avoid the spatial aliasing effect (Narita and Glassmeier, 2009; Sahraoui et al., 2010). Here, Taylor's hypothesis gives a reasonable estimate to set the frequency range in the analysis.

A fitting procedure is applied to the measured spectrum in two distinct domains. One is the plane spanned by the streamwise wavenumbers and the frequencies, and the other is the 3-D wave vector domain after integrating the 4-D

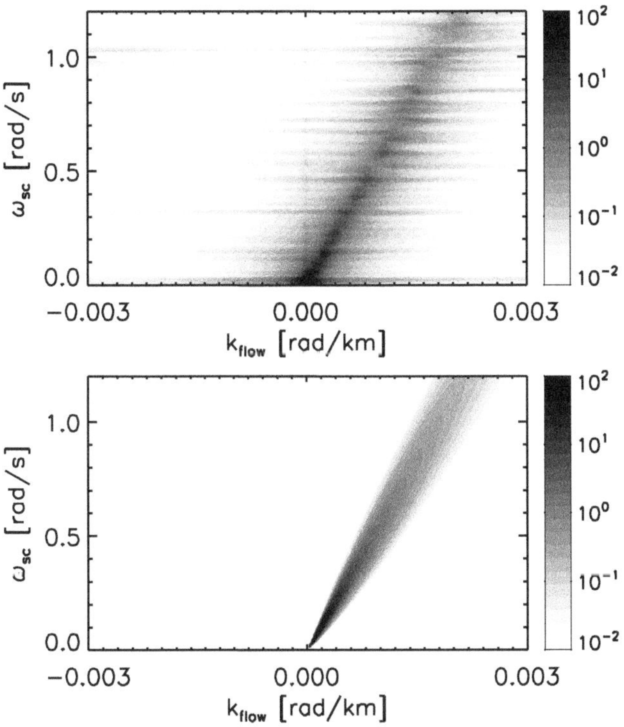

Fig. 3. Slice of the energy spectrum in the plane spanned by spacecraft-frame frequencies and streamwise wavenumbers derived by four Cluster spacecraft data (upper panel) and the reconstructed energy spectrum using the method of model fitting (lower panel). Best fitting was obtained for the parameter set $U = 547.6$ km (Doppler shift) and $V = 49.6$ km (Doppler broadening).

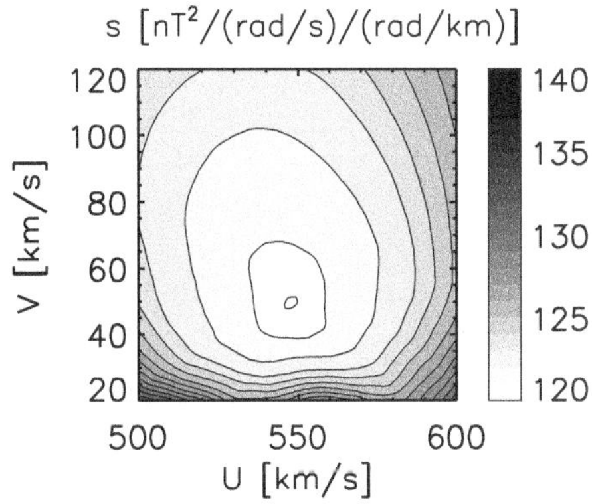

Fig. 4. Deviation between the measured spectrum and the model spectrum in the parameter space spanned by U (Doppler shift) and V (Doppler broadening).

Table 1. Key spectral shape parameters obtained from the Cluster data analysis for solar wind turbulence.

U	V	$a_{\perp 1}/a_{\parallel}$	$a_{\perp 2}/a_{\parallel}$
547.6 km s^{-1}	49.6 km s^{-1}	0.73	0.31

spectrum over the frequencies. The fitting procedure in the former domain determines the Doppler shift U and the broadening V, while the fitting in the latter domain determines the anisotropy coefficients. Fig. 3 displays the measured spectrum and the reconstructed spectrum using the set of the best-fitting parameters ($U = 547.6\,\mathrm{km\,s^{-1}}$, $V = 49.6\,\mathrm{km\,s^{-1}}$). The time series data of the ion bulk speed from Cluster-1 yields the standard deviation $\sqrt{\langle|\delta \boldsymbol{v}|^2\rangle} = 20.0\,\mathrm{km\,s^{-1}}$ in the temporal domain. The Doppler shift estimated in the spectral analysis agrees with the ion bulk speed within the accuracy better than 1 %. The estimated Doppler broadening agrees with the ion bulk speed fluctuation only moderately, i.e., the broadening V is larger than that estimated from the ion measurements by the factor of about 2.5. The Alfvén speed estimated from the mean magnetic field and the ion number density is about $38.9\,\mathrm{km\,s^{-1}}$, which is markedly close to the broadening V.

The reconstructed spectrum shows a difference from the measurement in that the conic shape is not clearly visible in the data. Possible explanations are (1) insufficient scale separation, and (2) finite wave propagation speed. In the former scenario, since the model spectrum assumes that fluctuations consist of large-scale and small-scale fields, the difference in the frequency dependence can be interpreted as a sign of insufficient scale separation. That is, the frequency broadening is no longer expressed by a constant Doppler broadening $V =$ const, but is a function of the wavenumber such as $V = f(k)$. In the present study, the parameters U and V are regarded as a constant, and these values are obtained by a least square method in the wavenumber-frequency domain (Fig. 4). In the latter scenario, since the model spectrum assumes that the frequency dependence represents solely Doppler shift and broadening and neglects the effect of finite wave phase speed, the difference in the spectrum may be interpreted as a sign of the existence of dispersion relations associated with, e.g., Alfvén waves, magnetosonic waves, or even dispersive waves. The intrinsic noise originating in the analysis technique is an unlikely source, since the MSR technique has been developed particularly to improve the signal-to-noise ratio of the conventional wave telescope technique significantly by a factor of about 10^5 in the spectrum as shown in Fig. 3 in Narita et al. (2011a) by incorporating the eigenvalue analysis into the wave telescope technique. The intrinsic noise associated with the analysis technique is negligible.

The energy spectrum in the 3-D wave vector domain was determined by integrating the 4-D spectrum over the spacecraft-frame frequencies. The anisotropy shape parameters ($a_{\parallel}$, $a_{\perp 1}$, and $a_{\perp 2}$) are evaluated in the three slices of the 3-D spectrum (planes at $k_{\parallel} = 0$, $k_{\perp 1} = 0$, and $k_{\perp 2} = 0$) with the help of the anisotropy angle θ_{ij} in the plane spanned by k_i and k_j (the indices i and j denote the three components of the wave vector). The anisotropy angle was introduced by

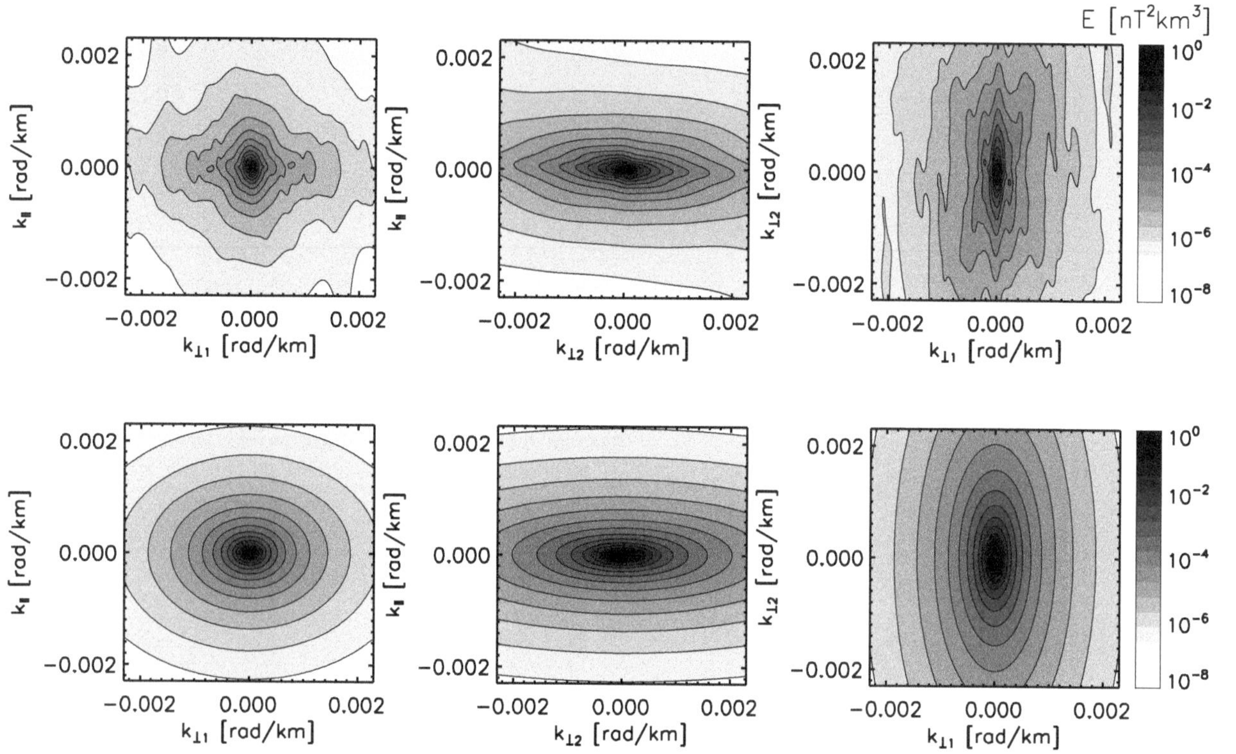

Fig. 5. Slices of the energy spectrum in the 3-D wave vector domain using Cluster spacecraft data in the solar wind (upper panels), and its reconstruction using the model spectrum (lower panels).

Shebalin et al. (1983) and its adapted form was used in the present analysis as

$$\tan^2\theta_{ij} = \frac{\sum_k k_j^2 E(k_i, k_j)}{\sum_k k_i^2 E(k_i, k_j)}. \tag{8}$$

The summation runs over all the wave vector components in the plane of interest. Fitting procedure using the anisotropy angle yields the set of optimal ratios of the anisotropy coefficients, $a_{\perp 1}/a_{\parallel} = 0.73$, and $a_{\perp 2}/a_{\parallel} = 0.31$. Fig. 5 displays the measured spectrum (upper panels) and the reconstructed spectrum using the best-fit parameters (lower panels).

In the $k_{\parallel}$-$k_{\perp 1}$ plane, the spectrum is not elliptic but exhibits two different kinds of anisotropy. At lower wavenumbers (below 0.0008 rad km^{-1}) the spectrum extends in the parallel direction, while at higher wavenumbers the extension turns to the perpendicular direction. This transition cannot be reproduced in the model spectrum, and the reconstructed spectrum is only moderately anisotropic. In the $k_{\parallel}$-$k_{\perp 2}$ plane, the spectrum is clearly anisotropic, elongated in the perpendicular direction. Approximation to ellipses can reproduce the spectral contours reasonably. In the $k_{\perp 1}$-$k_{\perp 2}$ plane, the spectrum is axi-asymmetric with detailed structures. The spectrum is elongated in the direction perpendicular to the mean flow, i.e., the spectral extension is in the direction of interplanetary electric field. These results support earlier studies using single and multiple spacecraft measurements that solar wind turbulence is anisotropic with respect to the wave vectors, and is axi-asymmetric around the direction of the large-scale magnetic field (Narita et al., 2010, 2011b; Chen et al., 2012).

4 Conclusion and discussion

The model energy spectrum is constructed in a simple fashion using analytic expression. The model is based on various assumptions such as incorporation of the random sweeping hypothesis, no wave propagation (or no dispersion relation), and elliptic approximation of the wave vector anisotropy. Still, the model has the potential to condense the energy spectrum of plasma turbulence (and hence space–time variation of turbulence) into a set of handful free parameters. Six parameters need to be determined in reconstructing the 4-D energy spectrum, but the shape of the spectrum itself requires four parameters only: the Doppler shift (U), the Doppler broadening (V), and the anisotropy ratios $a_{\perp 1}/a_{\parallel}$ and $a_{\perp 2}/a_{\parallel}$. The ratio $a_{\perp 2}/a_{\perp 1}$ reflects the axial asymmetry of the spectrum around the directions of the large-scale magnetic field. This key parameter set derived from the Cluster data in the solar wind is summarized in Table 1.

There are strengths and weaknesses in the model spectrum. Since the model reduces the energy spectrum into a set of free parameters, various realizations of plasma turbulence can be compared more easily and systematically. For example, the model can be applied to studies of beta dependence of plasma turbulence. Evolution of plasma turbulence can be studied by tracking time dependence of the parameter set. Quantitative comparison with numerically produced turbulence is possible. The model spectrum is constructed in the Eulerian frame of reference, and does not require the correction for the Doppler shift (instead, the concept of Doppler broadening has been introduced), which saves the computation time in data analysis. On the other hand, the assumptions used in the model are limiting the applicability, and the conditions of the plasma and magnetic field need to be carefully examined. The measured spectrum in the $k_\parallel$-$k_{\perp 1}$ plane exhibits two different kinds of anisotropy (parallel and perpendicular spectral extensions), which prefers the two-component picture plasma turbulence describing a superposition of fluctuations associated with the parallel wave vectors and that associated with the perpendicular wave vectors (Matthaeus and Ghosh, 1999). Highly structured wave vector anisotropy cannot be resolved or reproduced in the model spectrum. Also, the choice of the basis vectors needs to be exercised, as the axial asymmetry in the measured spectrum may not always be associated with the direction of the mean flow or the direction of interplanetary electric field.

The axial asymmetry in the $k_{\perp 1}$-$k_{\perp 2}$ plane is found to be stronger than anisotropy in the $k_\parallel$-$k_{\perp 1}$ on the analyzed time interval, and weaker than that in the $k_\parallel$-$k_{\perp 2}$. This feature was already found in the other case study using Cluster data (Narita et al., 2011b). The strong axial asymmetry cannot be explained by the spectral deformation induced by the irregular tetrahedral configuration, as spacecraft formed a nearly regular tetrahedron during the measurements. Possible origins of the axial asymmetry include the effect of radial expansion of the flow, the effect of interplanetary electric field, and the coronal field structure. The model spectrum can be extended in various ways to reconstruct the measured spectrum more accurately. For example, the parameter V may be treated as a function of the wavenumber. Another approach is to incorporate dispersion relations into the frequency dependence, which, however, requires the knowledge of the wave modes. Also, the presented model can be extended to describing the energy spectrum in the ion kinetic regime. For example, Comişel et al. (2013) introduced the Gaussian frequency distribution to measure the strength of sideband waves. The Doppler relation was also replaced by dispersion relations of linear modes assuming that linear modes co-exist in turbulent fluctuations. To conclude, model construction is an important task in plasma turbulence studies, and a suitable model will serve as a bridge connecting experimentally derived spectra and theoretical ones.

Acknowledgements. The research leading to these results has received funding from the European Community's Seventh Framework Programme under grant agreement 313038/STORM. The work conducted in Braunschweig is supported by Collaborative Research Center 963, *Astrophysical Flow, Instabilities, and Turbulence* of the German Science Foundation.

Edited by: A. C. L. Chian
Revied by: two anonymous referees

References

Balbus, S. A.: Enhanced angular momentum transport in accretion disks, Ann. Rev. Astron. Astrophys. 41, 555–597, doi:10.1146/annurev.astro.41.081401.155207, 2003.

Balogh, A., Carr, C. M., Acuña, M. H., Dunlop, M. W., Beek, T. J., Brown, P., Fornaçon, K.-H., Georgescu, E., Glassmeier, K.-H., Harris, J., Musmann, G., Oddy, T., and Schwingenschuh, K.: The Cluster magnetic field investigation: overview of in-flight performance and initial results, Ann. Geophys., 19, 1207–1217, doi:10.5194/angeo-19-1207-2001, 2001.

Bieber, J. W., Wannger, W., and Matthaeus, W. H.: Dominant two-dimensional solar wind turbulence with implications for cosmic ray transport, J. Geophys. Res., 101, 2511–2522, doi:10.1029/95JA02588, 1996.

Bruno, R. and Carbone, V.: The solar wind as a turbulence laboratory, Living Rev. Solar Phys., 10, 2, doi:10.12942/lrsp-2013-2, 2013.

Camporeale, E. and Burgess, D.: The dissipation of solar wind turbulent fluctuations at electron scales, Astrophys. J., 730, 114, doi:10.1088/0004-637X/730/2/114, 2011 (Correction, 735, 67, doi:10.1088/0004-637X/735/1/67, 2011).

Carbone, F., Sorriso-Valvo, L., Versace, C., Strangi, G., and Bartolino, R.: Anisotropy of spatiotemporal decorrelation in electrohydrodynamic turbulence, Phys. Rev. Lett., 106, 114502, doi:10.1103/PhysRevLett.106.114502, 2011.

Chang, O., Gary, S. P., and Wang, J.: Whistler turbulence at variable electron beta: Three-dimensional particle-in-cell simulations, J. Geophys. Res., Space Phys., 118, 2824–2833, doi:10.1002/jgra.50365, 2013.

Chen, C. H. K., Mallet, A., Schekochihin, A. A., Horbury, T. S., Wicks, R. T., and Bale, S. D.: Three-dimensional structure of solar wind turbulence, Astrophys. J., 758, 120, doi:10.1088/0004-637X/758/2/120, 2012.

Comişel, H., Verscharen, D., Narita, Y., and Motschmann, U.: Spectral evolution of two-dimensional kinetic plasma turbulence in the wavenumber-frequency domain, Phys. Plasmas, 20, 090701, doi:10.1063/1.4820936, 2013.

Elmegreen, B. G. and Scalo, J.: Interstellar turbulence I: Observations and processes, Ann. Rev. Astron. Astrophys., 42, 211–273, doi:10.1146/annurev.astro.41.011802.094859, 2004.

Escoubet, C. P., Fehringer, M., and Goldstein, M.: The Cluster mission, Ann. Geophys., 19, 1197–1200, doi:10.5194/angeo-19-1197-2001, 2001.

Gary, S. P., Chang, O., and Wang, J.: Forward cascade of whistler turbulence: Three-dimensional particle-in-cell simulations, Astrophys. J., 755, 142, doi:10.1088/0004-637X/755/2/142, 2012.

He, G.-W. and Zhang, J.-B.: Elliptic model for space-time correlation in turbulent shear flows, Phys. Rev. E, 73, 055303R, doi:10.1103/PhysRevE.73.055303, 2006.

Howes, G. G., Tenbarge, J. M., Dorland, W., Quataert, E., Schekochihin, A. A., Numata, R., and Tatsuno, T.:

Gyrokinetic simulations of solar wind turbulence from ion to electron scales, Phys. Rev. Lett., 107, 035004, doi:10.1103/PhysRevLett.107.035004, 2011.

Kolmogorov, A. N.: The local structure of turbulence in incompressible viscous fluid for very large Reynolds number, Dokl. Akad. Nauk. SSSR, 30, 299–303, 1941 (Reprinted in Proc. Roy. Soc. A 434, 9–13, doi:10.1098/rspa.1991.0075, 1991).

Kraichnan, R. H.: Kolmogorov's hypothesis and Eulerian turbulence theory, Phys. Fluids, 7, 1723–1734, 1964.

Julien, K. and Knobloch, E.: Magnetorotational instability: recent developments, Phil. Trans. R. Soc. A, 368, 1607–1633, doi:10.1098/rsta.2009.0251, 2010.

Matthaeus, W. H., Goldstein, M. L., and Roberts, D. A.: Evidence for the presence of quasi-two-dimensional nearly incompressible fluctuations in the solar wind, J. Geophys. Res., 95, 20, 673–683, doi:10.1029/JA095iA12p20673, 1990

Matthaeus, W. H., Gosh, S., Ougton, S., and Roberts, D. A.: Anisotropic three-dimensional MHD turbulence, J. Geophys. Res., 101, 7619–7630, doi:10.1029/95JA03830, 1996.

Matthaeus, W. H. and Ghosh, S.: Spectral decomposition of solar wind turbulence: Three-component model, AIP Conf. Proc., The solar wind nine conference, 5–9 October 1998, 471, 519–522, doi:10.1063/1.58688, 1999

Narita, Y. and Glassmeier, K.-H.: Spatial aliasing and distortion of energy distribution in the wave vector domain under multi-spacecraft measurements, Ann. Geophys., 27, 3031–3042, doi:10.5194/angeo-27-3031-2009, 2009.

Narita, Y., Glassmeier, K.-H., Sahraoui, F., and Goldstein, M. L.: Wave-vector dependence of magnetic-turbulence spectra in the solar wind, Phys. Rev. Lett., 104, 171101, doi:10.1103/PhysRevLett.104.171101, 2010.

Narita, Y., Glassmeier, K.-H., and Motschmann, U.: High-resolution wave number spectrum using multi-point measurements in space – The Multi-point Signal Resonator (MSR) technique, Ann. Geophys., 29, 351-360, doi:10.5194/angeo-29-351-2011, 2011a.

Narita, Y., Glassmeier, K.-H., Goldstein, M. L., Motschmann, U., and Sahraoui, F.: Three-dimensional spatial structures of solar wind turbulence from 10 000-km to 100-km scales, Ann. Geophys., 29, 1731–1738, doi:10.5194/angeo-29-1731-2011, 2011b.

Petrosyan, A., Balogh, A., Goldstein, M. L., Léorat, J., Marsch, E., Petrovay, K., Roberts, B., von Steiger, R., and Vial, J. C.: Turbulence in the solar atmosphere and solar wind, Space Sci. Rev., 156, 135–238, doi:10.1007/s11214-010-9694-3, 2010.

Rème, H., Aoustin, C., Bosqued, J. M., Dandouras, I., Lavraud, B., Sauvaud, J. A., Barthe, A., Bouyssou, J., Camus, Th., Coeur-Joly, O., Cros, A., Cuvilo, J., Ducay, F., Garbarowitz, Y., Medale, J. L., Penou, E., Perrier, H., Romefort, D., Rouzaud, J., Vallat, C., Alcaydè, D., Jacquey, C., Mazelle, C., d'Uston, C., Möbius, E., Kistler, L. M., Crocker, K., Granoff, M., Mouikis, C., Popecki, M., Vosbury, M., Klecker, B., Hovestadt, D., Kucharek, H., Kuenneth, E., Paschmann, G., Scholer, M., Sckopke, N., Seidenschwang, E., Carlson, C. W., Curtis, D. W., Ingraham, C., Lin, R. P., McFadden, J. P., Parks, G. K., Phan, T., Formisano, V., Amata, E., Bavassano-Cattaneo, M. B., Baldetti, P., Bruno, R., Chionchio, G., Di Lellis, A., Marcucci, M. F., Pallocchia, G., Korth, A., Daly, P. W., Graeve, B., Rosenbauer, H., Vasyliunas, V., McCarthy, M., Wilber, M., Eliasson, L., Lundin, R., Olsen, S., Shelley, E. G., Fuselier, S., Ghielmetti, A. G., Lennartsson, W., Escoubet, C. P., Balsiger, H., Friedel, R., Cao, J.-B., Kovrazhkin, R. A., Papamastorakis, I., Pellat, R., Scudder, J., and Sonnerup, B.: First multispacecraft ion measurements in and near the Earth's magnetosphere with the identical Cluster ion spectrometry (CIS) experiment, Ann. Geophys., 19, 1303–1354, doi:10.5194/angeo-19-1303-2001, 2001.

Sahraoui, F., Belmont, G., Goldstein M., and Rezeau, L.: Limitations of multi-spacecraft data techniques in measuring wavenumber spectra of space plasma turbulence, J. Geophys. Res., 115, A04206, doi:10.1029/2009JA014724, 2010.

Scalo, J. and Elmegreen, B. G.: Interstellar turbulence II: Implications and effects, Ann. Rev. Astron. Astrophys., 42, 275–316, doi:10.1146/annurev.astro.42.120403.143327, 2004.

Schmidt, R. O.: Multiple emitter location and signal parameter estimation, IEEE Trans. Ant., Prop., AP-34, 276–280, 1986.

Servidio, S., Carbone, V., Dmitruk, P., and Matthaeus, W. H.: Time decorrelation in isotropic magnetohydrodynamic turbulence, Europhys. Lett., 96, 55003, doi:10.1209/0295-5075/96/55003, 2011.

Shebalin, J. V., Matthaeus, W. H., and Montgomery, D.: Anisotropy in MHD turbulence due to a mean magnetic field, J. Plasma Phys, 29, 525–547, doi:10.1017/S0022377800000933, 1983.

Taylor, G. I.: The spectrum of turbulence, Proc. R. Soc. Lond. A 164, 476–490, doi:10.1098/rspa.1938.0032, 1938.

Tu, C.-Y., and Marsch, E.: MHD structures, waves and turbulence in the solar wind: Observations and theories, Space Sci. Rev., 73, 1–210, doi:10.1007/BF00748891, 1995.

Turner, A. J., Gogoberidze, G., Chapman, S. C., Hnat, B., and Müller, W.-C.: Nonaxisymmetric anisotropy of solar wind turbulence, Phys. Rev. Lett., 107, 095002, doi:10.1103/PhysRevLett.107.095002, 2011.

Valentini, F., Califano,F., and Veltri, P.: Two-dimensional kinetic turbulence in the solar wind, Phys. Rev. Lett. 104, 205002, doi:10.1103/PhysRevLett.104.205002, 2010.

Verscharen, D., Marsch, E., Motschmann, U., and Müller, J.: Kinetic cascade beyond magnetohydrodynamics of solar wind turbulence in two-dimensional hybrid simulations, Phys. Plasmas, 19, 022305 doi:10.1063/1.3682960, 2012.

von Stein, R., Glassmeier, K.-H., and Dunlop, M.: A configuration parameter for the Cluster satellites, Tech. Rep. 2/1992, Institut für Geophysik und Meteorologie der Technischen Universität Braunschweig, 1992.

Wilczek, M. and Narita, Y.: Wavenumber-frequency spectrum for turbulence from a random sweeping hypothesis with mean flow, Phys. Rev. E, 86, 066308, doi:10.1103/PhysRevE.86.066308, 2012.

Zhao, X. and He, G.-W.: Space-time correlations of fluctuating velocities in turbulent shear flows, Phys. Rev. E, 79, 046316, doi:10.1103/PhysRevE.79.046316, 2009.

Zimbardo, G., Greco, A., Sorriso-Valvo, L., Perri, S., Vörös, Z., Aburjania, G., Chargazia, K., and Alexandrova, O.: Magnetic turbulence in the geospace environment, Space Sci. Rev., 156, 89–134, doi:10.1007/s11214-010-9692-5, 2010.

Permissions

All chapters in this book were first published in NPG, by Copernicus Publications; hereby published with permission under the Creative Commons Attribution License or equivalent. Every chapter published in this book has been scrutinized by our experts. Their significance has been extensively debated. The topics covered herein carry significant findings which will fuel the growth of the discipline. They may even be implemented as practical applications or may be referred to as a beginning point for another development.

The contributors of this book come from diverse backgrounds, making this book a truly international effort. This book will bring forth new frontiers with its revolutionizing research information and detailed analysis of the nascent developments around the world.

We would like to thank all the contributing authors for lending their expertise to make the book truly unique. They have played a crucial role in the development of this book. Without their invaluable contributions this book wouldn't have been possible. They have made vital efforts to compile up to date information on the varied aspects of this subject to make this book a valuable addition to the collection of many professionals and students.

This book was conceptualized with the vision of imparting up-to-date information and advanced data in this field. To ensure the same, a matchless editorial board was set up. Every individual on the board went through rigorous rounds of assessment to prove their worth. After which they invested a large part of their time researching and compiling the most relevant data for our readers.

The editorial board has been involved in producing this book since its inception. They have spent rigorous hours researching and exploring the diverse topics which have resulted in the successful publishing of this book. They have passed on their knowledge of decades through this book. To expedite this challenging task, the publisher supported the team at every step. A small team of assistant editors was also appointed to further simplify the editing procedure and attain best results for the readers.

Apart from the editorial board, the designing team has also invested a significant amount of their time in understanding the subject and creating the most relevant covers. They scrutinized every image to scout for the most suitable representation of the subject and create an appropriate cover for the book.

The publishing team has been an ardent support to the editorial, designing and production team. Their endless efforts to recruit the best for this project, has resulted in the accomplishment of this book. They are a veteran in the field of academics and their pool of knowledge is as vast as their experience in printing. Their expertise and guidance has proved useful at every step. Their uncompromising quality standards have made this book an exceptional effort. Their encouragement from time to time has been an inspiration for everyone.

The publisher and the editorial board hope that this book will prove to be a valuable piece of knowledge for researchers, students, practitioners and scholars across the globe.

List of Contributors

G. V. Khazanov
NASA Goddard Space FlightCenter, Greenbelt, MD, USA

A. A. Tel'nikhin
Altai State University, Department of Physics and Technology, Barnaul, Russia

T. K. Kronberg
Altai State University, Department of Physics and Technology, Barnaul, Russia

R. A. Treumann
Department of Geophysics and Environmental Sciences, Munich University, Munich, Germany
International Space Science Institute, Bern, Switzerland

W. Baumjohann
Space Research Institute, Austrian Academy of Sciences, Graz, Austria

Y. Nariyuki
Faculty of Human Development, University of Toyama, 3190, Toyama City, Toyama 930-8555, Japan

T. Umeda
Solar-Terrestrial Environment Laboratory, Nagoya University, Nagoya, Aichi 464-8601, Japan

T. K. Suzuki
Department of Physics, Nagoya University, Furo-cho, Nagoya, Aichi 464-8602, Japan

T. Hada
Department of Earth System Science and Technology, Kyushu University, 6-1, Kasuga City, Fukuoka 816-8580, Japan

B. O. Ogunsua
Space Physics Laboratory, Department of Physics, Federal University of Technology, Akure, Nigeria

J. A. Laoye
Statistical and Nonlinear Physics Research group, Department of Physics, Olabisi Onabanjo University, Ago-Iwoye, Nigeria

I. A. Fuwape
Space Physics Laboratory, Department of Physics, Federal University of Technology, Akure, Nigeria

A. B. Rabiu
Space Physics Laboratory, Department of Physics, Federal University of Technology, Akure, Nigeria
National Space Research and Development Agency, Abuja, Nigeria

D. Falceta-Gonçalves
SUPA, School of Physics&Astronomy, University of St Andrews, North Haugh, St Andrews, Fife, KY16 9SS, UK
Escola de Artes, Ciencias e Humanidades, Universidade de Sao Paulo, Rua Arlindo Bettio 1000, CEP 03828-000 Sao Paulo, Brazil

G. Kowal
Escola de Artes, Ciencias e Humanidades, Universidade de Sao Paulo, Rua Arlindo Bettio 1000, CEP 03828-000 Sao Paulo, Brazil

E. Falgarone
LERMA/LRA, CNRS, Ecole Normale Supérieure and Observatoire de Paris, 24 rue Lhomond, 75231 Paris Cedex, France

A. C.-L. Chian
Observatoire de Paris, LESIA, CNRS, 92190 Meudon, France
National Institute for Space Research (INPE) and World Institute for Space Environment Research (WISER), P.O. Box 515, 12227-010 São José dos Campos-SP, Brazil
School of Mathematical Sciences, University of Adelaide, Adelaide, SA 5005, Australia

V. Jatenco-Pereira
Instituto de Astronomia, Geofísica e Ciências Atmosféricas (IAG/USP), Rua do Matão, 1226, 05508-090 São Paulo, SP, Brazil

N. Rubab
Department of Space Science, Institute of Space Technology (IST), Islamabad, Pakistan

C. Marsigli
ARPA-SIMC, HydroMeteoClimate Service of ARPA Emilia-Romagna, Bologna, Italy

A. Montani
ARPA-SIMC, HydroMeteoClimate Service of ARPA Emilia-Romagna, Bologna, Italy

T. Paccagnella
ARPA-SIMC, HydroMeteoClimate Service of ARPA Emilia-Romagna, Bologna, Italy

J. A. Wanliss
Presbyterian College, 503 South Broad Street, Clinton, SC 29325, USA

K. Shiokawa
Solar-Terrestrial Environment Laboratory, Nagoya University, Furo-cho, Chikusa-ku, Nagoya, Aichi, 464-8601, Japan

K. Yumoto
Presbyterian College, 503 South Broad Street, Clinton, SC 29325, USA
Solar-Terrestrial Environment Laboratory, Nagoya University, Furo-cho, Chikusa-ku, Nagoya, Aichi, 464-8601, Japan
Space Environment Research Center, Kyushu University, 53 6-10-1 Hakozaki, Higashi-ku Fukuoka, 812-8581, Japan

A. Cuzol
University of Bretagne-Sud, UMR 6205, LMBA, 56000 Vannes, France

E. Mémin
INRIA Rennes-Bretagne Atlantique, Rennes, France

G. S. Lakhina
Indian Institute of Geomagnetism, New Panvel (W), Navi Mumbai, India

B. T. Tsurutani
Jet Propulsion Laboratory, California Institute of Technology, Pasadena, CA, USA

A. C.-L. Chian
Paris Observatory-LESIA, Meudon, France and National Institute for Space Research (INPE), Sao Jose dos Campos, Brazil

T. Hada
Interdisciplinary Graduate School of Engineering Sciences, Kyushu University, Fukuoka, 816-8580, Japan

G. J. Morales
Physics and Astronomy Department, University of California, Los Angeles, CA 90095, USA

R. H. J. Grimshaw
Loughborough University, Department of Mathematical Sciences, Ashby Road, Loughborough, LE11 3TU, UK

G. S. Lakhina
Indian Institute of Geomagnetism, New Panvel (W), Navi Mumbai, India

B. T. Tsurutani
Jet Propulsion Laboratory, California Institute of Technology, Pasadena, CA, USA

D. V. Alexandrov
Department of Mathematical Physics, Ural Federal University, Lenin ave. 51, 620000 Ekaterinburg, Russia

I. A. Bashkirtseva
Department of Mathematical Physics, Ural Federal University, Lenin ave. 51, 620000 Ekaterinburg, Russia

L. B. Ryashko
Department of Mathematical Physics, Ural Federal University, Lenin ave. 51, 620000 Ekaterinburg, Russia

F. Uboldi
Independent researcher, Milan, Italy

A. Trevisan
CNR-ISAC, Bologna, Italy

A. Shapoval
IEPT RAS, Profsoyuznaya str. 84/32, 117 997 Moscow, Russia
Financial University, Leningradsky pr. 49, 125 167 Moscow, Russia
National Research University Higher School of Economics, 20 Myasnitskaya Ulitsa, 101 000 Moscow, Russia

J. L. Le Mouël
IPGP, 1 rue Jussieu, 75005 Paris, France

M. Shnirman
IEPT RAS, Profsoyuznaya str. 84/32, 117 997 Moscow, Russia
IPGP, 1 rue Jussieu, 75005 Paris, France

V. Courtillot
IPGP, 1 rue Jussieu, 75005 Paris, France

J. Guo
School of Mathematics and Physics, Qingdao University of Science and Technology, Qingdao, Shandong 266061, China
Sate Key Laboratory of Space Weather, Chinese Academy of Sciences, Beijing 100190, China

B. Yu
School of Mathematics and Physics, Qingdao University of Science and Technology, Qingdao, Shandong 266061, China

A. Osmane
University of New Brunswick, Physics Department, Fredericton, New Brunswick, Canada Aalto University, Radio Science and Engineering, Espoo, Finland

A. M. Hamza
University of New Brunswick, Physics Department, Fredericton, New Brunswick, Canada

S. Metref
CNRS - Univ. Grenoble Alpes, LGGE (UMR5183), 38041 Grenoble, France

E. Cosme
CNRS - Univ. Grenoble Alpes, LGGE (UMR5183), 38041 Grenoble, France

C. Snyder
National Center for Atmospheric Research, Boulder, Colorado, USA

P. Brasseur
CNRS - Univ. Grenoble Alpes, LGGE (UMR5183), 38041 Grenoble, France

O. E. Kurkina
Nizhny Novgorod State Technical University n.a. R. Alekseev, Nizhny Novgorod, Russia

A. A. Kurkin
Nizhny Novgorod State Technical University n.a. R. Alekseev, Nizhny Novgorod, Russia

E. A. Rouvinskaya
Nizhny Novgorod State Technical University n.a. R. Alekseev, Nizhny Novgorod, Russia

T. Soomere
Institute of Cybernetics, Tallinn University of Technology, Akadeemia tee 21, 12618 Tallinn, Estonia
Estonian Academy of Sciences, Kohtu 6, 10130 Tallinn, Estonia

C. Skandrani
NOVELTIS, Space and Remote Sensing Department, Space Weather Unit, rue du Lac 153, 31670 Labège, France

M. E. Innocenti
Center for Mathematical Plasma Astrophysics, Department of Mathematics, KULeuven - University of Leuven, Celestijnenlaan 200B, bus2400, 3001 Leuven, Belgium

L. Bettarini
Fluid and Plasma Dynamics group, Department of Physics, Université Libre de Bruxelles, Campus Plaine - CP231 Boulevard du Triomphe, 1050 Brussels, Belgium

F. Crespon
NOVELTIS, Space and Remote Sensing Department, Space Weather Unit, rue du Lac 153, 31670 Labège, France

J. Lamouroux
NOVELTIS, Space and Remote Sensing Department, Space Weather Unit, rue du Lac 153, 31670 Labège, France

G. Lapenta
Center for Mathematical Plasma Astrophysics, Department of Mathematics, KULeuven - University of Leuven, Celestijnenlaan 200B, bus2400, 3001 Leuven, Belgium

A. Matulka
Department of Applied Physics, UPC, Barcelona, Spain

P. López
Department of Applied Mathematics, UCM, Madrid, Spain

J. M. Redondo
Department of Applied Physics, UPC, Barcelona, Spain

A. Tarquis
Department of Applied Mathematics, UPM, Madrid, Spain

Y. Narita
Space Research Institute, Austrian Academy of Sciences, Schmiedlstr. 6, 8042 Graz, Austria
Institut für Geophysik und extraterrestrische Physik, Technische Universität Braunschweig, 38106 Braunschweig, Germany